Partition of Unity Methods

Partition of Unity Methods

Stéphane P. A. Bordas
University of Luxembourg, Luxembourg, UK

Alexander Menk
Robert Bosch GmbH, Germany

Sundararajan Natarajan
Indian Institute of Technology Madras, India

This edition first published 2024
© 2024 John Wiley & Sons Ltd

Registered Offices
John Wiley & Sons, Inc., 111 River Street, Hoboken, NJ 07030, USA
John Wiley & Sons Ltd, The Atrium, Southern Gate, Chichester, West Sussex, PO19 8SQ, UK

For details of our global editorial offices, customer services, and more information about Wiley products visit us at www.wiley.com.

Wiley also publishes its books in a variety of electronic formats and by print-on-demand. Some content that appears in standard print versions of this book may not be available in other formats.

A catalogue record for this book is available from the Library of Congress

Hardback ISBN: 9780470667088; ePub ISBN: 9781118535882; ePDF ISBN: 9781118535899

Cover Design: Wiley
Cover Image: Courtesy of Authors

Set in 9.5/12.5pt STIXTwoText by Integra Software Services Pvt. Ltd, Pondicherry, India
Printed and bound by CPI Group (UK) Ltd, Croydon, CR0 4YY

C9780470667088_161023

Contents

List of Contributors

Stéphane Cotin
Universite de Strasburg, Strasburg
France

Ravindra Duddu
School of Engineering, Vanderbilt University
Nashville, USA

Michel Duprez
Universite de Strasburg, Strasburg
France

Octavio Andrés González-Estrada
Escuela de Ingeniería Mecánica Colombia

Juan José Ródenas García
Universidad Politecnica De Valencia
Valencia, Spain

Robert Gracie
University of Waterloo, Ontario, Canada

Vanessa Lleras
IMAG, Univ Montpellier, CNRS, Montpellier, France

Alexei Lozinski
Laboratoire de Mathématiques, Université de
Franche-Comté, Besançon Cedex, France

Emilio Martínez-Pañeda
Imperical College, London, UK

Indra Vir Singh
Indian Institute of Technology Roorkee
Uttarakhand, India

Jon Trevelyan
Durham University, Durham, UK

Killian Vuillemot
Laboratoire de Mathématiques, Université de
Franche-Comté, Besançon Cedex, France

Preface

This book has been a moving target for the past 13 years (2009–2022). We are delighted to see its first edition published. We tell the story of extensions to the finite element method which are now globally accepted, and implemented in industrial simulation software. The book relies on the expertise of the authors and borrows additional know-how and experience from chapters contributed from leading experts in related methods. This makes this book the most complete account of enrichment methods, both in finite elements and boundary element methods. We also discuss the critically important topic of error estimation and adaptivity as well as practical applications.

This book has a long and complex history, typical of academic research. The idea for the book was born around 2007, when Alexander Menk, then funded by Bosch GmbH, was a PhD student with Stéphane Bordas in Glasgow. The original idea was to focus on the extended finite element method, but became a lot more ambitious as the authors investigated other discretization methods, industrialized their work, and collaborated with other research groups.

We start by an introduction on the origin of enriched methods, starting with global enrichment of finite element methods or specialized enrichments (e.g. for fracture mechanics), introduced in the 1970s. We explain how local enrichment methods took precedence, with the introduction of partition of unity methods in the 1990s. We give details on *a priori* error estimates (Cea's lemma), to explain how enrichment palliates limitations of polynomial approximations of non-smooth solutions.

By starting with the notion of partition of unity, we motivate how arbitrary functions can be exactly reproduced by the approximation space by multiplying this function with a partition of unity. We compare this approach to other methods, put forward within the context of meshfree methods, such as intrinsic enrichment of moving least squares.

Enrichment techniques require special treatments to support the generality of the enrichment functions used, which are not necessarily continuous, nor even polynomial. We therefore discuss and compare different options for numerical integration, in detail. The locality of enrichment implies the existence of interfaces between enriched and non-enriched regions, within the domain which can lead to decreased optimality in convergence rates, inaccuracies, and spurious oscillations in the solution close to the interface. We describe possibilities to overcome these difficulties.

After tackling these advanced topics, we discuss a wide variety of applications of enrichment schemes, including fracture mechanics, treatment of heterogeneities, and boundary layers.

In order to tackle special topics, we asked experts in free boundary problems, *a posteriori* error estimation, nonlinear material modeling, fracture simulations, and multiscale methods to provide the reader with up-to-date information on such important areas related to partition of unity enrichment. The book has a chapter on an exciting topic related to the enriched boundary element method for fracture and wave propagation. The book ends with a chapter on an exciting topic related to multiscale modeling of fracture.

This book is dedicated to our teachers and Professors. Stéphane Bordas thinks in particular about Monsieur Martin (math teacher at age 10 and 11), Madame Januel (math teacher at age 16), and Madame Goubet (math teacher in preparatory classes), as well as Monsieur Carsique (math teacher in preparatory classes). But, as when one teaches, two learn, this book is also dedicated to our students, in particular our PhD students, and to our team at large, who all contributed strongly to our education through their own work, research, questionings, philosophical or otherwise. We also thank our mentors Professor Ted Belytschko, Nenad Bićanić, Bhushan Karihaloo, Brian Moran, and Chris Pearce.

Acknowledgments

Writing this book on the partition of unity methods has been an ultra marathon, and as a long-distance runner, I realize the power of taking small steps. I would like to express my heartfelt gratitude to the individuals and organizations who have supported me throughout this 15-year journey.

First and foremost, I extend my appreciation to my mentors and educators who ignited my passion for excellence and instilled in me the importance of understanding the intricate details of any field. I am grateful to Monsieur Martin, Madame Januel, and Madame Goubet for their guidance during my formative years at Lycée Saint Louis in Paris. They showed me the beauty that lies in mastery and the endless quest for knowledge. Brian Moran, my PhD advisor, played a pivotal role by believing in me and introducing me to the field of computational mechanics, particularly solid mechanics. His unwavering support and guidance shaped my career.

The late Ted Belytschko remains a cherished role model whose leadership and influence continue to inspire me. I fondly remember the Friday group meetings at Northwestern University. Bhushan Karihaloo's trust in my abilities and his decision to entrust me with the leadership of the Institute of Mechanics and Advanced Materials in Cardiff were instrumental in my professional growth. Without his support, our achievements would have been delayed significantly. Thank you for helping me discover my inner potential. I extend my gratitude to the hundreds of co-authors on our research papers who have enriched my understanding of diverse research areas. Special thanks go to my PhD students who placed their trust in me and my research group. They showed me that *"when one teaches, two learn."* In particular, I thank Nguyen Vinh Phu, my first Master's Student, with whom I just co-signed a book on the material point method.

I thank Sundararajan Natarajan who has consistently showed me alternate ways into life and with whom I had long walks, bus rides, and discussions in Glasgow, Cardiff, and Chennai. I also thank Alex Menk, with whom Sundararajan and I co-sign this book. Alex was always an example of pragmatism mixed with mathematical rigor.

My heartfelt thanks go to the funding organizations, including the EU, EPSRC, and FNR (Luxembourg), for providing over €25 million in funding for our research. I am grateful to the universities that trusted me, including the University of Glasgow, EPFL, Cardiff University, and especially the University of Luxembourg, which allowed me to flourish at the heart of Europe.

I would like to express my deep appreciation to my children, Iphigénie, Augustin, Anatole, and Oscar Bordas, who have been constant sources of inspiration. Each of you has a unique quality that reminds me of the beauty and elegance in mathematics, science, and knowledge. Iphigénie, your growth mindset, your grit, competitiveness, and humility in the face of complexity inspire me daily. Augustin, your camaraderie, simplicity, and commitment to authenticity and beauty remind me of what truly matters. Anatole, your good humor, linguistic prowess, your drive to thrive, and your unwavering positive support have been my pillars. Oscar, your perpetual optimism, love for mathematics, chess, and logic along with your infectious positivity brighten every instant of my day. I am also grateful to their mother, Laurelle Demaurex, for her support during the early years of my research group and to

help me bring out the best of myself by challenging me in ways I did not anticipate I would be challenged. I also thank her father, Marc-Olivier Demaurex, for showing me the importance of reading between the lines. I also thank Beverly Johnston for daily reminders of the power of logic and sound reasoning over sophism.

Finally, I thank my parents (Pierre Bordas and Christiane Renault) and grandparents (Raymond Renault, Lucette Dusservaix, Suzanne Lebobe, and André Bordas) for their unwavering support and for introducing me to the world of hard work, mathematics, and science. They were the best pacers I could have hoped for in the race of life. I will never forget the advice given by my grandfather Raymond, his generosity, harmony, strength, and positivity which continue to live within us and within my son Oscar, who bears his name. As a conclusion, I leave you with a profound Sufi story that has resonated with me throughout my journey: "*You see in people what you yourself are. People are the same everywhere. The real problem is about you. Remember this.*"

In our pursuit of knowledge and excellence, we often look outward for answers and inspiration. However, this story reminds us that true understanding begins within ourselves. It is our own perspective, attitudes, and actions that shape our perception of the world and the people in it.

As we reflect on the acknowledgments and the journey chronicled in this book, may we remember that our interactions with others, our mentors, students, colleagues, and loved ones, are a reflection of our own inner world. By continually striving to improve ourselves and fostering a sense of humility, empathy, and gratitude, we can better navigate the complexities of life and contribute to the betterment of the world around us.

Thank you to everyone who has been a part of this remarkable journey. Your support, trust, and shared experiences have enriched my life and this book in immeasurable ways. With heartfelt gratitude and a commitment to ongoing growth, we would like to honor Ivo Babuška, who pioneered the partition of unity methods. His clarity and sagacity have deeply inspired me and motivated me to complete this work.

Stéphane P.A. Bordas

I would like to thank my wife Kathrin Lydia Menk and my children Julius and Magdalena for continuously supporting me with my work and for all the good times we had so far. I also would like to thank my parents and my brother for helping me become the person I am today. Last but not least I would like to thank Stéphane P. A. Bordas for giving me the opportunity to become active in academic research and Sundararajan Natarajan for all the fruitful discussions we had during our time in Glasgow.

Alexander Menk

I would like to thank my parents Geetha Natarajan and Natarajan, my wife Ramya Chandrasekaran, and son Vaibhav for continuously supporting me with my work and standing by me on numerous occasions. I admire Vaibhav's abundant source of energy, constant thirst to learn and do new things, and especially the attitude of carrying out tasks without the fear of failure and pursuing them until success is attained. I am thankful for my sister Parvathi's abundant love and affection. My thanks extend to Bhagirath, Nirmal, Mahalakshmi, and Chandrasekaran and Shantha Chandrasekaran for their support, trust, and encouragement. Special thanks to Young philosophers, Pranav, Ayush and Shubha. I would like to thank Stéphane P.A. Bordas for supporting me through different phases as a student, post-doctoral researcher and now as an academic researcher and Alexander Menk for all the good times we had during our time in Glasgow. Finally, I would like to thank the omnipresent for giving me the strength to swim through different hurdles.

Sundararajan Natarajan

We would like to thank our friends Professors Stéphane Cotin, Ravindra Duddu, Michel Duprez, Octavio Andrés González-Estrada, Juan José Ródenas García, Robert Gracie, Vanessa Lleras, Alexei Lozinski, Emilio Martinez-Pañeda, Indra Vir Singh, Jon Trevelyan, and Killian Vuillemot for sharing their expertise by contributing specialized topics on the partition of unity methods. We would like to thank our friends/students Chintan Jansari, Abir Elbeji, Dulce Canha, Saurabh Deshpande, Qiaoling Min, Aravind Kadhambariyil, Zhaoxiang Shen, Meryem Abbad Andaloussi, Geremy Loachamín Suntaxi, Vincent de Wit, Paris Papavasileiou, Sofia Farina, and Ehsan Mikaeili for extensive proofreading and for the numerous suggestions and corrections.

1

Introduction

Stéphane P. A. Bordas[1], Alexander Menk[2], and Sundararajan Natarajan[3]

[1] *University of Luxembourg, Luxembourg, UK*
[2] *Robert Bosch GmbH, Germany*
[3] *Indian Institute of Technology Madras, India*

Physical systems are often modeled using partial differential equations (PDEs). The exact solution or closed form or analytical solutions to these PDEs is only available in special cases for specific geometries. Numerical methods can be used to approximate the exact solution in more general settings. The result of a numerical simulation is rarely exact. Nonetheless, computer-based numerical simulation has revolutionized industrial product development throughout engineering disciplines. When comparing experiments and simulation with the aim of improving a simulation procedure to give more accurate results, it is necessary to understand the different sources of error. Figure 1.1 shows an overview of errors that occur at different stages of modeling and numerical simulation for a given numerical method.

One of these numerical methods is called the "finite element method" (FEM). It is most commonly used in structural mechanics, although the field of application is much broader. The historic origins of the FEM cannot be uniquely determined. Mathematicians and engineers seemed to develop similar methods simultaneously which laid the foundations for what is now popularly known as the FEM. In the mathematical community, the developments can be summarized as follows. In 1851, Schellbach obtained an approximate solution to Plateaus problem by using piecewise linear functions on a surface. Variational principles to solve partial differential equations were used by Ritz in 1909. Based on the Ritz method, Courant proposed a triangulation of a two-dimensional (2D)

Figure 1.1 Sources of error in simulation.

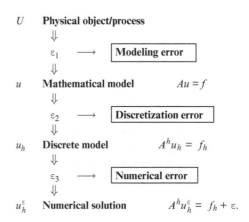

structure to solve the plane torsion problem. The first book providing a solid mathematical basis for the FEM is attributed to Babuška and Aziz (Babuška and Aziz, 1972). In the engineering community, the developments of FEM are motivated by physical analogies to describe continuous problems in a discrete fashion. Hrenikoff (Hrenikoff, 1941) combined trusses and beams to model plane elasticity problems. Turner, Clough, Martin, and Topp introduced plane elements in 1956. The term "*finite element*" was coined by Clough in 1960. The first book about the FEM written from an engineering perspective is attributed to Zienkiewicz (Zienkiewicz, 1971) in 1971. We will assume a certain familiarity of the reader with this method throughout the book, but we want to provide a short introduction to the FEM at this point, in order to introduce the main notations.

1.1 The Finite Element Method

Assume Ω to be a domain of $\mathbb{R}^d$, $d \in \{1, 2, 3\}$. Let us take a look at the following boundary value problem:

$$-\Delta\phi = f \quad \text{in} \quad \Omega \tag{1.1a}$$

$$\phi|_{\partial\Omega} = 0 \tag{1.1b}$$

where $\phi : \mathbb{R}^d \to \mathbb{R}$ is an unknown scalar field and $f : \mathbb{R}^d \to \mathbb{R}$.

Equation (1.1a) is known as Poisson's equation when $f \neq 0$ and as Laplace equation when $f \equiv 0$. The open domain Ω could be a region in $\mathbb{R}^d$, bounded by a $(d-1)$ dimensional surface $\partial\Omega$ whose outward normal is **n**. We are looking for a scalar function ϕ that fulfills Poisson's equation everywhere in Ω. On the domain boundary $\partial\Omega$, the function should be zero (c.f. Equation (1.1b)). In physics, this equation can be used to model a variety of phenomena, for example, an elasto-static rod under a torsional load, Newtonian gravity, electrostatics, diffusion, the motion of inviscid fluid, Schrödinger's equation in Quantum mechanics, the motion of biological organisms in a solution and can also be used in surface reconstruction.

Here, let us assume that the scalar function ϕ could for instance be a temperature distribution, T, that has come to an equilibrium. Then $-f$ describes the heat supply inside the domain. It is easy to interpret the equation physically under these assumptions. The heat flux is proportional to ∇T, the gradient of T. Because the system is in equilibrium, the sum of the heat flowing in and out of an infinitesimal subregion should be the same as the heat supplied by the heat sources inside that region. In other words, the divergence of the heat flux $\nabla(\nabla T)$ should be equal to $-f$ at any point, which is equivalent to $-\Delta T = f$. Assuming a constant temperature distribution at the domain boundary could be reasonable if a material with a high thermal conductivity is attached to the region of interest. To simplify things, we postulate in Equation (1.1b) that this constant temperature is zero. Please note that once this problem is solved, one can add an arbitrary constant to T and Equation (1.1a) is still fulfilled. If the problem is posed this way, then any solution T must be differentiable twice in Ω. This is a stronger condition on the choice of T and moreover in many situations the solution is not differentiable twice, although the underlying physics is the same.

Let us consider an example. Assume that the temperature does not vary in the z-direction. In that case, the problem can be posed in a 2D setting. Let the domain be the open unit square $\Omega = [0, 1]^2$. A scalar function defined on the unit square is shown in Figure 1.2. The function is piecewise constant. We take this function to be the heat supply $-f$. A discontinuous heat supply is a realistic assumption in several situations. One could imagine an electric current flowing through a metal. Then the heat is generated at every point inside the metal, but not outside. Assuming that the heat generation at some point is proportional to the electrical current, a function f containing jumps really is physically meaningful. Experimentally, one would measure a temperature distribution similar to the one shown in Figure 1.3. The x-component of the gradient of this temperature distribution is shown in Figure 1.4. It is easily observed that the gradient is not differentiable at certain points. Therefore, the temperature distribution in Figure 1.3 is not a solution of Equation (1.1a), although it is the correct solution from a physical point of view.

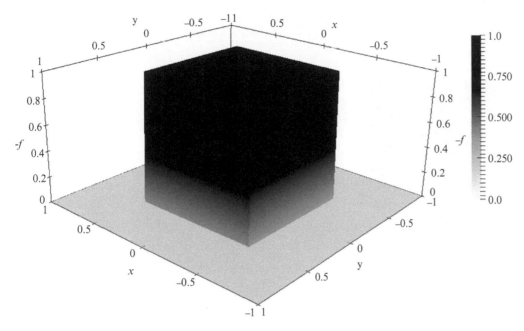

Figure 1.2 A piecewise constant function on the unit square as an example for $-f$.

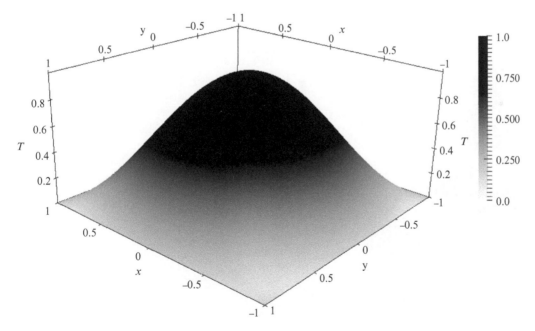

Figure 1.3 Temperature distribution.

This motivates the search for another problem description. Equation (1.1a) will subsequently be referred to as the classical formulation of the problem and a solution is called a classical solution. To obtain a new formulation, we multiply equation Equation (1.1a) by a scalar function v defined on the domain Ω and integrate over the whole domain to get:

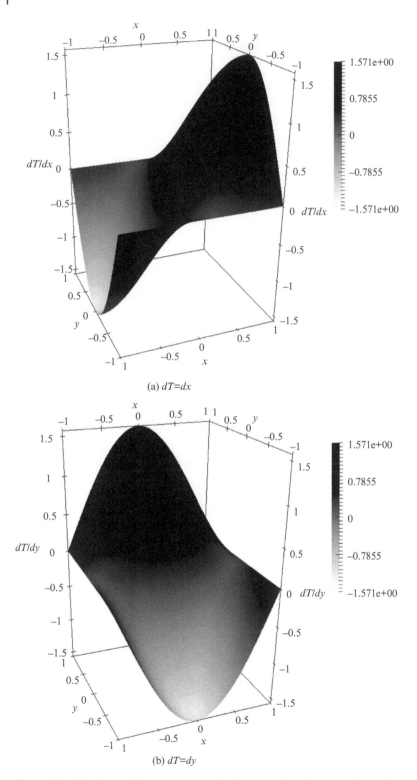

(a) *dT=dx*

(b) *dT=dy*

Figure 1.4 Gradient of the temperature distribution along the *x* and the *y* direction.

$$- \int_\Omega v \Delta T \, d\Omega = \int_\Omega v f \, d\Omega \qquad (1.2)$$

The function v is not completely arbitrary, but for now it suffices to assume certain nice properties such that we can perform the necessary integrations and differentiations in the following discussion. Applying partial integration to the left-hand side of Equation (1.2), we obtain:

$$\int_\Omega (\nabla v) \cdot \nabla T \, d\Omega = \int_\Omega v f \, d\Omega \qquad (1.3)$$

It is obvious that Equation (1.3) is fulfilled for any function v if T is a classical solution. Let us assume that there is a function space $\mathcal{V}$ which contains all the functions that are physically reasonable. By that we mean especially that the classical solution is in $\mathcal{V}$ if it exists and that all the functions in $\mathcal{V}$ vanish at the boundary[1]. Once such a space is known, the problem could be stated in the following abstract form, known as the weak form:

Weak form: Find $T \in \mathcal{V}$, such that for all $v \in \mathcal{V}$

$$a(T, v) = (v, f) \qquad (1.4)$$

where $a(T, v) = \int_\Omega (\nabla v) \cdot \nabla T \, d\Omega$ and $(v, f) = \int_\Omega v f \, d\Omega$ are the symmetric bilinear and linear forms, respectively.

⊙ **Example 1.** Find the weak form for the following strong form and identify the linear and bilinear form:

$$\kappa \frac{d^2 u(x)}{dx^2} - \lambda u(x) + f(x) = 0, \qquad 0 < x < 1$$

where κ, λ are constants independent of x and subject to the following Dirichlet boundary conditions: $u(0) = 1$, $u(1) = -2$.

⊙ **Example 2.** Repeat the above example, subjected to the following Dirichlet and Neumann boundary conditions: $u(0) = 0$, $\frac{du}{dx}(1) = -2$.

Using the weak form gives the function in Figure 1.3 a chance of being a solution to the problem because a solution of Equation (1.1a) needs only one time continuously differentiable. It remains to be checked under which circumstances a solution exists, and if this is a unique solution. Therefore, the term "physically meaningful," used when initially describing $\mathcal{V}$, needs to be defined more precisely. To do this, we need to address the integration and the differentiation in Equation (1.3). To motivate why the classical integration and differentiation operations are not useful for our purpose we take a first step toward the numerical solution of the problem. The function space $\mathcal{V}$ will generally have infinitely many dimensions. To compute an approximation to the exact solution one could try to solve the weak formulation in a finite dimensional subspace of $\mathcal{V}$, which we will denote by $\mathcal{V}_h \subset \mathcal{V}$. The solution associated with this restricted formulation is denoted by T_h. Without further analysis, it is not clear how T_h is related to the solution of Equation (1.3). But one can show that T_h minimizes the error over $\mathcal{V}_h$ in a particular norm called the "*energy norm.*" The problem then becomes:

---

1 The interested reader is referred to, e.g. Brenner and Scott (2002) for concepts of functional analysis relevant to finite element analysis

Find $T_h \in \mathcal{V}_h$ such that for all $v_h \in \mathcal{V}_h$:

$$a(T_h, v_h) = (v_h, f) \tag{1.5}$$

There exists a basis $\phi_1, ..., \phi_n$ for $\mathcal{V}_h$, that is, we can write the elements of $\mathcal{V}_h$ as:

$$v_h = \sum_I \hat{a}_I \phi_I \tag{1.6}$$

Then, certainly $T_h \in \mathcal{V}_h$ can also be written in this form:

$$T_h = \sum_I a_I \phi_I \tag{1.7}$$

To determine the unknown coefficients $\mathbf{a} = (a_1, ..., a_n)$ of T_h, we insert Equation (1.7) in Equation (1.5) and postulate that Equation (1.5) holds for all basis functions:

$$\int_\Omega (\nabla \phi_1)^T \nabla (\sum_I a_I \phi_I)\, d\Omega = \int_\Omega v_1 f\, d\Omega$$
$$\vdots \tag{1.8}$$
$$\int_\Omega (\nabla \phi_n)^T \nabla (\sum_I a_I \phi_I)\, d\Omega = \int_\Omega v_n f\, d\Omega$$

Because of the linearity of the integration operation, these equations are equivalent to postulating that Equation (1.5) holds for all $v_h \in \mathcal{V}_h$. Due to the linearity of the gradient operator and the integration operation, we can rewrite Equation (1.8) as:

$$\begin{bmatrix} \int_\Omega (\nabla \phi_1)^T \nabla \phi_1\, d\Omega & \cdots & \int_\Omega (\nabla \phi_1)^T \nabla \phi_n\, d\Omega \\ \vdots & & \vdots \\ \int_\Omega (\nabla \phi_n)^T \nabla \phi_1\, d\Omega & \cdots & \int_\Omega (\nabla \phi_n)^T \nabla \phi_n\, d\Omega \end{bmatrix} \begin{bmatrix} a_1 \\ \vdots \\ a_n \end{bmatrix} = \begin{bmatrix} \int_\Omega \phi_1 f\, d\Omega \\ \vdots \\ \int_\Omega \phi_n f\, d\Omega \end{bmatrix} \tag{1.9}$$

Therefore, to determine the unknown coefficients one has to solve a linear system of equations:

$$\mathbf{Ka} = \mathbf{f} \tag{1.10}$$

where

$$\mathbf{K} = \int_\Omega (\nabla v) \cdot \nabla v\, d\Omega$$

$$\mathbf{f} = \int_\Omega v f\, d\Omega$$

Different choices for $\mathcal{V}_h$ are possible. One could choose $\mathcal{V}_h$ to be the set of all global polynomials defined on Ω up to a certain order. In that case, all points in Ω are related to each other, and hence most of the entries of $\mathbf{K}$ will be non-zero. It is disadvantageous if the equation systems become large, because memory storage grows with the square of the system size N and solving large systems of equations using direct solver is computationally expensive ($O(N^3)$).

In practice, large equation systems are solved using iterative solvers. Iterative solvers multiply a vector with $\mathbf{K}$ at each iteration. This multiplication takes a large amount of time if $\mathbf{K}$ is fully populated. It is much easier to deal with sparse equation systems, in which most of the entries are zero. The zero entries do not have to be stored and can be neglected when evaluating the matrix-vector product. Now the question one would try to answer is:

Can we find function spaces $\mathcal{V}_h$ that result in sparse equation systems?

Assume that Ω is covered by a triangular mesh such as the one shown in Figure 1.5. The triangles will be called *elements* and the intersections of the element edges are called *nodes*. We could choose $\mathcal{V}_h$ to be the space of all

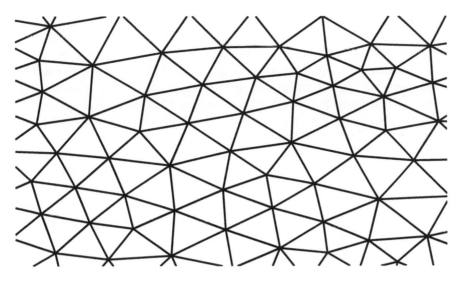

Figure 1.5 Mesh example.

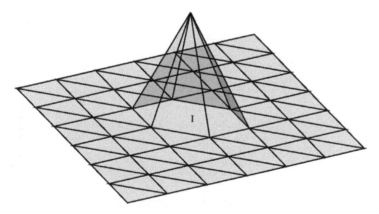

Figure 1.6 A typical basis function with its nodal support.

functions that are continuous in Ω, linear inside each element, and vanish at the boundary. A basis function $\phi_I \in \mathcal{V}_h$ is defined as follows:

- ϕ_I is 1 at the Ith interior node, i.e., $\phi_I(x_I) = 1$;
- ϕ_I is 0 at all other nodes; J, $\forall J$, $\phi_J(x_I) = 0$;
- for each element the function values can be obtained by a linear interpolation between the function values above.

Together, if chosen as described above, the basis functions v_I form a basis for the space $\mathcal{V}_h$. Figure 1.6 shows the basis function for a node I inside a domain covered by a triangular mesh. One could also cover the domain with elements with more than *three* sides. This has led to the "polygonal finite element methods." The shape functions over arbitrary elements are collectively known as "barycentric coordinates."

☞ *It is interesting to note that there is no unique way to represent the shape functions over polytopes. Any set of functions that satisfy the aforementioned properties is a candidature for basis functions.*

With a choice of such basis functions, we would have problems evaluating the elements of matrix **K**, since the functions v_h are not differentiable at the element boundaries. We could neglect this and evaluate the integration and the differentiation only for points inside the elements. The resulting matrix would then be sparse. Since the basis functions are zero in most elements, the same is true for their gradient. Therefore, most of the matrix entries in Equation (1.9) are also zero. From a computational point of view this is exactly what is done in the FEM. Please note that due to the linearity of the basis functions inside each element, the evaluation of the matrix entries can be done in a few computational steps. Since the gradient of the basis functions is constant inside each element, the integrals can be computed by evaluating the function only at the midpoint (Gauss quadrature of order one).

We now return to the discussion about the differentiation and the integration of terms in Equation (1.9). We need modified operations which, in some way, neglect subsets of Ω with measure zero, such as lines and points. With those operations we will be able to formulate the problem in a rigorous manner such that the basis functions we just discussed are part of space $\mathcal{V}_h$. For integration, this can be achieved by using the Lebesgue integral.

Lebesgue Integral:

The Lebesgue integral has the same value for functions that differ only on a point set of measure zero. The Lebesgue integral can also be used with functions that are undefined on such sets and still deliver meaningful values. We will denote the space of all infinitely differentiable functions that vanish on the boundary of Ω by $C_0^\infty(\Omega)$. Then, the weak derivative $\frac{dv}{dx}$ of a scalar function $v(x, y)$ is found if it fulfills for all $w \in C_0^\infty(\Omega)$:

$$-\int \frac{dw}{dx} v \, d\Omega = \int w \frac{dv}{dx} d\Omega$$

The derivative $\frac{dw}{dx}$ is defined in the classical sense. Similarly, other partial derivatives of higher order can be defined. The weak derivative, if it exists, is unique. If the classical derivative of a function exists, the weak derivative is the same (neglecting subsets of measure zero).

With our piecewise polynomial approximation functions, the weak derivative is not defined at the element boundaries, while inside the elements it coincides with the classical derivative. Together, the Lebesgue integral and the weak derivative make it possible to evaluate Equation (1.9) for our approximation functions. Moreover, we can now define the space $\mathcal{V}_h$. This space will be denoted by $H_0^1(\Omega)$ and a function v is an element of this space if:

- $\int_\Omega v^2 \, d\Omega$ can be evaluated in the Lebesgue sense and is smaller than infinity;
- the weak derivative of v exists;
- $\int_\Omega \left(\frac{dv}{dx}\right)^2 d\Omega$ and $\int_\Omega \left(\frac{dv}{dy}\right)^2 d\Omega$ can be evaluated in the Lebesgue sense and are smaller than infinity;
- v vanishes along the boundary.

☞   *The space of square-integrable functions is our starting point. Postulating that the weak partial derivatives exist and that they are square-integrable is necessary, otherwise one is not able to evaluate the weak form. Postulating that the functions vanish at the boundary is necessary to make sure that the essential boundary conditions are fulfilled.*

We motivated the definition of $H_0^1(\Omega)$ from a physical and from a computational point of view. But the definition has much more fundamental consequences from a mathematical point of view. One can define an inner product, such that this space becomes a Hilbert space. Hilbert spaces are inner product spaces in which geometrical operations like projections can be generalized. This makes a geometric interpretation of the FEM-procedure possible. The exact solution is a vector in this space. In this Hilbert space, the existence and the uniqueness of the exact solution can be proven. But more advanced tools from functional analysis are needed for this. Space $\mathcal{V}_h$ forms a hyperplane, and the numerical solution is the projection of the exact solution onto that plane. Therefore, in some sense the numerical solution is the best approximation one can get. To increase the accuracy of the numerical solution, one may, for example, increase the dimension of the hyperplane which will decrease the error. In the FEM, this is done by refining the mesh. However, since $H_0^1(\Omega)$ is infinite dimensional not every mesh refinement strategy results in convergence.

> ⊙ **Example 3.** Using triangular elements, write a program for an FEM-solver for the Poisson problem on a rectangular square and a right-hand side given by the function in Figure 1.2. Check how the solution behaves for different element sizes.
>
> ⊙ **Example 4.** Matlabs `minres()`-function is an iterative solver. Measure the time it takes to solve the FEM-equations from Exercise 1.1 directly. Compare this with the time `minres()` needs to solve the equation system. How do the computation times change if a sparse matrix is passed to `minres()`? How does this difference behave if the mesh is refined?

1.2   Suitability of the Finite Element Method

In this section, we discuss the convergence of the FEM. At the end of this section, you should be able:

- to assess what the convergence rate of the FEM depends on;
- to estimate the error;
- to comment on how it can possibly be improved.

To do this, we refer back to the weak form in the finite dimensional space (Equation (1.5)). We are interested in the relation of T_h to the exact solution T. Defining the error due to the finite element approximation $e = T - T_h$. This means that for all functions v_h in $\mathcal{V}_h^0$, $a(e, v_h) = 0$, or, equivalently, that the approximate solution T_h is the orthogonal projection of the exact solution u on $\mathcal{V}_h$. This is depicted geometrically in Figure 1.7, wherein, one can think of $e = T - T_h$ as a vectorial identity. The space where the exact solution, T, lives is represented artificially as a 3D space (in reality it is a space of infinite dimensions). The finite element subspace is a plane (dimension 2). The error, e, is a member of $\mathcal{V}$, and the finite element solution a member of $\mathcal{V}_h$. Considering the right triangle in the figure, it can be immediately seen that the length of T (squared) equals the length of T_h (squared) plus the length of e, squared, i.e., $a(T, T) = a(T_h, T_h) + a(e, e)^2$. This is an important property, if we add a function, which is not in the span of $\mathcal{V}_h$ to the basis of $\mathcal{V}_h$ and thus increase its dimension, the error will always decrease.

---

2 This identity is similar to the Pythagorean theorem in Euclidean spaces.

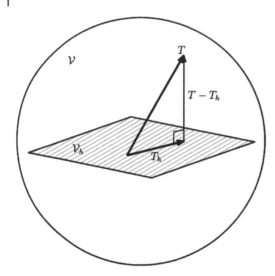

Figure 1.7 The orthogonal (with respect to bilinear form $a(\cdot,\cdot)$) projection of the exact solution u on the finite element space $\mathcal{V}_h \subset \mathcal{V}$ is the finite element solution u_h. In this figure, you can think of $e = u - u_h$ as a vectorial identity. In this figure, the space where the exact solution, u, lives is represented artificially as a three-dimensional space (in reality it is a space of infinite dimensions). The finite element subspace is a plane (dimension 2). The error, e, is a member of $\mathcal{V}$, and the finite element solution a member of $\mathcal{V}_h$. Considering the right triangle in the figure, it can be immediately seen that the length of u (squared) equals the length of u_h (squared) plus the length of e, squared, i.e., $a(u, u) = a(u_h, u_h) + a(e, e)$. This identity is similar to the Pythagorean theorem in Euclidean spaces. Interpretation of the error $e = u - u_h$ as a projection.

Galerkin Orthogonality $a(e, v_h) = 0$ is the Galerkin orthogonality relation for the error. From the Galerkin orthogonality, follows that the Galerkin approximation is the "*best approximant*" with respect to the associated bilinear form $a(\cdot,\cdot)$. This best approximation property means that the finite element solution is a least square fit of the exact solution in the sense of the bilinear form $a(\cdot,\cdot)$.

The success of the FEM is probably due to its rich mathematical analysis and its elegant framework. This facilitates accurate *a priori* and *a posteriori* estimates of the discretization error. Here, we briefly outline the *a priori* error estimates. Assume that $\| \cdot \|$ is a norm on the function space $\mathcal{V}_h$. And furthermore assume that the problem has a unique solution (i.e., appropriate essential boundary conditions are prescribed). Then, let $m \in \{\cdots\}$ be the order of the norm in which the error is measured, h a measure of the element size[3], p is the polynomial order (chosen as 1 in the example above), r the regularity of the exact solution T, Cea's Lemma states that the error of the finite element approximation in norm $\| \cdot \|$ satisfies the following inequality:

$$\|T - T_h\| \le c\, h^{\min(p+1-m,\,r-m)} \|u\|_r \tag{1.11}$$

In other words, up to a constant c which is independent of the mesh size, the solution T_h is the best possible solution in the chosen function space. If no statement can be made about the continuity of the exact solution, mesh refinement guarantees at least a linear convergence, provided the following is true:

$$\min(p + 1 - m, r - m) > 0 \tag{1.12}$$

The above condition can be rewritten as:

$$p + 1 - m > 0 \quad \text{and} \quad r - m > 0 \tag{1.13}$$

or, equivalently,

$$p > m + 1 \quad \text{and} \quad r > m \tag{1.14}$$

This means that that in order for the FEM to converge optimally in norm $\| \cdot \|_m$, we must both:

---

3 Typically the largest element in the mesh in Figure 1.8

Figure 1.8 The mesh size, *h*, is the diameter of the smallest circle enclosing the largest element in the mesh.

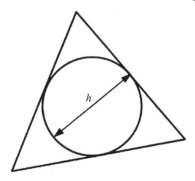

- select the polynomial order *p* larger than $(m - 1)$.
- make sure that the exact solution *u* is more regular than the order of the norm $\| \cdot \|_m$.

If the order of the continuity of the exact solution is high, increasing the polynomial order *p* might be a better strategy than refining the mesh. However, if the regularity of the solution is low, there is no use increasing the polynomial order *p*, because if *r* is small, then $(r - m)$ will remain the determining term in $\min(p + 1 - m, r - m)$. This is further discussed in detail in Section 1.3.

☞ A priori *estimation discussed above does not account for:*

- *numerical integration error to compute the bilinear and linear form;*
- *interpolating Dirichlet boundary conditions;*
- *approximating the boundary $\partial\Omega$ by piecewise polynomial functions.*

1.3 Some Limitations of the FEM

We have seen earlier that the FEM possesses the best approximation property. The FEM is optimal for problems that are self-adjoint, viz., elliptic or parabolic PDEs. This is because, in such cases, it is possible to show that there exists a quadratic functional, the minimum of which corresponds to the solutions of the PDEs governing the problem. However, in certain situations (even with self-adjoint equations) calculating a numerical approximation using the FEM can become a tedious task, especially when the problem involves features that lead to singular or discontinuous solutions. In this section, we discuss this in detail through (low value of regularity *r* in Equation (1.11)) two examples.

Problem with rough solutions Consider the domain shown in Figure 1.9. The domain has a re-entrant corner. Again, we consider a temperature distribution on this domain. The task is to find the equilibrium temperature distribution *T* in Ω, given the temperature distribution $\bar{T}$ on the boundary $\partial\Omega$. The governing differential equation is given by Equation (1.1a) with $f \equiv 0$ and the scalar function ϕ is assumed to be the temperature distribution. The problem can be transformed into polar coordinates with a radial component *r* and an angular component θ. The mapping between the Cartesian and the polar coordinates is given by:

$$(x(r, \theta), y(r, \theta)) = (r \cos \theta, r \sin \theta) \tag{1.15}$$

It is easily checked that using polar coordinates, Equation (1.1a) becomes:

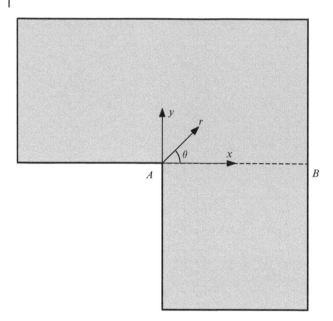

Figure 1.9 L-shaped domain.

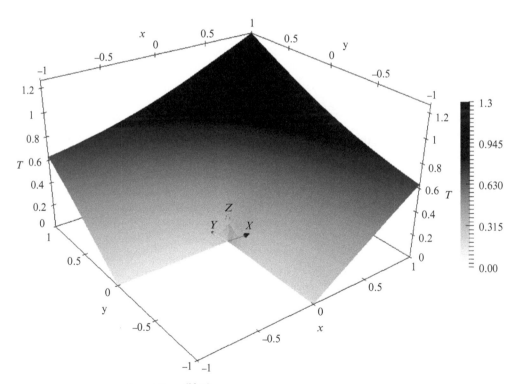

Figure 1.10 Function $\hat{T} = r^{2/3} \sin \frac{(2\theta+\pi)}{3}$.

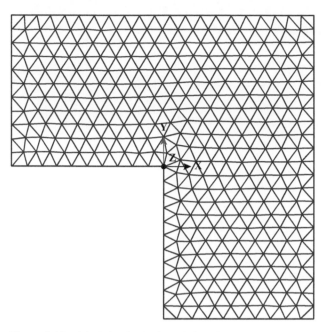

Figure 1.11 Mesh for the L-shaped domain.

$$\Delta_{r,\theta}T(r,\theta) := \frac{\partial^2 T(r,\theta)}{\partial^2 r} + \frac{1}{r}\frac{\partial T(r,\theta)}{\partial r} + \frac{1}{r^2}\frac{\partial^2 T(r,\theta)}{\partial^2 \theta} = 0 \tag{1.16}$$

Let us consider the following function:

$$\hat{T}(r,\theta) = r^{\frac{2}{3}}\sin\frac{(2\theta + \pi)}{3} \tag{1.17}$$

The function is visualized in Figure 1.10. In Ω, this function is a solution to the Laplace equation (see Equation (1.16)). Therefore, if we use in Equation (1.16), $\bar{T} = \hat{T}|_{\partial\Omega}$, then $\hat{T}$ becomes the (unique) solution to the Laplace equation. A numerical solution can be obtained in a similar way as previously described, although slight changes have to be made because the boundary conditions are not homogeneous. This has to be taken into account when the problem is formulated in its weak form using integration by parts.

To see why the numerical solution of this problem can become difficult, we consider $\hat{T}$ along a line segment $A-B$ in Ω given by $x \leq 0$ and $y = 0$ as shown in Figure 1.9. Since we are approximating the exact solution with piecewise linear functions in Ω, the numerical solution along the line $A - B$ is also a piecewise linear function. We choose a mesh as shown in Figure 1.11 to discretize the problem. In the preceding section, we already mentioned that the FEM solution can be interpreted as a projection onto a subspace of the function space in which the problem is formulated. We now try to motivate why, in this particular case, any function of the subspace is a bad approximation of the exact solution close to the re-entrant corner. To do this, we try to adjust the numerical solution T^h to obtain a good approximation along the line $A - B$. One possibility is to choose the nodal values of T^h to be the values of the exact solution at these locations.

Evaluation of T^h along the line $A - B$ in that case is shown in Figure 1.12 together with the exact solution and the error. Obviously, the error is large inside the elements close to the re-entrant corner, further away, the

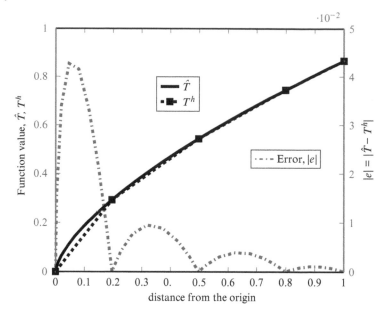

Figure 1.12 Approximation along the line $A - B$ (see Figure 1.9).

linear approximation resolves the exact solution much better. One could try to adjust the approximation in the vicinity of the re-entrant corner to reduce the error, but obviously there is a limit to this process. The problem is related to the gradient of the solution. In the vicinity of the re-entrant corner, the first derivative is unbounded. The x-component of the gradient of $\hat{T}$ is shown in Figure 1.13. Therefore, in this area, $\hat{T}$ deviates a lot from being a linear function and therefore it cannot be approximated by a straight line very well. One way to address this problem is to refine the mesh and use a larger number of elements. This would increase the ability to approximate the exact solution. On a smaller scale, however, similar problems would occur close to the re-entrant corner. In the presence of (weak) singularities, the approximation properties of piecewise linear functions (or more generally piecewise polynomial functions) are not sufficient. This certainly is a limitation of finite element approaches.

Conforming discretization Another limitation of the FEM is related to the mesh generation. To ensure a certain quality of the mesh, elements whose edges form sharp or obtuse angles should be avoided. Imagining a sharp triangle and the corresponding shape functions we may say that the exact solution cannot be resolved very well along the long edges. Therefore, it would be more efficient to generate a mesh with the same number of elements and equally sized element edges. When considering the deformation of a structure, an even more important reason for avoiding sharp element with large edge size ratios is a phenomena called locking. When bending or shearing occurs, sharp linear elements behave too stiff and the numerical solution is not acceptable. This is a geometric restriction on the mesh choice. Looking again at the mesh in Figure 1.11, we can see that, for this particular domain, generating a mesh with well-shaped elements is an easy task. The mesh can be generated by a very simple algorithm. But this is only possible if the geometry of the domain is simple as well and no other restrictions on the mesh choice are given.

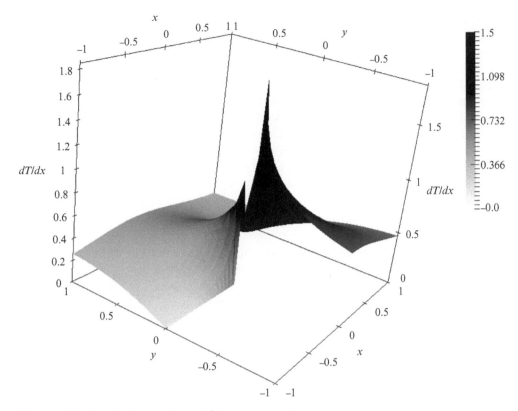

Figure 1.13 Magnitude of gradient of $\hat{T}$ in the x-direction.

Figure 1.14 Domains Ω^- and Ω^+.

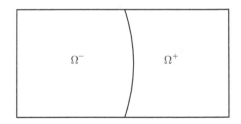

Assume that we want to generate a high-quality mesh for a domain Ω. Consider the geometry shown in Figure 1.14 with open domains Ω^- and Ω^+. In the following situations, problems may occur:

- Case1: The domain Ω is given by $\overline{\Omega^+}$;
- Case 2: The domain Ω is given by $\overline{\Omega^+ \cup \Omega^-}$ and the derivative of the exact solution is discontinuous along the interface;
- Case 3: The domain Ω is given by $\Omega^+ \cup \Omega^-$ and the exact solution or its derivatives are discontinuous along the interface between Ω^+ and Ω^-.

In all of the cases mentioned above, there would be the need to align the element edges with the interface between Ω^+ and Ω^-. In the first case, this is necessary in order to discretize the domain appropriately. If the derivative of the exact solution is discontinuous along the interface, the numerical solution should be able to represent such a weak discontinuity. In the framework discussed so far, the derivative of the numerical solution is constant inside each element and weak discontinuities can only form at the element edges. Thus, to represent a weak discontinuity numerically the element edges should align with the interface. If the exact solution is discontinuous along the interface, the substructures Ω^+ and Ω^- have to be meshed independently, that is, every subdomain is treated as an independent problem. Therefore, the numerical solutions in Ω^+ and Ω^- are not related to each other and the same is true for the nodal values associated with nodes from different domains.

However, in this particular case, aligning the element edges is not possible as the interface is curved. This is because the triangles elements considered here are formed by straight edges. One could try to generate a mesh whose edges resemble the interface sufficiently well and neglect the small error that is introduced. But as discussed earlier the elements should be well shaped. A possible mesh choice is shown in Figure 1.15. It is obvious that more sophisticated algorithms have to be used to generate these meshes. However, these algorithms are based on heuristic principles and one cannot guarantee that they generate high-quality meshes or that they generate any mesh at all for general geometries. The more complicated the geometries become, the more likely a mesh generator is to fail. The necessity to generate meshes which are subject to certain geometric restrictions is another drawback of the FEM. Both of the problems discussed here can be circumvented by using the so-called "*enrichment functions*" which will be discussed in Section 1.4.

1.4 The Idea of Enrichment

In both the examples considered in the previous section, we know in advance that the (weak) singularity and the discontinuous solution are likely to occur at the re-entrant corner and at the interface location, respectively. For the re-entrant corner, the strength of the singularity will be dependent on the interior angle. We showed this using a special choice of boundary conditions due to which we were able to write down the exact solution in an analytical form. However, even if we change the boundary conditions, (weak) singularities are likely to occur. This is due to the non-convexity of the domain. Similarly, in the example with two domains, especially case 3 mentioned in the previous section, a discontinuity along the interface between Ω^+ and Ω^- was assumed. This forced geometric restrictions onto the meshing procedure. In these examples, the regularity of the solution r is low and the term $(r - m)$ will be the dominant term (see Equation (1.11)). Due to this reason, the FEM will have suboptimal convergence rate. If the regularity of the solution is $r = \frac{3}{2}$, and we choose $m = 1$, then the convergence rate is $\frac{1}{2}$, i.e., the error decreases with the square root of the mesh size: $\|e\|_r = \|T - T^h\|_r < ch^{\min(\frac{3}{2}-1,p+1-1)} = ch^{\min(\frac{1}{2},p)}$. Once again notice that increasing p will not improve the convergence rate, as $\forall p \geq 1$, $\min(\frac{1}{2}, p) = \frac{1}{2}$.

> ☞ *Based on the geometry and the differential equation itself certain features of the solution are known before the numerical solution is determined. This* a priori *knowledge can be used to improve the numerical solution process.*

An obvious idea to enhance the approximation power of the FEM is to add "*suitable*" functions to the FEM space. These suitable functions represent the local nature of the solution. The direct consequence of this would be that we will be able to choose simpler meshes. But to do this we need to know something about the exact

Figure 1.15 Mesh for the domain Ω_+.

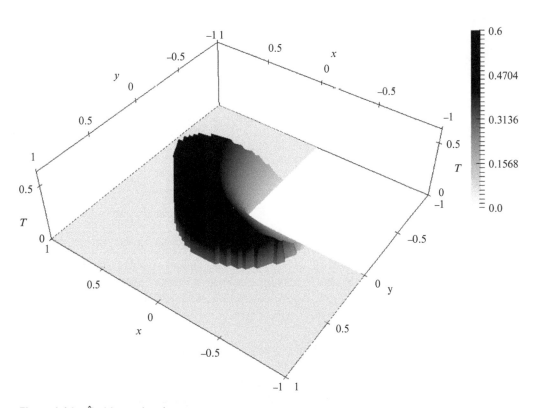

Figure 1.16 $\hat{T}$ with restricted support.

solution in advance. This is the basic idea of enrichment. For example, for the re-entrant corner, function $\hat{T}$ is a good candidate to generate functions. Using function $\hat{T}$ as globally to enrich the function space would not be appropriate since the function generally does not fulfill the essential boundary conditions. Also since its support is the whole domain its inner product with the shape functions will generally be non-zero, and this would

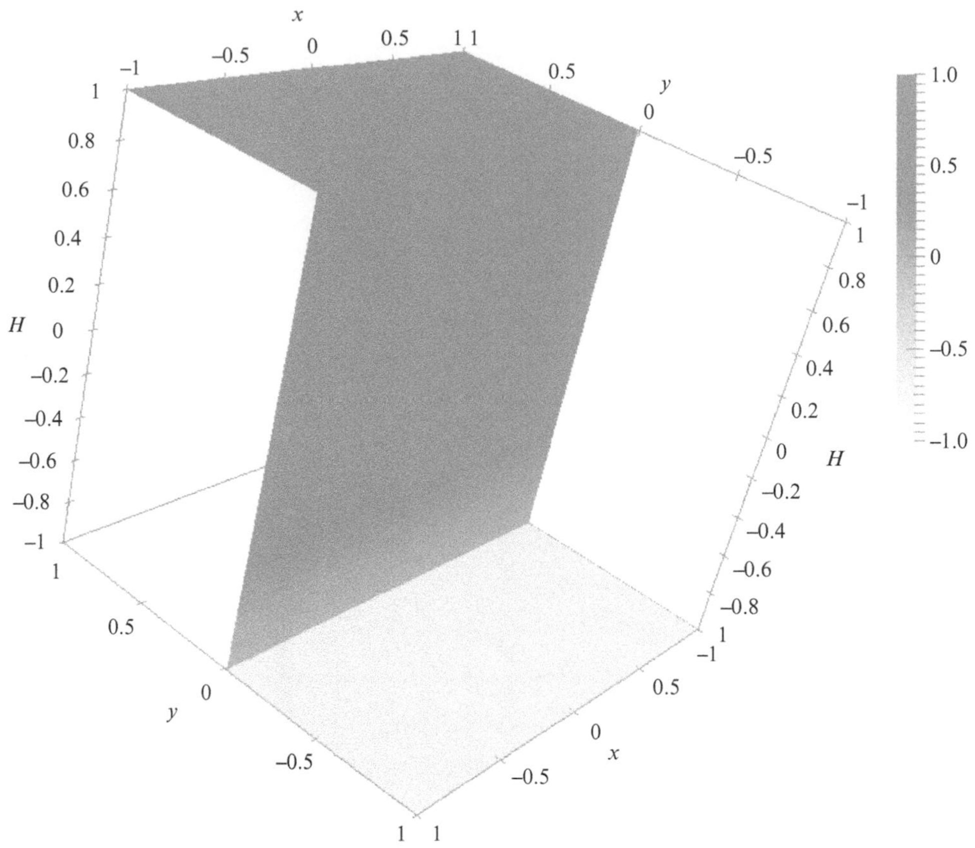

Figure 1.17 A discontinuous function.

increase the number of non-zero entries in the stiffness matrix significantly. However, the relevant part of $\hat{T}$ is the shape of the function in the vicinity of the re-entrant corner. A better approach would therefore be to set the function values of $\hat{T}$ to zero at some distance. But a transition zone must be set up to ensure continuity as discontinuous functions are not elements of $H^1(\Omega)$. A possible enrichment function could therefore be the one shown in Figure 1.16. Chapters 2 and 3 will provide details on general approaches to enrich finite element approximations.

We now return to mesh generation. A function that is discontinuous along the interface is shown in Figure 1.17. Following our recent thoughts, using such an enrichment function could enable the introduction of *a priori* knowledge about the solution within the approximation, but we do not want to use the whole function as an enrichment. Furthermore, the difference between function values in different subdomains may vary along the interface. Therefore, the final function space should be able to adjust the discontinuity locally. Thus, we have to split the function in Figure 1.17 somehow. In Figure 1.18, several functions are shown which are discontinuous along the interface. Their support is restricted to a small area, but their sum (Figure 1.19a) resembles the discontinuity partially. A different recombination of these functions can result in the function shown in Figure 1.19b. The functions are therefore able to locally represent the discontinuity, but also to change its shape along the interface. For example, functions whose derivative is discontinuous could be used to capture the discontinuity. A function whose derivative is discontinuous is shown in Figure 1.20. Again this function should be split into functions with smaller

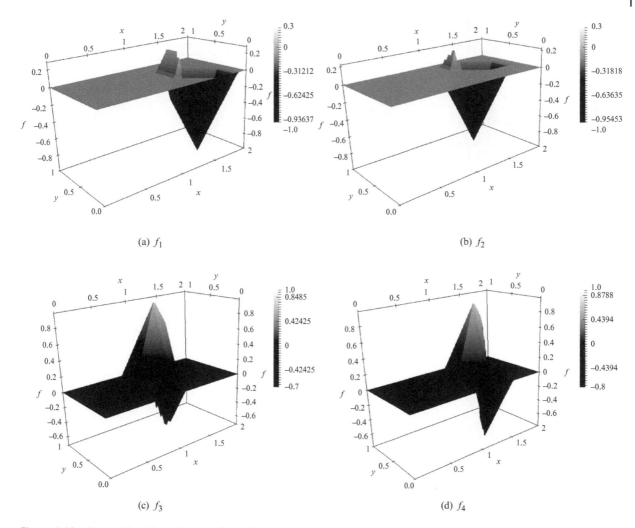

(a) f_1

(b) f_2

(c) f_3

(d) f_4

Figure 1.18 Several functions that are discontinuous along the interface.

support without loosing the ability to reproduce the weak discontinuity. In the first case, the domain boundary itself was curved. We already concluded that instead of meshing two adjacent domains independently of each other, we might add a strongly or weakly discontinuous function to the function space obtained from a non-conforming mesh. Similarly, we may, in this case, mesh a larger domain and separate the overlapping parts from the real domain by adding discontinuous enrichments.

1.5 Conclusions

In this chapter, we discussed the various sources of error in the numerical simulation, from setting up the problem to its numerical solution. We looked at *a priori* error estimation, which gives us, before carrying out a calculation, an upper bound (maximum value) for the error. We saw that the Galerkin finite element solution is optimal in

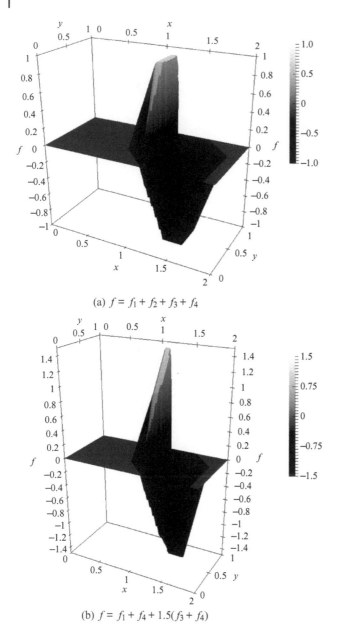

(a) $f = f_1 + f_2 + f_3 + f_4$

(b) $f = f_1 + f_4 + 1.5(f_3 + f_4)$

Figure 1.19 Recombination of the functions in Figure 1.18.

the sense of the bilinear form defining the boundary value problem. We learnt that the error of the finite element solution is governed by the degree of the continuity of the exact solution to be approximated, the polynomial order of the approximation, the mesh size, and the norm used to measure the error. We saw that the FEM is not well suited to solve problems with rough solutions. We later discussed how the approximation power of the FEM could possibly be improved by "*enriching*" the polynomial basis with appropriate functions. In the subsequent chapters (Chapters 2 and 3), we will discuss in greater detail how functions that contain *a priori* knowledge about

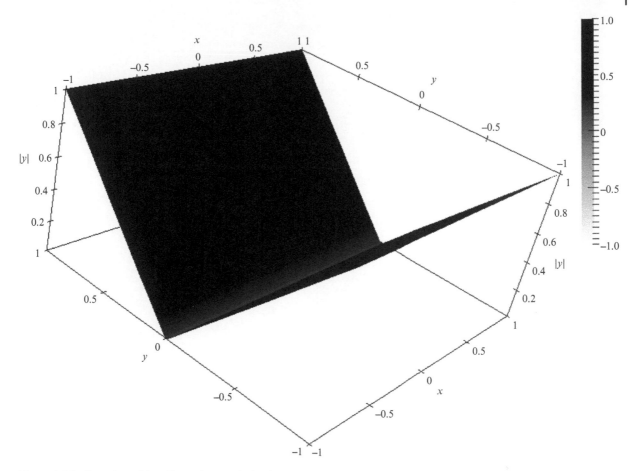

Figure 1.20 Function with a discontinuous derivative.

the solution can be split without loosing information. We will also discuss why phenomena like locking may not occur if enrichments are used.

References

Babuška, I. and Aziz, A.K. (1972). Lectures on mathematical foundations of the finite element method. Technical report, University of Maryland, College Park, Institute for Fluid Dynamics and Applied Mathematics.

Brenner, S.C. and Scott, L.R. (2002). *The Mathematical Theory of Finite Element Methods*. Springer.

Hrennikoff, A. (1941). Solution of problems of elasticity by the framework method. *Journal of Applied Mechanics* 8 (4): 169–175.

Zienkiewicz, O.C. (1971). *The Finite Element Method in Engineering Science*. McGraw-Hill Book Company.

2

A Step-by-Step Introduction to Enrichment

Stéphane P. A. Bordas[1], Alexander Menk[2], and Sundararajan Natarajan[3]

[1] *University of Luxembourg, Luxembourg, UK*
[2] *Robert Bosch GmbH, Germany*
[3] *Indian Institute of Technology Madras, India*

In the previous chapter, we discussed the approximation properties of the finite element method (FEM) and learnt that it is not ideally suited to treat problems involving "rough" solutions. This happens when the solution or its derivatives are of low-order continuity. Examples of problems with rough solutions include:

- partial differential equations (PDEs) with discontinuous coefficients (e.g., diffusion in multi-material domains, material interfaces, cracks, shocks, boundary layers, etc.);
- problems with singular solutions or steep gradients (e.g., cracks in linear elasticity, flow around obstacles, etc.).

We will require the following definitions.

> **Definition 1** A solution is said to be *strongly* discontinuous when the primary field is discontinuous. For example, a crack leads to strong discontinuities in the displacement field through the crack.
>
> **Definition 2** A solution is said to be *weakly* discontinuous when its first derivative is discontinuous. For example, perfect material interfaces lead to discontinuities in the strain or the flux field through the interface, i.e., discontinuities in the first (normal) derivative of the displacement or the temperature field.

From simple examples where the standard FEM is unable to reproduce certain features of the sought solution, we introduce a few enrichment schemes and solve the associated problems step by step by hand. In this book, enrichment is meant as a procedure to increase the approximation property of a given numerical approximation by adding specially selected bespoke functions, known to better approximate the solution. Based on this very broad definition, h, p, e (Bordas and Duflot, 2007; Bordas et al., 2008) and k (Hughes et al., 2005) refinement falls within the category of "enrichment schemes."

The role of this chapter is to provide no more than a succinct introduction to enrichment without delving into details that would confuse the reader at this point. Before providing simple examples of enrichment schemes for one-dimensional (1D) problems, we shall start by proposing a definition of the notion of enrichment through a historical perspective. By embedding it within a larger class of enrichment schemes, this discussion may be useful to understand the emergence of the particular type of enrichment which is the focus of this book, i.e., partition

Partition of Unity Methods, First Edition. Stéphane P. A. Bordas, Alexander Menk, and Sundararajan Natarajan.
© 2024 John Wiley & Sons Ltd. Published 2024 by John Wiley & Sons Ltd.

of unity enrichment for FEMs. A more rigorous definition of partition of unity enrichment itself is proposed in Chapter 3 and in the work of Babuška's group, in particular Babuška et al. (1994) and Melenk and Babuška (1996). In the following historical review, we consider only enrichment with specific *a priori* knowledge about the solution, as is the case with partition of unity enrichment, since this will be the central point of the book. We will therefore review, albeit succinctly and non-exhaustively:

- strong discontinuities (jumps in the primal unknown field) introduced at the element or node level;
- weak discontinuities (jumps in the first derivative of the primal unknown field) introduced at the element or node level;
- global enrichment based on *a priori* knowledge of the solution (near-tip fields in fracture mechanics, boundary layers, etc.);
- local enrichment based on *a priori* knowledge of the solution introduced intrinsically or extrinsically at the node or element level.

We denote by the term "local" any enrichment scheme which is based on a local partition of unity approach, as opposed to "global" enrichment schemes, where the enrichment is introduced independently of the underlying partition of unity. The *banded* structure of the matrix is retained in the local enrichment schemes, while it is larger in case of global enrichment scheme. Moreover, global enrichment schemes lead to potential storage difficulties and deterioration of the conditioning of the resulting matrix.

> ☞ **Intrinsic enrichment**: In this type of enrichment, there are no additional degrees of freedom. This term is used more particularly in the context of meshless methods such as intrinsic enrichment of the moving least square (MLS) basis as in Fleming et al. (1997), or weight function enrichment as in Duflot and Nguyen-Dang (2004a). It also appears within the realm of partition of unity methods, in particular with the intrinsic extended finite element method (XFEM) of Fries and Belytschko (2006). The same is true of hybrid Trefftz (1926) elements proposed by Trefftz where additional degrees of freedom are condensed at the element level. Intrinsic enrichment schemes also include most versions of elements with embedded discontinuities.

> ☞ **Extrinsic enrichment**: This type of enrichment is accompanied by additional degrees of freedom. Examples of this include the partition of unity method of Melenk and Babuška (1996) and its siblings, the XFEM of Moës et al. (1999) and generalized FEM (GFEM) of Strouboulis et al. (2000b). Methods such as partition of unity enriched meshfree methods, such as *hp*− clouds, enriched element-free Galerkin (EFG) also fall in this category.

In this chapter we introduce several key concepts for the first time:

- Enrichment functions for strong and weak discontinuities, localized gradients (details can be found in Section 2.2);
- Reproducing conditions for non-polynomial functions (details can be found in Chapter 3);
- Blending elements, spurious terms, and lack of partition of unity (details can be found in Chapter 4);
- Mesh-geometry interaction and level sets;
- Special numerical integration rules for enriched approximations (details can be found in Chapter 4);
- Imposition of essential (Dirichlet) boundary conditions on enriched nodes;
- Enriched right-hand side (e.g., nodal force vector).

2.1 History of Enrichment for Singularities and Localized Gradients

In the following, we consider more particularly discontinuous and "singular" enrichment, as those are the most commonly encountered in the literature. Using *a priori* knowledge about the solution in the approximation is a rather old idea. As an alternative to the Ritz method, the seminal work of Trefftz (1926), early in the second quarter of the twentieth century, where trial functions which satisfy the governing equations *a priori* are used within a variational formulation, is probably one of the earliest examples known to the authors of such an idea. Within the Ritz formalism, Fix (1969), in his "method of supplementary singular functions", also sought to improve convergence rates through enrichment and Fix and Strang were the main pioneers in this area. It will be helpful in the historical account below to separate the enrichment aiming at representing large gradients, oscillatory behavior, singularities, and boundary layers, known to represent well the solution of the boundary value problem, from the enrichments designed to capture discontinuities (strong or weak). The latter "discontinuous enrichments" also aim at improving the ability of a numerical method to reproduce the solution, but require rather special treatments, as will become clear below. The most salient points of the discussion are summarized in Table 2.1 for singular enrichment, and Table 2.2 for discontinuous enrichment.

2.1.1 Enrichment by the "Method of Supplementary Singular Functions"

A turning point in the treatment of singularities by FEMs occurred in the area of eigenvalue problems for elliptic problems with first results in the PhD thesis of George Fix at Harvard University (Fix (1968)) and a first paper in 1969 published in the *Journal of Mathematics and Mechanics* by George Fix and communicated by Garrett Birkhoff. The central idea of the method of "supplementary singular functions" is the following: Assume that the exact solution to the problem is u and contains a (known) singularity. The idea is to construct independent (singular) functions ψ_J able to represent the singularity and subtract from u a linear combination of those functions, chosen such that the resulting function is smooth and thus can be well approximated by (piecewise polynomial) finite element (FE) spaces. In this sense, the additional singular functions take care of approximating the solution near the singular points and the piecewise polynomial functions take over, far from the singularity. The key result of Fix (1969) is that, for a particular 2D elliptic problem (a particular case of which is the Helmholtz equation) defined on a domain containing re-entrant corners, the addition of suitably chosen singular functions to a p-degree, polynomial, Rayleigh-Ritz approximation leads to an error of order $h^2 p$ where h is the maximum characteristic mesh size. This work by Fix was followed by a sequence of papers, notably by Byskov (1970), Wait and Mitchell (1971), Fix et al. (1973), Benzley (1974), and Wait and Mitchell (1971), appearing between year 1970 and 1974.

In Byskov (1970), the author presented a version of the FEM including special "cracked elements" which addresses the difficulties encountered by the standard FEM in dealing with the stress singularities at the crack tip. In this formulation, the cracked element is a polygon containing a straight linear crack. An analytical solution respecting the traction boundary conditions in the element is used. The formulation does not provide continuity between the cracked element and neighboring elements such that a non-optimal convergence rate is achieved.

Wait and Mitchell (1971) applied the "method of supplementary singular functions" to a harmonic, mixed boundary value problem. This is a scalar Laplace problem with mixed boundary conditions solved on 2D problems with boundary singularities. The enrichment functions used are as derived by Lehman (1959) and Wigley (1969) are chosen to reproduce *all* the leading terms of the exact, singular, solution. The convergence results were provided in Fix (1969), showing that if all the basis functions are bilinear outside a neighborhood of the singular point and if the enrichment functions used are able to approximate the exact solution with second-order accuracy in another neighborhood of the singular point, then, a convergence rate of unity is obtained in the energy norm. This convergence rate is the same as that obtained by Birkhoff et al. (1968) for bilinear approximations in rectangular domains without singularities. This means that adding the singular functions to the FE approximation enables recovery of optimal convergence rates in the presence of singularities.

Table 2.1 A brief overview of enrichment techniques introduced since the 1970s to add *a priori* knowledge about the solution to the finite element space.

Published date (submission date)	References	Method	Local	Global
1926	Trefftz (1926)	Trial functions satisfying *a priori* the governing equations		✓
1969	Fix (1969)	Supplementary singular functions		✓
1970	Byskov (1970)	Stress intensity factors, cracked elements		
1970 (Dec. 1969)	Mote Jr (1971)	Global/local finite element method	✓	✓
1971 (Jul. 1970)	Wait and Mitchell (1971)	Singular Q4 elements for corner singularities		
1973 (Apr. 1972)	Fix et al. (1973)		✓	
1974 (Aug. 1973)	Benzley (1974)	Generalized Q4 elements that include a singularity at a corner node. Relies on the global local method of Mote Jr (1971)	✓	
1975 (Dec. 1973)	Henshell and Shaw(1975)	Quadrilateral element with corner singularity	✓	
1976	Barsoum (1976)	Finite element with a singularity	✓	
1994 (Mar. 1992)	Babuška et al. (1994)	Three special FEMs for second-order elliptic problems with rough coefficients. Hat functions used as partition of unity.	✓	
1995	Melenk (1995)	Generalization of Babuška et al. (1994) to any partition of unity, mathematical theory and applications	✓	
1996 (Jun. 1995)	Duarte and Oden (1996)	*hp* clouds	Meshless like enrichment based on clouds	✓
1996 (Dec.)	Duarte (1996)	*hp* clouds	Meshless like enrichment based on clouds	✓
1996 (Apr. 1996)	Melenk and Babuška(1996	Partition of unity method (PUM)	✓	
1997 (Jul. 1995)	Melenk and Babuška(1997	PUFEM (partition of unity FEM) oscillatory solutions, sharp gradients, rough coefficients	✓	
1997 (Jan. 1996)	Fleming et al. (1997)	Two methods to enrich the EFG globally: (1) similar to Benzley(1974) and Nash Gifford and Hilton (1978) with additional unknowns for the stress intensity factors. (2) with intrinsic enrichment of the MLS approximation		✓ method (2) can be localized using a ramp function
1997 (Dec. 1996)	Oden and Duarte (1997)	*hp* clouds. Clouds, cracks, and FEMs	✓	

1998 (Dec. 1996)	Oden et al. (1998)	*hp* clouds. Clouds, cracks, and FEMs where the clouds are based on a finite element partition of unity	✓	
1999 (Jul.1998)	Belytschko and Black(1999)	Enrichment of a hat function-based partition of unity with singular and discontinuous enrichment functions to simulate crack growth with minimal remeshing	✓	
1999 (Feb. 1999)	Moës et al. (1999)	Crack growth without remeshing	✓	
1999	Dolbow (1999)	First introduction of the name XFEM. Hat functions-based partition of unity for crack propagation problems (singularities and discontinuities)	✓	
2000 (Nov. 1998)	Strouboulis et al.(2000b); Strouboulis et al.(2001)	GFEM. Hat functions partition of unity enriched with non-polynomial functions. See special FEM3 in Babuška et al.(1994). Only a few enrichment functions are sufficient for problems with singularities. Introduction of numerically computed enrichment functions	✓	
2000 (Sep. 1999)	Sukumar et al. (2000)	XFEM for 3D crack growth without remeshing	✓	
2001 (Jul. 2000)	Strouboulis et al. (2001)	See Strouboulis et al. (2000b) and Strouboulis et al. (2001). Introduction of numerically computed enrichment functions.	✓	
2001 (Jan. 2001)	Karihaloo and Xiao(2001)	Hybrid crack element	✓	
2002 (Jul. 2001)	Ventura et al. (2002)	Extrinsic enrichment of the element-free Galerkin method (EFG) for discontinuities and singularities	✓	
2004 (Jul. 2001)	Duflot and Nguyen-Dang(2004a	Weight function enrichment (MLS)	✓	
2004 (Dec. 2002)	Rao and Rahman (2004)	Global enrichment of type Fleming et al. (1997) for linear and HRR crack tip fields		✓
2004 (Mar. 2003)	Lee et al. (2004)	Combined superimposed FEM and XFEM. Singularity brought by quarter-point elements	✓	
2006 (Mar. 2006)	Fries and Belytschko(2006)	Intrinsic XFEM, weight function enrichment (MLS) without additional degrees of freedom	✓	
2007	Rabczuk et al. (2007)	Extrinsically enriched EFG for cracks including asymptotic enrichment	✓	
2008 (Dec. 2006)	Bordas et al. (2008b)	Extrinsically enriched EFG for cracks without asymptotic enrichment	✓	

(Continued)

Table 2.1 (Continued)

Published date (submission date)	References	Method	Local	Global
2010 (Sep. 2007)	Bordas et al. (2010); Bordas et al. (2011); Vu-Bac et al. (2011)	Strain smoothing for enriched approximations	✓	
2010 (Dec. 2008)	Réthoré et al. (2010)	Combined hybrid crack element and XFEM	✓	
2011 (Jul. 2009)	Simpson and Trevelyan(2011a); Simpson and Trevelyan(2011b)	Enriched boundary element method for linear elastic fracture	✓	
2011 (Oct. 2010)	De Luycker et al. (2011)	Enriched isogeometric analysis	✓	

Table 2.2 Overview of enrichment techniques for discontinuities. Weight: stands for "enrichment of the weight function"; MLS: moving least squares; intrinsic enrichment Adapted from Nguyen et al. (2008) for details pertaining to enrichment in meshfree methods as well as associated implementation aspects.

Published date (submission date)	References	Method	Strong	Weak	Node level	Element level	PU
1987 (Apr. 1986)	Ortiz et al. (1987)	Finite element for localized failure	✓			✓	NA
1988 (Nov. 1987)	Belytschko et al. (1988)	Embedded localization zone		✓		✓	NA
1990 (Sep. 1989)	Dvorkin et al. (1990)	Embedded element with localization line	Strong	Weak	Node level	Element level	PU
1991	Shi (1991)	Manifold method	✓			✓	Hats
1991	Klisinski et al. (1991)	Inner softening band	✓			✓	NA
1992	Lotfi (1992)	Embedded crack	✓			✓	NA
1992	Shi (1992)	Numerical manifold method	✓			✓	NA
1993 (Mar. 1992)	Larsson et al. (1993)	Discontinuous displacement approximation	✓			✓	NA
1993	Simo et al. (1993)	Element with a strong discontinuity	✓			✓	NA
1993 (Mar. 1992)	Larsson et al. (1993)	Discontinuous displacement approximation	✓			✓	NA
1994	Simo and Oliver (1994)	Element with a strong discontinuity	✓			✓	NA
1995	Larsson et al. (1995)	Embedded cohesive crack	✓			✓	NA
1995 (Feb. 1994)	Lotfi and Shing (1995)	Embedded crack	✓			✓	NA
1996 (May 1995)	Oliver (1996)	Element with a strong discontinuity	✓			✓	NA
1996	Armero and Garikipati (1995)	Element with a strong discontinuity	✓			✓	NA
1996	Larsson and Runesson(1996)	Embedded localization band, regularized displacement discontinuity	✓			✓	NA
1998 (Nov. 1996)	Sluys and Berends (1998)	Discontinuous modeling of mode I and mode II cracking		✓	–	–	

(Continued)

Table 2.2 (Continued)

Published date (submission date)	References	Method	Strong	Weak	Node level	Element level	PU
1997 (Jan. 1996)	Fleming et al. (1997)	Two methods to enrich the EFG globally: (1) similar to Benzley (1974) and Nash Gifford and Hilton (1978) with additional unknowns for the stress intensity factors; (2) with intrinsic enrichment of the MLS approximation	Weight (visibility [Krysl and Belytschko, 1997]; diffraction [Organ et al., 1996])				MLS
1998 (Jun. 1996)	Krongauz and Belytschko (1998)	Enrichment of MLS functions with weak discontinuities		✓	✓		MLS
1999 (Jul. 1998)	Belytschko and Black (1999)	Enrichment of a hat function-based partition of unity with singular and discontinuous enrichment functions to simulate crack growth with minimal remeshing	✓		✓		Hats
1999 (Feb. 1999)	Moës et al.(1999)	Crack growth without remeshing	✓		✓		Hats
1999	J. Dolbow's PhD Dolbow (1999)	First introduction of the name XFEM. Hat functions-based partition of unity for crack propagation problems (singularities and discontinuities)	✓		✓		Hats
2002 (Jul. 2001)	Ventura et al. (2002)	Extrinsic enrichment of EFG and vector level sets	✓		✓		MLS
2003 (Aug. 2001)	Lin (2003)	Meshless finite cover method	✓	✓	✓		Hats
2003 (Jul. 2002)	Terada et al. (2003)	Finite cover method (see numerical manifold method)	✓	✓	✓		Hats
2003 (NA)	Remmers et al. (2003)	Cohesive segments	✓		✓		Hats
2004 (Mar. 2003)	Lee et al. (2004)	Combined superimposed FEM and XFEM. Singularity brought by quarter-point elements	✓		✓		Hats
2004 (Apr. 2004)	Rabczuk et al. (2004)	Cracking particles (see Remmers et al., 2003)	✓		✓		MLS
2007 (NA)	Rabczuk et al. (2007)	Continuous cracks with asymptotic enrichment	✓		✓		MLS
2008 (Dec. 2006)	Bordas et al.(2008b)	Continuous cracks without asymptotic enrichment	✓		✓		MLS
2011 (Jan. 2010)	Möes et al. (2011)	Thick level set approach		✓		Hats	
2012 (Oct. 2010)	Barbieri et al.(2012)	Enriched weight functions		✓		MLS	

Fix et al. (1973), within the framework of the "method of supplementary singular functions", showed numerically that cubic approximations are far more efficient than linear approximations when used in conjunction with singular enrichment (in the sense of accuracy per degree of freedom). In that paper, enrichment is either introduced on the whole domain, or on a subset of the domain and blended with the standard approximation over a transition/blending region. The results show that both variations lead to very similar results. The authors also report that cubic splines (with suitable modifications) are the most suitable cubic approximations to be used with singular functions. They also show that such enrichment techniques are more efficient than mesh refinement based on triangular elements. The authors explain the source of ill-conditioning of the matrix when singular enrichment is used (especially of high order), i.e., that the singular terms can be approximated by the standard FE shape functions, thereby leading to a nearly linearly dependent basis spanned by the FE shape functions and the additional singular functions. They propose a solution to alleviate this ill-conditioning and the lack of sparsity caused by the addition of enrichment, by using bordering techniques (Faddeev and Faddeeva, 1981).

The reference book by Strang and Fix (1973) presents, as early as 1973, a detailed account of the approximation properties of enriched approximations and the reader is referred to this excellent monograph for more details on the approximations described above. In particular, Section 8.2, p. 263 of the second edition deals with the principles of the method of supplementary singular functions.

2.1.2 Finite Element with a Singularity

Benzley (1974) proposed arbitrary quadrilateral FE with a singular corner node. The element ensures conformity across element edges and ensures optimal convergence[1]. The author writes that the method relies on "enriching" the standard FE space with the appropriate singular functions. The method relies on the local/global FEM of Mote Jr (1971). It is already observed in this paper that "[A] convergence criterion for enriched elements requires that the singular region not shrink to zero"[2]. Numerical integration of the non-polynomial terms is also discussed and a 7×7 Gauss quadrature scheme is used. The transition from the element with a singular node to the standard element is done by using a ramp function. A similar idea is used in the context of corrected/weighted XFEM approaches.

Barsoum (1974) and Henshell and Shaw (1975) seem to have co-discovered the idea of moving the mid-side nodes of isoparametric elements in order to generate the singularity required to resolve the stress field close to a crack tip in a linear elastic body. The authors propose a simple way of introducing singularities in isoparametric FEs, by displacing the mid-side node to a quarter of the element edge length from the node where the singularity is desired (see Figure 2.1). It is shown that these elements lead to comparable accuracy with other "special" crack tip elements introducing a singularity in the approximation[3].

⊙ **Example 1.** Consider a three-noded quadratic element with nodes at $x_1 = 0$, $x_2 = L$, and $x_3 = L/2$. Show that when node 3 (x_3) is specified at a distance of $L/4$ from node at x_1, the stress has a singularity of $1/\sqrt{x}$ at node 1.

---

1 In a partition of unity enrichment context, this notion of conformability is also key to ensure convergence and known as "blending."
2 The same notion exists in the XFEM in the sense that topological enrichment, where only the tip element is enriched with near-tip functions, is sub-optimal, while geometrical enrichment, where the tip enrichment lives over a region whose size is independent of the mesh size, leads to optimal convergence rate.
3 The fact that singularity could be obtained by placing mid-side nodes closer than 1/4 along the side of 2D isoparametric elements appears to have been first discovered by Jordan (1970). This was further investigated by Steinmueller (1974) in the context of mesh generation. This paper (Henshell and Shaw, 1975), fruit of the combined efforts of a fracture mechanics and mesh quality/generation researcher, is the first to report on the nature of singularity and its potential use in fracture mechanics.

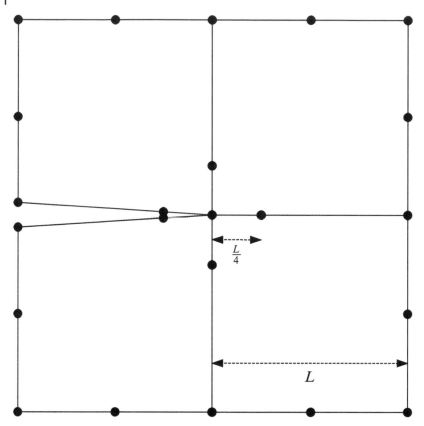

Figure 2.1 Two-dimensional quarter-point isoparametric element. Within the element, the stresses are singular for any angle θ.

A singularity mapping technique was used in Yamada et al. (1979) to represent the singularity at the crack tip. Based on this technique, a three-noded triangular element (capable of representing the singularity) is developed by degenerating the four-noded quadrilateral element. More interestingly, the technique was not restrictive to $\sqrt{r}$ singularity. Further to these initial developments, the quarter-node, or, more generally, distorted elements briefly defined above were generalized to higher dimension and higher order approximations. The review of these elements is not in the scope of this book, and the reader is referred to Shih et al. (1976) (crack growth), Foschi and Barrett (1976) and Saouma and Sikiotis (1986) (anisotropic cracks), Hibbitt (1977) (some properties of singular elements), Pu et al. (1978) (collapsed cubic isoparametric FE), Lynn and Ingraffea (1978) (transition elements), Hussain et al. (1981) (3D singular element) among many others. The work of Pageau et al. (1995) proposed an enriched approximation with a numerical solution for free-edge singular stress fields in anisotropic materials based on singularity mapping. This was further extended to various cases (Pageau and Biggers, 1995) including 3D cases in Pageau et al. (1995) and Pageau and Biggers (1996) and n-material wedges in Pageau et al. (1996).

☞ **Quarter-node elements versus "special" singular elements** The work of Henshell and Shaw (1975) led to a series of discussion articles commenting on these new "quarter-point isoparametric elements" for fracture. The reader can refer to Barsoum (1976, 1977), and Hibbitt (1977) for example. Some of the conclusions from those discussions are that triangular quarter-node elements are more accurate than their rectangular counterpart, that eight elements around the crack tip were generally sufficient in practical applications and that placing the quarter node at 1% of the crack length along the crack from the crack tip provided satisfactory results. It is also discussed in Barsoum (1976) that an advantage of quarter-node elements over "special" singular elements explicitly including the asymptotic fields is the possibility to involve constant strain terms. A disadvantage, however, is that while well suited for linear elastic fracture mechanics, these quarter-node elements cannot be extended to large-scale yielding, contrary to "special" singular elements (Barsoum, 1977; Shih, 1974; Tracey, 1976). Additionally, the quarter-node elements pass the patch test, contrary to the elements of Tracey (1971).

2.1.3 Partition of Unity Enrichment

FEMs using special non-polynomial shape functions were proposed in Babuška et al. (1994) for second-order problems with non-smooth (rough) coefficients and for which polynomial spaces are not adapted. In one of the methods proposed in Babuška et al. (1994), the "Special FEM3," the approximation functions are compactly (locally) supported. They are constructed as products of piecewise linear FE hat functions with a "special" (non-polynomial) function which is chosen to replicate known behaviors of the unknown solution.

Later, in the PhD thesis of Melenk (1995), the method was generalized and the mathematical theory detailed. The hat functions were replaced by an arbitrary, compactly supported, partition of unity and the method was first coined "partition of unity method (PUM)" in Melenk and Babuška (1996, 1997). These two publications give detailed theoretical and numerical results on the behavior of the PUM for problems with oscillatory or rough solutions as well as boundary layers.

From the PUM, several methods were born, which are all very similar. The first one is the generalized finite element method (GFEM), proposed by Strouboulis et al. (2000a,b, 2001), and where hat functions are used as partition of unity to enrich the FE space with non-polynomial functions. It is shown that only a few functions are sufficient in order to approximate singular solutions properly. The key novelty of these three papers, however, is the introduction of "numerically determined enrichment functions" obtained by the solution of local problems.

In parallel to the development of the GFEM, hat functions were used to construct discontinuous enrichment in Belytschko and Black (1999) and Moës et al. (1999) in the context of linear elastic crack growth with minimal remeshing. In the PhD thesis of Dolbow (1999) and subsequent papers, in particular Moës et al. (1999), this method was first named the extended finite element method (XFEM). Since then, it has continued to be associated with discontinuous partition of unity enrichment. Shortly after this, the XFEM was extended to 3D crack propagation (Sukumar et al. 2000), holes, arbitrarily branched cracks (Daux et al. 2000), and the fracture of Mindlin-Reissner plates (Dolbow et al. 2000).

The $h - p$ cloud method of Duarte and Oden (1996) is also a partition of unity method. Clouds are defined in Duarte and Oden (1996). The basic idea is to multiply these clouds, which form a partition of unity, by functions with good approximation properties for a given solution space. A simple example of clouds, in the context of the FEM, is the support of a node. The $h - p$ cloud method may be seen as a superclass of enriched FEMs and enriched meshless methods. The method was also extended to crack problems in Oden and Duarte (1997). In Oden et al. (1998), hat functions were used as clouds, similarly to the Special FEM3, to the GFEM, and the XFEM mentioned above.

☞ **Enrichment in meshfree methods** Further to the introduction of partition of unity enrichment in the context presented within the above paragraphs, the idea of enrichment became popular, and novel techniques were gradually introduced, to improve the approximation property of various discretization schemes. It is not in the scope of this book to review all these schemes in detail and we provide merely a summary. More details on enriched approximation schemes within meshless methods are provided in Nguyen et al. (2008). Note that around 1996, several papers dealing with the enrichment of so-called meshfree (or meshless) methods were published. In Krongauz and Belytschko (1998), an enriched method to capture weak discontinuities without a conforming mesh was proposed. Fleming et al. (1997) introduced the concept of enrichment of an MLS basis, in the context of the element-free Galerkin (EFG) method, which allowed the authors to propagate cracks with minimal alterations to the point distribution in the meshless method[a].

Duflot and Nguyen-Dang (2004a,b) and Duflot (2006) proposed, within the framework of meshfree methods based on MLS shape functions, the concept of "weight function enrichment" which was later revisited by Barbieri et al. (2012). The idea is to exploit the fact that the MLS shape functions inherit their continuity and properties from the weight functions. By modifying the latter, to better approximate the solution, Duflot was able to propagate cracks in three dimensions without remeshing, while capturing the discontinuities along the crack faces, and the singularities along the crack fronts. Rabczuk et al. (2007) extended the 2D work of Ventura et al. (2002) to extrinsically enriched EFG for 3D cracks in nonlinear materials. The method included asymptotic enrichment to close the cracks along their front. In contrast, Bordas et al. (2008b) proposed a simple extrinsically enriched EFG for 3D cracks without asymptotic enrichment, which simplified the implementation.

---

[a] See Nguyen et al. (2008) for details on enrichment within meshless formulations.

Rao and Rahman (2004) introduced enrichment of the same type as Fleming et al. (1997) for cracks in linear and elasto-plastic materials using enrichment based on the linear crack tip fields and on the Hutchinson-Rice-Rosengren fields for elasto-plasticity (Hutchinson, 1968), respectively. Lee et al. (2004) combined the superimposed FEM with XFEM, where the singularity part is introduced through quarter-point elements. Fries and Belytschko (2006) developed the "intrinsic XFEM" as a blend between a FE approximation and an intrinsic enrichment as introduced in meshless methods, where the basis vector, **p**, is enriched with *a priori* knowledge about the solution. The general idea of the method is to construct shape functions which are themselves able to reproduce the desired behavior, by constructing mesh-based MLS shape functions with suitably chosen basis vectors **p** and blending them together appropriately. The main advantages are that no additional degrees of freedom is required and that the conditioning number scales with the number of elements similarly to the standard FEM case. In a series of papers, Bordas et al. (2010), Bordas et al. (2011), and Vu-Bac et al. (2011) investigated the behavior of strain smoothing for enriched approximations and Réthoré et al. (2010) combined a hybrid crack element and the XFEM, where the region in the vicinity of the crack tip was modeled with analytical expressions, while the rest of the domain with FEs. The transition zone was modeled by using a cut-off function.

Simpson and Trevelyan (2011a) and Simpson and Trevelyan (2011b) introduced enrichment within boundary element methods for linear elastic fracture in two dimensions. De Luycker et al. (2011) showed how the recently introduced idea of isogeometric analysis could be enriched using similar methods as for standard FEs and applied the method to fracture mechanics.

☞ **Choice of local or global enrichment** In the method of supplementary singular function, the space is augmented with, say, a function of type x^{α}, which is a global shape function, i.e., defined on the whole domain, not premultiplied by local shape functions (e.g., those of a an FE basis). If this function is introduced over the whole domain, then, its presence leads to an extra "full row" and an extra "full column" in the stiffness matrix which is associated with this global shape function. This destroys the locality of the enrichment, the band structure of the matrix, leads to extra storage requirements, and deteriorates the conditioning number of the matrix. In the case where x^α is cut-off after a certain distance around the singular point, the "full" row and column are replaced by a "long" row and column, which also destroys the local structure of the stiffness matrix. In the particular case of a 1D partition of unity constituted of hat functions, the tridiagonal stiffness matrix is destroyed by any "global shape function." In the partition of unity method on the other hand, instead of augmenting the FEM space by x^α, we augment it by the shape function $N_I(x)x^\alpha$, where N_I is the hat function associated with the node x_I. Of course, we are increasing the dimension of the matrix, but its local structure is kept, i.e., the functions $N_I(x)x^\alpha$ extend only over the support of shape function N_I. Despite the fact that extra unknowns are required, and that thus the stiffness matrix is larger, the matrix is still banded, storage is no longer a problem, but conditioning may still be an issue since the enrichment functions can also be represented by the underlying partition of unity.

☞ Another interesting observation is that, in the case where the additional enrichment function (say x^α) is defined over the whole domain, i.e., it is not cut-off after a certain distance around the singularity, the method of supplementary singular functions is equivalent to partition of unity finite element methods (PUFEM/XFEM/GFEM). This is easily seen. Assume we augment the FEM space with $N_I(x)x^\alpha$ for all nodes x_I in the domain. In this case, the sum of the local enrichments writes $\sum_I N_I(x)x^\alpha = x^\alpha \sum_I N_I(x) = x^\alpha \cdot 1$. The last equality is obtained using the partition of unity property ($\sum_I N_I = 1$). This shows that function x^α also belongs to the enriched approximation space.

Consider a mesh of m nodes with standard hat functions $(N_I)_{1\leq I\leq m}$ defined at these nodes. Assuming that a function x^α is added to the approximation, the approximation space of the method of supplementary enrichment functions is $\mathcal{S}_{MSF} \equiv span(N_1, N_2, \ldots, N_m, x^\alpha)$, which is contained within the XFEM/GFEM approximation space, i.e., $\mathcal{S}_{XFEM} \equiv span(N_1, N_2, \ldots, N_m, N_1x^\alpha, N_2x^\alpha, \ldots, N_mx^\alpha)$, where it is assumed that all nodes are enriched.

This can be checked easily. Let $v \in \mathcal{S}_{MSF}$, then there exist real numbers $(a_I)_{1\leq I\leq m}$ and b such that $v = \sum_{I=1}^{m} a_I N_I + bx^\alpha$. And if we choose m coefficients $(b_J)_{1\leq J\leq m}$ to be all equal to b, we can write v as:

$$v = \sum_{I=1}^{m} a_I N_I + \sum_{J=1}^{m} b_J N_J x^\alpha$$

Therefore, given any $v \in \mathcal{S}_{MSF}$, we were able to write v as a linear combination of $(N_I)_{1\leq I\leq m}$ and $(N_J x^\alpha)_{1\leq J\leq m}$, which means that v is an element of $\mathcal{S}_{XFEM}$, and, in turn that $\mathcal{S}_{MSF} \subset \mathcal{S}_{XFEM}$. However, the converse is not true, i.e., $\mathcal{S}_{XFEM}$ is not contained in $\mathcal{S}_{MSF}$, since the dimension of $\mathcal{S}_{XFEM}$ is greater than that of $\mathcal{S}_{MSF}$: $\dim(\mathcal{S}_{XFEM}) = 2m > m + 1 = \dim(\mathcal{S}_{MSF})$.

2.1.4 Mesh Overlay Methods

Mote Jr (1971) introduced the idea of "global-local" FE. In this method, a *priori* known shape functions are added to enrich the local FEM. The formulation also includes singular elements in fracture mechanics. Tong et al. (1973) used a hybrid-element concept and complex variables to develop a super-element to be used in combination with the standard FE for problems in fracture mechanics. Belytschko et al. (1988) proposed the spectral overlay method for problems with high gradients. This is accomplished by superimposing the spectral approximation over the FE mesh in regions where high gradients are indicated by the solution. Based on this work, Fish (1992) proposed *s*-version of the FEM. This method consists of overlaying the FE mesh with a patch of higher order hierarchical elements in the regions of high gradients.

Hughes (1995) and Hughes et al. (1998) introduced the variational multiscale method (VMM) to solve problems involving multiscale phenomena. The idea is to decompose the displacement field into two parts: a coarse scale and a fine scale contribution. The fine scale contribution incorporates the local behavior, which is determined analytically. Hettich et al. (2008) combined XFEM with a multiscale approach to model failure of composites using a two-scale approach. Based on this work, Mergheim (2009) applied VMM to solve propagation of cracks at finite strains. Both scales are discretized with FEs.

2.1.5 Enrichment for Strong Discontinuities

Ortiz et al. (1987) proposed a method to model strain localization using isoparametric elements, which involves augmenting the FE approximation basis with localized deformation modes in the regions where localization is detected. The method was applied to study strain localization in both nearly incompressible and compressible solids. The additional degrees of freedom due to the localized deformation modes are eliminated at the element level by static condensation. Note that these localized modes do not form partition of unity.

By modifying the strain field, Belytschko et al. (1988) proposed a method to model localization zones. The jumps in strain are obtained by imposing traction continuity. Both these approaches use bifurcation analysis to detect the onset of localization. One of the distinct differences between these two methods is that the width of the localization band is smaller than the mesh parameter, h, in case of the method proposed by Belytschko et al. (1988), whereas the localization band width is of the same size as the mesh parameter in the method of Ortiz et al. (1987) and a very fine mesh is required to resolve the localization band. By using overlapping elements, Hansbo and Hansbo (2004) suggested a method to model strong and weak discontinuities. In their method, when a discontinuity surface cuts an element e, another element, $\bar{e}$, is superimposed and the standard FE displacement field is replaced by:

$$\mathbf{u}^h(\mathbf{x}) = \sum_{I \in e} N_I^e(\mathbf{x}) H(\mathbf{x}) \mathbf{q}_I^e + \sum_{J \in \bar{e}} N_J^{\bar{e}}(\mathbf{x})(1 - H(\mathbf{x})) \mathbf{q}_J^{\bar{e}} \tag{2.1}$$

where N_I are the standard FE shape functions, H is a step function, and $\mathbf{q}_I$ and $\mathbf{q}_J$ are the nodal variables associated with node I and node J, respectively. Areias and Belytschko (2006) showed that Hansbo and Hansbo's approximation basis is a linear combination of the XFEM basis and hence that both approximation spaces are identical. In the Hansbo-Hansbo method, the crack kinematics is obtained by overlapping elements and this does not introduce any additional degrees of freedom, see Figure 2.2. This has some advantages with respect to the implementation of the method in existing FE code and is also known as the "phantom node method."

In the discontinuous enrichment method (DEM) (Farhat et al. 2001), the FE approximation space is enriched within each element by additional non-conforming functions. This method also does not introduce additional degrees of freedom. The degrees of freedom associated with the enrichment functions are condensed at the element level prior to assembly. Lagrange multipliers are used to enforce the continuity across element boundaries and to apply Dirichlet boundary conditions. The idea of the method is summarized in Figure 2.3.

In the embedded finite element method (EFEM) (Lotfi and Shing 1995; Oliver et al. 2003), the enrichment is at the element level and is based on assumed enhanced strain. The additional degrees of freedom are condensed at element level leading to a global stiffness matrix with the same bandwidth as that of the FEM. Piecewise constant and linear crack opening can be modeled using EFEM. Oliver et al. (2006) presented a comparison between the XFEM and EFEM. Their study concluded the following:

- The convergence rates of both methods are similar;
- When implemented on the same element type, both methods yield identical results;
- The computational cost of XFEM was slightly greater than that of the EFEM;
- For relatively coarse meshes, EFEM obtained more accurate results when compared to that of the XFEM.

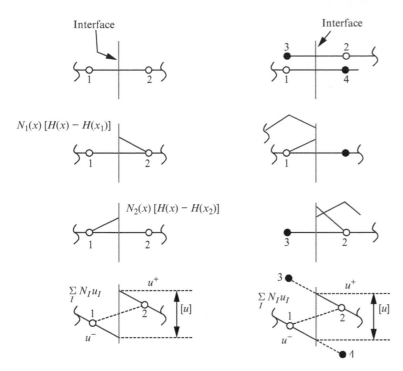

Figure 2.2 The representation of discontinuity: (a) standard XFEM and (b) Hansbo and Hansbo method to model discontinuities. Empty circles denote standard nodes and "filled" circles are called the phantom nodes.

Summary As can be seen from the literature, the concept of "enrichment" has been developed by mathematicians and engineers almost simultaneously. While the mathematicians aimed at providing a strong mathematical foundation and improving the approximation property, the engineers on the other hand aimed at accurate estimation of stress intensity factors (SIFs) by either enriching the approximation basis or by using quarter-point elements. From the above discussion, we notice that global enrichments have distinctive advantages, including the ability to reproduce special features known about the solution, but also drawbacks, such as:

- Deterioration of the conditioning number[4] of the system matrix.
- Increase in the bandwidth of the matrix, leading to increased storage and computational cost.
- In some versions of global enrichment, the number of additional unknowns may be proportional to the mesh parameter, h.

More importantly, high gradients or cracks or discontinuities are local phenomena and the emergence of partition of unity enrichment, which effectively localizes the action of the enrichment functions through their multiplication with a locally supported partition of unity, allowed to show that it is sufficient to enrich the FE approximation basis only in the vicinity of regions of interest to ensure optimal convergence. The novelty in the extended FEM itself is the idea of using discontinuous functions (strong discontinuities) or functions with discontinuous derivatives (weak discontinuities) as enrichment.

---

4 The condition number associated with the linear equation $Ax = b$ gives a bound on how inaccurate the solution x for small perturbations in b and is related to the ratio between the largest and the smallest eigenvalue of the system matrix.

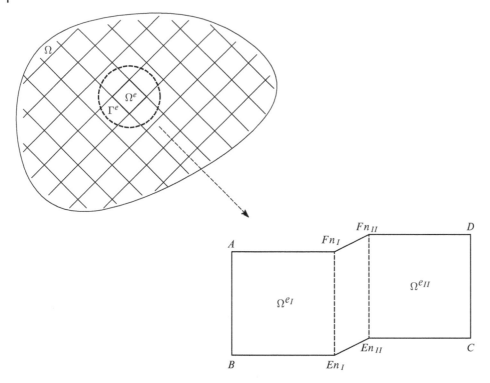

Figure 2.3 In the discontinuous enrichment method (DEM).

2.2   Weak Discontinuities for One-dimensional Problems

To do so, we shall first revisit the problem solution of the bi-material bar problem. This problem has discontinuous coefficients and the solution shows a kink at the location of the interface. Consider the problem of a bi-material bar as described in Figure 2.4. The bar is of length $L = 1$ and the interface is located at abscissa x_b along the bar. To simplify the derivations, we assume $x_b = L/2$. On the left of the interface, $x < x_b$, Young's modulus is E_1 and on the right of the interface, it is E_2. The cross-section area, A, is assumed constant all along the bar. A force F is applied on the right-hand side of the bar at $x = 1$, while the left-hand side $x = 0$ is fixed.

Strong form Letting u be the displacement field, ε the strain field, and σ the Cauchy stress in the bar, the equilibrium of the bar requires that the divergence of the stress vanishes in the absence of body force, i.e.:

$$\forall x \in [0, L] : \quad \frac{\mathrm{d}\sigma}{\mathrm{d}x}(x) = 0. \tag{2.2}$$

Figure 2.4   A bi-material bar subjected to force at the right end and the displacement is constrained at the left end.

The strain displacement relation writes:

$$\varepsilon(x) = \frac{du(x)}{dx}.$$ (2.3)

The constitutive relation writes:

$$\varepsilon(x) = \begin{cases} \frac{\sigma(x)}{E_1} & \text{if } x \le b, \\[2ex] \frac{\sigma(x)}{E_2} & \text{if } x > b. \end{cases}$$ (2.4)

Finally, the boundary conditions stating that the rod is fixed at $x = 0$ and that a force F is applied at $x = L$, and the continuity condition stating that the material interface at $x = b$ is perfect, i.e., that the stress on the left of the interface b^- equal the stress on the right of the interface b^+, close the system, mathematically:

$$u(0) = 0,$$
$$\sigma(L) = \frac{F}{A},$$
$$\sigma(b^-) = \sigma(b^+).$$ (2.5)

The solution to this problem can be derived analytically. The exact displacement is linear on either side of the interface and kinks at the interface. The exact strain field is constant on either side of the interface and discontinuous at the interface, while the stress field is constant throughout the bar. The solution is given in Table 2.3 and plotted in Figure 2.5.

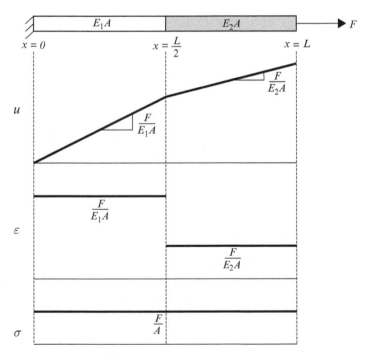

Figure 2.5 A bi-material bar subjected to force at the right end with its displacement is constrained at the left end. This is a problem with discontinuous coefficient.

Table 2.3 Bi-material problem – analytical solution for the displacement, strain, and stress along the length of the bar.

	$\forall x \in [0, b]$	$\forall x \in [b, L]$
$u(x)$	$\dfrac{Fx}{E_1 A}$	$\dfrac{Fx}{E_2 A} + \dfrac{FL}{2A}\left(\dfrac{1}{E_1} - \dfrac{1}{E_2}\right)$
$\varepsilon(x)$	$\dfrac{F}{E_1 A}$	$\dfrac{F}{E_2 A}$
$\sigma(x)$	$\dfrac{F}{A}$	$\dfrac{F}{A}$

Weak form The weak form for this problem can be developed by multiplying Equation (2.2) with a function v defined in the domain and integrating over the whole domain. The problem then becomes:

Find $u_h \in \mathcal{V}_h$ such that for all $v_h \in \mathcal{V}_h$:

$$a(u_h, v_h) = (v_h, f) \tag{2.6}$$

where $a(u_h, v_h) = \int_\Omega (\nabla u_h)^T E(x) \nabla v_h \, d\Omega$ and $(v_h, f) = \int_\Omega v_h f \, d\Omega + \dfrac{F}{A} v(L)$ are the symmetric bilinear and linear forms, respectively, $\dfrac{F}{A} v(L)$ is the force applied at $x = L$ and $E(x) = E_1$ for $x \le b$ and $E(x) = E_2$ for $x > b$. The trial (u^h) and the test function (v^h) are written as a linear combination of basis functions $\phi_1, \cdots, \phi_n$ as:

$$\forall x \in [0, L] : \quad u^h(x) = \sum_I N_I(x) u_I,$$

$$\forall x \in [0, L] : \quad v^h(x) = \sum_I N_I(x) v_I. \tag{2.7}$$

The weak form (Equation (2.6)) is then written as:

$$a\left(\sum N_I(x) u_I, \sum N_I(x) v_I\right) = \ell\left(\sum N_I(x) v_I\right) + \frac{F}{A} v^h(L). \tag{2.8}$$

Using the linearity of the bilinear form a and the linear form ℓ with respect to each argument, and the fact that the nodal unknowns u_I and nodal variations v_I are constants, and hence may be taken out of the bilinear and linear forms, the discretized weak form can be written as:

Weak form: Find $u_1, \cdots, u_n \in \mathbb{R}$, such that for all $v_1, \cdots, v_n \in \mathbb{R}$

$$u_1 a(N_1(x), N_1(x)) v_1 + u_1 a(N_1(x), N_2(x)) v_2 + \cdots + u_n a(N_n(x), N_n(x)) v_n$$

$$= v_1 \ell(N_1(x)) + v_2 \ell(N_2(x)) + \cdots v_{n-1} \ell(N_n(x)) + \frac{F}{A} v_n \tag{2.9}$$

Since the nodal variations $v_1, \cdots, v_n$ are arbitrary, the discretized weak form can be written as a system of equations as:

$$
\begin{bmatrix}
a\left(N_1(x), N_1(x)\right) & a\left(N_2(x), N_1(x)\right) & \cdots & a\left(N_n(x), N_1(x)\right) \\[2ex]
\vdots & \ddots & & \vdots \\[2ex]
a\left(N_n(x), N_2(x)\right) & a\left(N_2(x), N_2(x)\right) & \cdots & a\left(N_n(x), N_n(x)\right)
\end{bmatrix}
\begin{Bmatrix}
u_1 \\[2ex] u_2 \\[1ex] \vdots \\[1ex] u_n
\end{Bmatrix}
=
\begin{Bmatrix}
\ell\left(N_1(x)\right) \\[2ex]
\ell\left(N_2(x)\right) \\[1ex]
\vdots \\[1ex]
\ell\left(N_n(x)\right) + \dfrac{F}{A}
\end{Bmatrix}.
\tag{2.10}
$$

2.2.1 Conventional Finite Element Solution

In this section, we first derive the FEM solution to the bi-material problem by discretizing the weak form (Equation (2.6)) using two-noded linear FE. At the material interface, the solution must satisfy $\sigma(b^-) = \sigma(b^+)$. Thus, if $E_1 \neq E_2$, the strains will be different on each side of the material interface. We can distinguish two situations when meshing the bar with linear elements:

Case A If a node is located at the material interface (see Figure 2.6), the shape functions can represent the displacements that have a kink at the material interface. Thus, they can represent discontinuous strains. Consequently, the space in which the numerical solution is computed contains the exact solution space. In one dimension, it is trivial to select a mesh such that a node is located at the interface; however, in two or three dimensions, it is not as straightforward as the interface may have a complex shape and it would be necessary that the element edges/faces align with the interface.

Case B If none of the nodes coincide with the material interface (see Figure 2.6), the strain of the numerical solution is constant inside the element. This is because the shape functions are only able to represent the displacement by a linear function inside that element and the derivatives of the shape functions of the FE are continuous. Hence, they cannot accommodate discontinuities in the strain field. These functions are said to be incapable of reproducing the jump in the derivatives present at the interface.

The above two cases are demonstrated by deriving the FEM solution to the bi-material problem: (a) two elements for Case A and (b) one single element and two Gauss points to which the proper material properties are associated for Case B. This is a common technique to prevent the need to mesh the material interfaces.

Consider the bi-material bar of length L and cross-section A (see Figure 2.4). Let the domain be discretized with two noded linear elements (two elements for Case A and one element for Case B). For a single element with two nodes, the shape functions are given by:

$$
\begin{aligned}
N_1(x) &= \frac{L - x}{L} \\
N_2(x) &= \frac{x}{L}
\end{aligned}
\tag{2.11}
$$

The shape functions are depicted in Figure 2.7. The strain-displacement matrix, $\mathbf{B}$, takes the form:

$$
\mathbf{B} = \begin{bmatrix} N_{1,x}(x) & N_{2,x}(x) \end{bmatrix} = \begin{bmatrix} -1/L & 1/L \end{bmatrix}
\tag{2.12}
$$

Now, we can compute the stiffness matrix, $\mathbf{K} = \int \mathbf{B}^T \mathbf{D} \mathbf{B}$, with $\mathbf{D}$ a discontinuous function given by:

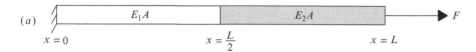

(*a*)

$x = 0$ $x = \dfrac{L}{2}$ $x = L$

(*b*)

(*c*)

Figure 2.6 Bi-material bar discretized with finite elements: (a) geometry and boundary conditions; (b) two finite elements with node 2 coinciding with the material interface (Case A); and (c) one single finite element (Case B). The "open" circles denote the location of Gauss points.

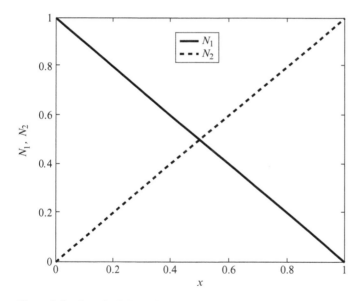

Figure 2.7 Standard shape function plot.

$$\mathbf{D} = \begin{cases} E_1 A & \text{if } x \in [0,b] \\[2ex] E_2 A & \text{if } x \in (b,L] \end{cases} \tag{2.13}$$

For Case A, the local stiffness matrix pertaining to an element is first computed and later assembled to the global stiffness matrix.

$$\mathbf{K} = \int_0^L \mathbf{B}^\mathrm{T} \mathbf{D} \mathbf{B} = \int_0^{L/2} \mathbf{B}^\mathrm{T} \mathbf{B} + \int_{L/2}^L \mathbf{B}^\mathrm{T} \mathbf{D} \mathbf{B}$$

$$= \int_0^{L/2} \begin{pmatrix} 1 & -1 & 0 \\ -1 & 1 & 0 \\ 0 & 0 & 0 \end{pmatrix} \frac{2E_1 A}{L} + \int_{L/2}^L \begin{pmatrix} 0 & 0 & 0 \\ 0 & 1 & -1 \\ 0 & -1 & 1 \end{pmatrix} \frac{2E_2 A}{L}$$

$$= \begin{pmatrix} 1 & -1 & 0 \\ -1 & 1 & 0 \\ 0 & 0 & 0 \end{pmatrix} \frac{2E_1 A}{L} + \begin{pmatrix} 0 & 0 & 0 \\ 0 & 1 & -1 \\ 0 & -1 & 1 \end{pmatrix} \frac{2E_2 A}{L}$$

$$= \begin{pmatrix} \frac{2E_1 A}{L} & -\frac{2E_1 A}{L} & 0 \\ -\frac{2E_1 A}{L} & \frac{2(E_1 + E_2)A}{L} & -\frac{2E_2 A}{L} \\ 0 & -\frac{2E_2 A}{L} & \frac{2E_2 A}{L} \end{pmatrix} \tag{2.14}$$

The system of equations can be written as ($\mathbf{Ku} = \mathbf{F}$):

$$\begin{pmatrix} \frac{2E_1 A}{L} & -\frac{2E_1 A}{L} & 0 \\ -\frac{2E_1 A}{L} & \frac{2(E_1 + E_2)A}{L} & -\frac{2E_2 A}{L} \\ 0 & -\frac{2E_2 A}{L} & \frac{2E_2 A}{L} \end{pmatrix} \begin{Bmatrix} u_1 \\ u_2 \\ u_3 \end{Bmatrix} = \begin{pmatrix} F_1 \\ F_2 \\ F_3 \end{pmatrix} \tag{2.15}$$

Considering the boundary conditions, $u_1 = 0$ and $F_3 = F$ and $F_2 = 0$ for equilibrium, we have:

$$u_2 = \frac{FL}{2E_1 A}$$

$$u_3 = \frac{FL(E_1 + E_2)}{2E_1 E_2 A} \tag{2.16}$$

The strain is given by:

$$\varepsilon(x) = \begin{cases} \dfrac{F}{2E_1 A} & \text{if } x \in [0, L/2] \\[2ex] \dfrac{F}{2E_2 A} & \text{if } x \in [L/2, L] \end{cases} \tag{2.17}$$

And finally the stress, is given as:

$$\sigma(x) = E\varepsilon(x) = \begin{cases} \dfrac{F}{A} & \text{if } x \in [0, L/2] \\[2ex] \dfrac{F}{A} & \text{if } x \in (L/2, L] \end{cases} \tag{2.18}$$

For Case B, Gauss-Legendre quadrature cannot be used to integrate discontinuous functions[5]. Here, we will calculate the integral analytically, as:

---

5 Dividing the integration domain in two parts where the function is continuous solves this problem. Later chapters of the book will show how to tackle this difficulty.

$$\mathbf{K} = \int_0^L \mathbf{B}^T \mathbf{DB} = \int_0^{L/2} \mathbf{B}^T \mathbf{B} + \int_{L/2}^L \mathbf{B}^T \mathbf{DB}$$

$$= \int_0^{L/2} \begin{pmatrix} 1 & -1 \\ -1 & 1 \end{pmatrix} \frac{E_1 A}{L^2} + \int_{L/2}^L \begin{pmatrix} 1 & -1 \\ -1 & 1 \end{pmatrix} \frac{E_2 A}{L^2}$$

$$= \begin{pmatrix} 1 & -1 \\ -1 & 1 \end{pmatrix} \frac{E_1 A}{2L} + \begin{pmatrix} 1 & -1 \\ -1 & 1 \end{pmatrix} \frac{E_2 A}{2L}$$

$$= \begin{pmatrix} 1 & -1 \\ -1 & 1 \end{pmatrix} \frac{A(E_1 + E_2)}{2L} \tag{2.19}$$

The system of equations can be written as ($\mathbf{Ku} = \mathbf{F}$):

$$\begin{pmatrix} 1 & -1 \\ -1 & 1 \end{pmatrix} \frac{A(E_1 + E_2)}{2L} \begin{Bmatrix} u_1 \\ u_2 \end{Bmatrix} = \begin{pmatrix} F_1 \\ F_2 \end{pmatrix} \tag{2.20}$$

Considering the boundary conditions, $u_1 = 0$ and $F_2 = F$, we have from the second equation:

$$u_2 = \frac{2LF}{A(E_1 + E_2)} \tag{2.21}$$

Therefore, the displacement field along the bar has the following equation (which does not coincide with the exact solution):

$$u(x) = \frac{2Fx}{A(E_1 + E_2)} \tag{2.22}$$

The strain is given by:

$$\varepsilon(x) = \frac{\mathrm{d}u(x)}{\mathrm{d}x} = \frac{2F}{A(E_1 + E_2)} \tag{2.23}$$

And finally the stress is given as

$$\sigma(x) = E\varepsilon(x) = \begin{cases} \dfrac{2FE_1}{A(E_1 + E_2)} & \text{if } x \in [0, L/2] \\[12pt] \dfrac{2FE_2}{A(E_1 + E_2)} & \text{if } x \in (L/2, L] \end{cases} \tag{2.24}$$

> ☞ *From the above derivation it is clear that unless the nodes are placed along the material interface, the standard FEM cannot reproduce the exact solution. If they are not, the discontinuity is smeared (smoothed) by the FEM, which degrades the convergence rate and the accuracy. The exact solution is not in the FE space and hence cannot be reproduced. Moreover, if local quantities, measured around the interface, are of interest, the accuracy is deteriorated. Conversely, by using two elements, one on either side of the discontinuity, the solution becomes linear in each element and it is possible to recover the exact solution.*

2.2.2 eXtended Finite Element Solution

In the previous section, we inferred that the FEM is not able to reproduce the strain discontinuity across the interface within a single element and that two elements are required. In this section, we will derive an enrichment technique which allows to represent the strain discontinuity at the interface even if none of the nodes is located at the interface. To do so, additional information is needed in the FE basis. For sake of brevity, we will assume that the bi-material bar is discretized one single two-noded FE with linear basis function (c.f. Figure 2.6c).

- Propose a suitable enrichment function which is continuous in an FE but whose first derivative is discontinuous at $b = L/2$;
- Discretize the weak form Equation (2.6);
- Derive the enriched FE equations;
- Impose Dirichlet boundary conditions;
- Solve for the displacement field along the bar, analyze and discuss the results.

The simplest enrichment function which satisfies the desired property was probably proposed in Krongauz and Belytschko (1998) as a V ramp function which is zero on the interface and linearly varying with distance from the interface. Such a function is shown in Figure 2.8 and writes as in Equation (2.25)[6]:

$$\forall x \in [0, L] : \quad V(x) \equiv \psi_b(x) \underbrace{\equiv}_{\text{def}} |x - b| = |x - L/2|. \tag{2.25}$$

The idea of partition of unity enrichment[7] is to multiply the standard shape functions of all nodes which are affected by the presence of interface, i.e., all nodes whose support is cut by the interface line, by the enrichment function, so as to ensure the reproduction of this enrichment function. In the present case, nodes 1 and 2 are such that their support is cut by the discontinuity. Letting N_1 and N_2 be the shape functions of nodes 1 and 2, respectively, the enriched shape function $\tilde{N}_1(x)$ for node 1 can be written as:

$$\forall x \in [0, L] : \quad \tilde{N}_1(x) = N_1(x)\psi_b(x) = \left(1 - \frac{x}{L}\right)|\psi_b(x)|. \tag{2.26}$$

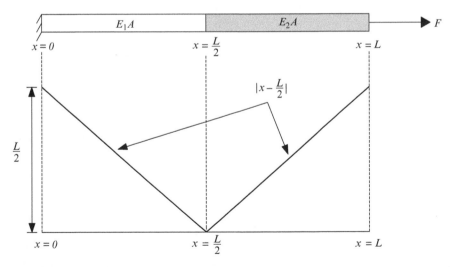

Figure 2.8 Enrichment function for a bi-material interface. This function is known as the "ramp function" or the "V" function.

---

6 This function may also be written conveniently as the distance function to the interface (c.f. Section 3.3.1), in other words the absolute value of the signed distance function (level set function).
7 A rigorous introduction to partition of unity enrichment is provided in Chapter 3.

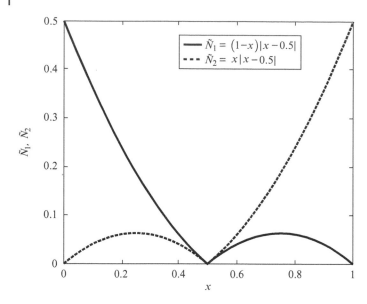

Figure 2.9 Bi-material enrichment functions.

Similarly, the enriched shape function associated with node 2, $\tilde{N}_2$, can be written as:

$$\forall x \in [0, L]: \quad \tilde{N}_2(x) = \left(\frac{x}{L}\right)(|\psi_b(x)|). \tag{2.27}$$

These two additional functions, referred to as enrichment functions, are shown in Figure 2.9. With these two additional functions, the element, made up of nodes 1 and 2, now has four shape functions: standard shape functions $(N_I)_{1 \leq I \leq 2}$ and enriched shape functions $(\tilde{N}_J)_{1 \leq J \leq 2}$. Associated with those four shape functions come four nodal unknowns: standard unknowns u_1, u_2 and enriched unknowns a_1, a_2. With the above definitions, the enriched trial and test functions write as the sum of the standard and the enriched approximations:

$$\forall x \in [0, L]: \quad u^h(x) = N_1(x)u_1 + N_2(x)u_2 + \tilde{N}_1(x)a_1 + \tilde{N}_2(x)a_2,$$
$$\forall x \in [0, L]: \quad v^h(x) = N_1(x)v_1 + N_2(x)v_2 + \tilde{N}_1(x)b_1 + \tilde{N}_2(x)b_2. \tag{2.28}$$

From Equation (2.28), the approximated strain field, which is, in one dimension, the first spatial derivative of the trial and test functions can be deduced: $\forall x \in [0, L]$:

$$\varepsilon^h(x) = \frac{du^h}{dx}(x)$$
$$\varepsilon^h(x) = \frac{dN_1(x)}{dx}u_1 + \frac{dN_2(x)}{dx}u_2 + \frac{d\tilde{N}_1(x)}{dx}a_1 + \frac{d\tilde{N}_2(x)}{dx}a_2, \tag{2.29}$$

and

$$\varepsilon^h_{\text{test}}(x) = \frac{dv^h}{dx}(x)$$
$$\varepsilon^h_{\text{test}}(x) = \frac{dN_1(x)}{dx}v_1 + \frac{dN_2(x)}{dx}v_2 + \frac{d\tilde{N}_1(x)}{dx}b_1 + \frac{d\tilde{N}_2(x)}{dx}b_2. \tag{2.30}$$

Discretization of the weak form The discretized weak form is obtained as in the non-enriched formulation of Section 2.2.1, by substituting the trial and the test (c.f. Equation (2.28)) functions into the weak form Equation (2.6) to obtain the corresponding discrete problem: find u_1, u_2, a_1, a_2, such that, for all $v_1, v_2, b_1, b_2 \in \mathbb{R}$, we have:

$$a\left(u^h(x), v^h(x)\right) = \ell(v^h(x)) + \frac{F}{A}v^h(L),$$

$$a\left(N_1(x)u_1 + N_2(x)u_2 + \tilde{N}_1(x)a_1 + \tilde{N}_2(x)a_2, N_1(x)v_1 + N_2(x)v_2 + \tilde{N}_1(x)b_1 + \tilde{N}_2(x)b_2\right) =$$

$$\ell\left(N_1(x)v_1 + N_2(x)v_2 + \tilde{N}_1(x)b_1 + \tilde{N}_2(x)b_2\right) + \frac{F}{A}v^h(L). \tag{2.31}$$

Considering the left-hand side first, and using the linearity of a with respect to its first argument:

$$a\left(N_1(x)u_1 + N_2(x)u_2 + \tilde{N}_1(x)a_1 + \tilde{N}_2(x)a_2, N_1(x)v_1 + N_2(x)v_2 + \tilde{N}_1(x)b_1 + \tilde{N}_2(x)b_2\right) =$$

$$a\left(N_1(x)u_1 + N_2(x)u_2, N_1(x)v_1 + N_2(x)v_2 + \tilde{N}_1(x)b_1 + \tilde{N}_2(x)b_2\right)$$

$$+ a\left(\tilde{N}_1(x)a_1 + \tilde{N}_2(x)a_2, N_1(x)v_1 + N_2(x)v_2 + \tilde{N}_1(x)b_1 + \tilde{N}_2(x)b_2\right). \tag{2.32}$$

Using again linearity with respect to the first argument of a, the left-hand side becomes:

$$u_1\, a\left(N_1(x), N_1(x)v_1 + N_2(x)v_2 + \tilde{N}_1(x)b_1 + \tilde{N}_2(x)b_2\right)$$

$$+u_2\, a\left(N_2(x), N_1(x)v_1 + N_2(x)v_2 + \tilde{N}_1(x)b_1 + \tilde{N}_2(x)b_2\right)$$

$$+a_1\, a\left(\tilde{N}_1(x), N_1(x)v_1 + N_2(x)v_2 + \tilde{N}_1(x)b_1 + \tilde{N}_2(x)b_2\right)$$

$$+a_2\, a\left(\tilde{N}_2(x), N_1(x)v_1 + N_2(x)v_2 + \tilde{N}_1(x)b_1 + \tilde{N}_2(x)b_2\right). \tag{2.33}$$

Using now the linearity of a with respect to its second argument, the left-hand side can be expanded into:

$$u_1\, a\left(N_1(x), N_1(x)\right)v_1 + u_1 a\left(N_1(x), N_2(x)\right)v_2$$

$$+u_1\, a\left(N_1(x), \tilde{N}_1(x)\right)b_1 + u_1 a\left(N_1(x), \tilde{N}_2(x)\right)b_2$$

$$+u_2\, a\left(N_2(x), N_1(x)\right)v_1 + u_2 a\left(N_2(x), N_2(x)\right)v_2$$

$$+u_2\, a\left(N_2(x), \tilde{N}_1(x)\right)b_1 + u_2 a\left(N_2(x), \tilde{N}_2(x)\right)b_2$$

$$+a_1\, a\left(\tilde{N}_1 x), N_1(x)\right)v_1 + a_1 a\left(\tilde{N}_1(x), N_2(x)\right)v_2$$

$$+a_1\, a\left(\tilde{N}_1(x), \tilde{N}_1(x)\right)b_1 + a_1 a\left(\tilde{N}_1(x), \tilde{N}_2(x)\right)b_2$$

$$+a_2\, a\left(\tilde{N}_2 x), N_1(x)\right)v_1 + a_2 a\left(\tilde{N}_2(x), N_2(x)\right)v_2$$

$$+a_2\, a\left(\tilde{N}_2(x), \tilde{N}_1(x)\right)b_1 + a_2 a\left(\tilde{N}_2(x), \tilde{N}_2(x)\right)b_2. \tag{2.34}$$

Using the linearity of ℓ, and $v^h(L) = v_2 + b_2|L - b|$, the right-hand side of Equation (2.31) writes:

$$\ell\left(N_1(x)v_1 + N_2(x)v_2 + \tilde{N}_1(x)b_1 + \tilde{N}_2(x)b_2\right) + \frac{F}{A}v^h(L),$$

$$= v_1\ell\left(N_1(x)\right) + v_2\ell\left(N_2(x)\right) + b_1\ell\left(\tilde{N}_1(x)\right) + b_2\ell\left(\tilde{N}_2(x)\right) + \frac{F}{A}\left(v_2 + b_2|L - b|\right)$$

$$= v_1\ell\left(N_1(x)\right) + v_2\left(\ell\left(N_2(x)\right) + \frac{F}{A}\right) + b_1\ell\left(\tilde{N}_1(x)\right) + b_2\left(\ell\left(\tilde{N}_2(x)\right) + |L - b|\right) \tag{2.35}$$

Equating the left-hand side Equation (2.34) to the right-hand side Equation (2.35) provides an expression for the discretized weak form, which reads: find $u_1, u_2, a_1, a_2 \in \mathbb{R}$ such that, for all $v_1, v_2, b_1, b_2 \in \mathbb{R}$, we have:

$$u_1 \, a\left(N_1(x), N_1(x)\right) v_1 + u_1 a\left(N_1(x), N_2(x)\right) v_2$$
$$+u_1 \, a\left(N_1(x), \tilde{N}_1(x)\right) b_1 + u_1 a\left(N_1(x), \tilde{N}_2(x)\right) b_2$$
$$+u_2 \, a\left(N_2(x), N_1(x)\right) v_1 + u_2 a\left(N_2(x), N_2(x)\right) v_2$$
$$+u_2 \, a\left(N_2(x), \tilde{N}_1(x)\right) b_1 + u_2 a\left(N_2(x), \tilde{N}_2(x)\right) b_2$$
$$+a_1 \, a\left(\tilde{N}_1 x), N_1(x)\right) v_1 + a_1 a\left(\tilde{N}_1(x), N_2(x)\right) v_2$$
$$+a_1 \, a\left(\tilde{N}_1(x), \tilde{N}_1(x)\right) b_1 + a_1 a\left(\tilde{N}_1(x), \tilde{N}_2(x)\right) b_2$$
$$+a_2 \, a\left(\tilde{N}_2 x), N_1(x)\right) v_1 + a_2 a\left(\tilde{N}_2(x), N_2(x)\right) v_2$$
$$+a_2 \, a\left(\tilde{N}_2(x), \tilde{N}_1(x)\right) b_1 + a_2 a\left(\tilde{N}_2(x), \tilde{N}_2(x)\right) b_2$$
$$= v_1 \ell\left(N_1(x)\right) + v_2 \left(\ell\left(N_2(x)\right) + \frac{F}{A}\right) + b_1 \ell\left(\tilde{N}_1(x)\right) + b_2 \left(\ell\left(\tilde{N}_2(x)\right) + |L - b|\right). \tag{2.36}$$

Enriched finite element equations Setting successively v_1, v_2, b_1, b_2 to one, and all other nodal variations to zero, we obtain a set of four equations in four unknowns, which writes as follows: find $u_1, u_2, a_1, a_2 \in \mathbb{R}$ such that:

$$u_1 \, a\left(N_1(x), N_1(x)\right) + u_2 a\left(N_2(x), N_1(x)\right) + a_1 a\left(\tilde{N}_1(x), N_1(x)\right) + a_2 a\left(\tilde{N}_2(x), N_1(x)\right)$$
$$= \ell\left(N_1(x)\right)$$
$$u_1 \, a\left(N_1(x), N_2(x)\right) + u_2 a\left(N_2(x), N_2(x)\right) + u_2 a\left(N_2(x), \tilde{N}_1(x)\right) + a_2 a\left(\tilde{N}_2(x), N_2(x)\right)$$
$$= \left(\ell\left(N_2(x)\right) + \frac{F}{A}\right)$$
$$u_1 \, a\left(N_1(x), \tilde{N}_1(x)\right) + u_2 a\left(N_2(x), \tilde{N}_1(x)\right) + a_1 a\left(\tilde{N}_1(x), \tilde{N}_1(x)\right) + a_2 a\left(\tilde{N}_2(x), \tilde{N}_1(x)\right)$$
$$= \ell\left(\tilde{N}_1(x)\right)$$
$$u_1 \, a\left(N_1(x), \tilde{N}_2(x)\right) + u_2 a\left(N_2(x), \tilde{N}_2(x)\right) + a_1 a\left(\tilde{N}_1(x), \tilde{N}_2(x)\right) + a_2 a\left(\tilde{N}_2(x), \tilde{N}_2(x)\right)$$
$$= \left(\ell\left(\tilde{N}_2(x)\right) + |L - b|\right). \tag{2.37}$$

This set of equations can be written in matrix form as follows: find $u_1, u_2, a_1, a_2 \in \mathbb{R}$ such that:

$$\left[\begin{array}{cc|cc} a\left(N_1(x), N_1(x)\right) & a\left(N_2(x), N_1(x)\right) & a\left(\tilde{N}_1(x), N_1(x)\right) & a\left(\tilde{N}_2(x), N_1(x)\right) \\ a\left(N_1(x), N_2(x)\right) & a\left(N_2(x), N_2(x)\right) & a\left(N_2(x), \tilde{N}_1(x)\right) & a\left(\tilde{N}_2(x), N_2(x)\right) \\ \hline a\left(N_1(x), \tilde{N}_1(x)\right) & a\left(N_2(x), \tilde{N}_1(x)\right) & a\left(\tilde{N}_1(x), \tilde{N}_1(x)\right) & a\left(\tilde{N}_2(x), \tilde{N}_1(x)\right) \\ a\left(N_1(x), \tilde{N}_2(x)\right) & a\left(N_2(x), \tilde{N}_2(x)\right) & a\left(\tilde{N}_1(x), \tilde{N}_2(x)\right) & a\left(\tilde{N}_2(x), \tilde{N}_2(x)\right) \end{array}\right] \left\{\begin{array}{c} u_1 \\ u_2 \\ a_1 \\ a_2 \end{array}\right\} =$$

$$\left\{\begin{array}{c} \ell\left(N_1(x)\right) \\ \ell\left(N_2(x)\right) + \dfrac{F}{A} \\ \ell\left(\tilde{N}_1(x)\right) \\ \ell\left(\tilde{N}_2(x)\right) + \underbrace{|L - b|}_{\psi_b(L)} \end{array}\right\}$$

$$\tag{2.38}$$

☞   *The matrix has been split into four parts in Equation (2.38). The top left block is the standard FE stiffness matrix for a two-noded element with linear shape functions. The bottom right block is an enriched block, acting only on the enriched degrees of freedom, and associated with the enriched shape functions. The top right and bottom left blocks are identical and represent mixed terms.*

☞   *Note that due to the body force, acting on an element with enriched nodes, an enriched load vector appears, i.e., the third and fourth rows of the right-hand side in Equation (2.38). Such terms also appear when Neumann boundary conditions are applied on the edges FEs whose nodes are enriched.*

☞   *The above derivation is easily generalized to the case where nodes are enriched with more than one enrichment function. In this case, additional enriched-enriched and mixed blocks appear in the stiffness matrix, for each enrichment function.*

The form in Equation (2.38) does not lend itself readily to a FE implementation. Below, we derive matrices which can be used to implement XFEM assuming that no body force is applied.

Implementation: $B^T D B$ formulation   To compute the strain at the element level, the derivatives of the shape functions are required, and are gathered in a strain displacement matrix **B**. The standard part of the strain displacement matrix writes $\mathbf{B}^{\mathrm{std}} = [\,N_{1,x}(x)\quad N_{2,x}(x)\,]$ and the enriched part of the strain displacement matrix writes $\mathbf{B}^{\mathrm{enr}}(x) = [\,\tilde{N}_{1,x}(x)\quad \tilde{N}_{2,x}(x)\,]$, so that the total strain displacement matrix writes:

$$\varepsilon^h(x) = \underbrace{\left[\; N_{1,x}(x)\quad N_{2,x}(x)\quad \tilde{N}_{1,x}(x)\quad \tilde{N}_{2,x}(x)\;\right]}_{\mathbf{B}(x)=[\mathbf{B}^{\mathrm{std}}(x)\quad \mathbf{B}^{\mathrm{enr}}(x)]}\left\{\begin{array}{c} u_1 \\ u_2 \\ a_1 \\ a_2 \end{array}\right\} \tag{2.39}$$

Or, remembering the definition of the enriched shape functions in Equation (2.25):

$$\varepsilon^h(x) = \underbrace{\left[\; N_{1,x}(x)\quad N_{2,x}(x)\quad \left(N_1(x)\psi_b(x)\right)_{,x}\quad \left(N_2(x)\psi_b(x)\right)_{,x}\;\right]}_{\mathbf{B}(x)=[\mathbf{B}^{\mathrm{std}}(x)\quad \mathbf{B}^{\mathrm{enr}}(x)]}\left\{\begin{array}{c} u_1 \\ u_2 \\ a_1 \\ a_2 \end{array}\right\} \tag{2.40}$$

$$\mathbf{B}(x) = [\mathbf{B}^{\mathrm{std}}(x)\quad \mathbf{B}^{\mathrm{enr}}(x)] = \left[\; N_{1,x}(x)\quad N_{2,x}(x)\quad \tilde{N}_{1,x}(x)\quad \tilde{N}_{2,x}(x)\;\right] \tag{2.41}$$

$$N_{1,x}(x) \;=\; \frac{\mathrm{d}\left(1-\frac{x}{L}\right)}{\mathrm{d}x} = \left(-\frac{1}{L}\right), \tag{2.42}$$

$$N_{2,x}(x) \;=\; \frac{\mathrm{d}\left(\frac{x}{L}\right)}{\mathrm{d}x} = \left(\frac{1}{L}\right) \tag{2.43}$$

$$\tilde{N}_{1,x}(x) = \frac{d}{dx}\left[\left(1 - \frac{x}{L}\right)(|x - b|)\right]$$

$$= \left(-\frac{1}{L}\right)(|x - b|) + \left(1 - \frac{x}{L}\right)\frac{d(|x - b|)}{dx}$$

$$\tilde{N}_{2,x}(x) = \frac{d}{dx}\left[\left(\frac{x}{L}\right)(|x - b|)\right]$$

$$= \left(\frac{1}{L}\right)(|x - b|) + \left(\frac{x}{L}\right)\frac{d(|x - b|)}{dx}. \tag{2.44}$$

Noting,

$$\frac{d|f(x)|}{dx} = sign(f(x))\frac{df(x)}{dx}. \tag{2.45}$$

The derivatives of the enriched shape functions in Equation (2.44) become:

$$\forall x \in [0, b]: \quad \tilde{N}_{1,x}(x) = \left(-\frac{1}{L}\right)(-(x - b)) + \left(1 - \frac{x}{L}\right)(-1) = \frac{2x}{L} - \frac{b}{L} - 1$$

$$\forall x \in [0, b]: \quad \tilde{N}_{2,x}(x) = \left(\frac{1}{L}\right)(-(x - b)) + \left(\frac{x}{L}\right)(-1) = -\frac{2x}{L} + \frac{b}{L} \tag{2.46}$$

and

$$\forall x \in]b, L]: \quad \tilde{N}_{1,x}(x) = \left(-\frac{1}{L}\right)(+(x - b)) + \left(1 - \frac{x}{L}\right)(1) = 1 - \frac{2x}{L} + \frac{b}{L}$$

$$\forall x \in]b, L]: \quad \tilde{N}_{2,x}(x) = \left(\frac{1}{L}\right)(+(x - b)) + \left(\frac{x}{L}\right)(1) = \frac{2x}{L} - \frac{b}{L}. \tag{2.47}$$

Letting $\mathbf{B}_1$ and $\mathbf{B}_2$:

$$\forall x \in [0, b]: \quad \mathbf{B}_1(x) = \begin{bmatrix} N_{1,x}(x) & N_{2,x}(x) & \tilde{N}_{1,x}(x) & \tilde{N}_{2,x}(x) \end{bmatrix} \tag{2.48}$$

$$\forall x \in]b, L]: \quad \mathbf{B}_2(x) = \begin{bmatrix} N_{1,x}(x) & N_{2,x}(x) & \tilde{N}_{1,x}(x) & \tilde{N}_{2,x}(x) \end{bmatrix}, \tag{2.49}$$

and recalling that $\psi_b(x) = |x - b|$, the distance function to the interface at $x = b$, $\mathbf{B}_1$ and $\mathbf{B}_2$ can be rewritten as:

$$\mathbf{B}_1(x) = \begin{bmatrix} -1 & 1 & (x - 1 - \psi_b(x)) & (-x + \psi_b(x)) \end{bmatrix} \tag{2.50a}$$

$$\mathbf{B}_2(x) = \begin{bmatrix} -1 & 1 & (-x + 1 - \psi_b(x)) & (x + \psi_b(x)) \end{bmatrix}. \tag{2.50b}$$

Defining $\mathbf{K}_1$ and $\mathbf{K}_2$ by:

$$\mathbf{K}_1 = \int_0^b \mathbf{B}_1^T(x)E_1 A\mathbf{B}_1(x)\, dx \tag{2.51}$$

$$\mathbf{K}_2 = \int_b^L \mathbf{B}_2^T(x)E_2 A\mathbf{B}_2(x)\, dx, \tag{2.52}$$

the stiffness matrix writes:

$$\mathbf{K} = \mathbf{K}_1 + \mathbf{K}_2. \tag{2.53}$$

$\mathbf{K}_1$ is computed by:

$$\mathbf{K}_1 = \int_0^b E_1 A \begin{bmatrix} 1 & -1 & a_1 & a_2 \\ -1 & 1 & a_3 & a_4 \\ a_1 & a_3 & a_6 & a_5 \\ a_2 & a_4 & a_5 & a_7 \end{bmatrix}(x)\mathrm{d}x \qquad (2.54)$$

where functions $a_1, a_2, a_3, a_4, a_5, a_6, a_7$ are given by:

$$a_1(x) = -x + 1 + \psi_b(x)$$
$$a_2(x) = x - \psi_b(x)$$
$$a_3(x) = x - 1 - \psi_b(x)$$
$$a_4(x) = -x + \psi_b(x)$$
$$a_5(x) = a_3(x)a_4(x)$$
$$a_6(x) = (a_3(x))^2$$
$$a_7(x) = (-x + \psi_b(x))^2.$$

After carrying out the integration, and noting:

$$\int |x|\mathrm{d}x = \frac{x|x|}{2}, \qquad (2.55)$$

$\mathbf{K}_1$ becomes:

$$\mathbf{K}_1 = \begin{bmatrix} 1/2 & -1/2 & 1/2 & 0 \\ -1/2 & 1/2 & -1/2 & 0 \\ 1/2 & -1/2 & 13/24 & -1/24 \\ 0 & 0 & -1/24 & 1/24 \end{bmatrix}. \qquad (2.56)$$

And similarly, $\mathbf{K}_2$ writes:

$$\mathbf{K}_2 = \int_b^L E_2 A \begin{bmatrix} 1 & -1 & b_1 & b_2 \\ -1 & 1 & b_3 & b_4 \\ b_1 & b_3 & b_6 & b_5 \\ b_2 & b_4 & b_5 & b_7 \end{bmatrix}(x)\mathrm{d}x \qquad (2.57)$$

where functions $b_1, b_2, b_3, b_4, b_5, b_6, b_7$ are given by:

$$b_1(x) = x - 1 + \psi_b(x)$$
$$b_2(x) = -x - \psi_b(x)$$
$$b_3(x) = -x + 1 - \psi_b(x)$$
$$b_4(x) = x + \psi_b(x)$$
$$b_5(x) = b_3(x)b_4(x)$$
$$b_6(x) = (b_3(x))^2$$
$$b_7(x) = (x + z\psi_b(x))^2.$$

Finally, the stiffness matrix writes:

$$
\mathbf{K} = \begin{bmatrix}
1.5 & -1.5 & 0.5 & -1 \\
-1.5 & 1.5 & -0.5 & 1 \\
0.5 & -0.5 & 0.625 & -0.125 \\
-1 & 1 & -0.125 & 1.125
\end{bmatrix}
\tag{2.58}
$$

Because an enriched node (node 2) is subjected to a non-zero Neumann (force) boundary condition, the discrete force corresponding to the enriched degree of freedom must be computed. The elemental force vector can be written as the sum of a body force contribution (integral over the whole element Ω^e) and a boundary contribution (integral over the Neumann boundary Γ_t^e):

$$
\mathbf{f}^e = \underbrace{\int_{\Omega^e} \mathbf{N}^{\mathrm{T}}(x) f(x) \, dx}_{\text{body force term } \mathbf{f}_\Omega^e} + \underbrace{\int_{\Gamma_t^e} \mathbf{N}^{\mathrm{T}}(x) \bar{t}(x)|_{\Gamma_t^e} \, dx}_{\text{Neumann term}}.
\tag{2.59}
$$

In the present case, there is no body force, and the applied force is a point force of magnitude F in the positive x direction at point $x = L$, i.e., a Dirac delta functional $\bar{t}|_{\Gamma_t^e}(x) = F\delta(x - L)$. Hence, using the sifting property of the Dirac delta distribution, the force vector for the element becomes:

$$\mathbf{f}^e = \mathbf{N}^{\mathrm{T}}(L)F,\tag{2.60}$$

which can be written, in matrix form (recalling that here $b = L/2$):

$$
\mathbf{f}^e = \mathbf{N}^{\mathrm{T}}(L)F = F \left\{ \begin{array}{c} N_1(x) \\ N_2(x) \\ \tilde{N}_1(x) \\ \tilde{N}_2(x) \end{array} \right\}_{x=L} = F \left\{ \begin{array}{c} 1 - \frac{x}{L} \\ \frac{x}{L} \\ \left(1 - \frac{x}{L}\right)|x - b| \\ \frac{x}{L}|x - b| \end{array} \right\}_{x=L} = \left\{ \begin{array}{c} 0 \\ 1 \\ 0 \\ 1/2 \end{array} \right\}
\tag{2.61}
$$

Enforcing boundary conditions Dirichlet boundary conditions must now be enforced. These are fixed displacement equals zero at the left end $u^h(0) = 0$.

☞ *As the enrichment functions do not go to zero at the nodes where Dirichlet boundary conditions are enforced, (in this case, node 1), enforcing these homogeneous Dirichlet boundary conditions cannot be done as is usual in FE, i.e., by simply setting the unknown degree of freedom at node 1 to zero, because the true solution at this node, which includes the part due to the enrichment function, is not equal to the nodal value u_1, but to $u_1 + a_1\tilde{N}_1(0)$. This is also known as a "lack of Kronecker delta" property and is characteristic of most approximations used in meshfree methods, in particular MLs. Any method to enforce constraints, such as Lagrange multipliers, the penalty method or Nitsche's method/Augmented Lagrangian techniques may be used. We use here the Lagrange multiplier method.*

The condition is derived as follows. We require $u^h(0) = u^h(x)|_{x=0} = 0$, so:

$$u^h(x)|_{x=0} = N_1(x)|_{x=0}u_1 + \underbrace{N_2(x)|_{x=0}}_{=0} u_2 + \tilde{N}_1(x)|_{x=0}a_1 + \underbrace{\tilde{N}_2(x)|_{x=0}}_{=0} a_2 = 0$$

$$u^h(x)|_{x=0} = \underbrace{N_1(x)|_{x=0}}_{=1} u_1 + \underbrace{\tilde{N}_1(x)|_{x=0}}_{1\times|0-b|=\frac{L}{2}=\frac{1}{2}} a_1 = 0, \tag{2.62}$$

hence $u^h(x)|_{x=0} = 0 = u_1 + (1/2)a_1$. This is a constraint equation linking the standard and enriched degrees of freedom of node 1. Using the Lagrange multiplier method (Hughes, 2000), we introduce an additional unknown in the set of equations, λ, which corresponds to the additional term $\int_{\Gamma_u} \lambda(u - \bar{u})$ in the variational formulation and can be interpreted as the force required to maintain the boundary condition. The complete system of equations with Lagrange multipliers can be written as:

$$\begin{bmatrix} 1.5 & -1.5 & 0.5 & -1 & 1 \\ -1.5 & 1.5 & -0.5 & 1 & 0 \\ 0.5 & -0.5 & 0.625 & -0.125 & 0.5 \\ -1 & 1 & -0.125 & 1.125 & 0 \\ 1 & 0 & 0.5 & 0 & 0 \end{bmatrix} \begin{Bmatrix} u_1 \\ u_2 \\ a_1 \\ a_2 \\ \lambda \end{Bmatrix} = \begin{Bmatrix} 0 \\ 1 \\ 0 \\ 1/2 \\ 0 \end{Bmatrix} \tag{2.63}$$

and the solution

$$\begin{Bmatrix} u_1 \\ u_2 \\ a_1 \\ a_2 \\ \lambda \end{Bmatrix} = \begin{Bmatrix} 0.125 \\ 0.875 \\ -0.250 \\ -0.250 \\ 1 \end{Bmatrix}. \tag{2.64}$$

☞ *As discussed above, the nodal values obtained here have no physical meaning. They do not correspond to the actual displacement values at the nodes because the enriched approximation does not vanish at the nodes. More work is therefore required to compute the actual displacement values at the nodes by accounting for the contribution of the enrichment. In Section 2.2.3, a shifting method is detailed, which allows to zero the enrichment function at all nodes, thereby suppressing this need for extra post-processing, and for special treatment of Dirichlet boundary conditions.*

Computation of true nodal displacements The enriched approximation given by Equation (2.28) writes:

$$\forall x \in [0, L], u^h(x) = N_1(x)u_1 + N_2(x)u_2 + \tilde{N}_1(x)a_1 + \tilde{N}_2(x)a_2 \tag{2.65}$$

Substituting the functional form for the shape functions and the computed unknown displacements, the above equation becomes, $\forall x \in [0, L]$:

$$u^h(x) = 0.125\left(1 - \frac{x}{L}\right) + 0.875\frac{x}{L} - 0.25\left(1 - \frac{x}{L}\right)|x - b| - 0.25\frac{x}{L}|x - b| \tag{2.66}$$

Displacement at the left end, i.e., at $x = 0$

$$
\begin{aligned}
u(x)|_{x=0} &= 0.125 - 0.25|0 - b| \\
&= 0.125 - 0.125 = 0
\end{aligned}
\tag{2.67}
$$

Displacement at the right end, i.e., at $x = L$

$$
\begin{aligned}
u(x)|_{x=L} &= 0.875 - 0.25|L - b| \\
&= 0.875 - 0.125 = 0.75
\end{aligned}
\tag{2.68}
$$

Displacement at the material interface, i.e., at $x = b = L/2$

$$
\begin{aligned}
u(x)|_{x=b} &= 0.125\left(1 - \frac{b}{L}\right) + 0.875\frac{b}{L} \\
&= 0.0625 + 0.4375 = 0.5.
\end{aligned}
\tag{2.69}
$$

Thus, the enriched method is therefore able to reproduce the exact solution at the nodes. A modified version of the approximation space with the same approximation properties is described next.

2.2.3 eXtended Finite Element Solution with Nodal Subtraction/Shifting

The purpose of this section is to revisit the bi-material problem in tension and show that a slightly modified enrichment scheme yields an identical solution. This modification reduces the extra effort that is otherwise required to compute the displacements at the nodes.

The standard FE shape functions for nodes 1 and 2 (c.f. Figure 2.6) of the element are given in Equation (2.11). We will now construct two additional shape functions, $\tilde{N}_1$, associated with node 1, and $\tilde{N}_2$ associated with node 2. As discussed previously, what is needed is an enrichment function which is continuous in the element and whose first derivative is discontinuous at $x_b = L/2$. As an additional requirement, we impose the constraint that the enrichment function constructed be zero at both nodes. Considering the signed distance function to the interface $\phi_b = x - x_b$, we define functions $\tilde{N}_1$ and $\tilde{N}_2$ as follows:

$$\forall x \in [0, L] : \quad \tilde{N}_1(x) = N_1(x)\left(|\phi_b(x)| - |\phi_b(x_1)|\right) \text{ and } \tilde{N}_2(x) = N_2(x)\left(|\phi_b(x)| - |\phi_b(x_2)|\right) \tag{2.70}$$

$|\phi_b(0)|$ is computed at the left end and is given by

$$|\phi_b(0)| = |x - b|_{x=0} = |0 - b| = |b|$$

and Equation (2.70) becomes:

$$\tilde{N}_1(x) = \left(1 - \frac{x}{L}\right)(|x - b| - |b|) \tag{2.71}$$

Similarly, $\tilde{N}_2$ can be written as:

$$\tilde{N}_2(x) = \left(\frac{x}{L}\right)(|x - b| - |L - b|). \tag{2.72}$$

We can use this new approximation to solve for the displacement field in the bar.

2.2.4 Solution

With these two additional functions, the L2 element now has four shape functions: standard shape functions $(N_I)_{1 \leq I \leq 2}$ and enriched shape functions $(\tilde{N}_I)_{1 \leq I \leq 2}$. Associated with those four shape functions come four nodal unknowns: standard unknowns u_1, u_2 and enriched unknowns a_1, a_2. With the above definitions, the enriched approximation writes:

$$\forall x \in [0, L] : \quad u^h(x) = N_1(x)u_1 + N_2(x)u_2 + \tilde{N}_1(x)a_1 + \tilde{N}_2(x)a_2. \tag{2.73}$$

☞ From Equation (2.73), it can be seen that the approximation in the element is actually the superposition of two approximations: that of the standard shape functions and that provided by the enriched shape functions, whose first derivative is discontinuous. Everything happens as if the element had four nodes with the four shape functions defined above. Pictorially, this can be seen as superimposing an element with two nodes and two enriched shape functions $\tilde{N}_1$ and $\tilde{N}_2$ on top of the standard L2 element with shape functions N_1 and N_2, as shown in Figure 2.2. This idea was developed in the context of strong discontinuities in Song et al. (2006) and is equivalent, as shown in Song et al. (2006), to the method proposed by Hansbo and Hansbo (2004).

From Equation (2.73), the strain field, i.e., the first derivative of the displacement approximation, can be deduced:

$$\varepsilon^h = \frac{du^h}{dx} \tag{2.74a}$$

$$\forall x \in [0, L] : \quad \varepsilon^h(x) = \frac{dN_1(x)}{dx}u_1 + \frac{dN_2(x)}{dx}u_2 + \frac{d\tilde{N}_1(x)}{dx}a_1 + \frac{d\tilde{N}_2(x)}{dx}a_2 \tag{2.74b}$$

In matrix form,

$$\varepsilon^h(x) = \begin{bmatrix} N_{1,x}(x) & N_{2,x}(x) & \tilde{N}_{1,x}(x) & \tilde{N}_{2,x}(x) \end{bmatrix} \begin{Bmatrix} u_1 \\ u_2 \\ a_1 \\ a_2 \end{Bmatrix} \tag{2.75}$$

To compute the strain at the element level, the derivatives of the shape functions are required, and those are gathered in a matrix:

$$\mathbf{B} = \begin{bmatrix} N_{1,x}(x) & N_{2,x}(x) & \tilde{N}_{1,x}(x) & \tilde{N}_{2,x}(x) \end{bmatrix} \tag{2.76}$$

$$N_{1,x}(x) = \frac{d\left(1 - \frac{x}{L}\right)}{dx} = \left(-\frac{1}{L}\right); \qquad N_{2,x}(x) = \frac{d\left(\frac{x}{L}\right)}{dx} = \left(\frac{1}{L}\right) \tag{2.77}$$

$$\tilde{N}_{1,x}(x) = \frac{d}{dx}\left[\left(1 - \frac{x}{L}\right)(|x - b| - |b|)\right] \tag{2.78a}$$

$$\tilde{N}_{1,x}(x) = \left(-\frac{1}{L}\right)(|x - b| - |b|) + \left(1 - \frac{x}{L}\right)\frac{d(|x - b| - |b|)}{dx}. \tag{2.78b}$$

Similarly, $\tilde{N}_{2,x}$ is given by:

$$\tilde{N}_{2,x}(x) = \frac{\mathrm{d}}{\mathrm{d}x}\left[\left(\frac{x}{L}\right)(|x - b| - |L - b|)\right] \tag{2.79a}$$

$$\tilde{N}_{2,x}(x) = \left(\frac{1}{L}\right)(|x - b| - |L - b|) + \left(\frac{x}{L}\right)\frac{\mathrm{d}(|x - b| - |L - b|)}{\mathrm{d}x}. \tag{2.79b}$$

The next step is to write the stiffness matrix for the element. This 4×4 matrix is obtained as in Section 2.2.3 and reads:

$$\mathbf{K} = \int_0^L \mathbf{B}^T(x)E(x)A(x)\mathbf{B}(x)\mathrm{d}x.$$

As can be seen from the expression above, the calculation of the stiffness matrix requires the integration of a function which is a polynomial expression on each of the subdomains $[0, b]$ and $[b, L]$. Because it is simple to integrate polynomials, the integration domain is split into those two subdomains. The stiffness matrix can therefore be split into two parts, the first, $\mathbf{K}_1$ for the contribution from the domain on the left of the interface $x < b$, the second, $\mathbf{K}_2$ pertaining to the right of the interface $x > b$:

$$\mathbf{K} = \mathbf{K}_1 + \mathbf{K}_2.$$

Matrices $\mathbf{K}_1$ and $\mathbf{K}_2$ are defined by:

$$\mathbf{K}_1 = \int_0^b \mathbf{B}_1^T(x)E_1A\mathbf{B}_1(x)\,\mathrm{d}x \tag{2.80a}$$

$$\mathbf{K}_2 = \int_b^L \mathbf{B}_2^T(x)E_2A\mathbf{B}_2(x)\,\mathrm{d}x, \tag{2.80b}$$

where $\mathbf{B}_1$ and $\mathbf{B}_2$ are given by:

$$\forall x \in [0, b] : \quad \mathbf{B}_1(x) = \left[\begin{array}{cccc} N_{1,x}(x) & N_{2,x}(x) & \tilde{N}_{1,x}(x) & \tilde{N}_{2,x}(x) \end{array}\right] \tag{2.81a}$$

$$\forall x \in [b, L] : \quad \mathbf{B}_2(x) = \left[\begin{array}{cccc} N_{1,x}(x) & N_{2,x}(x) & \tilde{N}_{1,x}(x) & \tilde{N}_{2,x}(x) \end{array}\right]. \tag{2.81b}$$

Noting,

$$\frac{\mathrm{d}|x|}{\mathrm{d}x} = sign(x) \qquad \int |x|\mathrm{d}x = \frac{x|x|}{2}$$

$\mathbf{B}_1$ and $\mathbf{B}_2$, after simple manipulations, can be expressed as:

$$\mathbf{B}_1(x) = \left[\begin{array}{cccc} -1 & 1 & (x - \psi_b(x) - 1/2) & (\psi_b(x) - 1/2 - x) \end{array}\right] \tag{2.82a}$$

$$\mathbf{B}_2(x) = \left[\begin{array}{cccc} -1 & 1 & (3/2 - x - \psi_b(x)) & (\psi_b(x) - 1/2 + x) \end{array}\right], \tag{2.82b}$$

and $\mathbf{K}_1$ can be written as:

$$\mathbf{K}_1 = \int_0^b E_1A \begin{bmatrix} 1 & -1 & p_1 & p_2 \\ -1 & 1 & p_3 & p_4 \\ p_1 & p_3 & p_6 & p_5 \\ p_2 & p_4 & p_5 & p_7 \end{bmatrix}(x)\mathrm{d}x \tag{2.83}$$

where the functions $(p_i(x))_{1\leq i\leq 7}$ are given by:

$$p_1(x) = \phi_b(x) + 1/2 - x$$
$$p_2(x) = x + 1/2 - \phi_b(x)$$
$$p_3(x) = x - \phi_b(x) - 1/2$$
$$p_4(x) = \phi_b(x) - x - 1/2$$
$$p_5(x) = p_3(x)p_4(x)$$
$$p_6(x) = (p_3(x))^2$$
$$p_7(x) = (\phi_b(x) - x - 1/2)^2$$

After carrying out the integration, $\mathbf{K}_1$ is given by:

$$\mathbf{K}_1 = \begin{bmatrix} 1/2 & -1/2 & 1/4 & 1/4 \\ -1/2 & 1/2 & -1/4 & -1/4 \\ 1/4 & -1/4 & 1/6 & 1/12 \\ 1/4 & -1/4 & 1/12 & 1/6 \end{bmatrix} \tag{2.84}$$

And similarly, $\mathbf{K}_2$ is given by:

$$\mathbf{K}_2 = \int_b^L E_2 A \begin{bmatrix} 1 & -1 & q_1 & q_2 \\ -1 & 1 & q_3 & q_4 \\ q_1 & q_3 & q_6 & q_5 \\ q_2 & q_4 & q_5 & q_7 \end{bmatrix}(x)\mathrm{d}x, \tag{2.85}$$

where the $(q_i)_{1\leq i\leq 7}$ are given by:

$$q_1(x) = x + \phi_b(x) - 3/2$$
$$q_2(x) = 1/2 - x - \phi_b(x)$$
$$q_3(x) = 3/2 - x - \phi_b(x)$$
$$q_4(x) = \phi_b(x) - 1/2 + x$$
$$q_5(x) = q_3(x)q_4(x)$$
$$q_6(x) = (q_3(x))^2$$
$$q_7(x) = (\phi_b(x) - 1/2 + x)^2.$$

The final stiffness matrix writes:

$$\mathbf{K} = \begin{bmatrix} 1.5 & -1.5 & -0.25 & -0.25 \\ -1.5 & 1.5 & 0.25 & 0.25 \\ -0.25 & 0.25 & 0.5 & 0.25 \\ -0.25 & 0.25 & 0.25 & 0.5 \end{bmatrix} \tag{2.86}$$

The Neumann boundary conditions impose that a force $F = 1$ is applied at node 2; hence, the force vector reads:

$$\mathbf{f} = \{ \begin{matrix} 0 & F & 0 & 0 \end{matrix} \}^T. \tag{2.87}$$

The assembled equations write, find the standard degrees of freedom $u_1, u_2 \in \mathbb{R}$, and the enriched degrees of freedom $a_1, a_2 \in \mathbb{R}$ such that:

$$\begin{bmatrix} 1.5 & -1.5 & -0.25 & -0.25 \\ -1.5 & 1.5 & 0.25 & 0.25 \\ -0.25 & 0.25 & 0.5 & 0.25 \\ -0.25 & 0.25 & 0.25 & 0.5 \end{bmatrix} \begin{Bmatrix} u_1 \\ u_2 \\ a_1 \\ a_2 \end{Bmatrix} = \begin{Bmatrix} 0 \\ F \\ 0 \\ 0 \end{Bmatrix} \tag{2.88}$$

Enforcing the essential (here Dirichlet) boundary conditions $u_1 = 0$ and recalling that F is a unit force, the final set of equations writes:

$$\begin{bmatrix} 1.5 & 0.25 & 0.25 \\ 0.25 & 0.5 & 0.25 \\ 0.25 & 0.25 & 0.5 \end{bmatrix} \begin{Bmatrix} u_2 \\ a_1 \\ a_2 \end{Bmatrix} = \begin{Bmatrix} 1 \\ 0 \\ 0 \end{Bmatrix} \tag{2.89}$$

and the solution is: $u_2 = 0.75$, $a_1 = -0.25$ and $a_2 = -0.25$.

2.3 Strong Discontinuities for One-dimensional Problem

This exercise deals with the problem of a 1D bar, fixed at both ends which has forces F and $-F$ applied at a particular point a along the bar. The bar is expected to split at this point. The displacements of the two portions of the bar are obtained using XFEM. The goal of the exercise is to compute the displacement field in both parts of the bar. Consider a 1D element with two nodes. The shape functions are given by:

$$N_1 = \frac{1}{L} \left(\frac{L}{2} - x \right)$$
$$N_2 = \frac{1}{L} \left(\frac{L}{2} + x \right) \tag{2.90}$$

The XFEM displacement approximation is given by:

$$\mathbf{u} = N_1 u_1 + N_2 u_2 + N_3 a_1 + N_4 a_2 \tag{2.91}$$

where u_1, u_2 are standard degrees of freedom and a_1, a_2 are enriched degrees of freedom associated with nodes 1 and 2. In this case, the Heaviside enrichment function $H(x)$ is used, such that:

$$H(x) = \begin{matrix} +1 & \forall x > a \\ -1 & \forall x < a \end{matrix} \tag{2.92}$$

The additional shape functions, N_3 and N_4, are given by:

$$N_3 = N_1(H(x) - H(x_1))$$
$$N_4 = N_2(H(x) - H(x_2)) \tag{2.93}$$

The terms $H(x) - H(x_1)$ and $H(x) - H(x_2)$ are referred to as the "shifted" enrichment functions which insure that the enrichment vanishes at the nodes. $H(x_1)$ indicates the value of $H(x)$ at node 1, while $H(x_2)$ refers to the value of $H(x)$ at node 2. The enriched shape functions can now be obtained for the two domains (the positive (+) domain which occurs to the right of the point "a" and the negative (−) domain which occurs to the left of the point "a").

$\forall x < a$

$$N(x) = [\; \frac{1}{L}\left(\frac{L}{2} - x\right) \quad \frac{1}{L}\left(\frac{L}{2} + x\right) \quad 0 \quad \frac{-2}{L}\left(\frac{L}{2} + x\right) \;] \tag{2.94}$$

$\forall x > a$

$$N(x) = [\; \frac{1}{L}\left(\frac{L}{2} - x\right) \quad \frac{1}{L}\left(\frac{L}{2} + x\right) \quad \frac{2}{L}\left(\frac{L}{2} - x\right) \quad 0 \;] \tag{2.95}$$

From these shape functions the B matrices (strain-displacement relation) can be obtained by differentiation, i.e.:

$\forall x < a$

$$B_n = [\; \frac{-1}{L} \quad \frac{1}{L} \quad 0 \quad \frac{-2}{L} \;] \tag{2.96}$$

$\forall x < a$

$$B_p = [\; \frac{-1}{L} \quad \frac{1}{L} \quad \frac{-2}{L} \quad 0 \;] \tag{2.97}$$

The stiffness matrix **K** for the problem is similarly calculated for the two independent regions to obtain K^- and K^+. These are then summed to give the combined stiffness matrix of the problem.

$$\mathbf{K} = \mathbf{K}^- + \mathbf{K}^+ \tag{2.98}$$

$$\mathbf{K}^- = \int_{\frac{-L}{2}}^{a} B_n^T EAB_n \, dx$$

$$\mathbf{K}^+ = \int_{a}^{\frac{L}{2}} B_p^T EAB_p \, dx \tag{2.99}$$

Therefore:

$$\mathbf{K}^- = \int_{-L/2}^{a} \frac{EA}{L^2}
\begin{bmatrix}
1 & 1 & 0 & 2 \\
-1 & 1 & 0 & -2 \\
0 & 0 & 0 & 0 \\
2 & -2 & 0 & 4
\end{bmatrix} dx$$

$$\mathbf{K}^+ = \int_{a}^{\frac{L}{2}} \frac{EA}{L^2}
\begin{bmatrix}
+1 & -1 & +2 & 0 \\
-1 & +1 & -2 & 0 \\
+2 & -2 & +4 & 0 \\
0 & 0 & 0 & 0
\end{bmatrix} dx \tag{2.100}$$

The system stiffness matrix, **K**, is then given by:

$$\mathbf{K} = \frac{EA}{L^2}
\begin{bmatrix}
(Z+V) & -(Z+V) & +2V & +2Z \\
-(Z+V) & (Z+V) & -2V & -2Z \\
+2Z & -2V & +4V & 0 \\
+2Z & -2Z & 0 & +4Z
\end{bmatrix} \tag{2.101}$$

where $V = 2\left(\frac{L}{2} - a\right)$ and $Z = 2\left(\frac{L}{2} + a\right)$. The force vector for the given problem is given by:

$$\mathbf{f} = \left\{ \begin{array}{c} R_1 \\ R_2 \\ -F \\ F \end{array} \right\} \tag{2.102}$$

Applying boundary conditions ($\mathbf{u}(x = -L/2) = 0, \mathbf{u}(x = L/2) = 0$):

$$\mathbf{K} = \left[\begin{array}{cc} 8\left(\frac{L}{2} - a\right) & 0 \\ 0 & 8\left(\frac{L}{2} + a\right) \end{array} \right]$$

$$\mathbf{f} = \left\{ \begin{array}{c} -F \\ F \end{array} \right\} \tag{2.103}$$

The displacement vector $\mathbf{u}$ is obtained:

$$\mathbf{u} = \left\{ \begin{array}{c} \frac{-8L^2 F\left(\frac{L}{2} + a\right)}{EA(16L^2 - 64a^2)} \\ \frac{8L^2 F\left(\frac{L}{2} - a\right)}{EA(16L^2 - 64a^2)} \end{array} \right\} \tag{2.104}$$

We saw that a simple discontinuous derivative enrichment is sufficient to allow the FEM to reproduce exactly a strain jump within an element. Similarly, introducing a discontinuous function in the FE space suffices to introduce a strong discontinuity in the displacement field. However, this flexibility comes at a price. In the standard FEM, interfaces are explicitly meshed and it is consequently clear where these interfaces are located within the solid since nodes and element edges are present along this interface. If an interface is modeled implicitly, using enrichment as described above, the geometrical position of the interface is no longer provided by the mesh connectivity and nodal coordinates. This leads to two complications:

- The geometry of the interface must be defined and updated in time. The interface can be geometrically described in several ways, which we will discuss in Section 3.3.1.
- The position of the interface relative to the mesh must be evaluated: this permits to define which nodes of the mesh are enriched, and which are not. This step, known as mesh geometry interaction (see Section 3.3.1), is usually performed by checking whether the support of a given node intersects the geometrical feature.

Probably the most widely used method to both describe the relative position of the interface and the mesh and to update this position in time is known as the level set method of Sethian (2000). This method is particularly well suited to closed interfaces, and to handling topological changes such as spawning, merging, and disappearance of interfaces.

A second important remark which can be made from the above is the need for special integration techniques for weak forms including weak discontinuities (e.g., material interfaces) or strong discontinuities (e.g., cracks). This is evident from the above bi-material bar example (Section 2.2), as the weak form had to be integrated separately on either side of the material interface, leading to two stiffness matrices which were subsequently added. More details will be provided in Chapter 4, but for now, it is sufficient for the reader to be aware that special care must be exercised when integrating weak forms with strong or weak discontinuities, so that sufficiently many Gauss points are present on either side of the discontinuity. While we will describe several other techniques in Section 4.2, the most widely used idea is to decompose both parts of the element intersected by the discontinuity into integration subcells and perform standard Gauss quadrature on those subcells. In Section 5.1, the case of linear elastic fracture mechanics is discussed, in which the enrichment functions are not polynomial and lead to the need to integrate singular functions.

⊙ **Example 2.** Write a Matlab-code to solve the one-dimensional bi-material bar problem shown in Figure 2.4 numerically with the FEM ($E_1 = 10$, $E_2 = 1$). Use piecewise linear shape functions and a discretization in which one of the nodes is at the material interface. The numerical solution should be the exact solution. Show this by refining the elements. What happens if the interface is located inside an element?

⊙ **Example 3.** Consider the one-dimensional bi-material problem subjected to cubic line force $q(x) = x^3$ with the following boundary conditions: $u(x = 0) = u(x = L) = 0$ with $E_1 = 1$ for $0 \le x \le 0.5$ and $E_2 = 10$ for $0.5 \le x \le 1$. Discus the choice of enrichment function and solve the problem within XFEM framework.

⊙ **Example 4.** Consider the following one-dimensional problem:

$$u_{,xx}(x) + b(x) = 0, \quad x \in [0,1]$$

with $u(0) = 0$ and $u(1) = 1$. The body force is given by:

$$b(x) = \begin{cases} (2\alpha^2 - 4[\alpha^2(x - 0.5)^2]) \, exp(-[\alpha(x - 0.5)]^2) & x \in [0.42, 0.58] \\ 0 & \text{otherwise} \end{cases}$$

The exact solution to this problem is $u(x) = x + exp(-[\alpha(x - 0.5)]^2)$. Write the weak form and solve using conventional FEM and XFEM with an appropriate enrichment function. Discuss the results.

2.4 Conclusions

The purpose of this chapter was to provide a simple introduction to the concept of enrichment. We started by reviewing, from a historical perspective, the notion of enrichment which dates back to the seventies when, already, known features of the solution had been introduced in the FE approximation. We differentiated these enrichment techniques from the more recent partition of unity enrichment of Babuška where the enrichment is localized to a small region around the features of interest. We then solved in detail problems for weak discontinuity in one dimension, which illustrated in very simple settings the notion of reproducibility of a function by an approximation. We remarked that the flexibility offered by enrichment, where elements need not conform with discontinuity surfaces, leads to the need to determine the interaction of the geometry of these interfaces with the underlying mesh and introduce the notion of level set representation of this geometry as a possible solution. Now that the scene is set, we will, in the following chapter, define the notion of partition of unity rigorously, and show how it is implemented in practice through a series of examples.

References

Areias, P. and Belytschko, T. (2006). Two-scale shear band evolution by local partition of unity. *International Journal for Numerical Methods in Engineering* 66 (5): 878–910.

Armero, F. and Garikipati, K. (1995). Recent advances in the analysis and numerical simulation of strain localization in inelastic solids. *Computational Plasticity Fundamentals and Applications* I 547–561.

Babuška, I., Caloz, G., and Osborn, J.E. (1994). Special finite element methods for a class of second order elliptic problems with rough coefficients. *SIAM Journal on Numerical Analysis* 31 (4): 945–981.

Barbieri, E., Petrinic, N., Meo, M., and Tagarielli, V.L. (2012). A new weight-function enrichment in meshless methods for multiple cracks in linear elasticity. *International Journal for Numerical Methods in Engineering* 90: 177–195.

Barsoum, R.S. (1974). Application of quadratic isoparametric finite elements in linear fracture mechanics. *International Journal of Fracture* 10 (4): 603–605.

Barsoum, R.S. (1976). On the use of isoparametric finite elements in linear fracture mechanics. *International Journal for Numerical Methods in Engineering* 10 (1): 25–37.

Barsoum, R.S. (1977). Triangular quarter-point elements as elastic and perfectly-plastic crack tip elements. *International Journal for Numerical Methods in Engineering* 11 (1): 85–98.

Bordas, S., Rabczuk, T., Hung, N.X., et al. (2010). Strain smoothing in FEM and XFEM. *Computers & Structures* 88 (23): 1419–1443.

Bordas, S., Natarajan, S., Kerfriden, P., et al. (2011). On the performance of strain smoothing for quadratic and enriched finite element approximations (XFEM/GFEM/PUFEM). *International Journal for Numerical Methods in Engineering* 86 (4–5): 637–666.

Belytschko, T., Fish, J., and Engelmann, B.E. (1988). A finite element with embedded localization zones. *Computer Methods in Applied Mechanics and Engineering* 70 (1): 59–89.

Belytschko, T. and Black, T. (1999). Elastic crack growth in finite elements with minimal remeshing. *International Journal for Numerical Methods in Engineering* 45 (5): 601–620.

Benzley, S.E. (1974). Representation of singularities with isoparametric finite elements. *International Journal for Numerical Methods in Engineering* 8 (3): 537–545.

Birkhoff, G., Schultz, M.H., and Varga, R.S. (1968). Piecewise Hermite interpolation in one and two variables with applications to partial differential equations. *Numerische Mathematik* 11 (3): 232–256.

Bordas, S. and Duflot, M. (2007). Derivative recovery and a posteriori error estimate for extended finite elements. *Computer Methods in Applied Mechanics and Engineering* 196 (35): 3381–3399.

Bordas, S., Duflot, M., and Le, P. (2008a). A simple error estimator for extended finite elements. *Communications in Numerical Methods in Engineering* 24 (11): 961–971.

Bordas, S., Rabczuk, T., and Zi, G. (2008b). Three-dimensional crack initiation, propagation, branching and junction in non-linear materials by an extended meshfree method without asymptotic enrichment. *Engineering Fracture Mechanics* 75 (5): 943–960.

Byskov, E. (1970). The calculation of stress intensity factors using the finite element method with cracked elements. *International Journal of Fracture* 6 (2): 159–167.

Daux, C., Moes, N., Dolbow, J., et al. (2000). Arbitrary branched and intersection cracks with the extended finite element method. *International Journal for Numerical Methods in Engineering* 48: 1731–1760.

De Luycker, E., Benson, D.J., Belytschko, T., et al. (2011). X-FEM in isogeometric analysis for linear fracture mechanics. *International Journal for Numerical Methods in Engineering* 87 (6): 541–565.

Dolbow, J.E. (1999). *An extended finite element method with discontinuous enrichment for applied mechanics*. PhD thesis, Northwestern university.

Dolbow, J., Moes, N., and Belytschko, T. (2000). Modeling fracture in Mindlin–Reissner plates with the extended finite element method. *International Journal of Solids and Structures* 37 (48): 7161–7183.

Duarte, C.A. (1996). *The hp cloud method*. PhD thesis, The University of Texas at Austin.

Duarte, C.A. and Oden, J.T. (1996). An hp-adaptive method using clouds. *Computer Methods in Applied Mechanics and Engineering* 139 (1): 237–262.

Duflot, M. (2006). A meshless method with enriched weight functions for three-dimensional crack propagation. *International Journal for Numerical Methods in Engineering* 65 (12): 1970–2006.

Duflot, M. and Nguyen-Dang, H. (2004a). A meshless method with enriched weight functions for fatigue crack growth. *International Journal for Numerical Methods in Engineering* 59 (14): 1945–1961.

Duflot, M. and Nguyen-Dang, H. (2004b). Fatigue crack growth analysis by an enriched meshless method. *Journal of Computational and Applied Mathematics* 168: 155–164.

Dvorkin, E.N., Cuitiño, A., and Gioia, G. (1990). Finite elements with displacement interpolated embedded localization lines insensitive to mesh size and distortions. *International Journal for Numerical Methods in Engineering* 30 (3): 541–564.

Faddeev, D.K. and Faddeeva, V.N. (1981). Computational methods of linear algebra. *Journal of Soviet Mathematics* 15: 531–560.

Farhat, C. Harari, I., and Franca, L.P. (2001). The discontinuous enrichment method. *Computer Methods in Applied Mechanics and Engineering* 190 (48): 6455–6479.

Fish, J. (1992). The s-version of the finite element method. *Computers & Structures* 43 (3): 539–547.

Fix, G. (1968). *Bounds and approximations for eigenvalues of self-adjoint boundary value problems.* Technical report.

Fix, G. (1969). Higher-order Rayleigh–Ritz approximations. *Journal of Mathematics and Mechanics* 18: 645–657.

Fix, G.J., Gulati, S., and Wakoff, G.I. (1973). On the use of singular functions with finite element approximations. *Journal of Computational Physics* 13 (2): 209–228.

Fleming, M., Chu, Y.A., Moran, B., et al. (1997). Enriched element-free Galerkin methods for crack tip fields. *International Journal for Numerical Methods in Engineering* 40 (8): 1483–1504.

Foschi, R.O. and Barrett, J.D. (1976). Stress intensity factors in anisotropic plates using singular isoparametric elements. *International Journal for Numerical Methods in Engineering* 10 (6): 1281–1287.

Fries, T.P. and Belytschko, T. (2006). The intrinsic XFEM: A method for arbitrary discontinuities without additional unknowns. *International Journal for Numerical Methods in Engineering* 68 (13): 1358–1385.

Gifford, L.N. and Hilton, P.D. (1978). Stress intensity factors by enriched finite elements. *Engineering Fracture Mechanics* 10 (3): 485–496.

Hansbo, A. and Hansbo, P. (2004). A finite element method for the simulation of strong and weak discontinuities in solid mechanics. *Computer Methods in Applied Mechanics and Engineering* 193 (33): 3523–3540.

Henshell, R.D. and Shaw, K.G. (1975). Crack tip finite elements are unnecessary. *International Journal for Numerical Methods in Engineering* 9 (3): 495–507.

Hettich, T., Hund, A., and Ramm, E. (2008 January). Modeling of failure in composites by X-FEM and level sets within a multiscale framework. *Computer Method Applied Mechanics* 197 (5): 414–424. doi: 10.1016/j.cma.2007.07.017.

Hibbitt, H.D. (1977). Some properties of singular isoparametric elements. *International Journal for Numerical Methods in Engineering* 11 (1): 180–184.

Hughes, T.J.R. (1995). Multiscale phenomena: Green's functions, the Dirichlet-to-Neumann formulation, subgrid scale models, bubbles and the origins of stabilized methods. *Computer Methods in Applied Mechanics and Engineering* 127 (1): 387–401.

Hughes, T.J.R. (2000). *The Finite Element Method: Linear Static and Dynamic Finite Element Analysis.* Dover Publications.

Hughes, T.J.R., Cottrell, J.A., and Bazilevs, Y. (2005). Isogeometric analysis: CAD, finite elements, NURBS, exact geometry and mesh refinement. *Computer Methods in Applied Mechanics and Engineering* 194 (39): 4135–4195.

Hughes, T.J.R., Feij'oo, G.R., Mazzei, L., and Quincy, J.B. (1998). The variational multiscale method–a paradigm for computational mechanics. *Computer Methods in Applied Mechanics and Engineering* 166 (1): 3–24.

Hussain, M.A., Coffin, L.F., and Zaleski, K.A. (1981). Three dimensional singular element. *Computers & Structures* 13 (5–6): 595–599.

Hutchinson, J.W. (1968). Plastic stress and strain fields at a crack tip. *Journal of the Mechanics and Physics of Solids* 16 (5): 337–342.

Jordan, W.B. (1970). *Plane isoparametric structural element.* Technical report, Schenectady, NY: Knolls Atomic Power Lab.

Karihaloo, B.L. and Xiao, Q.Z. (2001). Accurate determination of the coefficients of elastic crack tip asymptotic field by a hybrid crack element with p-adaptivity. *Engineering Fracture Mechanics* 68 (15): 1609–1630.

Klisinski, M., Runesson, K., and Sture, S. (1991). Finite element with inner softening band. *Journal of Engineering Mechanics* 117 (3): 575–587.

Krongauz, Y. and Belytschko, T. (1998). EFG approximation with discontinuous derivatives. *International Journal for Numerical Methods in Engineering* 41 (7): 1215–1233.

Krysl, P. and Belytschko, T. (1997). Element-free Galerkin method: Convergence of the continuous and discontinuous shape functions. *Computer Methods in Applied Mechanics and Engineering* 148 (3): 257–277.

Larsson, R., Runesson, K., and Akesson, M. (1995). Embedded cohesive crack models based on regularized discontinuous displacements. In: *Fracture Mechanics of Concrete Structures, Proceedings of FraMCoS-2* (ed. Wittmann F.H.), 899–911. Freiburg, Germany: Aedificatio Publishers.

Larsson, R. and Runesson, K. (1996). Element-embedded localization band based on regularized displacement discontinuity. *Journal of Engineering Mechanics* 122: 402.

Larsson, R., Runesson, K., and Ottosen, N.S. (1993). Discontinuous displacement approximation for capturing plastic localization. *International Journal for Numerical Methods in Engineering* 36 (12): 2087–2105.

Lee, S.H., Song, J.H., Yoon, Y.C., et al. (2004). Combined extended and superimposed finite element method for cracks. *International Journal for Numerical Methods in Engineering* 59 (8): 1119–1136.

Lehman. R.S. (1959). Developments at an analytic corner of solutions of elliptic partial differential equations. *Journal of Mathematics and Mechanics* 8: 727760.

Lin, J.S. (2003). A mesh-based partition of unity method for discontinuity modeling. *Computer Methods in Applied Mechanics and Engineering* 192 (11–12): 1515–1532

Lotfi, H.R. (1992). *Finite element analysis of fracture in concrete and masonry structures*. PhD thesis, University of Colorado, Boulder.

Lotfi, H.R. and Shing, P.B. (1995). Embedded representation of fracture in concrete with mixed finite elements. *International Journal for Numerical Methods in Engineering* 38 (8): 1307–1325.

Lynn, P.P. and Ingraffea, A.R. (1978).Transition elements to be used with quarter-point crack-tip elements. *International Journal for Numerical Methods in Engineering* 12 (6): 1031–1036.

Melenk, J.M. (1995). *On generalized finite element methods*. PhD thesis, The University of Maryland.

Melenk, J.M. and Babuška, I. (1996). The partition of unity finite element method: Basic theory and applications. *Computer Methods in Applied Mechanics and Engineering* 139 (1): 289–314.

Melenk, J.M. and Babuška, I. (1997). Approximation with harmonic and generalized harmonic polynomials in the partition of unity method. *Computer Assisted Mechanics and Engineering Sciences* 4: 607–632.

Mergheim, J. (2009). A variational multiscale method to model crack propagation at finite strains. *International Journal for Numerical Methods in Engineering* 80 (3): 269–289.

Moës, N., Dolbow, J., and Belytschko, T. (1999). Elastic crack growth in finite elements without remeshing. *International Journal for Numerical Methods in Engineering* 46: 131–150.

Moës, N., Stolz, C., Bernard, P.E., and Chevaugeon, N. (2011). A level set based model for damage growth: The thick level set approach. *International Journal for Numerical Methods in Engineering* 86 (3): 358–380.

Mote Jr., C.D. (1971). Global-local finite element. *International Journal for Numerical Methods in Engineering* 3 (4): 565–574.

Nguyen, V.P., Rabczuk, T., Bordas, S., and Duflot, M. (2008). Meshless methods: A review and computer implementation aspects. *Mathematics and Computers in Simulation* 79 (3): 763–813.

Oden, J.T. and Duarte, C.A. (1997). Clouds, cracks and FEM. *Recent Developments in Computational and Applied Mechanics* 302–321.

Oden, J.T. Duarte, C.A.M., and Zienkiewicz, O.C. (1998). A new cloud-based hp finite element method. *Computer Methods in Applied Mechanics and Engineering* 153 (1): 117–126.

Oliver, J. (1996). Modelling strong discontinuities in solid mechanics via strain softening constitutive equations. Part 2: Numerical simulation. *International Journal for Numerical Methods in Engineering* 39 (21): 3601–3623.

Organ, D., Fleming, M., Terry, T., and Belytschko, T. (1996). Continuous meshless approximations for nonconvex bodies by diffraction and transparency. *Computational Mechanics* 18 (3): 225–235.

Oliver, J., Huespe, A.E., and Samaniego, E. (2003). A study on finite elements for capturing strong discontinuities. *International Journal for Numerical Methods in Engineering* 56 (14): 2135–2161.

Oliver, J., Huespe, A.E., and Sanchez, P.J. (2006 January). A comparative study on finite elements for capturing strong discontinuities: E-FEM vs X-FEM. *Computer Method Applied Mechanics* 195 (37–40): 4732–4752.

Ortiz, M., Leroy, Y., and Needleman, A. (1987). A finite element method for localized failure analysis. *Computer Methods in Applied Mechanics and Engineering* 61 (2): 189–214.

Pageau, S.S. and Biggers, S.B. (1995). Finite element evaluation of free-edge singular stress fields in anisotropic materials. *International Journal for Numerical Methods in Engineering* 38 (13): 2225–2239.

Pageau, S.S. and Biggers, S.B. (1996). A finite element approach to three-dimensional singular stress states in anisotropic multi-material wedges and junctions. *International Journal of Solids and Structures* 33 (1): 33–47.

Pageau, S.S., Gadi, K.S., Biggers, S.B., and Joseph, P.F. (1996). Standardized complex and logarithmic eigensolutions for *n*-material wedges and junctions. *International Journal of Fracture* 77 (1): 51–76.

Pageau, S.S., Joseph, P.F., and Biggers Jr., S.B. (1995). A finite element analysis of the singular stress fields in anisotropic materials loaded in antiplane shear. *International journal for numerical methods in engineering* 38 (1): 81–97.

Pu, S.L., Hussain, M.A., and Lorensen, W.E. (1978). The collapsed cubic isoparametric element as a ingular element for crack problems. *International Journal for Numerical Methods in Engineering* 12 (11): 1727–1742.

Rabczuk, T., Belytschko, T., and Xiao, S.P. (2004). Stable particle methods based on Lagrangian kernels. *Computer Methods in Applied Mechanics and Engineering* 193 (12): 1035–1063.

Rabczuk, T., Bordas, S., and Zi, G. (2007). A three-dimensional meshfree method for continuous multiple-crack initiation, propagation and junction in statics and dynamics. *Computational Mechanics* 40 (3): 473–495.

Rao, B.N. and Rahman, S. (2004). An enriched meshless method for non-linear fracture mechanics. *International Journal for Numerical Methods in Engineering* 59 (2): 197–223.

Remmers, J.J.C., Borst, R., and Needleman, A. (2003). A cohesive segments method for the simulation of crack growth. *Computational Mechanics* 31 (1): 69–77.

Réthoré, J., Roux, S., and Hild, F. (2010). Hybrid analytical and extended finite element method (HAX-FEM): A new enrichment procedure for cracked solids. *International Journal for Numerical Methods in Engineering* 81 (3): 269–285.

Saouma, V.E. and Sikiotis, E.S. (1986). Stress intensity factors in anisotropic bodies using singular isoparametric elements. *Engineering Fracture Mechanics* 25 (1): 115–121.

Sethian, J. A. (1999). Fast marching methods. *Siam Review* 41 (2): 199–235.

Shi, G.H. (1991). Manifold method of material analysis. *Army Research Office Research Triangle Park NC*.

Shi, G.H. (1992). Modeling rock joints and blocks by manifold method. In: *Proceedings of 32nd US Symposium on Rock Mechanics*, 648.

Shih, C.F. (1974). Small-scale yielding analysis of mixed mode plane-strain crack problems. *ASTM STP* 560 (7): 187–210.

Shih, C.F., Lorenzi, H.G., and German, M.D. (1976). Crack extension modeling with singular quadratic isoparametric elements. *International Journal of Fracture* 12 (4): 647–651.

Simo, J. and Oliver, J. (1994). A new approach to the analysis and simulation of strain softening in solids. In: *Fracture and Damage in Quasi-brittle Structures* (ed. Bazant, Z.P. et al.), 25–39. London: E. and F.N. Spon.

Simo, J.C., Oliver, J., and Armero, F. (1993). An analysis of strong discontinuities induced by strain-softening in rate-independent inelastic solids. *Computational Mechanics* 12 (5): 277–296.

Simpson, R. and Trevelyan, J. (2011a). A partition of unity enriched dual boundary element method for accurate computations in fracture mechanics. *Computer Methods in Applied Mechanics and Engineering* 200 (1): 1–10.

Simpson, R. and Trevelyan, J. (2011b). Evaluation of J1 and J2 integrals for curved cracks using an enriched boundary element method. *Engineering Fracture Mechanics* 78 (4): 623–637.

Sluys, L.J. and Berends, A.H. (1998). Discontinuous failure analysis for mode-i and mode-ii localization problems. *International Journal of Solids and Structures* 35 (31): 4257–4274.

Song, J-H., Areias, P., and Belytschko, T. (2006). A method for dynamic crack and shear band propagation with phantom nodes. *International Journal for Numerical Methods in Engineering* 67 (6): 868–893.

Strang, W.G. and Fix, G.J. (1973). *Analysis of the Finite Element Method*. Prentice-Hall.

Steinmueller, G. (1974). Restrictions in the application of automatic mesh generation schemes by 'isoparametric'coordinates. *International Journal for Numerical Methods in Engineering* 8 (2): 289–294.

Strouboulis, T., Babuška, I., and Copps, K. (2000b). The design and analysis of the generalized finite element method. *Computer Methods in Applied Mechanics and Engineering* 181 (1): 43–69.

Strouboulis, T., Copps, K., and Babuška, I. (2000a). The generalized finite element method: An example of its implementation and illustration of its performance. *International Journal for Numerical Methods in Engineering* 47 (8): 1401–1417.

Strouboulis, T., Copps, K., and Babuška, I. (2001). The generalized finite element method. *Computer Methods in Applied Mechanics and Engineering* 190 (32): 4081–4193.

Sukumar, N., Moes, N., Moran, B., and Belytschko, T. (2000). Extended finite element method for three-dimensional crack modelling. *International Journal for Numerical Methods in Engineering* 48 (11): 1549–1570.

Terada, K., Asai, M., and Yamagishi, M. (2003). Finite cover method for linear and non-linear analyses of heterogeneous solids. *International Journal for Numerical Methods in Engineering* 58 (9): 1321–1346.

Tong, P., Pian, T.H.H., and Lasry, S.J. (1973). A hybrid-element approach to crack problems in plane elasticity. *International Journal for Numerical Methods in Engineering* 7 (3): 297–308.

Tracey, D.M. (1971). Finite elements for determination of crack tip elastic stress intensity factors. *Engineering Fracture Mechanics* 3 (3): 255–265.

Tracey, D.M. (1976). Finite element solutions for crack-tip behavior in small-scale yielding. *Journal of Engineering Materials and Technology* 98: 146.

Trefftz, E. (1926). Ein gegenstuck zum ritzschen verfahren. In: *Proceedings of the 2nd International Congress on Applied Mechanichs*, 131–137. Zurich. 1926.

Ventura, G., Xu, J.X., and Belytschko, T. (2002). A vector level set method and new discontinuity approximations for crack growth by EFG. *International Journal for Numerical Methods in Engineering* 54 (6): 923–944.

Vu-Bac, N., Nguyen-Xuan, H., Chen, L., et al. (2011). A node-based smoothed extended finite element method (NS-XFEM) for fracture analysis. *Computer Modeling in Engineering & Sciences (CMES)* 73 (4): 331–355.

Wait, R. and Mitchell, A.R. (1971). Corner singularities in elliptic problems by finite element methods. *Journal of Computational Physics* 8 (1): 45–52.

Wigley, N.M. (1969). On a method to subtract off a singularity at a corner for the Dirichlet or Neumann problem. *Mathematics of Computation* 23: 395–401.

Yamada, Y., Ezawa, Y., and Nishiguchi, I. (1979). Reconsiderations of singularity of crack tip elements. *International Journal for Numerical Methods in Engineering* 14: 1525–1544.

3

Partition of Unity Revisited

Stéphane P. A. Bordas[1], Alexander Menk[2], and Sundararajan Natarajan[3]

[1] University of Luxembourg, Luxembourg, UK
[2] Robert Bosch GmbH, Germany
[3] Indian Institute of Technology Madras, India

The previous chapter provided a first introduction to the concept of enrichment and introduced some key issues related to its use within a finite element framework. The goal of the present chapter is to take the concept one step further and analyze more in depth the notions of completeness, partition of unity, mesh-geometry interaction, and imposition of boundary conditions.

3.1 Completeness, Consistency, and Reproducing Conditions

The notion of partition of unity is easily introduced by defining the completeness of approximations, which is expressed in terms of the highest order of the polynomials that can be represented exactly.

Definition 1 [Reproducing condition] Let p be a function defined over Ω and a set of approximating functions N_J defined over a set of nodes $\mathbf{x}_J$ such that $p(\mathbf{x}_J) = u_J$. The N_J's reproduce p on Ω, if and only if:

$$\forall \mathbf{x} \in \Omega, \; N_J(\mathbf{x})u_J = N_J(\mathbf{x})p(\mathbf{x}_J) = p(\mathbf{x}) \tag{3.1}$$

For example, if the approximation is able to reproduce a constant function exactly, then it is "zero-order complete." If the approximation can reproduce linear functions exactly, it is first-order complete or linear complete and so on for higher order monomials. The term *completeness* is sometimes referred to as *reproducing conditions*.

Definition 2 [Constant and linear reproducing conditions in two dimensions] In two dimensions, the constant and linear reproducing conditions are given by:

$$\sum_J N_J(\mathbf{x}) = 1 \tag{3.2a}$$

$$\sum_J N_J(\mathbf{x}) \, x_J = x \quad \text{and} \quad \sum_J N_J(\mathbf{x}) \, y_J = y \tag{3.2b}$$

Reproducing conditions can be defined similarly for the derivatives of a function p above and the corresponding conditions are known as "derivative reproducing conditions."

Partition of Unity Methods, First Edition. Stéphane P. A. Bordas, Alexander Menk, and Sundararajan Natarajan.
© 2024 John Wiley & Sons Ltd. Published 2024 by John Wiley & Sons Ltd.

3.2 Partition of Unity

Functions that possess property Equation (3.2a) are called a partition of unity (PU), i.e., a partition of unity is a set of functions f_I defined on a domain Ω^{PU} such that:

$$\forall \mathbf{x} \in \Omega^{PU}, \sum_I f_I(\mathbf{x}) = 1. \tag{3.3}$$

> ☞ Isoparametric finite element shape functions meet this condition. Hence, isoparametric finite elements can be classified in the category of partition of unity methods.

Let us now remark that function p above needs not be a polynomial function, as is usually the case for standard finite element procedures as well as many meshless methods. Indeed, it is possible to choose p to be an arbitrary function defined on Ω^{PU}, which we shall denote by ψ to avoid confusion. Multiply Equation (3.3) by function ψ to obtain:

$$\forall \mathbf{x} \in \Omega^{PU}, \sum_I f_I(\mathbf{x})\psi(\mathbf{x}) = \psi(\mathbf{x}).$$

Based on this last expression, the concept of enrichment can be introduced very naturally, which is the topic of Section 3.3.

> ☞ Belytschko et al. (1996) have shown that a discretization has to be zero-order complete to guarantee conservation of linear momentum and linear complete to guarantee conservation of angular momentum. Conservation of linear momentum requires that the rate of change of linear momentum equals the total applied force such that the total change of linear momentum due to internal forces is zero. Thus, in the absence of external forces and body forces, conservation of linear momentum requires that:
>
> $$\frac{D}{Dt}\left(\sum_{I \in \mathcal{S}} m_I \mathbf{v}_I\right) = \sum_{I \in \mathcal{S}} m_I \dot{\mathbf{v}}_I = 0 \tag{3.4}$$
>
> where m_I are nodal masses and $\mathbf{v}$ the velocity field. From the linear momentum equation (without external and body forces) we know that:
>
> $$m_I \dot{\mathbf{v}}_I = -\sum_{J \in \mathcal{S}} \nabla \mathbf{N}_I(\mathbf{x}_J) \cdot \sigma(\mathbf{x}_J)\, w_J \tag{3.5}$$
>
> where $\mathbf{N}_I(\mathbf{x}_J)$ are shape functions and w_J are the quadrature weights. Substituting the right-hand side of Equation (3.5) into Equation (3.4) gives:
>
> $$\sum_{I \in \mathcal{S}} m_I \mathbf{v}_I = -\sum_{I \in \mathcal{S}}\sum_{J \in \mathcal{S}} \nabla \mathbf{N}_I(\mathbf{x}_J) \cdot \sigma(\mathbf{x}_J)\, w_J = -\sum_{J \in \mathcal{S}}\sum_{I \in \mathcal{S}} \nabla \mathbf{N}_I(\mathbf{x}_J) \cdot \sigma(\mathbf{x}_J)\, w_J = 0 \tag{3.6}$$
>
> that requires that $\sum_{I \in \mathcal{S}} \nabla \mathbf{N}_I(\mathbf{x}_J) = 0$ for arbitrary stress states that in turn requires zero-order complete shape functions.
>
>   Conservation of angular momentum requires that any change in angular momentum is exclusively due to external forces. Hence, let us show that the change of angular momentum in the absence of external forces vanishes. The time rate of change of angular momentum in the absence of external forces can be expressed as:
>
> $$\frac{D}{Dt}\left(\sum_I m_I \mathbf{v}_I \times \mathbf{x}_I\right) = \sum_I m_I\left(\dot{\mathbf{v}}_I \times \mathbf{x}_I + \underbrace{\mathbf{v}_I \times \mathbf{v}_I}_{=0}\right) = 0 \tag{3.7}$$

Here, $\times$ denotes the vector cross product. Substituting Equation (3.5) into Equation (3.7) leads to:

$$\frac{D}{Dt}\left(\sum_I m_I \mathbf{v}_I \times \mathbf{x}_I\right) = \sum_I \epsilon_{ijk}\left(\sum_J N_{I,m}(\mathbf{x}_J)\,\sigma_{mj}(\mathbf{x}_J)w_J\right)x_{Ik} \tag{3.8}$$

where ϵ_{ijk} is the permutation tensor and x_{Ik} refers to the kth component of node I. We sum over repeated indices. Equation (3.8) can be reformulated as:

$$\epsilon_{ijk}\sum_J \underbrace{\left(\sum_I N_{I,m}(\mathbf{x}_J)x_{Ik}\right)}_{\delta_{mk}}\sigma_{mj}(\mathbf{x}_J)w_J = \epsilon_{ijk}\delta_{mk}\sum_J \sigma_{mj}(\mathbf{x}_J)w_J$$

$$= \sum_J \underbrace{\epsilon_{ijm}\sigma_{mj}(\mathbf{x}_J)}_{=0}\,w_J = 0 \tag{3.9}$$

Note that we used the linear reproducing conditions of the derivatives of the approximation and the symmetry of the Cauchy stress tensor.

3.3 Enrichment

A partition of unity allows to introduce an arbitrary function $\psi(\mathbf{x})$ in the approximation space by splitting the approximation into a standard ^{s} and enriched ^{e} part: $u_h = u_h^s + u_h^e$ (Melenk and Babuška, 1996).

- The standard part of the approximation writes:

$$u_h^s(\mathbf{x}) = \sum_{I\in\mathcal{S}} N_I(\mathbf{x})u_I$$

where
– $\mathcal{S}$ is the set of all nodes in the mesh:
– N_I is the shape function associated with node I. The N_I's can be built using any approximation/interpolation scheme, e.g., Lagrange shape functions Belytschko and Black (1999), spectral elements Legay et al. (2005), moving least squares approximants Ventura et al. (2002) and Rabczuk et al. (2007), Non-Uniform Rational B-splines De Luycker et al. (2011), Ghorashi et al. (2011), and Nguyen et al. (2012);
– $\mathbf{x}_I$ are the coordinates of node I;
– u_I is the nodal unknown associated with node I.
- The enriched approximation field is the sum of a standard part and an enriched part:

$$u_h(\mathbf{x}) = \underbrace{\sum_{I\in\mathcal{S}} N_I(\mathbf{x})u_I}_{\text{standard part}} + \underbrace{\sum_{J\in\text{PU}} f_J(\mathbf{x})\psi(\mathbf{x})a_J}_{\text{enriched part}} \tag{3.10}$$

where
– PU is the set of nodes over which the partition of unity is built;
– function f_J is the Jth function of the partition of unity. The f_J's, similarly to the N_I, can be chosen among a wide class of approximations and interpolations;
– ψ is the enrichment function, which is chosen to represent as accurately as possible the behavior of the exact solution;
– the a_J's are the additional unknowns associated to function $f_J\psi$.

☞ Assuming that for all J, the additional unknowns $a_J = 1$, and that for all I, the standard nodal unknowns $u_I = 0$, we remark that the enriched approximation, built using the partition of unity f_I, reduces to the enrichment function ψ, i.e., the enriched approximation is able to *reproduce* the enrichment function.

Since the collection of the shape functions is a partition of unity, they can be used as a support for the additional function, which is what is generally done in practice. The new approximation then writes at any point $\mathbf{x}$ in the domain:

$$u_h(\mathbf{x}) = \sum_{I \in \mathcal{S}} N_I(\mathbf{x})u_I + \sum_{J \in \mathcal{S}^{enr}} N_J(\mathbf{x})\psi(\mathbf{x})a_J$$

where $\mathcal{S}^{enr}$ is the set of the so-called enriched nodes. The set of enriched nodes $\mathcal{S}^{enr}$ is included in the set of nodes $\mathcal{S}$. The set of enriched nodes is obtained by an operation known as "mesh-geometry interaction." This set depends both on the exact solution of the problem and on the type of enrichment. Concerning enrichment with discontinuities, the usual criterion for enrichment is that the support of the node be cut by the discontinuity line. For singular enrichment, such as that found in linear elastic fracture, the most common enrichment scheme is known as geometrical enrichment, where all nodes within a ball centered on the crack tip are enriched with the near-tip functions. Numerically, these enriched nodes can be easily found by looking at the signs of the nodal level set values in the support of a node: if at least two points in this support are such that the level set function has different signs, the nodal support is cut by the interface, and the node is enriched. In Figure 3.1, the set of enriched nodes $\mathcal{S}^{enr}$ are the nodes whose supports are cut by the discontinuity. From the enriched approximation:

$$u_h(\mathbf{x}) = \sum_{I \in \mathcal{S}} N_I(\mathbf{x})u_I + \sum_{J \in \mathcal{S}^{enr}} N_J(\mathbf{x})\psi(\mathbf{x})a_J \tag{3.11}$$

one can see that the value of u_h at an enriched node K in $\mathcal{S}^{enr}$ is given by:

$$
\begin{aligned}
u_h(\mathbf{x}_K) &= \sum_{I \in \mathcal{S}} N_I(\mathbf{x}_K)u_I + \sum_{J \in \mathcal{S}^{enr}} N_J(\mathbf{x}_K)\psi(\mathbf{x}_K)a_J \\
&= \sum_{I \in \mathcal{S}} \delta_{IK}u_I + \sum_{J \in \mathcal{S}^{enr}} \delta_{JK}\psi(\mathbf{x}_K)a_J \\
&= u_K + \psi(\mathbf{x}_K)a_K.
\end{aligned}
\tag{3.12}
$$

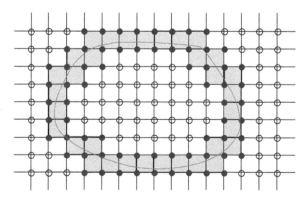

Figure 3.1 Set of standard nodes, $\mathcal{S}^{std}$, set of enriched nodes $\mathcal{S}^{enr}$, enriched elements, and support of the partition of unity.

☐ Ω^{PU}, enriched elements

○   Set S^{std}, of standard nodes

●   Set S^{enr}, of enriched nodes

This last expression is not the standard expected relation where the nodal unknown u_K is the physical unknown at the node[1]. The approximation lacks the "Kronecker delta" property.

$$u_h(\mathbf{x}) = \sum_{I \in \mathcal{S}} N_I(\mathbf{x}) u_I + \sum_{J \in \mathcal{S}^{enr}} N_J(\mathbf{x}) \big(\psi(\mathbf{x}) - \psi(\mathbf{x}_J) \big) a_J \tag{3.13}$$

Using this last approximation, the expected property is reached: $u_h(\mathbf{x}_K) = u_K$. This property is useful to enforce boundary conditions and facilitate post-processing, as introduced in Section 2.2.3 and detailed in Section 3.3.3. The necessary steps involved in the implementation of the extended finite element method (XFEM) are as follows:

1. **Description of geometry of enrichment features** The interface or discontinuity can be represented explicitly by line segments or implicitly by using the level set method (LSM) (Osher and Sethian, 1988).
2. **Selection of enriched nodes** In case of the local enrichment, only a subset of the nodes closer to the region of interest is enriched. The nodes to be enriched can be selected by using an area criterion or from the nodal values of the level set function.
3. **Choice of enrichment functions** Depending on the physics of the problem, different enrichment functions can be used.
4. **Integration** A consequence of adding custom tailored enrichment functions to the finite element approximation basis, which are not necessarily smooth polynomial functions (for example, $\sqrt{r}$ in case of the linear elastic fracture mechanics [LEFM]), is that special care has to be taken in numerically integrating over the elements that are intersected by the discontinuity surface. The standard Gauss quadrature cannot be applied in elements enriched by discontinuous terms, because Gauss quadrature implicitly assumes a polynomial approximation.
5. **Imposition of boundary conditions** Another consequence of augmenting enriched basis functions to standard finite element shape function is that special care needs to be given while imposition essential boundary conditions if the part of the boundary contains enriched elements.

3.3.1 Description of Geometry of Enrichment Features

We have seen that enrichment permits treating the discontinuities which are arbitrarily oriented with respect to a fixed background mesh. Because the (moving) discontinuity lines are not explicitly discretized, their position (possibly evolving in time) needs to be located with respect to the background mesh. Here, we will discuss a few tools to define fixed interfaces and capture moving interfaces relative to a fixed background mesh without any explicit tracking or meshing. The problem consists of tracking the motion of an interface Γ, as a function of time t, knowing its initial position $\Gamma(t = 0)$ and assuming it evolves in a direction that is normal to itself with speed F. The problem at hand is to describe the position of the domain Ω_B relative to the background mesh $\mathcal{T}$. Assuming that no analytical form is available for Γ, the first option that comes to mind is to place a set of points along Γ (i.e., discretize it) and locate their position in space relative to $\mathcal{T}$. If Γ evolved in time t, these points could be moved according to a prescribed velocity field defined on the interface and the procedure could be repeated.

Aforementioned difficulty of tracking the evolving interfaces explicitly is alleviated by LSM, wherein the interfaces are defined implicitly. The LSM introduced in Osher and Sethian (1988) is an interface capturing method (as opposed to interface tracking) which is widely used to describe the evolution of surfaces without requiring their explicit description by points. It is more particularly adapted to representing closed boundaries and strong topological changes such as merging, cusping, self contact, etc. The LSMs have also been adapted to describe open interface problems such as cracks within the extended finite element framework (Gravouil et al., 2002).

The LSM defines Γ implicitly using a scalar valued function $\phi : \Omega \to \mathbb{R}$ which maps the space where the interface lives on the real line. The ideas is that given the value of ϕ at any point in the domain, one can deduce

---

1 See also Section 2.2.2 for numerical illustration in case of one-dimensional problem.

the position of this point relative to the interface by defining ϕ such that the interface is one of its iso-lines. At any time, t, the interface is thus obtained by cutting the graph of ϕ at different heights.

With the above preamble, we now define the concept of level set. A level set function ϕ defining a curve $\Gamma = \partial \Omega_A$ is a function from $\Omega = \Omega_A \bigcup \Omega_B$, such that: $\Omega_A = \{\mathbf{x} \in \Omega / \phi(\mathbf{x}) > 0\}$, $\Omega_B = \{\mathbf{x} \in \Omega / \phi(\mathbf{x}) < 0\}$, and $\Gamma = \{\mathbf{x} \in \Omega / \phi(\mathbf{x}) \equiv 0\}$. The interface Γ is the zero iso-contour of ϕ. Note that this function can be time dependent, the zero iso-contour of $\phi(\mathbf{x}, t)$ is then the position of the interface at time t (Osher and Sethian, 1988). A natural choice for ϕ is the signed distance function to the interface of interest. The signed distance function associates to each point $\mathbf{x} \in \Omega$, its distance to the interface. Γ with different signs depending on whether $\mathbf{x}$ is inside or outside the domain enclosed by Γ, see Figure 3.2. The signed distance function has all properties of a level set representation of the interface Γ. The distance d from a point $\mathbf{x}$ to the interface Γ is given by:

$$d = ||\mathbf{x} - \mathbf{x}_\Gamma|| \tag{3.14}$$

where $\mathbf{x}_\Gamma$ is the normal projection of $\mathbf{x}$ on Γ. The signed distance function ϕ is a level set function defined by:

$$\phi(\mathbf{x}) = \min_{\tilde{\mathbf{x}} \in \Gamma} ||\mathbf{x} - \tilde{\mathbf{x}}|| \ \mathrm{sign}\Big(\mathbf{n} \cdot (\mathbf{x} - \tilde{\mathbf{x}})\Big) \tag{3.15}$$

Figure 3.3 shows the graph in $\mathbb{R}^3$ of the signed distance function to a circle and the graph of the signed distance function for two circles in a plane.

$\phi \equiv 0 \longrightarrow$

d

x

$\mathbf{n}$

$\mathbf{x}_\Gamma$

$\phi < 0$

$\phi > 0$

Γ

Figure 3.2 Signed distance function.

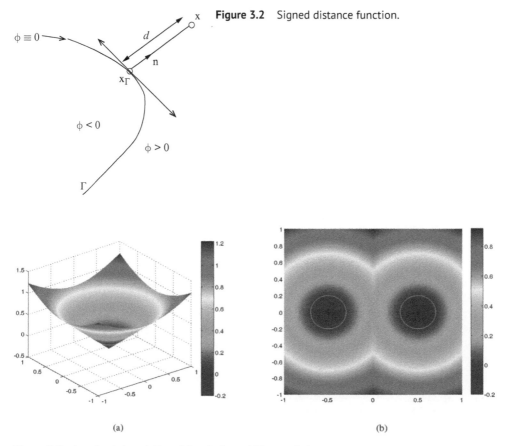

(a)

(b)

Figure 3.3 Level set description: (a) a circle and (b) two circles.

In general, however, ϕ cannot be easily defined analytically, and its numerical value must then be evaluated at each node of the mesh. This process is known as initialization of the level set function. Then, the underlying shape functions associated with the mesh $\mathcal{T}$ can be used to obtain the value of the level set function anywhere within the domain using the nodal values through simple interpolation. This is known as the discretization of the level set. Of course, the smoothness of the resulting (discretized) curve or surface is directly related to that of the shape functions constructed on the mesh. In the context discretization based on a finite element mesh, where N_I is the shape function associated to a node I and the set of nodes in the mesh $\mathcal{T}$ is $\mathcal{S}$, the discretized level set is given by:

$$\phi(\mathbf{x}) = \sum_{I \in \mathcal{S}} N_I(\mathbf{x})\phi_I \tag{3.16}$$

where ϕ_I is the value of the level set function at node I. The level set approach can be used to model the shape of the computational domain or to delineate the boundary between two phases, for example, in composite materials. Consider the example of a solid defined by domain Ω_A, included in the domain Ω called the computational domain (see Figure 3.4). The shape of the solid may be defined by the zero iso-contour of the level set function ϕ. The elements which have at least one ϕ positive node are activated for the computation. For instance, a node which is ϕ-negative and which belongs to an element cut by the zero level set is taken into account in the computation even if it is outside of the solid. However, there are advantages and drawbacks of using implicit boundary definition with the finite element framework. Some of them include:

- In order of impose boundary conditions, the use of a mesh which does not conform to the boundary leads to now well-known difficulties in the imposition of essential boundary conditions, which are shared by the whole class of immersed boundary techniques. This is discussed in more detail in Section 3.3.3.
- The elements cut by the zero level set have to be split into two parts for integration. The part which is outside of the solid is not considered during integration, since it contains no material. When dealing with composite material problems, the material properties are assigned to Gauss points depending on the sign of the level function. This process is made simple because the location of the interface is known from the nodal values of the level set function. Of course, the accuracy of the description of the interface is determined by the ability of the underlying shape functions used for its interpolation.
- The advantage of this method includes the possibility to describe the boundary of the domain/phases analytically. Moreover, when performing shape or topology optimization, no remeshing is necessary.

So far, we have discussed about how the LSM applies to the description of closed interfaces. Next, we will see how the ideas developed so far can be extended to treat open surfaces such as cracks. The first papers in this area that use level set to describe cracks include: Gravouil et al. (2002); Moës et al. (2002); Stolarska et al. (2001). The most common description of cracks is an explicit description. In two dimensions, a crack would typically be represented by a broken line and by a union of triangular elements in three dimensions.

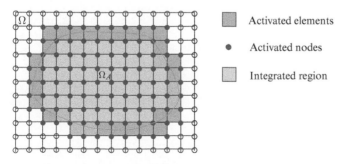

Figure 3.4   Solid shape represented by a level set.

Contrary to the explicit definition, Stolarska et al. (2001) proposed to use level sets to describe and update crack geometries implicitly. In two dimensions (three dimensions), a crack is an open curve (surface) which propagates through its tips (front). A single level set function is insufficient to completely define an open curve (surface). Consider for example, the case of an open circular surface (e.g., a penny shape crack). A level set function is sufficient to describe the plane containing the crack (for example, such a function could take the value zero on the crack plane, be positive on one side of the plane and negative on the other side), but a single level set function is insufficient to define the circular crack front, i.e., the boundary of the open surface. One way to describe the geometry of this open surface completely is to employ a second level set function. In addition to the necessity to define the crack front, following points characterize the description of a crack geometry by level set functions:

- The history of the crack must be preserved as it grows, i.e., the update of the level set functions must not modify the crack interior (crack surface). This means that the zero level set must not be updated behind the front to take into account the fact that once a material point is cracked, it remained cracked.
- The crack does not usually grow in plane; hence, the level set functions are not updated with the speed of an interface in the direction normal to itself but with the speed function computed along the front, depending on the value of the stress intensity factors

For above reasons, it is not possible to use the same technique to describe and propagate cracks that was described for closed interfaces. A crack is defined by two level set functions: a normal and a tangent level set (Stolarska et al., 2001):

- The normal level set function ϕ is the signed distance to the union of the crack and its extension to its front.
- The tangent level set function ψ is the signed distance function to a surface that passes through the crack boundary and normal to the crack.

The crack surface is the set of points that belong to the zero level set of ϕ and for which ψ is negative. The crack front is defined as the intersection of the two zero level set. See Figure 3.5, where the crack is represented by a thick line, the tangent extensions as a dashed line, and the surface normal to the crack at the front (tip) as a dotted line.

For the case of LEFM, two sets of functions are used: a jump function to capture the displacement jump across the crack faces and asymptotic branch functions that span the asymptotic crack tip fields. The enriched approximation for LEFM takes the following form:

$$\mathbf{u}_h(\mathbf{x}) = \sum_{I \in \mathcal{S}^{fem}} N_I(\mathbf{x}) u_I + \sum_{J \in \mathcal{S}^c} N_J(\mathbf{x}) \big(\psi(\mathbf{x}) - \psi(\mathbf{x}_J) \big) \mathbf{a}_J + \sum_{K \in \mathcal{S}^f} N_K(\mathbf{x}) \sum_{\alpha=1}^{n} \Xi_\alpha(r, \theta) \mathbf{b}_K^\alpha \tag{3.17}$$

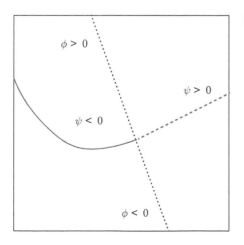

Figure 3.5 Crack definition by two level set functions.

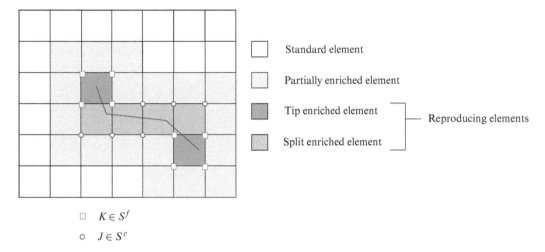

$\square \quad K \in S^f$

$\circ \quad J \in S^c$

Figure 3.6 A typical finite element mesh with an arbitrary crack. "Circled" nodes are enriched with ψ and "squared" nodes are enriched with near-tip asymptotic fields. "Reproducing elements" are the elements whose all the nodes are enriched.

where S^c is the set of nodes whose shape function support is cut by the crack interior ("circled" nodes in Figure 3.6) and S^f is the set of nodes whose shape function support is cut by the crack tip ("squared" nodes in Figure 3.6). ψ and Ξ_α are the enrichment functions chosen to capture the displacement jump across the crack surface and the singularity at the crack tip and n is the total number of asymptotic functions. $\mathbf{a}_J$ and $\mathbf{b}_K^\alpha$ are the nodal degrees of freedom corresponding to functions ψ and Ξ_α, respectively. n is the total number of near-tip asymptotic functions and (r, θ) are the local crack tip coordinates. In Equation (3.17), the displacement field is global, but the support of the enriched functions are local because they are multiplied by the nodal shape functions. The local enrichment strategy introduces the following four types of elements apart from the standard elements:

- *Split elements* are elements completely cut by the crack. Their nodes are enriched with the discontinuous function ψ.
- *Tip elements* either contain the tip, or are within a fixed distance independent of the mesh size, r_{enr} of the tip, if geometrical enrichment is used. All nodes belonging to a tip element are enriched with the near-tip asymptotic fields.
- *Tip-blending elements* are elements neighboring tip elements. They are such that some of their nodes are enriched with the near-tip and others are not enriched at all.
- *Split-blending elements* are elements neighboring split elements. They are such that some of their nodes are enriched with the strongly or weakly discontinuous function and others are not enriched at all.

3.3.2 Choice of Enrichment Functions

In this section, the common types of enrichment functions used in the XFEM in the context of solid mechanics are discussed.

Weak discontinuities For weak discontinuities, two types of enrichment functions are used in the literature. One choice for the enrichment function is the absolute value of the level set function:

$$\psi_{\text{abs}}(\mathbf{x}) = |\phi(\mathbf{x})| \tag{3.18}$$

The other choice was proposed by Moës et al. (2003) as:

$$\psi(\mathbf{x}) = \sum_{I \in \mathcal{N}^c} |\phi_I| N_I(\mathbf{x}) - | \sum_{I \in \mathcal{N}^c} \phi_I N_I(\mathbf{x})| \tag{3.19}$$

The main advantage of the above enrichment function is that this enrichment function is non-zero only in the elements that are intersected by the discontinuity surface. Figure 3.7 shows the two choices of enrichment functions in the one-dimensional (1D) case. For 2D and 3D problems, the enrichment function proposed by Moës et al. (2003) is a ridge centered on the interface and has zero value on the elements which are not crossed by the interface.

Strong discontinuities and Singularity   As discussed earlier, within LEFM framework, to represent the crack independent of the background mesh, one would need two enrichment functions, viz., to capture the jump across the crack face and to capture the singularity ahead of the crack tip. To model a strong discontinuity, popular choices of enrichment functions include Heaviside function and sign of the level set function, defined by:

$$H(\mathbf{x}) = \begin{cases} +1 & \mathbf{x} \text{ above the crack face} \\ 0 & \mathbf{x} \text{ below the crack face} \end{cases} \tag{3.20}$$

and

$$\text{sign}(\phi(\mathbf{x})) = \begin{cases} +1 & \mathbf{x} \text{ above the crack face} \\ -1 & \mathbf{x} \text{ below the crack face} \end{cases} \tag{3.21}$$

These enrichment functions have been proposed in Belytschko and Black (1999) and Moës et al. (1999). Iarve (2003) proposed a modification to the Heaviside function that eliminates the need to sub-triangulate for the purpose of numerical integration (see Section 3.3.4). Figure 3.8 shows the plot of the Heaviside function for a straight and a kinked crack.

For isotropic materials, the near-tip function, Ξ_α, in Equation (3.17) is given by:

$$\{\Xi_\alpha\}_{1 \leq \alpha \leq 4}(r, \theta) = \sqrt{r}\left\{\sin\frac{\theta}{2}, \cos\frac{\theta}{2}, \sin\theta\sin\frac{\theta}{2}, \sin\theta\cos\frac{\theta}{2}\right\}. \tag{3.22}$$

where (r, θ) are the crack tip polar coordinates. Figure 3.9 shows the near-tip asymptotic fields. Note that the first function in Equation (3.22) is discontinuous across the crack surface. For orthotropic materials, the asymptotic functions given by Equation (3.22) has to be modified because the material property is a function of material

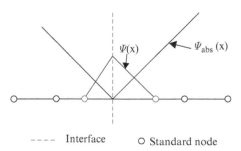

Figure 3.7   Weak discontinuity: different choices of enrichment functions.

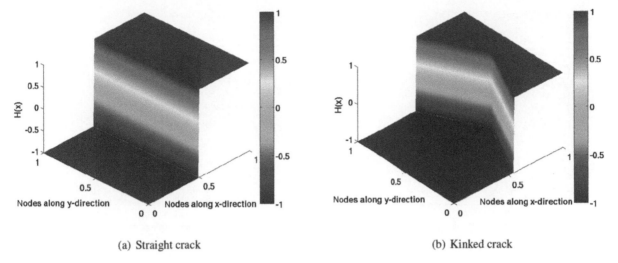

(a) Straight crack

(b) Kinked crack

Figure 3.8 Heaviside function to capture strong discontinuities.

orientation. Asadpoure and Mohammadi (2007) proposed special near-tip functions for orthotropic materials as:

$$\{\Xi_\alpha\}_{1\leq\alpha\leq4}(r,\theta) = \sqrt{r}\left\{\cos\frac{\theta_1}{2}\sqrt{g_1(\theta)}, \cos\frac{\theta_2}{2}\sqrt{g_2(\theta)}, \sin\frac{\theta_1}{2}\sqrt{g_1(\theta)}, \sin\frac{\theta_2}{2}\sqrt{g_2(\theta)}\right\} \tag{3.23}$$

where (r,θ) are the crack tip polar coordinates. The functions $g_i(i=1,2)$ and $\theta_i(i=1,2)$ are given by:

$$g_j(\theta) = \left(\cos^2\theta + \frac{\sin^2\theta}{e_j^2}\right), \quad j=1,2$$

$$\theta_j = \arctan\left(\frac{\tan\theta}{e_j}\right). \quad j=1,2 \tag{3.24}$$

where $e_j(j=1,2)$ are related to material constants, which depend on the orientation of the material (Asadpoure and Mohammadi, 2007). Figure 3.10 shows the near-tip asymptotic fields in case of orthotropic material with material orientation, $\theta=45°$.

Cohesive cracks In cohesive cracks, the stresses and the strains are no longer singular and the step enrichment alone is suitable for the entire crack. Due to this, the XFEM approximation cannot treat crack tips or fronts that lie within the element and thus the crack is virtually extended to the next element edge. Wells and Sluys (2001) used the step function as an enrichment function to treat cohesive cracks. Moës and Belytschko (2002) used the following enrichment function for 2D cohesive crack tips:

$$\{\Xi_\alpha\}_{1\leq\alpha\leq3}(r,\theta) = \left\{r\sin\frac{\theta}{2}, \sqrt[3]{r}\sin\frac{\theta}{2}, r^2\sin\frac{\theta}{2}\right\} \tag{3.25}$$

Based on the generalized Heaviside function, Zi and Belytschko (2003) proposed a new crack tip element for modeling cohesive crack growth. This method overcame the difficulty and provided a method for the crack tips to lie within the element. Rabczuk et al. (2008) developed a crack tip element for the phantom-node method to model cracks independent of the underlying mesh. The main idea is to use reduced integrated finite elements with hourglass control. With this method, the crack tip can be modeled within an element. This was an advancement to the earlier work by Song et al. (2006) who used the phantom-node method to simulate dynamic crack growth and

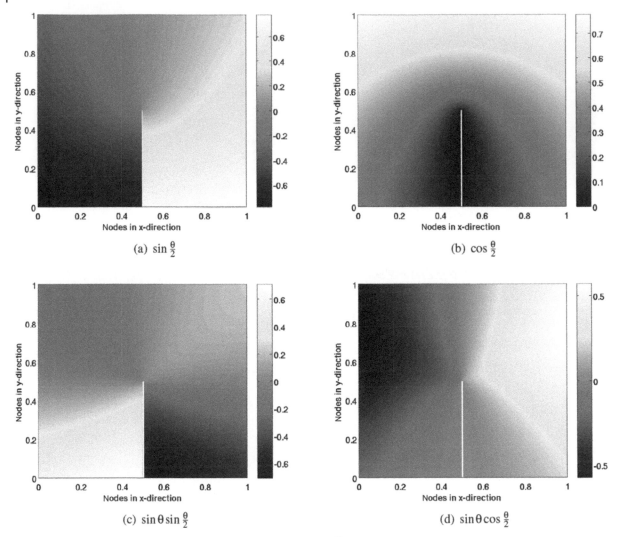

(a) $\sin\frac{\theta}{2}$

(b) $\cos\frac{\theta}{2}$

(c) $\sin\theta\sin\frac{\theta}{2}$

(d) $\sin\theta\cos\frac{\theta}{2}$

Figure 3.9 Near-tip asymptotic fields. Note that the function $\sin\frac{\theta}{2}$ is discontinuous across the crack surface. The white line denotes the discontinuity line. The crack tip is located at (0.5,0.5).

shear band propagation. In their approach, the crack has to cross an entire element. Very recently, an isogeometric approach was combined with XFEM to model cohesive cracks in Verhoosel et al. (2010).

Dynamic cracks Réthoré et al. (2005) simulated dynamic propagation of arbitrary 2D cracks using an enrichment strategy for time-dependent problems. By adding phantom nodes and superposing elements on the original mesh, Song et al. (2006) proposed a new method for modeling dynamic crack and shear band propagation. From the numerical studies, it was observed that the method exhibits almost no mesh dependence once the mesh is sufficiently refined to resolve the relevant physics of the problem. This method was based on the Hansbo and Hansbo (2004) approach. XFEM has been applied to concurrent continuum-atomistic simulations of cracks and dislocations by Gracie et al. (2009) and Aubertin et al. (2009, 2010).

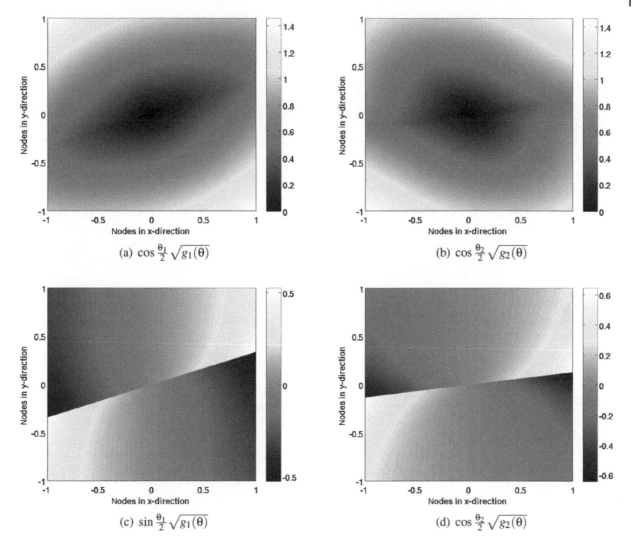

Figure 3.10 Near-tip asymptotic fields for orthotropic material with material orientation, $\theta = 45°$. The crack tip is located at (0.5,0.5).

Dislocations Ventura et al. (2005) applied the PUM concept for the solution of edge dislocations. As in cracks, the dislocation solutions are characterized by discontinuities and singular points. Gracie et al. (2007) used the following approximation for modeling the dislocations:

$$\mathbf{u}^h(\mathbf{x}) = \sum_{I \in \mathcal{N}^{fem}} N_I(\mathbf{x})\mathbf{q}_I + \mathbf{b} \cdot \sum_{J \in \mathcal{N}^e} N_J(\mathbf{x}) \cdot \psi_{\text{step}}(\mathbf{x}) \tag{3.26}$$

where **b** is the Burgers vector, which is a known quantity and closed-form solutions are used for enrichment function. Note that Equation (3.26) does not introduce additional unknowns. At the core of the dislocation, the step function ψ_{step} is modified and the rest of the dislocation is modeled by a step function. Gracie et al. (2009) computed the enrichment functions by coupling the XFEM with atomistic simulations. This eliminated the need for closed-form solutions, which are available only for isotropic materials. Tables 3.1 give a summary of different enrichment functions proposed in the literature for different classes of problems.

3.3.3 Imposition of boundary conditions

The imposition of Neumann boundary conditions is straightforward in the XFEM. The corresponding integrals have to be evaluated to obtain the force vector. In many situations the imposition of Dirichlet boundary conditions is straightforward too. If the surface at which Dirichlet conditions are prescribed is discretized by standard finite elements, the situation in the XFEM is the same as in the classical finite element method (FEM). But if enriched elements are present, a special treatment is necessary. In the following, we will discuss these situations and algorithmic approaches toward them.

There are situations, in which the treatment of Dirichlet conditions in the FEM context is particularly cumbersome. Imagine that the boundary of a structure has a complex shape. Meshing such boundaries is not a trivial task and meshing algorithms might run into problems. However, XFEM-like approaches can also be used to treat complex boundary shapes and the corresponding Dirichlet conditions. This will also be discussed in the following.

Treatment of Dirichlet boundary conditions in the presence of enriched elements What makes the imposition of Dirichlet conditions extremely simple in the FEM (once the structure is meshed) is the fact that the displacement in a certain direction at a certain node is determined only by one particular degree of freedom. Thus, we can ensure that the Dirichlet conditions at this node are fulfilled by setting the value in the displacement vector accordingly. This results in an additional force term when solving the equation system. If we do this for all nodes along the Dirichlet boundary, the Dirichlet conditions are fulfilled at all nodes and the prescribed displacements are automatically interpolated along the boundary. But now let us assume that we introduce a crack by Heaviside-enrichments and displacements are prescribed along the crack faces. The displacement along the crack faces always depends on more than one degree of freedom.

Let us take a look at a 1D example. Assume that a 1D element with domain $[0, 1]$ and linear shape functions is enriched with the Heaviside-function ϕ, whose jump is located in the center of the element. The displacement can be written as:

$$u = u_1 N_1 + u_2 N_2 + u_3 (N_1(\phi - 1)) + u_4 (N_2(\phi + 1)) \tag{3.27}$$

Depending on whether we approach the element center from the left (u^+) or from the right (u^-), the displacement becomes:

$$u^+ = u_1 N_1 + u_2 N_2 + u_4 (N_2(\phi + 1)) \tag{3.28a}$$

$$u^- = u_1 N_1 + u_2 N_2 + u_3 (N_1(\phi - 1)) \tag{3.28b}$$

If we prescribe a certain value for the displacement for u^+ or u^-, we get an additional constraint which depends on more than one degree of freedom. There are different ways to deal with those constraints. Both can be derived nicely if we interpret the structural problem as an energy minimization problem:

$$min \| \frac{1}{2} \boldsymbol{u}^T \boldsymbol{K} \boldsymbol{u} - \boldsymbol{u}^T \boldsymbol{f} \| \tag{3.29}$$

Minimization problems can be solved by setting the gradient to zero, which yields our standard system of equations:

$$\boldsymbol{K}\boldsymbol{u} - \boldsymbol{f} = 0 \tag{3.30}$$

Table 3.1 Choice of enrichment functions.

Kind of problem	Displacement	Strain	Enrichment
Crack body	Discontinuous	-	Heaviside: $\psi(\mathbf{x}) = \text{sign}(\phi(\mathbf{x}))$
Isotropic materials	Discontinuous for $\theta = \pm\pi$, $\sqrt{r}$ order	High gradient, $\frac{1}{\sqrt{r}}$ order	$\sqrt{r}\sin\frac{\theta}{2}$, $\sqrt{r}\cos\frac{\theta}{2}$, $\sqrt{r}\sin\frac{\theta}{2}\sin\theta$, $\sqrt{r}\cos\frac{\theta}{2}\sin\theta$.
Orthotropic materials (Asadpoure and Mohammadi, 2007)	Discontinuous	High gradient	$\sqrt{r}\cos(\frac{\theta_1}{2})\sqrt{g_1(\theta)}$, $\sqrt{r}\cos(\frac{\theta_2}{2})\sqrt{g_2(\theta)}$, $\sqrt{r}\cos(\frac{\theta_2}{2})\sqrt{g_2(\theta)}$, $\sqrt{r}\sin(\frac{\theta_1}{2})\sqrt{g_1(\theta)}$.
Cohesive crack (Moës and Belytschko, 2002)	Discontinuous	High gradient	$r\sin(\frac{\theta}{2})$, $r^{\frac{3}{2}}\sin(\frac{\theta}{2})$, $r^2\sin(\frac{\theta}{2})$.
Reissner-Mindlin plate (Dolbow et al., 2000)	Discontinuous	High gradient	
Kirchhoff-Love plate (Lasry et al., 2010)	Discontinuous	High gradient	
Incompressible materials (Legrain et al., 2008)	Discontinuous	High gradient	$\frac{1}{\sqrt{r}}\sin(\frac{\theta}{2})$, $\frac{1}{\sqrt{r}}\cos(\frac{\theta}{2})$.
Plastic materials (Elguedj et al., 2006)	Discontinuous	High gradient	
Thermo-elastic material (Duflot, 2008)	Discontinuous	High gradient	
Crack aligned to the interface of a material interface (Huynh and Belytschko, 2009)	Discontinuous	High gradient	$\sqrt{r}\cos(\epsilon\log r)e^{-\epsilon\theta}\sin\frac{\theta}{2}$, $\sqrt{r}\cos(\epsilon\log r)e^{-\epsilon\theta}\cos\frac{\theta}{2}$, $\sqrt{r}\cos(\epsilon\log r)e^{\epsilon\theta}\sin\frac{\theta}{2}$, $\sqrt{r}\cos(\epsilon\log r)e^{\epsilon\theta}\cos\frac{\theta}{2}$, $\sqrt{r}\sin(\epsilon\log r)e^{-\epsilon\theta}\sin\frac{\theta}{2}$, $\sqrt{r}\sin(\epsilon\log r)e^{\epsilon\theta}\sin\frac{\theta}{2}$, $\sqrt{r}\cos(\epsilon\log r)e^{\epsilon\theta}\sin\frac{\theta}{2}\sin\theta$, $\sqrt{r}\cos(\epsilon\log r)e^{\epsilon\theta}\cos\frac{\theta}{2}\sin\theta$, $\sqrt{r}\sin(\epsilon\log r)e^{\epsilon\theta}\sin\frac{\theta}{2}\sin\theta$, $\sqrt{r}\sin(\epsilon\log r)e^{\epsilon\theta}\cos\frac{\theta}{2}\sin\theta$, $\sqrt{r}\cos(\epsilon\log r)e^{-\epsilon\theta}\sin\frac{\theta}{2}\sin\theta$, $\sqrt{r}\sin(\epsilon\log r)e^{\epsilon\theta}\sin\frac{\theta}{2}\sin\theta$.

(Continued)

Table 3.1 (Continued)

Kind of problem	Displacement	Strain	Enrichment
Piezoelastic materials (Béchet et al., 2009; Wells et al., 2002)	Discontinuous	High gradient	
Magnetoelectroelastic materials (Rojas-Díaz et al., 2009)	Discontinuous	High gradient	
Complex cracks (Kaufmann et al., 2009)	Discontinuous	High gradient	$\Delta H(\xi) = 0$
Hydraulic fracture (Lecampion, 2009)	Discontinuous	High gradient	
Crack tip in polycrystals (Menk and Bordas, 2011a)	Discontinuous	High gradient	$\Re[r(\mathbf{x})\Psi(\Theta(\mathbf{x}))]$
Bi-material	Continuous	Discontinuous	Ramp: $\psi(\mathbf{x}) = \lvert\phi(\mathbf{x})\rvert$
Dislocations (Gracie et al., 2008a, 2007; Ventura et al., 2005)	Discontinuous	Discontinuous	
Topology optimization (Belytschko et al., 2003)	Continuous	Discontinuous	Ramp: $\psi(\mathbf{x}) = \lvert\phi(\mathbf{x})\rvert$
Stokes flow (Lecampion, 2009)	–	–	
Solidification (Chessa et al., 2002)	–	–	Ramp: $\psi(\mathbf{x}) = \lvert\phi(\mathbf{x})\rvert$.
Fluid-structure interaction (Gerstenberger and Wall, 2008)	Continuous	Discontinuous	Heaviside: $\psi(\mathbf{x}) = \operatorname{sign}(\phi(\mathbf{x}))$.

In the presence of additional constraints there is the well-known Lagrange multiplier approach which can be used to solve the problem[2]. Let us say we want to prescribe a displacement $u^+ = 0$. Using Equation (3.28a), this can be written as:

$$\boldsymbol{D}\boldsymbol{u} = 0 \tag{3.31}$$

where $\boldsymbol{D}$ can be obtained simply by evaluating the shape and the enrichment functions at the element center:

$$\boldsymbol{D} = [0.5\ 0.5\ 0\ 0.5] \tag{3.32}$$

Then, the Lagrange multiplier approach yields the following minimization problem:

$$min\|\frac{1}{2}\boldsymbol{u}^T\boldsymbol{K}\boldsymbol{u} - \boldsymbol{u}^T\boldsymbol{f} + \lambda\,(\boldsymbol{D}\boldsymbol{u})\| \tag{3.33}$$

λ is an additional degree of freedom. Upon setting the gradient to zero, we obtain:

$$\begin{bmatrix} \boldsymbol{K} & \boldsymbol{D}^T \\ \boldsymbol{D} & 0 \end{bmatrix}\begin{bmatrix} \boldsymbol{u} \\ \lambda \end{bmatrix} = \begin{bmatrix} \boldsymbol{f} \\ 0 \end{bmatrix} \tag{3.34}$$

Looking at Equation (3.34), we can understand why the Lagrange multiplier approach works. The last row of the equation system ensures $\boldsymbol{D}\boldsymbol{u} = 0$. Furthermore, it is ensured that the gradient $Ku - f$ of the elastic energy points in the same direction as D. Since D is perpendicular to the hyperplane spanned by $Du = 0$, this is equivalent to optimality. However, the stiffness matrix is not positive definite any more, it has negative eigenvalues. This is a disadvantage, if the number of unknowns become large, since a smaller number of solvers is available for such problems.

Another approach toward the treatment of additional constraints is the penalty method. Again we consider the minimization problem, but with an additional term:

$$min\|\frac{1}{2}\boldsymbol{u}^T\boldsymbol{K}\boldsymbol{u} + \rho\frac{1}{2}\boldsymbol{u}^T\boldsymbol{D}^T\boldsymbol{D}\boldsymbol{u} - \boldsymbol{u}^T\boldsymbol{f} + \lambda\,(\boldsymbol{D}\boldsymbol{u})\| \tag{3.35}$$

For increasing values of ρ the solution of this problem converges to the solution of the Lagrange approach. This is easy to see. If $\boldsymbol{D}\boldsymbol{u} \neq 0$, then the value of the objective function is large since ρ is large. On the other hand, those values of the objective function for which $\boldsymbol{D}\boldsymbol{u} = 0$ are not affected. Therefore, the minimum of the objective function is shifted away from those vectors $\boldsymbol{u}$ that do not fulfill the Dirichlet boundary conditions. One might also say that those values are "penalized," hence the name of the method. The solution is again computed by setting the gradient to zero:

$$\left(\boldsymbol{K} + \rho\boldsymbol{D}^T\boldsymbol{D}\right)\boldsymbol{u} - \boldsymbol{f} = 0 \tag{3.36}$$

in the above equation, ρ must be finite in an actual computation. Therefore, the boundary conditions are never fulfilled exactly. This error becomes negligible if ρ is a large number, but then another problem occurs. Large values of ρ have a negative effect on the conditioning of the matrix $\boldsymbol{K} + \rho\boldsymbol{D}^T\boldsymbol{D}$. Thus, ρ must be chosen carefully. Ji and Dolbow (2004) apply the penalty and the Lagrange methods to enforce Dirichlet constraints within the framework of the XFEM. The problem considered is a two-phase body, the interface is modeled by enrichments and Dirichlet conditions are prescribed at the interface. Ji and Dolbow (2004) reported problems with the matrix conditioning when using the penalty method. They also reported oscillating solutions when using the Lagrange approach; however, these difficulties could finally be resolved, the interested reader is referred to the original work.

---

2 See Section 2.2.2, Chapter 2 for numerical illustration.

Treatment of complex boundary shapes and corresponding Dirichlet conditions by XFEM approaches In certain situations the meshing of the boundary can give rise to problems. Consider the domain shown in Figure 3.11. The mesh shown in the background is easy to construct. One could use those elements that intersect with the domain to discretize the problem. But then the boundary would not be represented very well in the simulation. There are different approaches to deal with this situation. One approach would be to use more sophisticated meshing techniques, which can take the shape of the boundary into account. Another one was described by Kumar et al. (2008). They developed an approach which has a close connection to the enrichment procedure in the XFEM. We will make some simplifying assumptions, i.e., zero displacements are prescribed along the whole boundary of the domain. For the treatment of more general cases the interested reader is referred to the original work. An implicit description of the boundary must be available for this approach. A function $D(x)$ is necessary, which is zero along the boundary. Furthermore, it should have a non-zero gradient along the boundary and be positive inside the structural domain. The signed distance function would be an example of such a function. Then the discrete space is chosen as:

$$u = \sum N_i D u_i \tag{3.37}$$

All displacements u are zero along the boundary and therefore the Dirichlet conditions are automatically fulfilled, even for the mesh in Figure 3.11. The corresponding function space can be used in the weak form, where integration is performed only over the structural domain. Thus, meshing can be done using a set of uniform elements which do not conform to the boundary. It is useful to choose a function D which is 1 over large parts of the domain. In that case, D can be neglected for most elements when the element stiffness matrices are calculated. So the signed distance function would be a good function to start with, but it should be modified away further inside the domain.

Let us demonstrate this with a little example. Consider a 1D element with linear shape functions. Assume that the domain boundary (of the 1D domain) is located at the center of this element, so the element is not able to represent the boundary very well. This situation is similar to the 2D example considered before, in both examples the meshes do not conform to the boundary. Let D be the signed distance function. If the elements domain is $[0, 1]$, we obtain the following functions:

$$N_1 = 1 - x$$
$$N_2 = x$$
$$D = x - 0.5$$
$$N_1 D = (x - 0.5)(1 - x)$$
$$N_2 D = (x - 0.5)x \tag{3.38}$$

Figure 3.11 Structural domain with non-conforming background mesh.

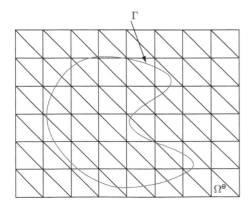

Γ

Ω^e

Figure 3.12 Sub-triangles used in numerical integration. Solid line represents the line of discontinuity and "dashed" lines denote the boundaries of the subcells.

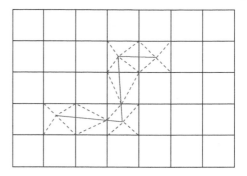

It is easy to see that any linear combination of $N_1 D$ and $N_2 D$ vanishes at $x = 0.5$, so the Dirichlet boundary conditions are automatically fulfilled.

3.3.4 Numerical Integration of the Weak Form

One potential solution for numerical integration is to partition the elements into subcells (triangles for example) aligned to the discontinuous surface in which the integrands are continuous and differentiable (Belytschko and Black, 1999). Figure 3.12 shows a possible sub-triangulation commonly used in the XFEM. The purpose of subdividing into triangles is solely for the purpose of numerical integration and does not introduce new degrees of freedom. Although the generation of quadrature subcells does not alter the approximation properties, it inherently introduces a "mesh" requirement. The steps involved in this approach are as follows:

- Split the element into subcells with the subcells aligned to the discontinuity surface. Usually the subcells are triangular.
- Numerical integration is performed with the integration points from triangular quadrature.

The subcells must be aligned to the crack or to the interface and this is costly and less accurate if the discontinuity is curved (Cheng and Fries, 2010). Figure 3.13 illustrates necessary steps to determine the coordinates of the Gauss points over the enriched elements.

3.4 Numerical Examples

3.4.1 One-Dimensional Multiple Interface

Consider a 1D bar with two material interfaces. The material interfaces are located at $x = x_a = 1/4$ and $x = x_b = 3/4$, as shown in Figure 3.14. The left edge of the bar is clamped and a unit force $F = 1$ is prescribed at the right end. Young's moduli for $x \in [0, x_a)$, $x \in (x_a, x_b)$, and $x > x_b$ are 1 N/m², 2 N/m², and 1 N/m², respectively. The total length of the bar is $L = 1$. The XFEM displacement approximation at a point $x \in [0, 1]$ writes:

$$u(x) = \begin{bmatrix} N_1(x) & N_2(x) & N_3(x) & N_4(x) & N_5(x) & N_6(x) \end{bmatrix} \begin{Bmatrix} u_1 \\ u_2 \\ a_1 \\ a_2 \\ b_1 \\ b_2 \end{Bmatrix} \qquad (3.39)$$

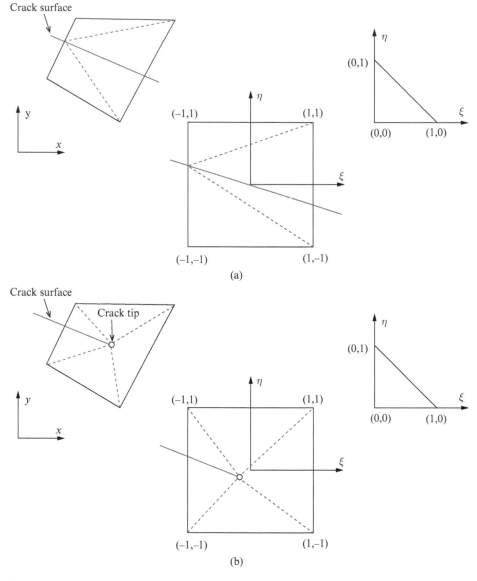

Figure 3.13 Physical element (a) intersected by the crack surface and (b) containing the crack tip is subdivided into subcells (triangular subcells in this case). A quadrature rule on a standard triangular domain is used for the purpose of integration.

where u_1, u_2 are standard degrees of freedom and a_1, a_2, b_1, b_2 are enriched degrees of freedom associated with nodes 1 and 2. N_1, N_2 are the standard finite element shape functions, N_3, N_4, N_5 and N_6 are the enriched shape functions, given by, $\forall x \in [0, 1]$:

$$N_3(x) = N_1(x)(|\phi_1(x)| - |\phi_1(x_1)|)$$
$$N_4(x) = N_2(x)(|\phi_1(x)| - |\phi_1(x_2)|)$$
$$N_5(x) = N_1(x)(|\phi_2(x)| - |\phi_2(x_1)|)$$
$$N_6(x) = N_2(x)(|\phi_2(x)| - |\phi_2(x_2)|) \tag{3.40}$$

Figure 3.14 One-dimensional bar with multiple interfaces, where x_a and x_b are the location of the material interfaces and L_1, L_2, and L_3 are the length of each segment. A unit force is prescribed at the right end.

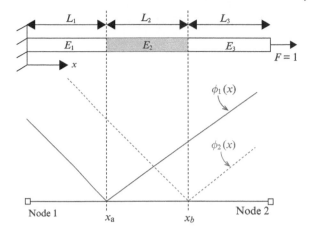

where $\phi_1(x)$ and $\phi_2(x)$ are level set functions for the two material interfaces. $\phi_1(x_i), \phi_2(x_i), i = 1, 2$ are the values of the level set function evaluated at nodes 1 and 2. $\phi_1(x) = x - x_a$ is the level set function in one dimension and x_a is the location of the first interface from the left end and $\phi_2(x) = x - x_b$ is the level set function for the second interface. An absolute value of the level set function, given by Equation (2.25), is used as an enrichment function (see Figure 3.14 for a schematic representation). As the solution's first derivative is discontinuous at two locations along the length of the bar, $x_a = 1/4$ and $x_b = 3/4$, the domain is split into three regions:

$$\int_0^L f(x)\,dx = \int_0^{x_a} f(x)\,dx + \int_{x_a}^{x_b} f(x)\,dx + \int_{x_b}^L f(x)\,dx. \tag{3.41}$$

The total stiffness matrix for the problem under consideration is given by:

$$\mathbf{K} = \mathbf{K}_1 + \mathbf{K}_2 + \mathbf{K}_3, \tag{3.42}$$

where $\mathbf{K}_1, \mathbf{K}_2$, and $\mathbf{K}_3$ are given by:

$$\mathbf{K}_1 = \int_0^b \mathbf{B}_1^T E_1 A\, \mathbf{B}_1\, dx$$

$$\mathbf{K}_2 = \int_b^L \mathbf{B}_2^T E_2 A\, \mathbf{B}_2\, dx$$

$$\mathbf{K}_3 = \int_b^L \mathbf{B}_3^T E_3 A\, \mathbf{B}_3\, dx, \tag{3.43}$$

and $\mathbf{B}_i, i = 1, 2, 3$ are the strain-displacement matrices. The terms in Section 3.4.1 are computed as outlined earlier for the bi-material problem. The assembled system of equations is given by:

$$\begin{bmatrix} 1.5 & 0.25 & 0 & 0 & -0.25 \\ 0.25 & 0.5 & 0 & 0.0208 & -0.2708 \\ 0 & 0 & 0.3750 & 0.3542 & 0.0208 \\ 0 & 0.0208 & 0.3542 & 0.3750 & 0 \\ -0.25 & -2708 & 0.0208 & 0 & 0.5 \end{bmatrix} \begin{Bmatrix} u_2 \\ a_1 \\ a_2 \\ b_1 \\ b_2 \end{Bmatrix} = \begin{Bmatrix} 1 \\ 0 \\ 0 \\ 0 \\ 0 \end{Bmatrix} \tag{3.44}$$

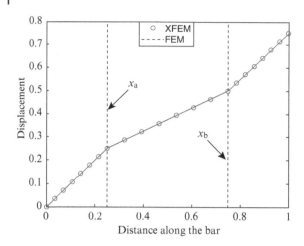

Figure 3.15 Displacement along the length of the bar. Three elements are used in case of the FEM, while one element is used in case of the XFEM. x_a and x_b denote the material interfaces.

with boundary conditions $u_1 = 0$, $F_1 = 1$. The solution is given by $u_2 = 0.75$, $a_1 = -0.25$, $a_2 = -0.25$, $b_1 = 0.25$, $b_2 = 0.25$. The displacement function is given by:

$$\forall x \in [0, 1], \quad u(x) = 0.75N_2(x) - 0.25N_3(x) - 0.25N_4(x) + 0.25N_5(x) + 0.25N_6(x) \tag{3.45}$$

Figure 3.15 shows the displacement along the length of the bar for the XFEM and for the conforming FEM. In the case of the FEM, three elements are used to solved the problem. Thanks to the local PU and enrichment functions, the XFEM with one element is able to capture the weak discontinuities along the length of the bar. Note that the number of unknowns in case of the XFEM is 6 (two standard degrees of freedom and four enriched degrees of freedom), while in the case of the FEM, its only 4 (one for each node). Although the number of unknowns in case of the XFEM is increased, the advantage is that the mesh does not conform to the material interface. Hence, the influence of the location of the material interface on the solution can be solved without changing the underlying mesh by using appropriate values of the location of the interfaces x_a and x_b while evaluating the integrals given by Section 3.4.1.

3.4.2 Two-Dimensional Circular Inhomogeneity

In this example, the enriched finite element solutions for the elasto-static response of a circular material inhomogeneity under radially symmetric loading, as shown in Figure 3.16, are examined within the framework of the XFEM. The effect of modifying the approximation in the blending elements is studied. Plane strain conditions are assumed. The material properties are constant within each domain, Ω_1 and Ω_2, but there is a material discontinuity across the interface, $\Gamma_1(r = a)$. The Lamé constants in Ω_1 and Ω_2 are: $\lambda_1 = \mu_1 = 0.4$ and $\lambda_2 = 5.7692$, $\mu_2 = 3.8461$, respectively. These correspond to $E_1 = 1$, $\nu = 0.25$ and $E_2 = 10$, $\nu_2 = 0.3$. A linear displacement field: $u_1 = x_1, u_2 = x_2$ ($u_r = r, u_\theta = 0$) on the boundary $\Gamma_2(r = b)$ is imposed (Fries, 2008; Chessa et al., 2003). The governing equation in polar coordinates is given by:

$$\frac{d}{dr}\left[\frac{1}{r}\frac{d}{dr}(ru_r)\right] = 0 \tag{3.46}$$

The exact displacement solution is written as:

$$u_r(r) = \begin{cases} \left[\left(1 - \frac{b^2}{a^2}\right)\beta + \frac{b^2}{a^2}\right]r, & 0 \le r \le a, \\ \left(r - \frac{b^2}{r}\right)\beta + \frac{b^2}{r}, & a \le r \le b, \end{cases}$$

$$u_\theta(r) = 0. \tag{3.47}$$

Figure 3.16 Bi-material boundary value problem.

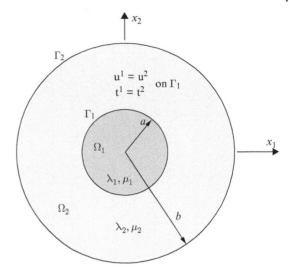

where

$$\beta = \frac{(\lambda_1 + \mu_1 + \mu_2)b^2}{(\lambda_2 + \mu_2)a^2 + (\lambda_1 + \mu_1)(b^2 - a^2) + \mu_2 b^2} \tag{3.48}$$

The radial (ε_{rr}) and hoop ($\varepsilon_{\theta\theta}$) strains are given by:

$$\varepsilon_{rr}(r) = \begin{cases} \left(1 - \frac{b^2}{a^2}\right)\beta + \frac{b^2}{a^2}, & 0 \le r \le a, \\ \left(1 + \frac{b^2}{a^2}\right)\beta - \frac{b^2}{a^2}, & a \le r \le b, \end{cases}$$

$$\varepsilon_{\theta\theta}(r) = \begin{cases} \left(1 - \frac{b^2}{a^2}\right)\beta + \frac{b^2}{a^2}r, & 0 \le r \le a, \\ \left(1 - \frac{b^2}{r^2}\right)\beta + \frac{b^2}{r^2}r, & a \le r \le b. \end{cases} \tag{3.49}$$

and the radial (σ_{rr}) and hoop ($\sigma_{\theta\theta}$) stresses are given by:

$$\sigma_{rr}(r) = 2\mu\varepsilon_{rr} + \lambda(\varepsilon_{rr} + \varepsilon_{\theta\theta}),$$
$$\sigma_{\theta\theta}(r) = 2\mu\varepsilon_{\theta\theta} + \lambda(\varepsilon_{rr} + \varepsilon_{\theta\theta}). \tag{3.50}$$

where the appropriate Lamé constants are to be used in the evaluation of the normal stresses.

For the present numerical study, a square domain of size $L \times L$ with $L = 2$ is considered, where the outer radius is chosen to be $b = 2$ and inner radius $a = 0.4 + \epsilon$. The parameter ϵ is set to be 10^{-3} to avoid the level set function to be exactly zero at the node. This is done because if the material interface conforms to the mesh, then there is no need to enrich the approximation. Along the outer boundary, closed-form displacements are imposed. The XFEM displacement approximation for this problem is given by Equation (3.13), where $\psi(\mathbf{x})$ is chosen as the signed distance function (c.f. Equation (3.15)). Figures 3.17a and 3.17b show a typical structured four-noded quadrilateral mesh employed for this study and a deformed shape for the applied boundary conditions, respectively. Meshes with characteristic element sizes of $h \in \{0.2, 0.1, 0.05, 0.025\}$ are used. In order to assess the performance of the method,

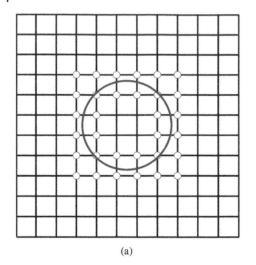
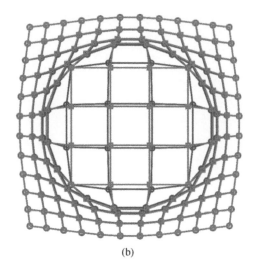

(a)   (b)

Figure 3.17 Two-dimensional circular inhomogeneity: (a) domain discretized with structured four-noded quadrilateral elements, "blue" line represents the interface, "red" circle represents the enriched nodes and (b) "red" shows the deformed shape and "blue" represents the solution obtained from XFEM and analytical approach, respectively.

we use the relative error in the displacement norm (L_2 norm) and the energy norm (H_1 semi-norm), defined by:

$$||\mathbf{u} - \mathbf{u}^h||_{L_2\Omega} = \frac{\sqrt{\int\limits_{\Omega}[(\mathbf{u} - \mathbf{u}^h) \cdot (\mathbf{u} - \mathbf{u}^h)]\ \mathrm{d}\Omega}}{\sqrt{\int\limits_{\Omega}\mathbf{u}^h \cdot \mathbf{u}^h\ \mathrm{d}\Omega}} \tag{3.51a}$$

$$||\mathbf{u} - \mathbf{u}^h||_{H_1(\Omega)} = \frac{\sqrt{\int\limits_{\Omega}[(\varepsilon - \varepsilon^h)\mathbf{D}(\varepsilon - \varepsilon^h)]\ \mathrm{d}\Omega}}{\sqrt{\int\limits_{\Omega}\varepsilon\ \mathbf{D}\ \varepsilon\ \mathrm{d}\Omega}} \tag{3.51b}$$

where **D** is the material constitutive matrix. Figure 3.18 shows the rate of convergence in the displacement norm (L_2 norm) and in the energy norm (H_1 semi-norm). It is opined that with mesh refinement, the results converge; however, the rate of convergence is suboptimal in both the norms. This is attributed to the problem of inaccurate representation of the solution in the partially enriched elements, also commonly referred to as blending problem. This has received considerable attention among researchers and we shall be discussing this again in Chapter 4 with some possible remedy.

3.4.3 Infinite Plate with a Center Crack Under Tension

In this example, we shall consider an example where XFEM is used to model a strong discontinuity and singularity, viz., cracks. For this consider an infinite plate containing a straight crack of length a and loaded by a remote uniform stress field σ, as shown in Figure 3.19. Along ABCD, the closed-form near-tip displacements are imposed. All simulations are performed with $a = 100$ mm and $\sigma = 10^4$ N/mm^2 on a square mesh with sides of length 10 mm. The closed-form solution in terms of polar coordinates in a reference frame (r, θ) centered at the crack tip is defined by:

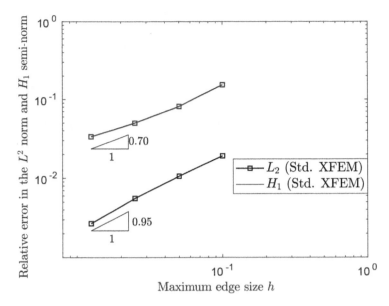

Figure 3.18 Bi-material circular inhomogeneity problem: convergence of the relative error in the L_2 norm and H_1 semi-norm.

Figure 3.19 Infinite plate with a center crack under remote tension: geometry and loads.

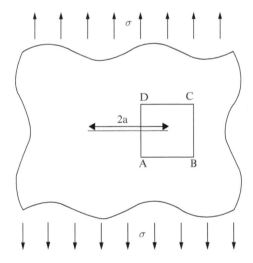

$$\sigma_x(r,\theta) = \frac{K_I}{\sqrt{r}}\cos\frac{\theta}{2}\left(1 - \sin\frac{\theta}{2}\sin\frac{3\theta}{2}\right)$$

$$\sigma_y(r,\theta) = \frac{K_I}{\sqrt{r}}\cos\frac{\theta}{2}\left(1 + \sin\frac{\theta}{2}\sin\frac{3\theta}{2}\right)$$

$$\sigma_{xy}(r,\theta) = \frac{K_I}{\sqrt{r}}\sin\frac{\theta}{2}\cos\frac{\theta}{2}\cos\frac{3\theta}{2} \tag{3.52}$$

The closed-form near-tip displacement field is written as:

$$u_x(r, \theta) = \frac{2(1 + \nu)}{\sqrt{2\pi}} \frac{K_I}{E} \sqrt{r} \cos\frac{\theta}{2} \left(2 - 2\nu - \cos^2\frac{\theta}{2}\right)$$

$$u_y(r, \theta) = \frac{2(1 + \nu)}{\sqrt{2\pi}} \frac{K_I}{E} \sqrt{r} \sin\frac{\theta}{2} \left(2 - 2\nu - \cos^2\frac{\theta}{2}\right) \tag{3.53}$$

In the two previous expression, $K_I = \sigma\sqrt{\pi a}$ denotes the stress intensity factor (SIF), ν is Poisson's ratio, and E is Young's modulus.

☞ The computation of stress intensity factors using the domain form of the interaction integral is presented here. The energy release rate for general mixed-mode problems in two dimensions is written as:

$$G = \frac{1}{\tilde{E}}(K_I^2 + K_{II}^2) \tag{3.54}$$

where K_I and K_{II} are the mode I and mode II SIF's, respectively, and $\tilde{E}$ is defined as:

$$\tilde{E} = \begin{cases} \dfrac{E}{1 - \nu^2} & \text{plane strain} \\ E & \text{plane stress} \end{cases} \tag{3.55}$$

Consider a crack in two dimensions. Let Γ be a contour encompassing the crack tip and $\mathbf{n}$ be the unit normal vector (see Figure 3.20). The contour integral J is defined as:

$$J = \lim_{\Gamma_o \to 0} \int_{\Gamma_o} \left[W dx_2 - T_i \frac{\partial u}{\partial x_1} d\Gamma \right] = \lim_{\Gamma_o \to 0} \int_{\Gamma_o} \left[W dx_2 - \sigma_{ij} \frac{\partial u}{\partial x_1} n_j d\Gamma \right] \tag{3.56}$$

where $\Gamma_o = \Gamma \cup C_o \cup C^+ \cup C^-$ (see Figure 3.20). Physically, the J integral can be interpreted as the energy flowing per unit crack advance, i.e., it is equivalent to the energy release rate. Two states of the cracked body are considered, viz., the current (denoted as state 1) and the auxiliary state (referred to as state 2). The J integral given by Equation (3.56) is written as a superposition of these two states as:

$$J^{(1+2)} = \lim_{\Gamma_o \to 0} \int_{\Gamma_o} \left[\frac{1}{2}(\sigma_{ij}^{(1)} + \sigma_{ij}^{(2)})(\varepsilon_{ij}^{(1)} + \varepsilon_{ij}^{(2)})\delta_{1j} - (\sigma_{ij}^{(1)} + \sigma_{ij}^{(2)})\frac{\partial(u_i^{(1)} + u_i^{(2)})}{\partial x_1} \right] n_j \, d\Gamma \tag{3.57}$$

By rearranging the terms in Equation (3.57), we get:

$$J^{(1+2)} = J^{(1)} + J^{(2)} + I^{(1+2)} \tag{3.58}$$

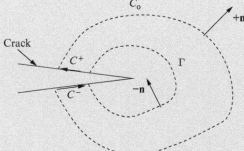

Figure 3.20 Contours and domain for computation of mixed mode stress intensity factors.

where $I^{(1+2)}$ is called the interaction integral for the states 1 and 2 and is given by:

$$I^{(1+2)} = \lim_{\Gamma_o \to 0} \int_{\Gamma_o} \left[\frac{1}{2}(\sigma_{ij}^{(1)}\varepsilon_{ij}^{(2)} + \sigma_{ij}^{(2)}\varepsilon_{ij}^{(1)}) - \sigma_{ij}^{(1)}\frac{\partial u_i^{(2)}}{\partial x_1} - \sigma_{ij}^{(2)}\frac{\partial u_i^{(1)}}{\partial x_1} \right] n_j \, d\Gamma \qquad (3.59)$$

or,

$$I^{(1+2)} = \lim_{\Gamma \to 0} \left\{ \int_{\Gamma \cup C_o \cup C^+ \cup C^-} \left[W^{(1,2)}\delta_{1j} - \sigma_{ij}^{(1)}\frac{\partial u_i^{(2)}}{\partial x_1} - \sigma_{ij}^{(2)}\frac{\partial u_i^{(1)}}{\partial x_1} \right] q \, n_j \, ds \right\} \qquad (3.60)$$

For a combined state, Equation (3.54) can be rewritten as:

$$J^{(1+2)} = J^{(1)} + J^{(2)} + \frac{2}{\tilde{E}}(K_I^{(1)}K_{II}^{(2)} + K_I^{(2)}K_{II}^{(1)}) \qquad (3.61)$$

By comparing Equation (3.58) and Equation (3.61), we get:

$$I^{(1+2)} = \frac{2}{\tilde{E}}(K_I^{(1)}K_{II}^{(2)} + K_I^{(2)}K_{II}^{(1)}) \qquad (3.62)$$

In order to facilitate the numerical implementation of the interaction integral, it is advantageous to recast the contour integrals into equivalent domain integrals. The equivalent domain form of the interaction integral is obtained by defining an appropriate test or weighting function and applying the divergence theorem. In this study, the weighting function is chosen to take a value of unity on an open set containing the crack tip and vanishes on an outer prescribed contour. By applying divergence theorem, to Equation (3.60), the interaction integral in domain form is given by:

$$I^{(1+2)} = \int_{\Omega} \left[-W^{(1,2)}\delta_{1j} + \sigma_{ij}^{(1)}\frac{\partial u_i^{(2)}}{\partial x_1} - \sigma_{ij}^{(2)}\frac{\partial u_i^{(1)}}{\partial x_1} \right] \frac{\partial q}{\partial x_j} \, d\Omega \qquad (3.63)$$

Note that, in the above derivation, it is assumed that the crack faces are straight and traction free.

For this problem, the governing equation in the absence of inertia and body force is given by:

$$\boldsymbol{\nabla} \cdot \boldsymbol{\sigma} = \mathbf{0} \quad \text{in} \quad \Omega \qquad (3.64)$$

with $\boldsymbol{\sigma} \cdot \mathbf{n} = 0$ on Γ_t, $\mathbf{u} = \hat{\mathbf{u}}$ on Γ_u, $\boldsymbol{\sigma} = \mathbf{D}\boldsymbol{\varepsilon}$ and $\boldsymbol{\varepsilon} = \frac{1}{2}\left(\nabla\mathbf{u} + \nabla\mathbf{u}^{\mathrm{T}}\right)$, where $\mathbf{D}$ is the elastic constitutive matrix, $\boldsymbol{\sigma}$ and $\boldsymbol{\varepsilon}$ are the Cauchy stress and small strain tensor, respectively, and $\mathbf{u}$ is the displacement field. In this case, the XFEM approximation for the displacement is given by Equation (3.13).

In this example, the domain is discretized with structured four-noded quadrilateral elements. As discussed in Section 3.3.1, XFEM framework leads to five different categories of elements, viz., standard elements, split elements, tip elements, split-blending elements, and tip-blending elements. The elements intersected by the discontinuous surface are subdivided into triangles for the purpose of numerical integration. For standard elements, 4 Gauss points are used; for split-blending and tip-blending elements, 8 Gauss points are used; for split elements 6 points per triangle are used; and 13 points per triangle are used for tip elements. Figure 3.21 shows a representative mesh and the location of Gauss points used for this study. To study the convergence behavior, a sequence of mesh, viz., 10×10, 20×20, 40×40, 80×80, and 160×160 is considered. The relative error in the L_2 norm and H_1 semi-norm is shown in Figure 3.22. It is seen that with mesh refinement, the error converges, however, not at an optimal rate of 2 and 1 in L_2 norm and H_1 semi-norm, respectively. This suboptimal behavior can be attributed to the following reasons: (i) only the tip element is enriched, which asymptotically reduces the XFEM approximation space to the standard approximation space and (ii) no blending correction is performed. This limits the optimal convergence rate to 0.5 in the energy norm in the presence of singularity.

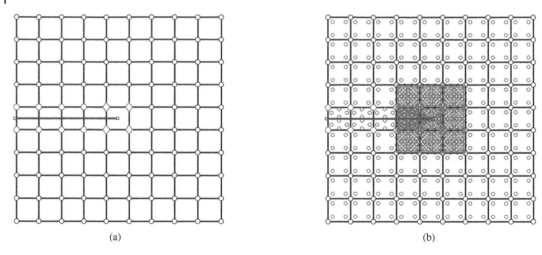

Figure 3.21 Plate with an edge crack: (a) representative back ground mesh with discontinuity independent of the mesh, black filled circles are standard nodes, red circles are nodes enriched with Heaviside enrichment, and red diamonds are nodes enriched with near-tip asymptotic functions and (b) location of Gauss points.

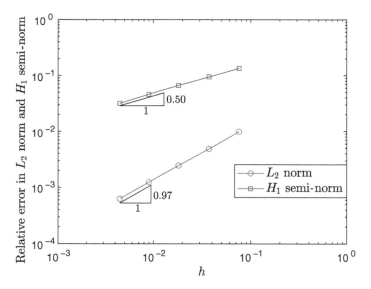

Figure 3.22 Convergence of the relative error in the L_2 norm and H_1 semi-norm for a plate with an edge crack in tension.

3.5 Conclusions

The purpose of this chapter was to revisit the idea of partition of unity. We started by defining the completeness, consistency, and reproducing conditions, followed by the concept of partition of unity and enrichment. The key ingredients in the implementation of the XFEM are discussed in detail. We then presented representative examples for weak/strong discontinuity and singularity. The convergence of the relative error in the L_2 norm and H_1 norm were presented and it is inferred that the XFEM framework presented in this chapter yields accurate results, alleviates the need for a conforming mesh, however, converges suboptimally. The reason for suboptimal rate of convergence was highlighted. In Chapter 4, we will discuss the advancements to the XFEM.

References

Asadpoure, A. and Mohammadi, S. (2007). Developing new enrichment functions for crack simulation in orthotropic media by the extended finite element method. *International Journal for Numerical Methods in Engineering* 69 (10): 2150–2172.

Aubertin, P., Réthoré J., and de Borst, R. (2009). A coupled molecular dynamics and extended finite element method for dynamic crack propagation. *International Journal for Multiscale Computational Engineering* 81: 72–88.

Aubertin, P., Réthoré, J., and de Borst, R. (2010). Dynamic crack propagation using a combined molecular dynamics/extended finite element approach. *International Journal for Multiscale Computational Engineering* 2: 221–235.

Bechet, E., Scherzer, M., and Kuna, M. (2009). Application of the X-FEM to the fracture of piezoelectric materials. *International Journal for Numerical Methods in Engineering* 77: 1535–1565.

Belytschko, T. and Black, T. (1999). Elastic crack growth in finite elements with minimal remeshing. *International Journal for Numerical Methods in Engineering* 45 (5): 601–620.

Belytschko, T., Krongauz, Y., Fleming, M., et al. (1996). Smoothing and accelerated computations in the element free galerkin method. *Journal of Computational and Applied Mathematics* 74 (1): 111–126.

Belytschko, T., Xiao, S.P., and Parimi, C. (2003). Topology optimization with implicit functions and regularization. *International Journal for Numerical Methods in Engineering* 57 (8): 1177–1196.

Cheng, K.W. and Fries, T.P. (2010). Higher-order XFEM for curved strong and weak discontinuities. *International Journal for Numerical Methods in Engineering* 82 (5): 564–590.

Chessa, J., Smolinski, P., and Belytschko, T. (2002). The extended finite element method (XFEM) for solidification problems. *International Journal for Numerical Methods in Engineering* 53 (8): 1959–1977. doi: 10.1002/nme.386.

Chessa, J., Wang, H., and Belytschko, T. (2003). On the construction of blending elements for local partition of unity enriched finite elements. *International Journal for Numerical Methods in Engineering* 57 (7): 1015–1038.

De Luycker, E., Benson, D.J., Belytschko, T., et al. (2011). X-FEM in isogeometric analysis for linear fracture mechanics. *International Journal for Numerical Methods in Engineering* 87 (6): 541–565.

Dolbow, J., Moes, N., and Belytschko, T. (2000). Modeling fracture in Mindlin–Reissner plates with the extended finite element method. *International Journal of Solids and Structures* 37 (48): 7161–7183.

Duflot, M. (2008). The extended finite element method in thermoelastic fracture mechanics. *International Journal for Numerical Methods in Engineering* 74 (5): 827–847.

Elguedj, T., Gravouil, A., and Combescure, A. (2006). Appropriate extended functions for X-FEM simulation of plastic fracture mechanics. *Computer Methods in Applied Mechanics and Engineering* 195 (7): 501–515.

Fries, T.P. (2008). The intrinsic XFEM for two-fluid flows. *International Journal for Numerical Methods in Fluids* 60 (4): 437–471.

Gerstenberger, A. and Wall, W.A. (2008). An eXtended finite element method/Lagrange multiplier based approach for fluid-structure interaction. *Computer Methods in Applied Mechanics and Engineering* 197 (19): 1699–1714.

Ghorashi, S.S., Valizadeh, N., and Mohammadi, S. (2011). Extended isogeometric analysis for simulation of stationary and propagating cracks. *International Journal for Numerical Methods in Engineering* 89 (9): 1069–1101.

Gracie, R., Oswald, J., and Belytschko, T. (2008a). On a new extended finite element method for dislocations: Core enrichment and nonlinear formulation. *Journal of the Mechanics and Physics of Solids* 56 (1): 200–214.

Gracie, R., Oswald. J., and Belytschko, T. (2009). Concurrently coupled atomistic and XFEM models for dislocations and cracks. *International Journal for Numerical Methods in Engineering* 78: 354–378.

Gracie, R., Ventural, G., and Belytschko, T. (2007). A new fast finite element method for dislocations based on interior discontinuities. *International Journal for Numerical Methods in Engineering* 69: 423–441.

Gravouil, A, Moës, N., and Belytschko, T. (2002). Non-planar 3d crack growth by the extended finite element and level sets. Part II: Level set update. *International Journal for Numerical Methods in Engineering* 53 (11): 2569–2586.

Hansbo, A. and Hansbo, P. (2004). A finite element method for the simulation of strong and weak discontinuities in solid mechanics. *Computer Methods in Applied Mechanics and Engineering* 193 (33): 3523–3540.

Huynh, D.B.P. and Belytschko, T. (2009). The extended finite element method for fracture in composite materials. *International Journal for Numerical Methods in Engineering* 77 (2): 214–239.

Iarve, E.V. (2003). Mesh independent modelling of cracks by using higher order shape functions. *International Journal for Numerical Methods in Engineering* 56: 869–882.

Ji, H. and Dolbow, J.E. (2004). On strategies for enforcing interfacial constraints and evaluating jump conditions with the extended finite element method. *International Journal for Numerical Methods in Engineering* 61 (14): 2508–2535.

Kaufmann, P., Martin, S., Botsch, M., et al. (2009). Enrichment textures for detailed cutting of shells. *In ACM Transactions on Graphics (TOG)* 28: 50. ACM.

Kumar, A.V., Padmanabhan, S., and Burla, R. (2008). Implicit boundary method for finite element analysis using nonconforming mesh or grid. *International Journal for Numerical Methods in Engineering* 74 (9): 1421–1447.

Lasry, J., Pommier, J., Renard, Y., and Salaün, M. (2010). eXtended finite element methods for thin cracked plates with Kirchhoff-love theory. *International Journal for Numerical Methods in Engineering* 84 (9): 1115–1138.

Lecampion, B. (2009). An extended finite element method for hydraulic fracture problems. *Communications in Numerical Methods in Engineering* 25 (2): 121–133.

Legay, A., Wang, H.W., and Belytschko, T. (2005). Strong and weak arbitrary discontinuities in spectral finite elements. *International Journal for Numerical Methods in Engineering* 64 (8): 991–1008.

Legrain, G., Moes, N., and Huerta, A. (2008). Stability of incompressible formulations enriched with X-FEM. *Computer Methods in Applied Mechanics and Engineering* 197 (21): 1835–1849.

Melenk, J.M. and Babuška, I. (1996). The partition of unity finite element method: Basic theory and applications. *Computer Methods in Applied Mechanics and Engineering* 139 (1): 289–314.

Menk, A. and Bordas, S. (2010). Numerically determined enrichment functions for the extended finite element method and applications to bi-material anisotropic fracture and polycrystals. *International Journal for Numerical Methods in Engineering* 83 (7): 805–828.

Moës, N. and Belytschko, T. (2002). Extended finite element method for cohesive crack growth. *Engineering Fracture Mechanics* 69 (7): 813–833.

Moës, N., Cloirec, M., Cartraud, P., and Remacle, J.F. (2003). A computational approach to handle complex microstructure geometries. *Computer Methods in Applied Mechanics and Engineering* 192 (28): 3163–3177.

Moës, N., Gravouil, A., and Belytschko, T. (2002). Non-planar 3d crack growth by the extended finite element and level sets. Part I: Mechanical model. *International Journal for Numerical Methods in Engineering* 53 (11): 2549–2568.

Nguyen, V.P., Simpson, R.N., Bordas, S., and T. Rabczuk. (2012). An introduction to isogeometric analysis with Matlab implementation: FEM and XFEM formulations. *Arxiv preprint arXiv*: 1205–2129.

Osher, S. and Sethian, J.A. (1988). Fronts propagating with curvature-dependent speed: Algorithms based on Hamilton-Jacobi formulations. *Journal of Computational Physics* 79 (1): 12–49.

Rabczuk, T., Bordas, S., and Zi, G. (2007). A three-dimensional meshfree method for continuous multiple-crack initiation, propagation and junction in statics and dynamics. *Computational Mechanics* 40 (3): 473–495.

Rabczuk, T., Zi, G., Gerstenberger, A., and Wall, W.A. (2008). A new crack tip element for the phantom-node method with arbitrary cohesive cracks. *International Journal for Numerical Methods in Engineering* 75: 577–599.

Réthoré, J., Gravouil, A., and Combescure, A. (2005). An energy-conserving scheme for dynamic crack growth using the extended finite element method. *International Journal for Numerical Methods in Engineering* 63: 631–659.

Rojas-Díaz, R., Sukumar, N., Sáez, A., and García-Sánchez, F. (2009). Crack analysis in magnetoelectroelastic media using the extended finite element method. In *International Conference on Extended Finite Element Methods*.

Song, J.-H., Areias, P., and Belytschko, T. (2006). A method for dynamic crack and shear band propagation with phantom nodes. *International Journal for Numerical Methods in Engineering* 67 (6): 868–893.

Stolarska, M., Chopp, D.L., Moës, N., and Belytschko, T. (2001). Modelling crack growth by level sets in the extended finite element method. *International Journal for Numerical Methods in Engineering* 51: 943–960.

Ventura, G., Moran, B., and Belytschko, T. (2005). Dislocations by partition of unity. *International Journal for Numerical Methods in Engineering* 62:1463–1487.

Ventura, G., Xu, J.X., and Belytschko, T. (2002). A vector level set method and new discontinuity approximations for crack growth by EFG. *International Journal for Numerical Methods in Engineering* 54 (6): 923–944.

Verhoosel, C.V., Scott, M.A., Borst, R., and Hughes, T.J.R. (2010). An isogeometric analysis to cohesive zone modeling. *International Journal for Numerical Methods in Engineering* 10.1002/nme.3061. doi: 10.1002/nme.3061.

Wells, G.N. and Sluys, L.J. (2001). A new method for modelling cohesive cracks using finite elements. *International Journal for Numerical Methods in Engineering* 50: 2667–2682.

Wells, G.N., Sluys, L.J., and De Borst, R. (2002). Simulating the propagation of displacement discontinuities in a regularized strain-softening medium. *International Journal for Numerical Methods in Engineering* 53 (5): 1235–1256. ISSN 1097-0207.

Zi, G. and Belytschko, T. (2003). New crack-tip elements for (XFEM) and applications to cohesive cracks. *International Journal for Numerical Methods in Engineering* 57: 2221–2240.

4

Advanced Topics

Stéphane P. A. Bordas[1], Alexander Menk[2], and Sundararajan Natarajan[3]

[1] *University of Luxembourg, Luxembourg, UK*
[2] *Robert Bosch GmbH, Germany*
[3] *Indian Institute of Technology Madras, India*

The previous chapters aimed to convey an understanding of the general concept of the extended finite element method (XFEM) and its advantages and drawbacks. In this chapter, different aspects and advancements of the XFEM will be analyzed in greater detail. We start by discussing some of the more advanced integration techniques for the enriched elements. Although it is possible to integrate enrichment functions sufficiently accurately with traditional Gaussian quadrature performed on subcells, the techniques discussed in this chapter are more efficient in certain situations. We continue by discussing the choice of the enriched domain and a correction to unwanted artifacts introduced by the enrichments. This is done by taking the same one-dimensional (1D) example discussed in Chapter 2. Both the topics are closely related to the convergence of the XFEM itself. For large system of equations, iterative methods are an ideal choice to obtain a numerical solution. The condition number of the equations plays an important role if the convergence and the stability of these methods are considered. We, therefore, finish the chapter by discussing preconditioning techniques for the XFEM framework which improve the condition number.

4.1 Size of the Enrichment Zone

We already discussed that the crack tip enrichments are necessary to represent the weak singularity in the vicinity of the crack tip. We also discussed that optimal convergence rates cannot be guaranteed with the finite element method (FEM) if the exact solution contains a weak singularity. One may hope that with the XFEM optimal convergence rates can be achieved, since the enrichments enhance the approximation properties. But it turns out that it is not always the case. If optimal convergence rates are desired, the size of the enriched area plays a significant role.

In the standard version of the XFEM, only the nodes of the element containing the crack tip are enriched with the crack tip enrichments. There is a simple argument why this is not sufficient to obtain an optimal convergence rate. Suppose that the domain $\hat{\Omega}$ is the structural domain $\Omega \subset \mathbb{R}^2$, but without the elements enriched by the crack tip enrichments. If h is the mesh parameter, then in the standard XFEM, there is a constant c such that ch^2 is greater than the difference in area between $\hat{\Omega}$ and Ω.

Now, let us assume that optimal convergence rates can be achieved by the standard XFEM. We define the error e as the minimum error in the L^2-norm of all possible solutions. Please note that e may be smaller than the L^2-error of the XFEM solution. Then, for some constant C_1, the error e is bounded by: $e \leq C_1 h^2$. Now suppose that $e_{\hat{\Omega}}$ is the

Partition of Unity Methods, First Edition. Stéphane P. A. Bordas, Alexander Menk, and Sundararajan Natarajan.
© 2024 John Wiley & Sons Ltd. Published 2024 by John Wiley & Sons Ltd.

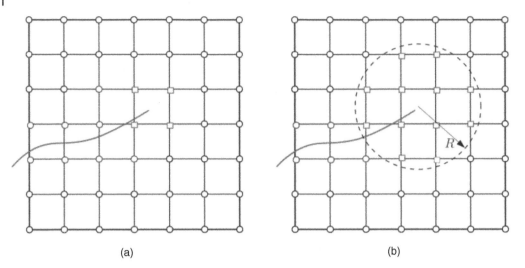

(a)　　　　　　　　　　　　　　(b)

Figure 4.1 Schematic representation of different enrichment strategies within the xfem framework: (a) Topological enrichment, where only the element containing the crack tip is enriched and (b) Geometrical enrichment (or) fixed area of enrichment, where all the nodes within the radius of circle R are enriched with near-tip asymptotic functions. The solid line represents the crack, circled nodes are enriched with Heaviside function, and squared nodes are enriched with near-tip asymptotic functions.

minimum error in the L^2- norm of all possible solutions, but restricted to the domain $\hat{\Omega}$ (that is, the error is only integrated over $\hat{\Omega}$). Obviously, we have:

$$e_{\hat{\Omega}} \leq e \leq C_1 h^2 \tag{4.1}$$

Since the difference in area between $\hat{\Omega}$ and Ω is smaller than ch^2 and all the displacements are bounded, there is another constant C_2, such that:

$$||e_{\hat{\Omega}} - e|| \leq C_1 h^2 \tag{4.2}$$

But taking Equation (4.1) and Equation (4.2) together, we can conclude that it is possible to approximate the exact solution with an error which decreases quadratically during mesh refinement, even if no crack tip enrichments are used. Of course this cannot be true. Therefore, the assumption that the standard XFEM yields an optimal convergence rate must be wrong. The important part in the argument above is that the area difference between $\hat{\Omega}$ and Ω tends zero quadratically. So the least thing we have to make sure when looking for a method with an optimal convergence rate is that the enriched area cannot be bound by ch^2 for some constant c during mesh refinement.

Assume that we enrich all nodes which are inside a circle of radius r around the crack tip using crack tip enrichments. This situation is shown in Figure 4.1. Then, the enriched area cannot be bound by something like ch^2. This is called the geometric enrichment or fixed-area enrichment (Béchet et al., 2005). It turns out that we can actually guarantee optimal convergence rates if geometrical enrichment is used.

4.2 Numerical Integration

4.2.1 Polar Integration

The singular functions commonly used in the XFEM requires a high density of integration points. By using polar integration, Laborde et al. (2005) and Béchet et al. (2005) showed that a concentration of integration points in the vicinity of the singularity improves the results significantly. This process eliminates the singular terms from the quadrature. The idea is to subdivide into triangles such that each triangle has one node at the singularity. Instead of using triangular quadrature, tensor-product type Gauß points in a quadrilateral reference element are mapped into each triangle such that two nodes of the quadrilateral coincide at the singularity node of each triangle. Figure 4.2 shows the procedure. The proposed technique is well suited for point singularities (for example, in two dimensions), but in the case of three dimensions, such a mapping is not trivial (Laborde et al.,2005; Béchet et al.,2005). Park et al. (2008) extended the mapping method introduced by Nagarajan and Mukherjee (1993) in two dimensions to three dimensions to integrate singular integrands. The 2D mapping is given by:

$$T_M : (\rho, \theta) \mapsto (\xi, \eta) \tag{4.3}$$

where $\xi = \rho \cos^2 \theta$ and $\eta = \rho \sin^2 \theta$. The inverse mapping (T_M^{-1}) transforms a right triangle into a rectangle.

4.2.2 Equivalent Polynomial Integration

Obviously, the division of the enriched elements into subcells is an additional effort. This division is necessary if enrichments containing strong or weak discontinuities are to be integrated exactly. Since these functions are not polynomial functions, using Gaussian quadrature can only give an approximation. However, these functions may be polynomials on certain subdomains, thus Gaussian quadrature on these subdomains would return the exact value as discussed earlier.

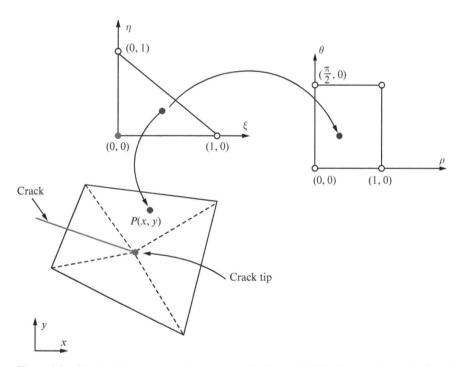

Figure 4.2 Physical element containing a singularity is subdivided such that each triangular subdivision contains the singular point as one of its nodes. Each sub-triangle is mapped onto a standard triangle in the area coordinate system (ξ, η) with crack tip assigned as the origin of the coordinate system. Next the triangle is transformed into a rectangular coordinate (ρ, θ) system.

If the enrichments were replaced by polynomial functions, subdivision would not be necessary. But generally one would expect that replacing the enrichments by polynomials would have an influence on the solution probably eliminating the advantages introduced by the enrichments in the first place. Ventura et al. (2009) showed that it is possible to find polynomials, which can replace the enrichment functions without changing the entries of the stiffness matrix, thus the displacement vector is the same. We would like to demonstrate this idea using the Heaviside function as an example. For Heaviside enrichment, the stiffness matrix of an element takes a rather simple form:

$$\mathbf{K}_{el} = \begin{bmatrix} \int \mathbf{B}^T \mathbf{DB} d\Omega & H \int \mathbf{B}^T \mathbf{DB} d\Omega \\ H \int \mathbf{B}^T \mathbf{CB} d\Omega & H^2 \int \mathbf{B}^T \mathbf{DB} d\Omega \end{bmatrix} = \begin{bmatrix} \int_\Omega \mathbf{B}^T \mathbf{DB} d\Omega & \int_\Omega H \mathbf{B}^T \mathbf{DB} d\Omega \\ \int_\Omega H \mathbf{B}^T \mathbf{DB} d\Omega & \int_\Omega \mathbf{B}^T \mathbf{DB} d\Omega \end{bmatrix} \tag{4.4}$$

where H is the Heaviside function which is 1 in the domain Ω_+ and -1 in the domain Ω_-. Ventura et al. (2009) proposed to use a polynomial $\tilde{H}$ instead of H, such that:

$$\int_\Omega H \mathbf{B}^T \mathbf{DB} d\Omega = \int_\Omega \tilde{H} \mathbf{B}^T \mathbf{DB} d\Omega \tag{4.5}$$

Let us assume that the domains Ω_+ and Ω_- are separated by a straight line in the element. In this example, we will restrict ourselves to the case of a triangular element with piecewise linear shape functions. Using a description of the geometry in natural coordinates, the situation is shown in Figure 4.3. If the shape functions are piecewise linear, the term $\mathbf{B}^T \mathbf{DB}$ is constant. Therefore, the equivalence condition Equation (4.5) becomes:

$$\int_\Omega \tilde{H} d\Omega = \int_\Omega H d\Omega = \int_{\Omega_+} 1 d\Omega - \int_{\Omega_-} 1 d\Omega \tag{4.6}$$

It is easily checked that a suitable $\tilde{H}$ is given by:

$$\tilde{H} = 2\tilde{\eta}\tilde{\nu} - 1 \tag{4.7}$$

Since the components of $\tilde{H}\mathbf{B}^T \mathbf{DB}$ are polynomials of order 1, they can be integrated exactly using Gaussian quadrature without dividing the element into subcells. This idea can be transferred to the 3D case, enrichments containing weak discontinuities and other element types.

4.2.3 Conformal Mapping

Based on Schwarz-Christoffel conformal mapping for arbitrary polygons, Natarajan et al. (2009, 2010) proposed a numerical integration scheme for elements intersected by a discontinuity. It eliminates the need to sub-triangulate the polygonal region intersected by the discontinuity or to sub-triangulate the polygonal element.

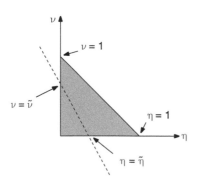

Figure 4.3 Triangular element cut by a straight crack.

☞ A conformal maps transform any pair of curves intersecting at a point in the region so that the image curves intersect at the same angle (Figure 4.4).

Figure 4.4 A rectangular grid (top) and its image under a conformal map f (bottom). It can be seen that f maps pairs of lines intersecting at 90° to pairs of curves intersecting at 90°. In general, a conformal map preserves oriented angles.

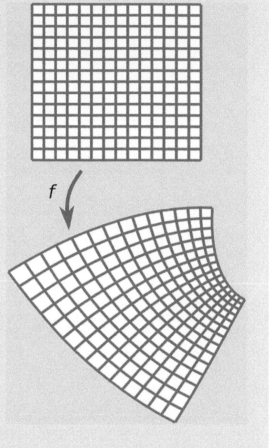

The Schwarz-Christoffel transformation is based on the following postulate: Let U be the interior of a polygon Γ having vertices $w_1, w_2, \cdots, w_n$ and interior angles $\pi\alpha_1, \pi\alpha_2, \cdots, \pi\alpha_n$ in counter-clockwise order. Let f be any conformal map from the unit disk D to U, then the Schwarz-Christoffel formula for a disk is given by:

$$w = f(z) = A + C \int_0^z \prod_{k=1}^n \left(1 - \frac{\varsigma}{z_k}\right)^{\alpha_k - 1} d\varsigma \tag{4.8}$$

where A and C are complex constants ($C \neq 0$). Here $z = x + iy$ correspond to a point in the complex plane and $w = \xi + i\eta$ is its corresponding map in the complex polygonal plane. The function f maps the unit disk in the complex plane conformally onto a polygonal plane U. An inverse mapping also exists (Trefethen, 1979, 1980) and since the map is conformal, the positivity of the Jacobian is ensured.

In the conventional XFEM, the numerical integration over the elements intersected by the discontinuity is done by subdividing the elements into subdomains (triangular subdomains in this case) (see Section 3.3.4). This poses two problems:

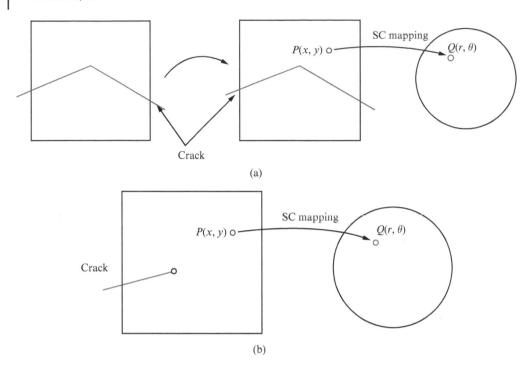

Figure 4.5 Integration over an element with discontinuity (solid red line): (a) with kinked discontinuity, representing a split element and (b) strong discontinuity, representing the tip element. In both cases, the sub-polygon is mapped conformally onto the unit disk using Schwarz-Christoffel conformal mapping.

- The sub-triangulation introduces a "mesh" requirement (c.f. Section 3.3.4, Figure 3.12).
- Involves a two-level isoparametric mapping.

By employing the Schwarz-Christoffel conformal mapping, the above two steps can be circumvented. By conformally mapping each part of the element intersected by the discontinuous surface, the "mesh" requirement is suppressed. Figure 4.5 illustrates the above idea for the split and the tip element. The quadrature points over the circle can be obtained from either Midpoint rule or Gauss-Chebyshev rule (De and Bathe, 2001; Peirce, 1957).

4.2.4 Strain Smoothing in XFEM

Stabilized conformal nodal integration (SCNI) was constituted to suppress the instabilities arising in nodally integrated mesh-free methods by avoiding the computation of the derivatives of the shape functions, which vanish at the nodes. Liu et al. (2007) extended the idea of SCNI to finite element approximations and named the resulting method the smoothed finite element method (SFEM). Within the framework of SCNI and SFEM, the strain field used to construct the stiffness matrix is written as the divergence of a spatial average of the standard (compatible) strain field – i.e., symmetric gradient of the displacement field. The smoothed strain field $\bar{\varepsilon}_{ij}^{h}$, used to compute the stiffness matrix, is computed by a weighted average of the standard strain field ε_{ij}^{h}. At a point $\mathbf{x}_c$ in an element Ω^h:

$$\bar{\varepsilon}_{ij}^{h}(\mathbf{x}_c) = \int_{\Omega^h} \varepsilon_{ij}^{h}(\mathbf{x}) \Phi(\mathbf{x} - \mathbf{x}_c) d\mathbf{x}, \tag{4.9}$$

where Φ is a smoothing function that generally satisfies the following properties (Yoo et al., 2004):

$$\Phi \geq 0 \quad \text{and} \quad \int_{\Omega^h} \Phi(\mathbf{x}) d\mathbf{x} = 1. \tag{4.10}$$

One possible choice of Φ is given by:

$$\Phi = \frac{1}{A_c} \quad \text{in} \quad \Omega_C \quad \text{and} \quad \Phi = 0 \quad \text{elsewhere} \tag{4.11}$$

where A_c is the area of the subcell. To use Equation (4.9), the subcell containing the point $\mathbf{x}_c$ must first be located in order to compute the correct value of the weight function Φ. The discretized smoothed strain field is computed through the smoothed discretized gradient operator or the smoothed strain-displacement operator, $\tilde{\mathbf{B}}$, defined by for a schematic representation of the construction:

$$\tilde{\varepsilon}^h(\mathbf{x}_c) = \tilde{\mathbf{B}}_c(\mathbf{x}_c)\mathbf{q} \tag{4.12}$$

where $\mathbf{q}$ are the unknown displacements coefficients defined at the nodes of the finite element, as usual. The smoothed element stiffness matrix for element e is computed by the *sum of the contributions of the subcells*[1]:

$$\tilde{\mathbf{K}}^e = \sum_{C=1}^{nc} \int_{\Omega_C} \tilde{\mathbf{B}}_C^{\mathrm{T}} \mathbf{D} \tilde{\mathbf{B}}_C \, d\Omega = \sum_{C=1}^{nc} \tilde{\mathbf{B}}_C^{\mathrm{T}} \mathbf{D} \tilde{\mathbf{B}}_C \int_{\Omega_C} d\Omega = \sum_{C=1}^{nc} \tilde{\mathbf{B}}_C^{\mathrm{T}} \mathbf{D} \tilde{\mathbf{B}}_C A_C \tag{4.13}$$

where nc is the number of the smoothing cells of the element. The strain-displacement matrix $\tilde{\mathbf{B}}_c$ is constant over each Ω_C and is of the following form:

$$\tilde{\mathbf{B}}_C = \begin{bmatrix} \tilde{\mathbf{B}}_{C1} & \tilde{\mathbf{B}}_{C2} & \tilde{\mathbf{B}}_{C3} & \tilde{\mathbf{B}}_{C4} \end{bmatrix} \tag{4.14}$$

where for all shape functions $I \in \{1, \ldots, 4\}$, the 3×2 submatrix $\tilde{\mathbf{B}}_{CI}$ represents the contribution to the strain-displacement matrix associated with shape function I and cell C and writes:

$$\forall I \in \{1, 2, \ldots, 4\}, \forall C \in \{1, 2, \ldots nc\} \tilde{\mathbf{B}}_{CI} = \int_{\Gamma_C} \begin{bmatrix} n_x & 0 \\ 0 & n_y \\ n_y & n_x \end{bmatrix} (\mathbf{x}) N_I(\mathbf{x}) d\Gamma \tag{4.15}$$

since Equation (4.15) is computed on the boundary of Ω_C, one Gauss point is sufficient for an exact integration (in the case of a bilinear approximation):

$$\tilde{\mathbf{B}}_{CI}(\mathbf{x}_c) = \frac{1}{A_c} \sum_{b=1}^{nb} \begin{pmatrix} N_I\left(\mathbf{x}_b^G\right) n_x & 0 \\ 0 & N_I\left(\mathbf{x}_b^G\right) n_y \\ N_I\left(\mathbf{x}_b^G\right) n_y & N_I\left(\mathbf{x}_b^G\right) n_x \end{pmatrix} l_b^C \tag{4.16}$$

where $\mathbf{n} = (n_x, n_y)$ is the outward normal to the smoothing cell, Ω_C, n_b is the number of edges of the subcell, $\mathbf{x}_b^G$ and l_b^C are the center point (Gauss point) and the length of Γ_b^C, respectively. Figure 4.6 shows a possible subdivision of a quadrilateral element into subcells and the location of the integration points. The general procedure for the SFEM consists of the following steps:

The SmXFEM uses a similar approximation for the displacement field as the XFEM, given by Equation (3.10) (c.f. Chapter 3, Section 3.3). As in the SFEM, subcells are employed to smooth the strain field and calculate the stiffness matrix. All elements are divided into number of subcells, which can vary from element to element. Typically, elements that are split by a discontinuity (weak or strong) will be divided into only two subcells, one on either

---

1 The subcells Ω_C form a partition of the element Ω^h.

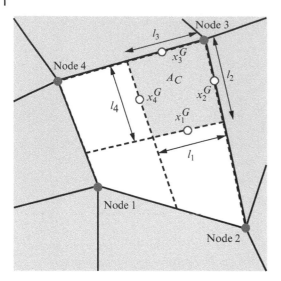

Figure 4.6 SFEM: description of the subcells and the integration points. The boundary of the subcell is denoted by dashed lines.

side of the discontinuity. Elements in which non-polynomial approximations are present are divided into a larger number of subcells, similar to the case of the standard XFEM.

4.2.4.1 Computation of the Stiffness Matrix

Let us now derive the enriched stiffness matrix of the SmXFEM, assuming all nodes in $\mathcal{N}^{\text{enr}}$ are enriched with function ρ. Denote by ε^h the discretized enriched strain field deduced by differentiation of Equation (3.10) is given by:

$$\varepsilon^h(\mathbf{x}) = \sum_{I \in \mathcal{N}^{\text{fem}}} B_{I\text{fem}}(\mathbf{x})\mathbf{u}_I + \sum_{J \in \mathcal{N}^{\text{enr}}} B_{J\text{enr}}(\mathbf{x})\mathbf{a}_J = [\mathbf{B}_{\text{fem}}|\mathbf{B}_{\text{enr}}][\mathbf{q}]$$

$$\varepsilon^h(\mathbf{x}) = [\mathbf{B}_{\text{xfem}}][\mathbf{q}] \tag{4.17}$$

where $\mathcal{N}^{\text{fem}}$ is the set of all nodes in the finite element mesh, $\mathcal{N}^{\text{enr}}$ is the set of nodes that are enriched with the Heaviside function and near-tip asymptotic fields. The $\mathbf{B}_{\text{xfem}}$ matrix in Equation (4.17) includes two terms $\mathbf{B}_{\text{fem}}$ and $\mathbf{B}_{\text{enr}}$ corresponding to the standard nodes and the enriched nodes. The $\mathbf{B}_{\text{fem}}$ term contains the first derivatives of the standard finite element shape functions:

$$\mathbf{B}_{\text{fem}} = \begin{bmatrix} \frac{\partial N_I}{\partial x} & 0 \\ 0 & \frac{\partial N_I}{\partial y} \\ \frac{\partial N_I}{\partial y} & \frac{\partial N_I}{\partial x} \end{bmatrix} \tag{4.18}$$

The $\mathbf{B}_{\text{enr}}$ term is composed of the first derivatives of the product of the finite element shape functions with the enrichment functions:

$$\mathbf{B}_{\text{enr}} = \begin{bmatrix} \frac{\partial}{\partial x}[N_J(\rho(\mathbf{x}) - \rho(\mathbf{x}_J))] & 0 \\ 0 & \frac{\partial}{\partial y}[N_J(\rho(\mathbf{x}) - \rho(\mathbf{x}_J))] \\ \frac{\partial}{\partial y}[N_J(\rho(\mathbf{x}) - \rho(\mathbf{x}_J))] & \frac{\partial}{\partial x}[N_J(\rho(\mathbf{x}) - \rho(\mathbf{x}_J))] \end{bmatrix} \tag{4.19}$$

The smoothed strain field at an arbitrary point $\mathbf{x}_c$ is defined as for the standard SFEM:

$$\bar{\varepsilon}^h_{ij}(\mathbf{x}_c) = \int_\Omega \varepsilon^h_{ij}(\mathbf{x})\Phi(\mathbf{x} - \mathbf{x}_c)\, d\mathbf{x} \tag{4.20}$$

where Φ is a smoothing function defined exactly as in the SFEM:

$$\Phi \geq 0 \quad \text{and} \quad \int_\Omega \Phi(\mathbf{x}) \, d\mathbf{x} = 1,$$

$$\Phi(\mathbf{x} - \mathbf{x}_c) = \begin{cases} 1/A_c, & \mathbf{x} \in \Omega_C \\ 0, & \mathbf{x} \notin \Omega_C \end{cases} . \tag{4.21}$$

Substituting Equation (4.17) into Equation (4.20) and using Equation (4.18)–(4.21), we obtain:

$$\tilde{\varepsilon}^h(\mathbf{x}_c) = \int_\Omega \mathbf{B}_{\text{xfem}} \mathbf{q} \, \Phi(\mathbf{x} - \mathbf{x}_c) \, d\mathbf{x} = \tilde{\mathbf{B}}_{\text{xfem}} \mathbf{q}, \tag{4.22}$$

where the smoothed matrix $\tilde{\mathbf{B}}_{\text{xfem}}$ in Equation (4.22) is defined by:

$$\tilde{\mathbf{B}}_{\text{xfem}} = \frac{1}{A_c} \int_{\Omega_C} \mathbf{B}_{\text{xfem}}(\mathbf{x}) \, d\mathbf{x}. \tag{4.23}$$

The $\tilde{\mathbf{B}}_{\text{xfem}}$ in Equation (4.23) includes two terms: $\tilde{\mathbf{B}}_{\text{fem}}$ and $\tilde{\mathbf{B}}_{\text{enr}}$ corresponding to the standard nodes and enriched nodes. The $\tilde{\mathbf{B}}_{I\text{fem}}$ term is given by:

$$\tilde{\mathbf{B}}_{I\text{fem}} = \frac{1}{A_c} \int_{\Omega_C} \begin{bmatrix} \frac{\partial N_I}{\partial x} & 0 \\ 0 & \frac{\partial N_I}{\partial y} \\ \frac{\partial N_I}{\partial y} & \frac{\partial N_I}{\partial x} \end{bmatrix} d\Omega. \tag{4.24}$$

By using the divergence theorem and noting $\mathbf{n} = (n_x, n_y)$ the outward normal to the smoothing cell Ω_C and Γ_c, Equation (4.24) can be written as:

$$\tilde{\mathbf{B}}_{I\text{fem}} = \frac{1}{A_c} \int_{\Gamma_c} \begin{bmatrix} n_x N_I & 0 \\ 0 & n_y N_I \\ n_y N_I & n_x N_I \end{bmatrix} d\Gamma. \tag{4.25}$$

Performing the same operations for $\tilde{\mathbf{B}}_{J\text{enr}}$, we obtain:

$$\tilde{\mathbf{B}}_{J\text{enr}} = \frac{1}{A_c} \int_{\Omega_C} \begin{bmatrix} \frac{\partial}{\partial x}[N_J(\rho(\mathbf{x}) - \rho(\mathbf{x}_J))] & 0 \\ 0 & \frac{\partial}{\partial y}[N_J(\rho(\mathbf{x}) - \rho(\mathbf{x}_J))] \\ \frac{\partial}{\partial y}[N_J(\rho(\mathbf{x}) - \rho(\mathbf{x}_J))] & \frac{\partial}{\partial x}[N_J(\rho(\mathbf{x}) - \rho(\mathbf{x}_J))] \end{bmatrix} d\Omega. \tag{4.26}$$

Using the divergence theorem, we obtain:

$$\tilde{\mathbf{B}}_{J\text{enr}} = \frac{1}{A_c} \int_{\Gamma_c} \begin{bmatrix} n_x[N_J(\rho(\mathbf{x}) - \rho(\mathbf{x}_J))] & 0 \\ 0 & n_y[N_J(\rho(\mathbf{x}) - \rho(\mathbf{x}_J))] \\ n_y[N_J(\rho(\mathbf{x}) - \rho(\mathbf{x}_J))] & n_x[N_J(\rho(\mathbf{x}) - \rho(\mathbf{x}_J))] \end{bmatrix} d\Gamma. \tag{4.27}$$

The smoothed enriched stiffness matrix for subcell C, $\tilde{\mathbf{K}}^C_{\text{xfem}}$, is computed by:

$$\tilde{\mathbf{K}}^C_{\text{xfem}} = \int_{\Omega_C} \underbrace{\tilde{\mathbf{B}}_c^T \mathbf{D} \tilde{\mathbf{B}}_c}_{=\text{constant}} \, d\Omega = \tilde{\mathbf{B}}_c^T \mathbf{D} \tilde{\mathbf{B}}_c A_c \tag{4.28}$$

where $\widetilde{\mathbf{B}}_C \equiv \widetilde{\mathbf{B}}_{\text{xfem}}$ and A_c is the area of the subcell. The smoothed enriched element stiffness matrix $\widetilde{\mathbf{K}}^e_{\text{xfem}}$ is the sum of the $\widetilde{\mathbf{K}}^C_{\text{xfem}}$, for all subcells, C:

$$\widetilde{\mathbf{K}}^e = \sum_{C=1}^{nc} \widetilde{\mathbf{B}}^T_C \mathbf{D} \widetilde{\mathbf{B}}_C \int_{\Omega_C} \mathrm{d}\Omega = \sum_{C=1}^{nc} \widetilde{\mathbf{B}}^T_C \mathbf{D} \widetilde{\mathbf{B}}_C A_c \tag{4.29}$$

where $C \in \{1, 2, \cdots nc\}$ is the number of subcell Ω_C, $A_c = \int_{\Omega_C} \mathrm{d}\Omega$ is the area of the subcell Ω_C. Note that in Equation (4.29), the stiffness matrix is rewritten as the sum of the contributions from the individual subcell, because all the entries in matrix $\check{\mathbf{B}}_c$ are constants over each subcell Ω_C – each of these entries are line integrals calculated along the boundaries of the subcells.

> ☞ It is clear from Equation (4.27) and Equation (4.29) that the derivatives of the shape functions are not needed to compute the stiffness matrix. Moreover, when $\sqrt{r}$ enrichment is used, the singular terms normally present in the stiffness matrix integrand disappear.

4.3 Blending Elements and Corrections

Enrichments are used to model certain characteristic features of the solution which cannot be reproduced by the FEM-shape functions. These features are usually local features, i.e., they appear only in some subregion of the computational domain. Therefore, in order to represent these features, only nodes in that region or its vicinity have to be enriched.

In Sections 2.2 and 2.3, we looked at two enrichment schemes capable of reproducing weak (material interfaces) and strong (cracks) discontinuities within an element for which both nodes are enriched. We saw that several enrichment functions could be used to this effect and that the key feature which qualifies or disqualifies such an enrichment is its order of continuity along the interface. Thus far, however, we have only considered one-element problems where all nodes were enriched. Yet, in the general case, enrichment is kept local, and only a few elements in the mesh are affected, i.e., only a few nodes are enriched. Typically, for strong and weak discontinuities, only the nodes whose support are cut by the discontinuity are enriched. Consequently, the problem of the transition between enriched and non-enriched elements arises. These elements are such that at least one of their nodes is not enriched. Due to the partition of unity property of the shape functions, the enrichments can be reproduced exactly in an element if all nodes are enriched. But since generally not all nodes are enriched, there are some elements in which the enrichment cannot be reproduced. These elements are called "*blending elements.*"

4.3.1 Blending Between Different Partitions of Unity

The regions of high strain gradients or discontinuities are local phenomena. To capture such local phenomena, enrichment functions are added to the finite element approximation basis locally. Around the enriched region, Ω^{PU}, which is the support of the partition of unity, there is a partially enriched area where the elements do not have all their nodes enriched (see Figure 4.7). Consider, for example, a four-noded bilinear quadrilateral element. The approximation in a four-noded element which has less than four enriched nodes, three for instance, is given by:

$$u_i(\mathbf{x}) = \sum_{I=1}^{4} N_I(\mathbf{x}) U_{Ii} + \sum_{J=1}^{3} N_J(\mathbf{x}) \Psi(\mathbf{x}) A_{Ji} \tag{4.30}$$

Figure 4.7 Partially enriched elements around the enriched area.

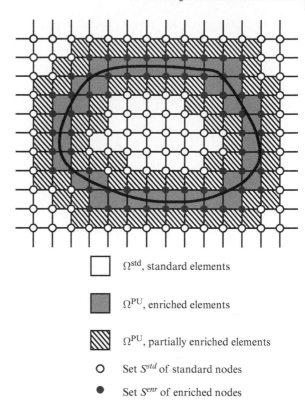

$\square$ Ω^{std}, standard elements

■ Ω^{PU}, enriched elements

▨ Ω^{PU}, partially enriched elements

○   Set S^{std} of standard nodes

●   Set S^{enr} of enriched nodes

In this element, the function $\Psi(\mathbf{x})$ cannot be reproduced by taking $U_{Ii} = 0$ and $A_{Ji} = 1$, because (N_1, N_2, N_3) is not anymore a partition of unity, in other terms:

$$\sum_{J=1}^{3} N_J(\mathbf{x}) \neq 1$$

The fact that the additional function cannot be reproduced in these elements is not important, since these elements do not contain the discontinuity. The main point is that it may introduce spurious terms in the approximation which induces error in the numerical solution. These spurious terms can be automatically corrected by the standard part of the approximation if the order of the standard part is more or equal to the order of the partition of unity times the enrichment (Legay et al., 2005). A few possible combinations are given in Table 4.1.

Before we look at various approaches that are proposed to address the blending problems, let us look at the error introduced in the numerical solution due to the partially enriched elements.

4.3.2 Interpolation Error in Blending Elements

In this section, we evaluate an error estimate for the blending element. We carry out the derivation in a general way assuming that a linear partition of unity is used to introduce the enrichment function. We then particularize to the case of a weakly discontinuous enrichment (i.e., discontinuous gradient).

Consider a domain $[0, L]$ discretized by equi-spaced two-noded linear elements of length h, such that the nodes of the element number I have coordinates, $x_I = Ih$ and $x_{(I+1)} = (I + 1)h$. Let V be the enrichment function and assume that element I is a blending element such that its left node (I) is enriched, but not its right node $(I + 1)$ (c.f.

Table 4.1 Standard combination of enriched and standard shape functions, and resulting blending difficulties.

Standard shape functions $N_I(\mathbf{x})$	Partition of unity $f_i(\mathbf{x})$	Enrichment $\psi(\mathbf{x})$	order of $f_i(\mathbf{x})\psi(\mathbf{x})$	Spurious terms
4-node element	4-node element	Heaviside		
order 1	order 1	order 0 (constant)	1	No
4-node element	4-node element	Ramp		
order 1	order 1	order 1	2	Yes
9-node element	4-node element	Ramp		
order 2	order 1	order 1	2	No

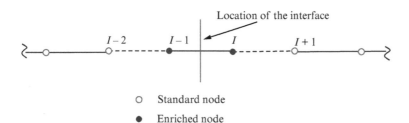

Figure 4.8 Finite element discretization using two-noded linear elements.

Figure 4.8). In this partially enriched element of length $h = x_{I+1} - x_I$, the approximation is composed of a standard part (linear) and an enriched part, but the enrichment function only multiplies the shape function N_I; hence, the approximation in the blending element is incomplete and unable to reproduce the enrichment function:

$$u_h^{e=I}(x) = u_h^{\text{linear}} + u_h^{\text{blending}} = N_I(x)u_I + N_{I+1}u_{I+1} + N_I(x)V(x)a_I \qquad (4.31)$$

Consider the blending element defined by nodes I and $I+1$, call it element 3. Let $\xi = \dfrac{x - x_3}{h}$ be the local coordinate in element 3. The shape functions write $N_I(\xi) = 1 - \xi$ and $N_{I+1}(\xi) = \xi$. The enrichment function associated with node 3 can be expressed as:

$$V(x) - V(x_3) = |x - x_b| - |x_3 - x_b| = x - x_b - (x_3 - x_b) = x - x_3 \qquad (4.32)$$

or, in terms of the local coordinates $V(x(\xi)) - V(x_3) = \xi h$ and Equation (4.31) can be recast in the following form where the spurious term, which cannot be "canceled" by the standard part of the approximation, is clearly visible:

$$u_h^{e=I}(x) = \underbrace{\left(1 - \frac{x - Ih}{h}\right)u_I + \frac{x - Ih}{h}u_{I+1}}_{\text{standard (linear) approximation}} + \underbrace{\left(1 - \frac{x - Ih}{h}\right)V(x)a_I}_{\text{spurious quadratic term}}. \qquad (4.33)$$

We will now examine the effect of this spurious term on the approximation power in element I. For any point $x \in [x_I, x_{I+1}]$, denote by $e(x) = u(x) - u_h(x)$ the interpolation error between the exact solution u and XFEM interpolant (which we shall name u_h) at any point x in element I (we have dropped the superscript $^{e=I}$ for legibility). Because the error vanishes at both nodes since u_h is assumed interpolatory, there exists a point $\bar{x}$ in the element such that the error is extremum, and hence $\frac{de}{dx}(\bar{x})$ which we note $e_{,x}(\bar{x})$ vanishes. Writing a Taylor series expansion about point $\bar{x}$ yields:

$$\forall x \in [x_I, x_{I+1}], e(x) = e(\bar{x} + x - \bar{x})$$

$$= e(\bar{x}) + (x - \bar{x})e_{,x}(\bar{x}) + \frac{(x - \bar{x})^2}{2!}e_{,xx}(\bar{x}) + o((x - \bar{x})^3)$$

$$= e(\bar{x}) + \frac{(x - \bar{x})^2}{2!}e_{,xx}(\bar{x}) + o((x - \bar{x})^3). \tag{4.34}$$

Using the Lagrange form of the second-order remainder of the Taylor series expansion, there exists a point $\tilde{x} \in [x, \bar{x}]$ such that the remainder of the Taylor series writes:

$$\frac{(x - \bar{x})^2}{2!}e_{,xx}(\bar{x}) + \mathcal{O}((x - \bar{x})^3) = \frac{(x - \bar{x})^2}{2!}e_{,xx}(\tilde{x}). \tag{4.35}$$

Since Equation (4.34) is valid for any point x in element I, let us choose for x the position of the node closest to $\bar{x}$. Without loss of generality, let $x = x_I{}^2$ so that, combining Equation (4.34) with Equation (4.35), we obtain:

$$\exists \tilde{x} \in [x_I, \bar{x}], e(x_I) = e(\bar{x}) + \frac{(x_I - \bar{x})^2}{2!}e_{,xx}(\tilde{x}). \tag{4.36}$$

But since we assumed that the approximation u_h is interpolatory, the error vanishes at the nodes, hence $e(x_I) = u(x_I) - u_h(x_I) = 0$ which provides an expression for the error at point $\bar{x}$, i.e., the maximum error in element I:

$$\exists \tilde{x} \in [x_I, \bar{x}], e(\bar{x}) = -\frac{(x_I - \bar{x})^2}{2!}e_{,xx}(\tilde{x}). \tag{4.37}$$

The final step is to find an upper bound for $e(\bar{x})$ as a function of h:

$$\left|e(\bar{x})\right| \leq \left|\frac{(x_I - \bar{x})^2}{2!}e_{,xx}(\tilde{x})\right| \leq \frac{(x_I - \bar{x})^2}{2!}\left|\max_{[x_I,\bar{x}]} e_{,xx}(\tilde{x})\right|. \tag{4.38}$$

Remarking that $|x_I - \bar{x}| \leq \frac{h}{2}$, we obtain the following upper bound:

$$\left|e(\bar{x})\right| \leq \left|\frac{(\frac{h}{2})^2}{2!}\max_{[x_I,\bar{x}]} e_{,xx}(\tilde{x})\right| = \frac{h^2}{8}\left|\max_{[x_I,\bar{x}]} e_{,xx}(\tilde{x})\right|. \tag{4.39}$$

Equation (4.39) provides an upper bound for the maximum error $e(\bar{x})$. But the exact error $e(x)$ being unknown, the right-hand side is not known explicitly. The next step is thus to obtain an upper bound for the right-hand side of Equation (4.39) in terms of the second derivative of the exact solution, the mesh size h, the nodal coefficients, and the enrichment function. Differentiating twice Equation (4.31), and noting that the standard part of the approximation is linear, hence its second derivative vanishes, we get:

---

2 Note that this implies that $x - x_I \leq \frac{h}{2}$ which will be used later.

$$e_{,xx}(x) = u_{,xx}(x) - u^h_{,xx}(x) = u_{,xx}(x) - u^{\text{blend}}_{h,xx}(x)$$

$$= u_{,xx}(x) - a_I(N_I(x)V(x))_{,xx}$$

$$= u_{,xx}(x) - a_I \left(N_{I,xx}(x)V(x) + 2N_{I,x}(x)V_{,x}(x) + N_I(x)V_{,xx}(x) \right). \tag{4.40}$$

Recalling that N_I is linear in x and that $N_{I,x} = -\frac{1}{h}$:

$$e_{,xx}(x) = u_{,xx}(x) - u^h_{,xx}(x) = u_{,xx}(x) - u^{\text{blend}}_{h,xx}(x)$$

$$= u_{,xx}(x) - a_I(N_I(x)V(x))_{,xx}$$

$$= u_{,xx}(x) - a_I \left(\frac{-2}{h}V_{,x}(x) + N_I(x)V_{,xx}(x) \right),$$

$$= u_{,xx}(x) - a_I \left(\frac{-2}{h}V_{,x}(x) + \left(1 - \frac{x - Ih}{h} \right) V_{,xx}(x) \right). \tag{4.41}$$

This final equality provides a general expression for the error in the blending element assuming linear shape functions and will be revisited later to estimate the approximation error in blending elements for general enrichment functions.

Particularizing to the case of a linear enrichment function $V(x) = |x - x_b| = x - x_b$ (since in the present case, the interface position is such that $x_b < x$), we obtain[3] Equation (4.39) and Equation (4.41) provide a final upper bound for the maximum error in the blending element:

$$= u_{,xx}(x) + 2a_I \frac{1}{h}. \tag{4.43}$$

Recalling the estimate Equation (4.39) providing an upper bound for the exact error $e(\tilde{x}) \leq \frac{h^2}{8} \left| \max_{[x_I,\tilde{x}]} e_{,xx}(\tilde{x}) \right|$:

$$\left| e(\tilde{x}) \right| \leq \frac{h^2}{8} \max_{[x_I,\tilde{x}]} \left| u_{,xx}(x) + 2a_I \frac{1}{h} \right| \leq \frac{h^2}{8} \left(\max_{[x_I,\tilde{x}]} \left| u_{,xx}(x) \right| + \left| 2a_I \frac{1}{h} \right| \right). \tag{4.44}$$

Rewriting the last expression in Equation (4.44), we can separate the contribution of the spurious bending term from the standard finite element approximation:

$$\left| e(\tilde{x}) \right| \leq \underbrace{\frac{h^2}{8} \max_{[x_I,\tilde{x}]} \left| u_{,xx}(x) \right|}_{\text{standard FE}} + \underbrace{\frac{1}{4} \left| a_I \right| h}_{\text{blending}}. \tag{4.45}$$

Note that the first term is the usual finite element error estimate providing a convergence rate of 2 (h^2 term) in the displacement norm. $\max_{[x_I,\tilde{x}]} \left| u_{,xx}(x) \right|$ is a constant independent of the mesh size measuring the "steepness" of the second derivative of the solution to be approximated. The second term however introduces an error term

---

3 A material interface is present at location $x = x_b \in [x_{I-1}, x_I]$, i.e., in the element immediately to the left of element I as shown in Figure 4.8 (see also Section 2.2.1 for numerical illustration.)

Recall V, the ramp function permitting to introduce a jump in the gradient of the unknown function, first introduced by Krongauz and Belytschko (1998) (see Equation (2.25)) defined on the whole length of the bar $[0, L]$:

$$\forall x \in [0, L], V(x) \equiv \psi_b(x) \underbrace{\equiv}_{\text{def}} |x - x_b|. \tag{4.42}$$

Assume that no shifting is used (see Section 2.2.2 for details).

of order h which, in the blending elements, will limit the convergence rate to 1 in the displacement norm. While there are only a few blending elements in the domain, this suffices to reduce the overall convergence rate of the solution, as was indicated by the numerical examples in Section 3.4.

☞ The methodology described above to obtain the estimate Equation (4.45) can be generalized to the case of an arbitrary enrichment function. The following two sections focus, in turn, on the case of an arbitrary polynomial enrichment function and an arbitrary (smooth) enrichment function.

☞ A similar result can be shown in the case of an arbitrary (but smooth) enrichment function by performing a Taylor series expansion of this enrichment at an arbitrary point within the blending element and using the results obtained for a polynomial enrichment. For more details, the reader is referred to Chessa and Belytschko (2003) and Strang and Fix (1973).

4.3.3 Addressing Blending Phenomena

As discussed in the previous section, blending phenomena introduce error in the numerical solution. This pathological behavior has been studied in great detail. Here, we present some of the approaches employed to address the blending issue.

4.3.3.1 Enhanced Strain Formulation

A technique to deal with the spurious terms in the blending elements was proposed by Chessa et al. (2003). Inside a blending element the displacements caused by possible recombinations of the enrichments form a function space. We can define a basis for those displacements with support restricted to the domain of the blending element. If we were allowed to use the basis for those spurious terms in addition to the shape functions and the enrichments, we could use them to cancel out the spurious terms inside the blending elements.

But we are generally not allowed to use them. They may be discontinuous along the element edges, thus we would introduce unwanted displacement jumps. But there is a condition which, if fulfilled, allows us to use additional strains in our element formulation. A function ε_{enh} with support restricted to the area of the blending element Ω_{bl} can be used additionally to solve the structural problem if:

$$\int_{\Omega_{bl}} \varepsilon_{enh} d\Omega = 0 \tag{4.46}$$

This is called the orthogonality condition. If fulfilled, the strains ε_{enh} are orthogonal to any constant stress field inside the element. The idea is to choose a basis for ε_{enh} such that the strains arising from the spurious terms in the blending elements can be canceled out. We would like to demonstrate this idea using an example. Consider the enrichment constructed using the 1D abs-function as shown on the top in Figure 4.9. The left element would be a blending element. We can see the quadratic term introduced by the enrichments. Now consider the function shown in the middle labeled u_{sp}. This is the restriction of the enrichment to the blending element. The corresponding strains are shown at the bottom of Figure 4.9. Obviously, the orthogonality condition is fulfilled. Thus, the strains $\varepsilon(u_{sp})$ can be used as enhanced strains ε_{enh} in the element definition which would add a degree of freedom. Furthermore, the strain arising from the enrichment functions in the blending element could be canceled out. However, in more complex situations in two or three dimensions, it might not be obvious how to construct function spaces that fulfill the orthogonality condition and allow the cancellation of the spurious terms.

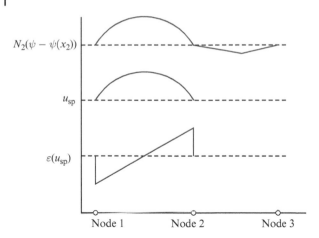

Figure 4.9 Avoiding problem in blending elements using enhanced strains.

4.3.3.2 Discontinuous Galerkin

Another possibility to avoid spurious terms due to the blending elements was proposed by Gracie et al. (2008b). The method was named DG-XFEM, with DG being short for discontinuous Galerkin. In this method, the contribution of the blending elements to the stiffness matrix is neglected. Let us call the union of all fully enriched elements of a particular enrichment function as its enriched domain. Then, during integration, the enrichment function outside its enriched domain is set to zero. As a consequence, the displacement fields may be discontinuous at the boundaries of the enriched domains. The continuity of the displacement field and the traction along the boundaries of the enriched domains is enforced in a weak sense using a penalty method. The term "weak sense" can be interpreted as follows: The discontinuity multiplied by the shape functions integrated along the element boundary is zero.

4.3.3.3 Corrected XFEM

Fries (2008) suggested that by multiplying the enrichment function with a ramp function, the enrichment vanishes in the elements where only some of the nodes are enriched. The corrected or weighted XFEM approximation is given by:

$$\mathbf{u}^h(\mathbf{x}) = \underbrace{\sum_{I \in \mathcal{N}^{\text{fem}}} N_I(\mathbf{x})\mathbf{q}_I}_{\text{Std. FE approx}} + \underbrace{\sum_{J \in \mathcal{N}^{\text{fem}}} N_J(\mathbf{x}) \cdot \underbrace{R(\mathbf{x})}_{\text{Ramp Function}} \cdot [\psi(\mathbf{x}) - \psi(\mathbf{x}_J)]\mathbf{a}a_J}_{\text{Enrichment}} \tag{4.47}$$

where $R(\mathbf{x})$ is the ramp function that equals 1 over enriched elements and linearly decreases to 0 in the blending elements.

☞ Consider the 1D bi-material bar problem (c.f. Section 2.2.3). To study the influence of partially enriched elements, let the domain be discretized with 3 two-noded 1D finite elements, as shown in Figure 4.10, with the material interface located in the second element. The elements on either side of the element containing the interface is partially enriched. The displacement approximation for the partially enriched element to the left of the material interface is given by:

Figure 4.10 Bi-material bar discretized with 3 two-noded 1D finite elements.

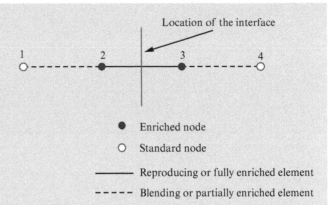

$$u(x) = N_1(x)u_1 + N_2(x)u_2 + N_2(x)\psi(x)a_2$$

And the strain approximation within the element is given by:

$$\varepsilon_x = \frac{du(x)}{dx} = \frac{dN_1(x)}{dx}u_1 + \frac{dN_2(x)}{dx}u_2 + \frac{d}{dx}\Big[N_2(x)\psi(x)a_2\Big]$$

The term $\frac{d}{dx}\Big[N_2(x)\psi(x)a_2\Big]$ in the above equation causes the zig-zag behavior as shown in Figure 4.11. With a small modification as proposed by Fries (2008), it can be seen from Figure 4.11 that the zig-zag behavior in the blending elements is removed. And with the increase in the number of elements, the performance is improved.

☞ Corrected or weighted XFEM shares similarities with the work of Benzley (1974), who proposed to use a function that equals 1 on boundaries adjacent to "enriched" elements and equals 0 on boundaries adjacent to "standard" elements, i.e., the value of the function decreases from 1 to 0 in the blending.

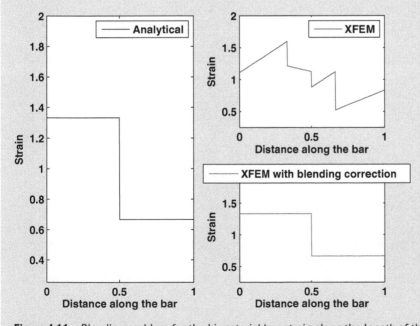

Figure 4.11 Blending problem for the bi-material bar: strain along the length of the bar.

4.3.3.4 Fast Integration

Ventura et al. (2009) suggested a nonlinear form for the ramp function $R(\mathbf{x})$. Given a point $\mathbf{x} \in \Omega$, let $d = d(\mathbf{x})$ be the signed distance function from $\mathbf{x}$ to Γ. The function $R(d)$ is defined by:

$$R(d) = \begin{cases} 1 & 0 \leq |d| \leq d_i \\ (1-g)^n & d_i < |d| < d_e \\ 0 & |d| > d_e \end{cases} \tag{4.48}$$

where d_i and d_e are chosen such that $R = 1$ for $|d| \leq d_i$ and $R = 0$ for any $|d| > d_e$, n is a positive integer and g is the linear ramp function given by:

$$g = \frac{|d| - d_i}{d_e - d_i} \tag{4.49}$$

> ☞ Ventura et al. (2009) used a nonlinear form for the ramp function. This method shares some similarities with the work of Strang and Fix (1973) who used polynomials to merge the coefficients of singular functions smoothly into zero on the boundary of standard elements.

4.4 Preconditioning Techniques

If a system of equations become large, a direct solution is not efficient any more. Iterative solvers can be used instead. Especially for sparse matrices the solution time can be reduced significantly, since matrix-vector products (which are the basic operations used by iterative solvers) are cheap from a computational point of view. Their performance is highly dependent on the condition number of the stiffness matrix.

For the FEM several criteria to evaluate the quality of a mesh are known. Following these criteria, one can not only hope for good approximation properties, but also for well-conditioned stiffness matrices. For example, it is known that the condition number grows to infinity if the element size tends to zero. Therefore, one would try to omit extremely small elements in the finite element mesh, at least if they are not necessary to resolve the exact solution appropriately.

For the XFEM such criteria do not exist. In fact, one can show that the enrichment schemes explained so far can result in arbitrarily ill-conditioned matrices (Menk and Bordas, 2011a). Generally, this happens if the enrichments become almost linearly dependent.

We attempt to solve:

$$\underbrace{\mathbf{P}^T \mathbf{K} \mathbf{P}}_{\tilde{\mathbf{K}}} \underbrace{\mathbf{P}^{-1} \mathbf{a}}_{\tilde{\mathbf{a}}} = \underbrace{\mathbf{P}^T \mathbf{f}}_{\tilde{\mathbf{f}}} \tag{4.50}$$

Once this equation system is solved to obtain $\tilde{\mathbf{a}}$, one can calculate the real displacements by:

$$\mathbf{a} = \mathbf{P}\tilde{\mathbf{a}} \tag{4.51}$$

The matrix $\mathbf{P}$ is a preconditioner. It should be chosen such that the condition number of the transformed stiffness matrix $\tilde{\mathbf{K}}$ is smaller than the $\mathbf{K}$. Furthermore, the effort necessary for its computation should be small. The advantage is that an iterative solver would need a smaller number of iterations to get an acceptable result. This can justify the additional effort, especially if several solutions for different vectors $\mathbf{f}$ have to be calculated.

The explicit form of $\tilde{\mathbf{K}}$ is not calculated. When iterative solution algorithms are used the explicit form is not needed, in every iteration step it suffices to calculate a matrix-vector product with each of the matrices which $\tilde{\mathbf{K}}$ is composed of.

4.4.1 The First Preconditioner Proposed for the XFEM

Béchet et al. (2005) proposed a preconditioner specially tailored to the XFEM, which stabilizes the enrichments by applying Cholesky-decompositions to certain submatrices of the stiffness matrix. These submatrices are formed by the degrees of freedom associated with each enriched node. This can be understood as a local stabilization, because the problem of almost linearly dependent enrichment functions is eliminated for each node.

Assume that $\mathbf{K}^i$ is the submatrix of degrees of freedom associated with the ith node. We define:

$$\mathbf{D} := \sqrt{\text{diag}(\mathbf{K}^i)} \tag{4.52}$$

[4]Let the Cholesky-decomposition of $\mathbf{K}^i$ be given by:

$$\mathbf{K}^i = \mathbf{C}\mathbf{C}^T \tag{4.53}$$

Then a preconditioner for $\mathbf{K}^i$ is given by:

$$\mathbf{P}^i := \mathbf{C}^{-1}\mathbf{D} \tag{4.54}$$

The whole preconditioner is formed by assembling these matrices for all enriched nodes.

4.4.2 A domain Decomposition Preconditioner for the XFEM

The problems due to ill-conditioning introduced by almost linearly dependent enrichment could be resolved if a matrix decomposition, such as the Cholesky-decomposition, is applied to the submatrix formed by the enriched degrees of freedom. The inverse of the Cholesky-factor could then be used to form a preconditioner. The disadvantage of such matrix decomposition algorithms is that the computation time as well as the memory consumption depends cubically on the matrix size. Although in most applications the number of enriched degrees of freedom is much smaller than the number of standard degrees of freedom, there are a lot of situations in which the number of enriched degrees of freedom is still very large (e.g., Menk and Bordas, 2011a). If the structure would be decomposed into several smaller disconnected domains, the submatrix associated with the enriched degrees of freedom would be a block diagonal matrix. In that case, a Cholesky-decomposition has to be applied only to each one of these smaller blocks. This would decrease the numerical effort. But how to apply this idea to a structure that is actually connected? In that case, the structure has to be split into several subdomains and continuity conditions must be added. It often suffices to apply this idea only to the enrichment functions. In the following, a possible implementation is described more precisely. The degrees of freedom of the XFEM stiffness matrix can be ordered such that:

$$\mathbf{K} = \begin{bmatrix} \mathbf{K}_{FEM,FEM} & \mathbf{K}_{X,FEM} \\ \mathbf{K}_{FEM,X} & \mathbf{K}_{X,X} \end{bmatrix} \tag{4.55}$$

$\mathbf{K}_{FEM,FEM}$ is the submatrix formed by the standard degrees of freedom and $\mathbf{K}_{X,X}$ the one formed by the enriched degrees of freedom. Let us assume that the domain Ω is decomposed into several non-overlapping subdomains

---

4 Do not confused **D** with the constitutive tensor, for which this notation is usually used in the book.

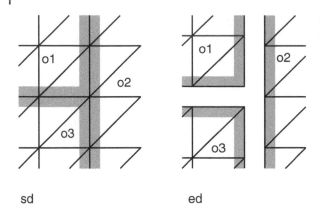

sd   ed

Figure 4.12 Example of a domain decomposition, the nodal enrichments are split in to the domain decomposition, the FEM shape functions remain unchanged.

Ω_i. Each subdomain is formed by a union of elements. The matrix $\mathbf{K}_{X,X}$ is evaluated as if these domains were disconnected. The matrix $\mathbf{K}_{FEM,FEM}$ is evaluated as usual, that is, treating the domain Ω as one connected domain. An example is given in Figure 4.12. A domain is decomposed into three domains. Each of the subdomains is formed by a union of elements. The number of standard degrees of freedom does not change, since the decomposition has no effect on the corresponding part of the stiffness matrix. The number of enriched degrees of freedom increases due to the nodes at the boundary of the subdomains. Since the domains are disconnected, the submatrix of enriched degrees of freedom can be written as a block-diagonal matrix:

$$\mathbf{K}_{X,X} = \begin{bmatrix} \mathbf{K}_{X,X}^{\Omega_1} & 0 & \\ 0 & \mathbf{K}_{X,X}^{\Omega_2} & \\ & & \ddots \end{bmatrix} \tag{4.56}$$

$\mathbf{K}_{X,X}^{\Omega_i}$ are the enriched degrees of freedom of all the enrichments that are non-zero inside the domain Ω_i. However, without further restrictions on the function space this new system would give an erroneous solution, since the enrichment functions might be discontinuous at the boundaries of the subdomains. We therefore use an additional matrix $\mathbf{Y}$ to ensure continuity of the displacements (at least in the absence of cracks):

$$\mathbf{K} = \begin{bmatrix} \mathbf{K}_{FEM,FEM} & \mathbf{K}_{X,FEM} & 0 \\ \mathbf{K}_{FEM,X} & \mathbf{K}_{X,X} & \mathbf{Y}^T \\ 0 & \mathbf{Y} & 0 \end{bmatrix} \tag{4.57}$$

To demonstrate how $\mathbf{Y}$ is constructed we again consider Figure 4.12. Assume that the ith node is at the boundary of two subdomains (e.g., it is connected to Ω_1 and Ω_2 but not Ω_3). Also assume that there is one enriched degree of freedom associated with this node in the standard XFEM formulation. Then due to the domain decomposition there would be two enriched degrees of freedom associated with this node in the stiffness matrix in Equation (4.57), one associated with Ω_1, the other one with Ω_2. If the vector $\mathbf{a}^{X_i}$ contains those degrees of freedom, then a condition for the continuity of the nodal enrichment can be written as:

$$\begin{bmatrix} 1 & -1 \end{bmatrix} \mathbf{a}^{X_i} = 0 \tag{4.58}$$

If the node is the one in Figure 4.12 which is connected to all three domains, then the condition becomes:

$$\begin{bmatrix} 1 & -1 & 0 \\ 0 & 1 & -1 \end{bmatrix} \mathbf{a}^{X_i} = 0 \tag{4.59}$$

The whole matrix $\mathbf{Y}$ is then constructed by ensuring such conditions for all enriched nodes that are connected to two or more subdomains. More precisely, for every nodal enrichment whose support is contained in n distinct subdomains $n - 1$ rows must be added to the matrix $\mathbf{Y}$. They are constructed using the $(n - 1) \times n$ matrix:

$$
\begin{bmatrix}
1 & -1 & 0 & & \cdots & 0 \\
0 & 1 & -1 & 0 & \cdots & 0 \\
& & & \vdots & & \\
0 & \cdots & 0 & 1 & -1 & 0 \\
0 & \cdots & & 0 & 1 & -1
\end{bmatrix}
\tag{4.60}
$$

To construct the new rows of $\mathbf{Y}$ the columns of Equation (4.60) are placed in an empty matrix. Their position is determined by the degrees of freedom that are assigned to the nodal enrichment after performing the domain decomposition. Each block $\mathbf{K}_{X,X}^{\Omega_i}$ in Equation (4.56) can be decomposed by a Cholesky-decomposition with Cholesky factor $\mathbf{C}_i$:

$$
\mathbf{K}_{X,X}^{\Omega_i} = \mathbf{C}_i^T \mathbf{C}_i
\tag{4.61}
$$

We define the preconditioner for $\mathbf{K}_{X,X}$ as:

$$
\mathbf{P}_X = \begin{bmatrix}
\mathbf{C}_1^{-1} & 0 & \\
0 & \mathbf{C}_2^{-1} & \\
& & \ddots
\end{bmatrix}
\tag{4.62}
$$

This preconditioner transforms $\mathbf{K}_{X,X}$ to the unit matrix and therefore would remove the small eigenvalues:

$$
\mathbf{P}_X^T \mathbf{K}_{X,X} \mathbf{P}_X = \mathbf{I}
\tag{4.63}
$$

Please note that the inverse of the Cholesky-factors should not be calculated. It suffices to provide a routine for the iterative solver that calculates the matrix-vector product with $\mathbf{P}_X^T$ and $\mathbf{P}_X$. Therefore, in each iteration step several small equation systems with the Cholesky-factors can be solved instead by performing forward and backward substitution.

A preconditioner for the whole system is then given by:

$$
\mathbf{P} = \begin{bmatrix}
\mathbf{P}_{FEM} & 0 & \\
0 & \mathbf{P}_X & 0 \\
& 0 & \mathbf{I}
\end{bmatrix}
\tag{4.64}
$$

$\mathbf{P}_{FEM}$ can be any preconditioner for the standard degrees of freedom. The final equation system may however still be ill-conditioned because $\tilde{\mathbf{K}}$ contains the matrix $\mathbf{Y}\mathbf{P}_X$ whose rows can be almost linearly dependent. Applying an LQ-decomposition to $\mathbf{Y}\mathbf{P}_X$ we see that:

$$
\mathbf{Q} = \mathbf{L}^{-1} \mathbf{Y} \mathbf{P}_X
\tag{4.65}
$$

$\mathbf{L}$ is a lower triangular matrix and $\mathbf{Q}$ is a matrix with orthogonal rows. The final preconditioner then takes the form:

$$
\mathbf{P} = \begin{bmatrix}
\mathbf{P}_{FEM} & 0 & \\
0 & \mathbf{P}_X & 0 \\
& 0 & \mathbf{L}^{-1}
\end{bmatrix}
\tag{4.66}
$$

Thus, the transformed equation system becomes:

$$
\tilde{\mathbf{K}} = \begin{bmatrix}
\tilde{\mathbf{K}}_{FEM,FEM} & \tilde{\mathbf{K}}_{X,FEM} & 0 \\
\tilde{\mathbf{K}}_{FEM,X} & \mathbf{I} & \mathbf{Q}^T \\
0 & \mathbf{Q} & 0
\end{bmatrix}
\tag{4.67}
$$

A comment must be made about the Cholesky-decomposition in Equation (4.61). When using domain decomposition in combination with the FEM the stiffness matrix of a subdomain may become indefinite if appropriate boundary conditions are missing for this particular subdomain. This is because the rigid body motions of the subdomain are described by eigenvectors of the subdomains stiffness matrix. The Cholesky-decomposition is likely to fail in that case.

However, if the Cholesky-decomposition is applied only to the enriched degrees of freedom, the Cholesky-decomposition can in most cases be performed without problems: Due to the nodal shifting, the enrichments are zero at all nodes. Thus, rigid body motions of a subdomain cannot be described by the corresponding enrichments alone.

But there may be other situations in which the submatrix $\mathbf{K}_{X,X}^{\Omega_i}$ becomes indefinite. A simple way to deal with the problem of indefiniteness would be to perform the Cholesky-decomposition for a stabilized version of this matrix. Assume that the columns of $\mathbf{Z}$ contain an orthogonal basis of the null space. A stabilized version of $\mathbf{K}_{X,X}^{\Omega_i}$ is given by:

$$\mathbf{K}_{X,X}^{\Omega_i} + \lambda \mathbf{Z}\mathbf{Z}^T \quad \text{with} \quad \lambda > 0 \tag{4.68}$$

In an algorithm one would always try to do a Cholesky-decomposition first. If the decomposition fails or if extremely small entries appear on the diagonal, the matrix $\mathbf{Z}$ must be determined. The impact of the preconditioner on the equation systems will be addressed using the structure shown in Figure 4.13. A material interface separates material 1 and material 2. A vertical line through the center of the structure is also shown in the figure. The material interface is not quite vertical, its distance toward the vertical line at the boundary of the structure is characterized by the distance ϵ. The materials are both isotropic materials with material constants shown in Table 4.2. Zero displacements are prescribed at the boundary on the left and a constant traction on the right.

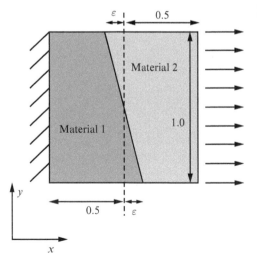

Figure 4.13 Structure with slanting material interface.

Table 4.2 Material properties for the slanting material interface problem

	Material 1	Material 2
E	10	20
ν	0.3	0.3

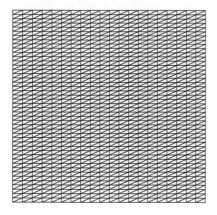

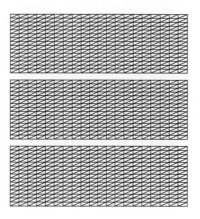

(a) Mesh used to discretize the material interface problem

(b) Domain decomposed into three subdomains

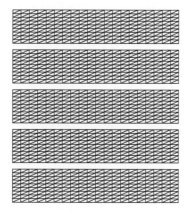

(c) Domain decomposed into five subdomains

Figure 4.14 Mesh used to discretize the material interface problem and different domain decompositions.

A simple mesh is used to discretize this structure consisting of 1800 equally sized triangular elements. The mesh and different domain decompositions are shown in Figure 4.14. The mesh around the center of the structure is shown in Figure 4.15. The blending elements and the fully enriched elements are indicated. To perform preconditioning for the FEM-part of the stiffness matrix, we choose $\mathbf{P}_{FEM}$ to be the diagonal matrix such that the diagonal elements of $\mathbf{\hat{K}}_{FEM,FEM}$ are 1. The condition number of the stiffness matrices has been evaluated for different values of ϵ. The results are shown in Figure 4.16. The condition number of the preconditioned FEM-submatrix is slightly below those obtained for the preconditioned version of the whole equation system. For the preconditioned version of XFEM the choice of domain decomposition as well as the value of ϵ have no effect on the condition number of the equation system. The condition number of the standard version, however, grows for decreasing values of ϵ. If $\epsilon = 10^{-7}$ the difference between the standard and the preconditioned version is three orders of magnitude. The relative error of the equation systems after different iteration steps and $\epsilon = 10^{-7}$ is shown in Figure 4.17. Clearly, the solver converges much faster if the preconditioned version is used. The difference between the standard and the preconditioned version is due to the different condition numbers observed in Figure 4.16.

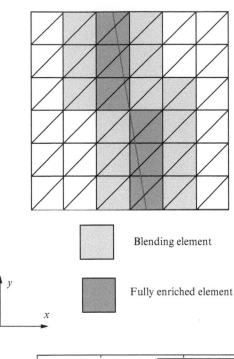

Figure 4.15 Mesh in the center of the structure with the material interface, represented by solid red line.

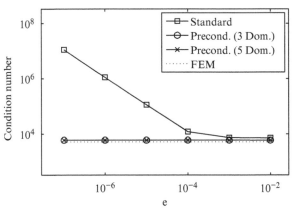

Figure 4.16 Condition number evaluated for different values of ϵ (material interface problem).

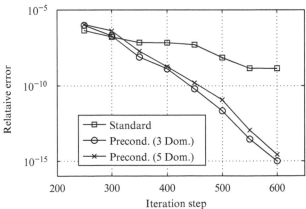

Figure 4.17 Relative error of the MINRES solver after different iteration steps (material interface problem, $\epsilon = 10^{-7}$).

References

Bechet, E., Minnebo, H., Moes, N., and Burgardt, B. (2005). Improved implementation and robustness study of the X-FEM for stress analysis around cracks. *International Journal for Numerical Methods in Engineering* 64 (8): 1033–1056.

Benzley, S.E. (1974). Representation of singularities with isoparametric finite elements. *International Journal for Numerical Methods in Engineering* 8 (3): 537–545.

Chessa, J. and Belytschko, T. (2003). An enriched finite element method and level sets for axisymmetric two-phase flow with surface tension. *International Journal for Numerical Methods in Engineering* 58 (13): 2041–2064.

De, S. and Bathe, K.J. (2001). The method of finite spheres with improved numerical integration. *Computers and Structures* 79 (22):2183–2196.

Fries, T.P. (2008). The intrinsic XFEM for two-fluid flows. *International Journal for Numerical Methods in Fluids* 60 (4): 437–471.

Gracie, R., Wang, H., and Belytschko, T. (2008b). Blending in the extended finite element method by discontinuous Galerkin and assumed strain methods. *International Journal for Numerical Methods in Engineering* 74 (11): 1645–1669.

Krongauz, Y. and Belytschko, T. (1998). EFG approximation with discontinuous derivatives. *International Journal for Numerical Methods in Engineering* 41 (7): 1215–1233.

Laborde, P., Pommier, J., Renard, Y., and Salaün, M. (2005 September). High-order extended finite element method for cracked domains. *International Journal for Numerical Methods in Engineering* 64 (3): 354–381. doi: 10.1002/nme.1370.

Legay, A., Wang, H.W., Belytschko, T. (2005). Strong and weak arbitrary discontinuities in spectral finite elements. *International Journal for Numerical Methods in Engineering* 64 (8): 991–1008.

Liu, G.R., Nguyen, T.T., Dai, K.Y., and Lam, K.Y. (2007). Theoretical aspects of the smoothed finite element method (SFEM). *International Journal for Numerical Methods in Engineering* 71: 902–930.

Menk, A. and Bordas, S. (2011a). A robust preconditioning technique for the extended finite element method. *International Journal for Numerical Methods in Engineering* 85 (13): 1609–1632.

Natarajan, S., Bordas, S., and Roy Mahapatra, D. (2009). Numerical integration over arbitrary polygonal domains based on Schwarz-Christoffel conformal mapping. *International Journal for Numerical Methods in Engineering* 80 (1): 103–134.

Natarajan, S., Mahapatra, D.R., and Bordas,S. (2010). Integrating strong and weak discontinuities without integration subcells and example applications in an XFEM/GFEM framework. *International Journal for Numerical Methods in Engineering* 83 (3): 269–294.

Nagarajan, A. and Mukherjee, S. (1993). A mapping method for numerical evaluation of two-dimensional integrals with $1/r$ singularity. *Computational Mechanics* 12: 19–26.

Park, K., Paulino, G.H., Pereira, J.P., and Duarte, C.A. (2008). Integration of singular enrichment functions in the generalized/extended finite element method for three-dimensional problems. *International Journal for Numerical Methods in Engineering* 78: 1220–1257.

Peirce, W.H. (1957). Numerical integration over the planar annulus. *Journal of the Society of Industrial and Applied Mathematics* 5 (2): 66–73.

Strang, W.G. and Fix, G.J. (1973). *Analysis of the Finite Element Method*. Prentice-Hall.

Trefethen, L.N. (1979). Numerical computation of the Schwarz-Christoffel transformation. *Technical report, Department of Computer Science*, Stanford University.

Trefethen, L.N. (1980). Numerical conformal mapping. *SIAM Journal on Scientific and Statistical Computing* 1 (1): 82–102.

Ventura, G., Gracie, R., and Belytschko, T. (2009). Fast integration and weight function blending in the extended finite element method. *International Journal for Numerical Methods in Engineering* 77 (1): 1–29.

Yoo, J.W., Moran, B., and Chen, J.S. (2004). Stabilized conforming nodal integration in the natural-element method. *International Journal of Numerical Methods in Engineering* 60: 861–890.

5

Applications

Stéphane P. A. Bordas[1], Alexander Menk[2], and Sundararajan Natarajan[3]

[1] *University of Luxembourg, Luxembourg, UK*
[2] *Robert Bosch GmbH, Germany*
[3] *Indian Institute of Technology Madras, India*

In this chapter, we provide some examples of applications of extended finite element method (XFEM) for problems with weak/strong discontinuities and singularities, viz., linear elastic homogeneous isotropic fracture mechanics; anisotropic fracture mechanics and automatic numerical enrichment; simulation of creep and crack initiation and propagation in solder joints; fatigue crack in two dimensions and plate subjected to thermo-mechanical loading. For the first four examples, consider a linear homogeneous body with an internal discontinuity occupying $\Omega \subset \mathbf{R}^2$ is subjected to mechanical boundary conditions. The boundary Γ of Ω accommodates the following decomposition: $\Gamma = \Gamma_t \bigcup \Gamma_u$, where Dirichlet and Neumann boundary conditions are specified. The governing equation for the linear elasto-static problem in the absence of inertia and body forces is given by:

$$\nabla \cdot \sigma = \mathbf{0} \quad \text{in} \quad \Omega \tag{5.1}$$

supplemented with the following boundary conditions are on the external boundary:

$$\sigma \cdot \mathbf{n} = \hat{\mathbf{t}} \quad \text{on} \quad \Gamma_t \tag{5.2a}$$

$$\mathbf{u} = \hat{\mathbf{u}} \quad \text{on} \quad \Gamma_u \tag{5.2b}$$

$$\sigma \cdot \mathbf{n} = 0 \quad \text{on} \quad \Gamma_c \tag{5.2c}$$

where Γ_c represents the crack surface, $\mathbf{u}$ and σ are the displacement and stress field, respectively, $\mathbf{n}$ is the unit outward normal and variables with $\hat{\ }$ are the prescribed quantities. The Cauchy stress tensor is related to the displacement field by:

$$\sigma = \mathbf{D} \cdot \varepsilon \tag{5.3}$$

where $\varepsilon = \nabla \mathbf{u}$ is the small strain tensor. In this case, the XFEM approximation for the displacement is given by (c.f. Equation (5.14)):

$$\mathbf{u}_h(\mathbf{x}) = \sum_{I \in \mathcal{S}^{fem}} N_I(\mathbf{x}) u_I + \sum_{J \in \mathcal{S}^c} N_J(\mathbf{x}) \Big(H(\mathbf{x}) - H(\mathbf{x}_J) \Big) \mathbf{a}_J + \sum_{K \in \mathcal{S}^f} N_K(\mathbf{x}) \sum_{\alpha=1}^{n} \Xi \alpha(r, \theta) \mathbf{b}_K^\alpha \tag{5.4}$$

5.1   Linear Elastic Fracture in Two Dimensions with XFEM

In Section 3.4, we discussed how the XFEM framework could be used to represent the discontinuities independent of the background mesh and still capture the necessary physics of the problem through the process of enrichment. In this section, we look at a more general case of a plate with an inclined crack and a crack in the presence of inclusions. Unless mentioned otherwise, the computational domain is discretized with structured four-noded quadrilateral elements. As discussed in Section 3.3.1, XFEM framework leads to five different categories of elements, viz., standard elements, split elements, tip elements, split-blending elements, and tip-blending elements. The elements intersected by the discontinuous surface are subdivided into triangles for the purpose of numerical integration. For standard elements, 4 Gauss points are used; for split-blending and tip-blending elements, 8 Gauss points are used; for split elements 6 points per triangle are used; and 13 points per triangle are used for tip elements. For blending elements, corrected XFEM approach is employed.

5.1.1   Inclined Crack in Tension

Consider a plate with an angled crack subjected to a far field uni-axial stress field, σ (see Figure 5.1). In this example, mode I and mode II SIFs, K_I and K_{II}, respectively, are obtained as a function of the crack angle β. For the loads shown, the analytical stress intensity factors (SIFs) are given by (Sih, 1973):

$$K_I = \sigma\sqrt{\pi a}\cos\beta\cos\beta, \quad K_{II} = \sigma\sqrt{\pi a}\cos\beta\sin\beta. \tag{5.5}$$

A structured mesh (100×100) is used for the study and the crack is represented independent of the background finite element mesh. The material properties employed for this study are: Young's modulus, $E = 1000$ N/mm^2 and Poisson's ratio, $\nu = 0.3$. The tensile load magnitude is $\sigma = 1$ N/mm^2. The influence of the crack angle, β, on the SIFs is shown in Figure 5.2 and it is seen that the numerical results from the proposed techniques are comparable with the analytical solution.

5.1.2   Example of a Crack Inclusion Interaction Problem

In this example, we study in this section the interaction between a crack and an inclusion. Such examples occur, for example, in particle reinforced composites. Both the strong discontinuity associated with the crack and the weak

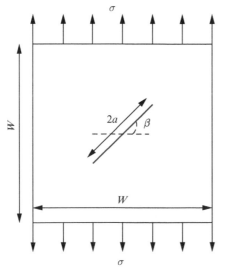

Figure 5.1   Inclined crack in tension.

discontinuity associated with the inclusion are represented independent of the background mesh using level sets and appropriate enrichment functions. The enriched approximation for this problem takes the following form:

$$
\begin{cases}
\mathbf{u}^h(\mathbf{x}) = \sum_{I \in \mathcal{N}^{\text{fem}}} N_I(\mathbf{x})\mathbf{q}_I + \\
\underbrace{\sum_{J \in \mathcal{N}^c} N_J(\mathbf{x})H(\mathbf{x})\mathbf{a}_J + \sum_{K \in \mathcal{N}^f} N_K(\mathbf{x}) \sum_{\alpha=1}^{4} B_\alpha(\mathbf{x})\mathbf{b}_K^\alpha}_{\text{for cracks}} + \underbrace{\sum_{L \in \mathcal{N}^{\text{inc}}} N_L(\mathbf{x})\Psi(\mathbf{x})\mathbf{c}_L}_{\text{for inclusions}}
\end{cases}
\tag{5.6}
$$

5.1.3 Effect of the Distance Between the Crack and the Inclusion

Figure 5.3 shows a two-dimensional plate of length $2L(= W)$ and width W with a horizontal crack of length $2a$ in the neighborhood of a circular inclusion of radius R. The following parameters are chosen for the current study: $W = 100; a/W = 0.01; b/a = (3, 4, 6, 8, 10); R = 2a; E_1 = 1, E_2 = 10^4; \nu_1 = \nu_2 = 0.35; \sigma_o = 1; K_o = \sigma_o\sqrt{\pi a}$.

Figure 5.2 Variation of stress intensity factors K_I and K_{II} with crack angle, β.

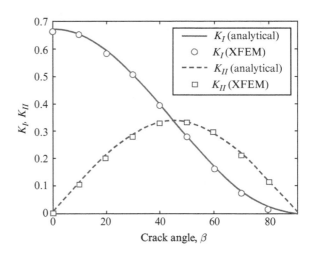

Figure 5.3 A plate with a horizontal crack in the neighborhood of a circular inclusion: geometry and boundary conditions.

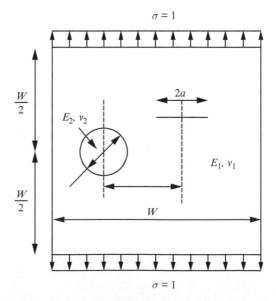

Table 5.1 Normalized SIFs at left tip (A) for a plate with a crack and a circular inclusion under far field tension σ_o.

b/a	K_I			K_{II}		
	Ref. Wang and Chou (2001)	Ref. Yu et al. (2009)	Present	Ref. Wang and Chou (2001)	Ref. Yu et al. (2009)	Present
3	0.5810	0.5908	0.5916	0.0636	0.0652	0.0592
4	0.8199	0.8174	0.8185	−0.0661	−0.0671	−0.0646
6	0.9506	0.9505	0.9517	−0.0368	−0.0373	−0.0359
8	0.9787	0.9785	0.9798	−0.0173	−0.0179	−0.0172
10	0.9878	0.9877	0.9889	−0.0091	−0.0101	−0.0117

Table 5.2 Normalized SIFs at left tip (A) for a plate with a crack and a circular inclusion under far field tension σ_o.

b/a	K_I			K_{II}		
	Ref. Wang and Chou (2001)	Ref. Yu et al. (2009)	Present	Ref. Wang and Chou (2001)	Ref. Yu et al. (2009)	Present
3	0.7995	0.8011	0.7916	−0.0733	−0.0711	−0.0684
4	0.9068	0.9065	0.8957	−0.0560	−0.0568	−0.0555
6	0.9684	0.9687	0.9572	−0.0252	−0.0255	−0.0245
8	0.9842	0.9844	0.9727	−0.0125	−0.0129	−0.0124
10	0.9901	0.9903	0.9924	−0.0069	−0.0076	−0.0069

Tables 5.1 and 5.2 show the variation of the crack tip SIF (normalized SIF)[1] as a function of the distance to the center of the circular inclusion. It can be seen that as the distance between the crack and the inclusion increases, the effect of the inclusion on the crack tip SIF decreases. In other words, the crack tip SIF increases with increasing distance from the inclusion. The results show good agreement with the results available in the literature (Wang and Chou, 2001; Yu et al., 2009).

Next, the crack tip shielding in the presence of an inclusion is studied. When the inclusion is perfectly bonded to the matrix, the crack deflection is a prominent toughening mechanism. The crack deflection depends on the inclusion size, the inclusion eccentricity, the number of inclusions surrounding the crack, the orientation of crack, to name a few. A rigid inclusion in a relatively compliant matrix shields the crack tip as it is approached. Also, shielding depends on the aforementioned factors. The geometry and boundary conditions are shown in Figure 5.4. The radius of the inclusion, $R = W/20$, is considered, where W is the width of the plate. The variation of the energy release rate (ERR), G, with respect to the crack length and the elastic moduli is studied. The ratio of the elastic moduli E_p/E_m is varied from 2 to 8 as in the literature (Bush, 1997; Kitey et al., 2006), where E_p is the Young's modulus of the inclusion and E_m is the Young's modulus of the matrix. The ERR, G, is calculated using the SIFs by the following equation:

$$G = \frac{\kappa + 1}{8\mu}(K_I^2 + K_{II}^2)$$ (5.7)

---

1 Normalized SIF $= K_{\text{numerical}}/K_o$

Figure 5.4 A plate with a horizontal edge crack in the neighborhood of a circular inclusion: geometry and boundary conditions.

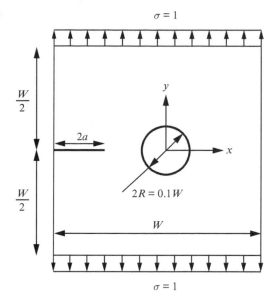

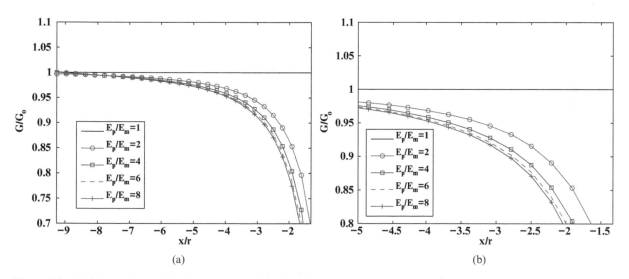

Figure 5.5 Variation of non-dimensionalized ERR for various ratio of elastic moduli, E_p/E_m, where E_p is the Young's modulus of the inclusion and E_m is the Young's modulus of the matrix. It can be seen that the crack senses the inclusion at a distance of ≈ 5 times the radius of the inclusion (Bush, 1997; Kitey et al., 2006).

where the Kolosov constant, κ, is $(3 - 4\nu)$ for plane strain and $(3 - \nu)/(1 + \nu)$ for plane stress. The Poisson's ratio for the inclusion and the matrix are assumed to be $\nu = 0.33$ and 0.17, respectively. The variation of the non-dimensional ERR, G/G_o, for different crack lengths and the elastic moduli is shown in Figure 5.5, where G_o is the ERR for the matrix without the inclusion. With an increasing ratio of the elastic moduli, the non-dimensional ERR decreases. From Figure 5.5, it can be seen that the crack senses the rigid inclusion in front of it at least from a distance of ≈ 5 times the radius of the inclusion (Bush, 1997; Kitey et al., 2006). The observed variation of the ERR agrees with the results reported in the literature.

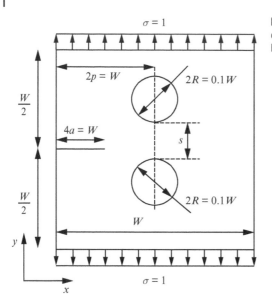

Figure 5.6 A plate with a horizontal edge crack in the neighborhood of symmetrically located two circular inclusions, geometry and boundary conditions.

The effect of a pair of circular inclusions in the crack path on the ERR is studied next. A symmetrically located pair of inclusions with respect to the crack, as shown in Figure 5.6, leads to crack growth under mode I conditions. The length of the crack is $a = W/4$ and the inclusions are located at $p/W = 0.25$ from the crack tip. The separation distance 's' between the inclusions is varied between $W/8$ and W. Figure 5.7 shows the variation of the ERR as a function of the crack length for various position of the inclusions. It can be seen that by increasing the crack length, the ERR decreases indicating the crack shielding and maximum shielding occurs when the crack tip is approximately at a distance of $W/2$ in front of the center of the inclusion. With further increase in the crack length, the crack shielding decreases and approaches unity when the crack length approaches p/W (the center of the inclusions). And with further increase, an amplification of the ERR can be seen. This is because, when the crack length, a, is greater than p/W, the effect of the inclusion below the crack is small, as the crack experiences the majority of the far field tension. The amplification decreases with further increase in the crack length. The observed crack shielding/amplification effects agree well with the results reported in the literature (Bush, 1997; Kitey et al., 2006; Li and Chudnovsky, 2003).

5.2 Numerical Enrichment for Anisotropic Linear Elastic Fracture Mechanics

The enrichment functions used to approximate the asymptotic behavior of the displacement fields in the vicinity of a crack tip in an isotropic material have already been discussed. These are examples of enrichments that can be used to approximate displacement fields that contain weak singularities. Weak singularities may arise at crack tips, junctions, or re-entrant corners. For some cases it is possible to determine suitable enrichment functions analytically. In the case of general strain singularities this is not always possible. The shape of the singularities is dependent on the surrounding materials and the geometry of the structure.

However, it is often possible to determine the shape of a weak singularity numerically using *a priori* knowledge about the materials and the geometry. Consider the structure shown in Figure 5.8. A notch is formed by several infinitely long material wedges. Each wedge is described by an anisotropic linear elastic stress-strain relationship. The geometry of a great class of two-dimensional problems which exhibit weak singularities can be described this

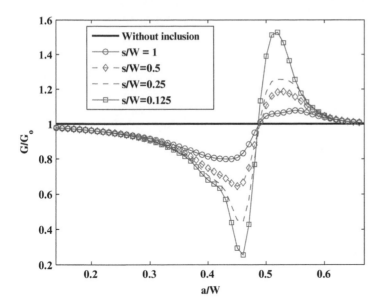

Figure 5.7 Crack tip shielding and amplification factors due to pair of symmetrically situated inclusions in the crack path.

Figure 5.8 Notch formed by different anisotropic material wedges.

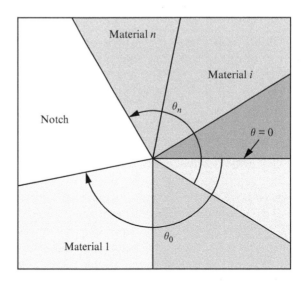

way in the vicinity of the singularities. In the case of a crack in an anisotropic material for instance, the notch angle would be zero and there would only be one type of material.

The key idea is to separate the determination of the asymptotic fields from the solution of the full structural problem. The equilibrium equations for the problem in Figure 5.8 can be written in polar coordinates. We will assume that the components of the weakly singular displacement fields behave like r^{λ} in the radial direction. If λ is known, the problem simplifies to a one-dimensional problem. Usually λ will not be known. Then an iterative procedure has to be applied which solves a one-dimensional problem in every step. Such a procedure was described for instance by Li et al. (2001). The procedure was extended by Menk and Bordas (2011a) to the case where the

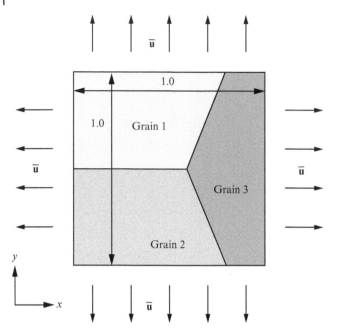

Figure 5.9 Polycrystalline structure with prescribed displacements at the boundary.

Table 5.3 Elastic constants for a copper grain

$D_{1111}, D_{2222}, D_{3333}$	168.4 GPa
$D_{1122}, D_{1133}, D_{2233}, D_{2211}, D_{3311}, D_{3322}$	121.4 GPa
$D_{2323}, D_{3131}, D_{1212}$	75.4 GPa

notch angle is zero and Material 1 and Material n are connected. The numerical effort necessary to obtain the shape of the singularity sufficiently well is usually small. To see the potential of the numerical enrichments we consider the structure shown in Figure 5.9. Three different grains form a structure. Many metals and metal alloys are polycrystals, i.e., they are formed by a number of small crystals also called grains. The grains often show an anisotropic elastic behavior depending on their orientation. Copper crystals for instance are strongly anisotropic, the stiffness varies up to a factor of two depending on the orientation. The elastic behavior of copper grains is given by the constants shown in Table 5.3. For a rotated grain the elastic constants can be obtained by applying the correct transformation. Assume that the grains of the structure in Figure 5.9 are copper grains where the orientation of each grain is described by the Euler angles in Table 5.4. To solve this problem numerically with the XFEM , we interpret the structural problem in the center of the structure as a junction formed by three infinitely long material wedges with perfect bonds between them. Determining the weakly singular fields with the previously mentioned methods, we obtain weakly singular enrichment functions. The resulting enrichment functions were used together with the abs-enrichment to model the grain boundaries in the corrected XFEM approach. Non-conforming meshes and triangular elements were used to discretize the polycrystalline structure. The von Mises stresses for the finite element method (FEM) (i.e., the same approach without using enrichments) and the XFEM are compared in Figures 5.10a and 5.10b. Although the dimension of the FEM equation systems is larger, the XFEM

Table 5.4 Euler angles describing crystal rotations for the different grains in the polycrystalline structure.

	Grain 1	Grain 2	Grain 3
ϕ_x	0°	0°	36°
ϕ_y	45°	120°	0°
ϕ_z	0°	-22.5^o	22.5^o

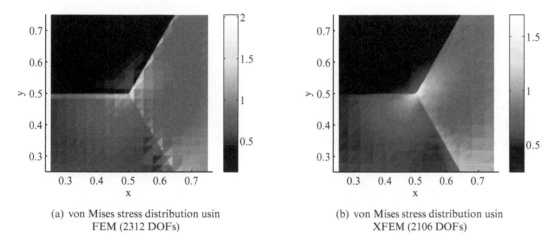

(a) von Mises stress distribution usin FEM (2312 DOFs)

(b) von Mises stress distribution usin XFEM (2106 DOFs)

Figure 5.10 von Mises stress distribution for the polycrystalline structure.

solution is obviously much smoother which is an indicator for good approximation properties of the numerical solution. Indeed, if the numerical error is evaluated for the XFEM, it is much lower than the one of the FEM. Furthermore, optimal convergence rates can be restored in the XFEM (Menk and Bordas, 2011a).

5.3 Creep and Crack Growth in Polycrystals

One of the numerical problems in which remeshing is particularly cumbersome is the simulation of anisotropic polycrystalline structures involving crack growth. Polycrystalline structures are formed by different crystals each of which can have a different orientation. Due to the anisotropy of the crystals the mechanical properties of the structure as a whole depend on those crystal orientations. To adequately capture the strain fields with the FEM the element edges should align with the crystal boundaries. In the presence of cracks the element edges should also align with the cracks and the crack tips. Furthermore, the mesh should be refined in the vicinity of crack tips or possible weak singularities.

This geometric restriction on the mesh choice may require a very sophisticated meshing strategy if in addition a certain mesh quality is required. Using enrichments, geometrical features such as crystal boundaries and cracks can be introduced mesh-independently. The element edges do not have to align with the geometrical features, but the ability to represent the influence of these features in the numerical solution adequately is not lost. This allows for a very simple meshing strategy. Due to its simplicity the simulation can be automated completely and no user intervention is necessary. As an example, we consider Figure 5.11. The initial mesh is shown in Figure 5.11a. It con-

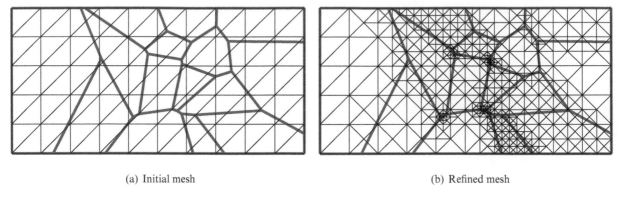

(a) Initial mesh　　　　　　　　　　　　　　　　　　(b) Refined mesh

Figure 5.11 Mesh generation for a polycrystalline structure (crystal boundaries are indicated by blue lines).

sists of equally sized triangular elements. Menk and Bordas (2011b) proposed a refinement technique which produces the mesh shown in Figure 5.11b. The mesh is refined along short crystal boundaries and inside small crystals.

Suppose a polycrystalline structure is given and a background mesh as shown in Figure 5.11a has been generated. The background mesh is now refined at certain locations. To refine a particular element, it is split along the line formed by the center of the longest element edge and the node opposite to this edge. This introduces two new elements. If the element splitting produces a hanging node in a neighboring element, the neighboring element is also refined by splitting the longest edge. This is done recursively to eliminate all hanging nodes. Please note that by using this scheme we always obtain elements with a good aspect ratio.

The code example 5.3.1 demonstrates the implementation of the element refinement using recursive function calls and a refinement level. The refinement level is initially set to 1 for all elements.

Remark 5.3.1. *Element refinement implemented using recursive function calls and the idea of a refinement level*

```
refineElement(element)
   elAttached=findElAttached(element)
   if refinementLevel(element)=refinementLevel(elAttached)
      split(element)
      split(elAttached)
   else
      refineElement(elAttached)
      refineElement(element)
   end
return
```

The function `refineElement()` takes the element that should be refined as an input argument. `findElAttached()` determines the element which is attached to the longest element edge of `element`. This element is stored in `elAttached`. If the two elements have the same refinement level, splitting both elements will not introduce hanging nodes, since both elements are connected via their longest edge. In that case, the first part of the `if`-statement is executed and the element is successfully refined. Otherwise the function `refineElement()` is called recursively for both elements again.

The mesh in Figure 5.11b can be obtained by applying this element refinement strategy to all the elements cut by a crystal boundary or inside a crystal until a certain criterion is reached (e.g., until a certain number of elements intersect with a crystal domain).

An application of the ideas mentioned above is the calculation of crack growth in solder joints. Solder joints form a mechanical as well as an electrical connection between the circuit board and components (e.g., chips and resistors) in electronic devices. If theses joints fail the device fails because the electrical connectivity is interrupted. Different failure mechanisms are known. Considering automotive electronics the most important damage mechanism is thermomechanical fatigue. Frequent temperature changes as they occur in the vicinity of the engine cause joint damage due to the thermal mismatch of the components and the board. It is of great importance to predict the lifetime of these joints under varying temperature profiles.

One of the most commonly used solder alloys in the electronics industry is the lead-free tin-silver-copper alloy. After the soldering process the joint is formed only by a few grains (Lehman et al., 2004). The grains exhibit highly anisotropic mechanical behavior. Recrystallization occurs if the devices are exposed to varying temperature profiles. If ball grid array (BGA) assemblies are considered, the joints form little balls. The part of these balls that is connected to the component is known to recrystallize (Limaye et al., 2006; Sundelin et al., 2006; Vandevelde et al., 2007). The number of grains in the recrystallized area is larger. Menk and Bordas (2011b) proposed a two-dimensional model for these joints which captures the phenomena of grain anisotropy and recrystallization. Consider Figure 5.12a. The boundary of the structure in Figure 5.12a describes the shape of a solder joint cross-section which forms if BGA devices are soldered onto a circuit board. The flat part on the top is attached to a copper pad. The cavity on the lower part of the joint encloses another copper pad. Each of the gray areas represents a

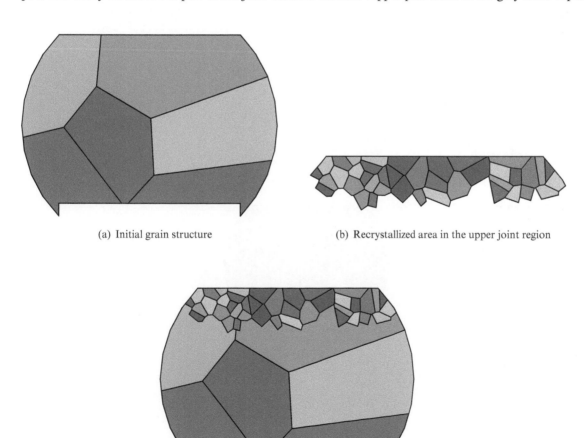

(a) Initial grain structure

(b) Recrystallized area in the upper joint region

(c) Final model of the joint

Figure 5.12 Generation of random recrystallized grain structures for a solder joint.

grain with a random orientation. The grain structure was generated randomly by using a Voronoi tessellation. The number of source points was chosen such that the number of grains roughly matches experimental observations (Sundelin et al., 2008).

A recrystallized structure consisting of smaller grains forming in the upper part of the joint is shown in Figure 5.12b. Again a Voronoi tesselation was used in which the number of source points was chosen such that experimental observations can be reproduced to some extent. The final joint model in Figure 5.12c is obtained by superimposing the recrystallized area onto the initial grain structure.

To capture the variety of possible grain structures in real solder joints, a series of these random grain structures should be generated. Crack growth calculations should then be performed for each of these structures. An automated meshing algorithm like the one explained above is highly desirable. Otherwise user intervention would be necessary for each grain structure and at each stage of crack growth. This would prevent the methodology from becoming an industrial standard.

To test the applicability of the idea to solder joint lifetime prediction Menk and Bordas (2011b) generated six random grain structures as shown in Figure 5.13. These grain structures were chosen to represent possible joint configurations for a certain joint location in a plastic ball grad array (PBGA) assembly. A two-dimensional model of the assembly is shown in Figure 5.14. Experimental crack growth data for the third joint from the left as shown in Figure 5.14 was gathered by Tunga (2008). Therefore, this joint position was chosen to conduct numerical studies on the crack growth using the joint structures above. Menk and Bordas determined constitutive laws for the grains

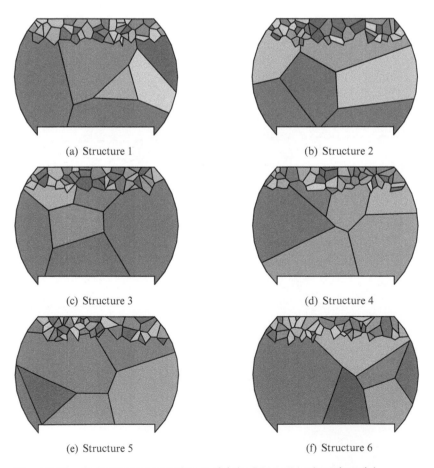

(a) Structure 1   (b) Structure 2

(c) Structure 3   (d) Structure 4

(e) Structure 5   (f) Structure 6

Figure 5.13 Grain structures used to model the BGA ball in the submodel.

in an inverse procedure. Failure of the boundaries was correlated with the accumulated creep deformation in their vicinity. In the experiments, cracks grow during several thousands of thermal cycles. But using simplifying assumptions not all of these cycles have to be simulated. The geometry of the device and the boundary conditions including the thermal profile were chosen to match the experiments conducted by Tunga (2008). Crack growth patterns for structure 1 are shown in Figure 5.15. A comparison of the crack length with the experimental data

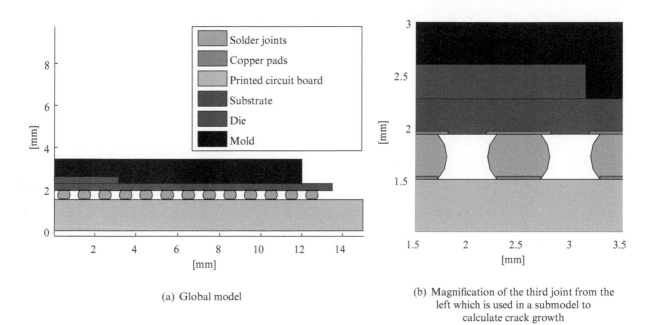

(a) Global model

(b) Magnification of the third joint from the left which is used in a submodel to calculate crack growth

Figure 5.14   Model of the PBGA 676 package soldered onto a circuit board.

Figure 5.15   Crack development in structure 1 from Figure 5.13 (cracks are marked by red lines).

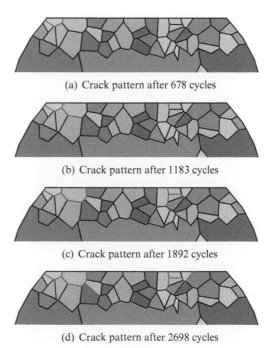

(a) Crack pattern after 678 cycles

(b) Crack pattern after 1183 cycles

(c) Crack pattern after 1892 cycles

(d) Crack pattern after 2698 cycles

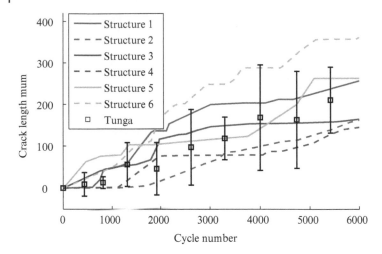

Figure 5.16 Crack lengths calculated for the polycrystalline solder joint models compared with data from the literature.

as gathered by Tunga is shown in Figure 5.16. Further investigation shows that the average crack length and the standard deviation of the crack length in the experiments can be represented numerically by using such a procedure. The key issue in implementing this strategy in an industrial environment is the automation of the underlying algorithms. In this case, the problem was solved by using enrichments in combination with a meshing strategy that guarantees a successful mesh generation.

5.4 Fatigue Crack Growth Simulations

Consider a plate with an edge crack of length a, inclined at an angle β, measured clockwise from the horizontal position. The geometry and boundary conditions are shown in Figure 5.17. The direction of the crack propagation is determined based on the maximum hoop stress criterion as given in Equation (5.8a).

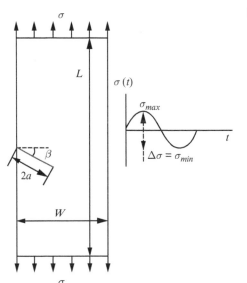

Figure 5.17 Finite plate with an edge crack subjected to tensile stress.

$$\theta_c = 2 \arctan \left(\frac{K_I - \sqrt{K_I^2 + 8K_{II}^2}}{4K_{II}} \right) \tag{5.8a}$$

$$K_{Ieq} = K_I \cos^3 \frac{\theta_c}{2} - 3K_{II} \cos^2 \frac{\theta_c}{2} \sin \frac{\theta_c}{2} \tag{5.8b}$$

$$\frac{da}{dN} = C \left(\Delta K_{Ieq} \right)^n \tag{5.8c}$$

where θ_c is the angle at which the crack is likely to propagate and ΔK_{Ieq} is the change in the equivalent SIF due to the differential change in the crack length, a with respect to the number of life, N. This direction corresponds to the direction in which the local shear stress approaches zero. For mixed mode problems, instead of pure mode I SIF (K_I), an equivalent mode I SIF (K_{Ieq}) is used which is evaluated as per the maximum hoop stress criterion (c.f. Equation (5.8b)). The crack growth rate is based on the generalized Paris's law as given by Equation (5.8c). The resulting crack increment is updated in the predicted direction via updating the level sets. In this example, the influence of crack angle on the fatigue life is studied. In the first case, the crack is assumed to be straight, i.e., $\beta = 0°$. The width and the length of the plate are 90 mm and 108 mm, respectively, and the thickness of the plate is taken as 6 mm. The initial crack length is assumed to be $a = 45$ mm and the plate is subjected to a cyclic tensile stress, $\sigma_{max} = 29.63$ N/mm^2 and $\sigma_{min} = 14.815$ N/mm^2 with Young's modulus and Poisson's ratio as 200,000 N/mm^2 and 0.3, respectively. The computational domain is discretized with four-noded structured quadrilateral elements with the mesh consisting of 46×54 nodes. For the study, the crack increment is fixed as 2 mm at each step of the crack growth. Figure 5.18 shows the crack growth rates computed by the XFEM and the results from the XFEM is compared with the experimental results reported in Ma et al. (2006). It is opined that the crack growth rate predicted by the XFEM are in close agreement with the results reported in the literature. Next, the influence of the crack angle β on the fatigue life is studied. For this study, we set $\beta = 40°$. The dimensions of the plate are 100 mm × 200 mm and the thickness of the plate is 1 mm. A cyclic tensile stress, $\sigma_{max} = 40$ N/mm^2 and $\sigma_{min} = 0$ N/mm^2, is applied. The material properties are as follows: Elastic modulus, $E = 74,000$ N/mm^2, Poisson's ratio, $\nu = 0.3$, critical fracture toughness, $K_{Ic} = 1897.36$ N/mm$^{3/2}$, Paris's exponent $n = 3.32$, and Paris constant $C = 2.087136 \times 10^{-13}$ (m/cycles)(Pa$\sqrt{m}$)$^{-n}$. In this case, the domain is discretized with 75×150 four-noded quadrilateral elements. The fatigue life obtained with the present XFEM framework is shown in Figure 5.19. The fatigue life obtained with the present formulation is 137,391 for a crack increment of $da = 1.33$ mm and 146,589 for a crack increment of $da = $

Figure 5.18 Crack growth rate curve for a plate with an edge crack subject to tensile stress.

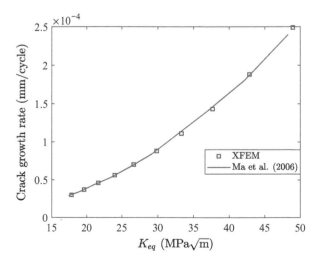

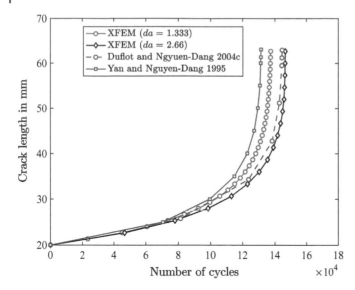

Figure 5.19 Crack growth rate curve for a plate with an edge crack subject to tensile stress.

2.666 mm. This is in fair agreement with the results obtained in the literature based on boundary element method and meshless method (Duflot and Nguyen-Dang, 2004; Yan and Nguyen-Dang, 1995).

5.5 Rectangular Plate with an Inclined Crack Subjected to Thermo-Mechanical Loading

Next, we present an example of coupled thermo-mechanical problem. In this case, the linear homogeneous body with an internal discontinuity occupying $\Omega \subset \mathbf{R}^2$ is subjected to both mechanical and thermal boundary conditions. The boundary, Γ, of Ω accommodates the following decomposition: $\Gamma = \Gamma_t \bigcup \Gamma_u$ and $\Gamma = \Gamma_T \bigcup \Gamma_q$, where Dirichlet and Neumann boundary conditions are specified. The governing equations for the linear thermo-elastic problem in the absence of inertia, body forces, and heat generation are given by:

$$\nabla \cdot \mathbf{q} = 0 \quad \text{in} \quad \Omega \tag{5.9a}$$

$$\boldsymbol{\nabla} \cdot \boldsymbol{\sigma} = \mathbf{0} \quad \text{in} \quad \Omega \tag{5.9b}$$

The above equations are supplemented with the following boundary conditions on the external boundary:

$$\varphi = \hat{\varphi} \quad \text{on} \quad \Gamma_T \tag{5.10a}$$

$$\mathbf{q} \cdot \mathbf{n} = \hat{\mathbf{q}} \quad \text{on} \quad \Gamma_q \tag{5.10b}$$

$$\boldsymbol{\sigma} \cdot \mathbf{n} = \hat{\mathbf{t}} \quad \text{on} \quad \Gamma_t \tag{5.10c}$$

$$\mathbf{u} = \hat{\mathbf{u}} \quad \text{on} \quad \Gamma_u \tag{5.10d}$$

and

$$\boldsymbol{\sigma} \cdot \mathbf{n} = 0 \quad \text{on} \quad \Gamma_c \tag{5.11a}$$

$$\mathbf{q} \cdot \mathbf{n} = 0 \quad \text{on} \quad \Gamma_c \quad \text{if crack is adiabatic} \tag{5.11b}$$

$$\varphi = \hat{\varphi} \quad \text{on} \quad \Gamma_c \quad \text{if crack is isothermal} \tag{5.11c}$$

where Γ_c represents the crack surface, $\varphi(\mathbf{x}) = T(\mathbf{x}) - T_o(\mathbf{x})$ is the increase in the temperature with respect to the strain-free reference configuration, $\mathbf{u}$ and $\boldsymbol{\sigma}$ are the displacement and stress field, respectively, $\mathbf{n}$ is the unit outward normal, $\mathbf{q}$ is the heat flux, and variables with $\hat{\ }$ are the prescribed quantities. The heat flux and the Cauchy stress tensor are related to the temperature and displacement field by:

$$\mathbf{q} = -\kappa \nabla \varphi \tag{5.12a}$$

$$\boldsymbol{\sigma} = \mathbf{D} \cdot \left(\boldsymbol{\varepsilon} - \boldsymbol{\varepsilon}^{\text{th}}\right) \tag{5.12b}$$

where κ is the thermal conductivity matrix, $\boldsymbol{\varepsilon}^{\text{th}} = \alpha\varphi[1\ 1\ 0]^{\text{T}}$ is the thermal strain and $\boldsymbol{\varepsilon} = \nabla\mathbf{u}$ is the small strain tensor. In this case, the XFEM approximation for the displacement is given by (c.f. Equation (5.14)):

$$\mathbf{u}_h(\mathbf{x}) = \sum_{I \in \mathcal{S}^{fem}} N_I(\mathbf{x})u_I + \sum_{J \in \mathcal{S}^c} N_J(\mathbf{x})\Big(H(\mathbf{x}) - H(\mathbf{x}_J)\Big)\mathbf{a}_J + \sum_{K \in \mathcal{S}^f} N_K(\mathbf{x})\sum_{\alpha=1}^{n} \Xi\alpha(r,\theta)\mathbf{b}_K^{\alpha} \tag{5.13}$$

and the temperature field is given by:

$$\varphi_h(\mathbf{x}) = \sum_{I \in \mathcal{S}^{fem}} N_I(\mathbf{x})\varphi_I + \sum_{J \in \mathcal{S}^c} N_J(\mathbf{x})\Big(H(\mathbf{x}) - H(\mathbf{x}_J)\Big)\mathbf{c}_J + \sum_{K \in \mathcal{S}^f} N_K(\mathbf{x})\Xi_\alpha(r,\theta)\mathbf{d}_K^{\alpha} \tag{5.14}$$

where $\alpha = 1,2$ for adiabatic and isothermal cracks. For the definition of $\Xi_\alpha(r,\theta)$ refer to Equation (3.22). The influence of temperature on the stress field in the vicinity of the crack is studied next.

Consider a rectangular plate with an inclined center crack subjected to thermo-mechanical boundary conditions (see Figure 5.20). For this study, the following are the thermal boundary conditions: adiabatic boundary conditions along edges $A{-}D$ and $B{-}C$ and the edges $A{-}B$ and $C{-}D$ are subjected to $-T$ and T, respectively. The displacement in y direction is restrained at B and the displacement in both x and y is constrained at A. No other mechanical boundary conditions are applied. The crack is assumed to be adiabatic. The following dimensions of the plate are considered: $L = 1$, $W = 1$. The influence of the crack length on the thermal SIF is numerically studied. The SIFs are normalized with $\alpha(T - To)E\sqrt{L}$, where T_o is the ambient temperature. The above boundary conditions induce a pure sliding mode at the crack tips, i.e., mode II. Based on a progressive refinement, a structured quadrilateral mesh consisting of 50×100 elements is used for this study. The radius of the domain of influence is taken as one-third the crack length. Figure 5.20b shows the temperature distribution for a horizontal crack, $\beta = 0°$. Table 5.5

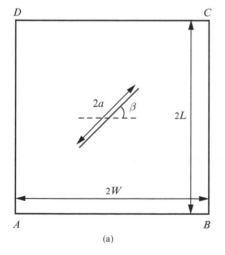

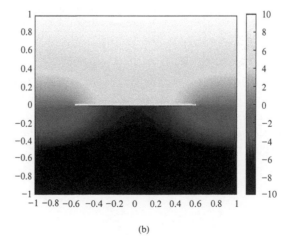

(a) (b)

Figure 5.20 Square plate with a center crack: (a) geometry and thermo-mechanical boundary conditions and (b) temperature profile.

Table 5.5 Square plate with a center horizontal crack subjected to thermo-mechanical boundary conditions. The crack is assumed to be adiabatic.

a/W	K_{II}	
	Ref. (Prasad and Aliabadi, 1994)	xfem
0.2	0.054	0.0512
0.3	0.096	0.0961
0.4	0.141	0.1430
0.5	0.191	0.1920
0.6	0.245	0.2447

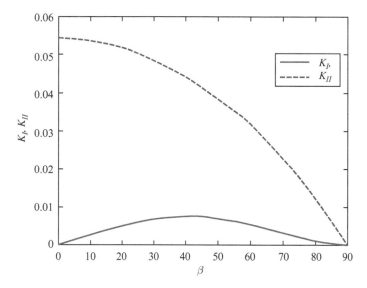

Figure 5.21 Influence of the crack orientation β on the mode I and mode II SIFs for a rectangular plate with a center crack. The crack length to width ratio is $a/W = 0.3$. The crack is assumed to be adiabatic.

presents the normalized SIF from the present formulation and is compared with the results obtained by the dual boundary element method (Prasad and Aliabadi, 1994). It is seen that the results are in close agreement. To study, the influence of the crack orientation, the following dimensions of the plate are considered: $L/W = 2$ and the SIFs are normalized with $\alpha T_1 E(W)^{1.5}/L$. Figure 5.21 shows the variation of the mode I and mode II SIFs with the crack orientation β for a rectangular plate with a center crack. When the crack is horizontal, i.e., $\beta = 0°$, the boundary conditions induce a pure mode II and when the orientation $\beta = 90°$, the temperature field is not influenced by the presence of crack.

References

Bush, M.B. (1997). The interaction between a crack and a particle cluster. *International Journal of Fracture* 88: 215–232.

Duflot, M. and Nguyen-Dang, H. (2004). A meshless method with enriched weight functions for fatigue crack growth. *International Journal for Numerical Methods in Engineering* 59 (14): 1945–1961.

Kitey, R., Phan, A.V., Tippur, H.V., and Kaplan, T. (2006). Modeling of crack growth through particulate clusters in brittle matrix by symmetric-Galerkin boundary element method. *International Journal of Fracture* 141 (1): 11–25.

Lehman, L.P., Athavale, S.N., Fullem, T.Z., *et al.* (2004). Growth of Sn and intermetallic compounds in Sn-Ag-Cu solder. *Journal of Electronic Materials* 33 (12): 1429–1439.

Li, J., Zhang, X.B., and Recho, N. (2001). Stress singularities near the tip of a two-dimensional notch formed from several elastic anisotropic materials. *International Journal of Fracture* 107 (4): 379–395.

Li, R. and Chudnovsky, A. (2003). Energy analysis of crack interaction with an elastic inclusion. *International Journal of Fracture* 63: 247–261.

Limaye, P., Vandevelde, B., Vandepitte, D., and Verlinden, B. (2006). Crack growth rate measurement and analysis for WLCSP SnAgCu solder joints. *Circuits Assembly* 17 (2): 68.

Ma, S., Zhang, X., Recho, N., and Li, J. (2006). The mixed-mode investigation of the fatigue crack in CTS metallic specimen. *International Journal of Fatigue* 28: 1780–1790.

Menk, A. and Bordas, S. (2011a). A robust preconditioning technique for the extended finite element method. *International Journal for Numerical Methods in Engineering* 85 (13): 1609–1632.

Menk, A. and Bordas, S. (2011b). Crack growth calculations in solder joints based on microstructural phenomena with X-FEM. *Computational Materials Science* 50 (3): 1145–1156.

Prasad, N.N.V. and Aliabadi, M.H. (1994). The dual boundary element method for thermoelastic crack problems. *International Journal of Fracture* 66: 255–272.

Sih, G.C. (1973). Energy-density concept in fracture mechanics. *Engineering Fracture Mechanics* 5: 1037–1040.

Sundelin, J.J., Nurmi, S.T., and Lepisto, T.K. (2008). Recrystallization behaviour of SnAgCu solder joints. *Materials Science and Engineering: A* 474 (1): 201–207.

Sundelin, J.J., Nurmi, S.T., Lepisto, T.K., and Ristolainen, E.O. (2006). Mechanical and microstructural properties of SnAgCu solder joints. *Materials Science and Engineering: A* 420 (1): 55–62.

Tunga, K.R. (2008). *Study of tin-silver-copper alloy reliability through material microstructure evolution and laser moire interferometry*. *ProQuest*. Dissertation of thesis.

Vandevelde, B., Gonzalez, M., et al. (2007). Thermal cycling reliability of SnAgCu and SnPb solder joints: A comparison for several IC-packages. *Microelectronics Reliability* 47 (2): 259–265.

Wang, Y.B. and Chou, K.T. (2001). A new boundary element method for mixed boundary value problems involving cracks and holes: Interactions between rigid inclusions and cracks. *International Journal of Fracture* 110: 387–406.

Yan, A. and Nguyen-Dang, H. (1995). Multiple-cracked fatigue crack growth by BEM. *Computational Mechanics* 16: 273–280.

Yu, H., Wu, L., Guo, L., et al. (2009). Investigation of mixed-mode stress intensity factors for nonhomogeneous materials using an interaction integral method. *International Journal of Solids and Structures* 46 (20): 3710–3724.

6

Recovery-Based Error Estimation and Bounding in XFEM

Octavio Andrés González-Estrada[1], Juan José Ródenas García[2] and Stéphane P. A. Bordas[3]

[1] *Universidad Industrial de Santander, Colombia*
[2] *Universidad Politécnica de Valencia, Valencia, Spain*
[3] *Univesity of Luxembourg, Luxembourg, UK*

6.1 Introduction

Engineering structures, in particular in aerospace engineering, are intended to operate with flawless components, especially for safety critical parts. However, there is always a possibility that cracking will occur during operation, risking catastrophic failure, and associated casualties.

The mission of damage tolerance assessment (DTA) is to assess the influence of these defects, cracks, and damage on the ability of a structure to perform safely and reliably during its service life. An important goal of DTA is to estimate the fatigue life of a structure, i.e., the time during which it remains safe given a pre-existing flaw.

DTA relies on the ability to accurately predict crack paths and growth rates in complex structures. Since the simulation of three-dimensional (3D) crack growth is either not supported by commercial software, or requires significant effort and time for the analysts and is generally not coupled to robust error indicators, reliable, quality-controlled, industrial DTA is still a major challenge in engineering practice.

The extended finite element method (XFEM) (Moës et al., 1999) is now one of many successful numerical methods to solve fracture mechanics problems. As shown in the rest of the book, the particular advantage of the XFEM, which relies on the partition of unity (PU) property (Melenk and Babuška, 1996) of finite element (FE) shape functions, is its ability to model cracks without the mesh conforming to their geometry. This allows the crack to split the background mesh arbitrarily, thereby leading to significantly increased freedom in simulating crack growth.

This feat is achieved by adding new degrees of freedom (dof) to (i) describe the discontinuity of the displacement field across the crack faces, within a given element, and (ii) reproduce the asymptotic fields around the crack tip. Thanks to the advances made in the XFEM during the last years, the method is now considered to be a robust and accurate means of analyzing fracture problems, has been implemented in commercial codes (ABAQUS, 2009; CENAERO, 2011), and is industrially in use for DTA of complex 3D structures (Bordas and Moran, 2006; Duflot, 2007; Wyart et al., 2007).

While XFEM allows to model cracks without (re)meshing the crack faces as they evolve and yields exceptionally accurate stress intensity factors (SIFs) for 2D problems, Bordas and Moran (2006); Duflot (2007); Wyart et al. (2007) showed that for realistic 3D structures, a "very fine" mesh is required to accurately capture the complex 3D stress field and obtain satisfactory SIFs, the main drivers of linear elastic crack propagation.

Consequently, based on the mesh used for the stress analysis, a new mesh offering sufficient refinement throughout the whole potential path of the crack must be constructed *a priori*, i.e., before the crack path is known.

Practically, this is done by running preliminary analyses on coarse meshes to obtain an approximative crack path and heuristically refining the mesh around this path to increase accuracy.

Typically, this refinement does not rely on sound error measures; thus, the heuristically chosen mesh is in general inadequately suited and can cause large inaccuracies in the crack growth path, especially around holes, which can lead to non-conservative estimates of the safe life of the structure.

Thus, it is clear that although XFEM simplifies the treatment of cracks compared to the standard finite element method (FEM) by lifting the burden of a geometry-conforming mesh, it still requires iterations and associated user intervention and it employs heuristics which are detrimental to the robustness and accuracy of the simulation process.

It would be desirable to minimize the changes to the mesh topology, and as a result user intervention, while ensuring the stress fields and SIFs are accurately computed at each crack growth step. Herein, we present different approaches to control the discretization error committed by enriched FE approximations, decrease human intervention in DTA of complex industrial structures, and enhance confidence in the results by providing enriched FEMs with sound error estimators which will guarantee a predetermined accuracy level and suppress recourse to manual iterations and heuristics.

The error assessment procedures used in FE analysis are well known and can be usually classified into different families as seen in Ainsworth and Oden (2000): residual-based error estimators, recovery-based error estimators, dual techniques, etc. Explicit and implicit residual-based error estimators have a strong mathematical basis and have been frequently used to obtain lower and upper bounds of the error (see the seminal works from Babuška and Miller (1987); Babuška and Rheinboldt (1979) for explicit methods and Ainsworth and Oden (1993); Babuška and Rheinboldt (1978); Bank and Weiser (1985); Ladevèze and Leguillon (1983) for implicit ones). Recovery-based error estimates were first introduced by Zienkiewicz and Zhu (1987) and are often preferred by practitioners because they are robust and simple to use and provide an enhanced solution. Further improvements were made with the introduction of new recovery processes such as the superconvergent patch recovery (SPR) technique proposed by Zienkiewicz and Zhu (1992a,b). Dual techniques based on the evaluation of two different fields, one compatible in displacements and another equilibrated in stresses, have also been used by Pereira et al. (1999) to obtain bounds of the error. In this work, we are going to focus on recovery-based techniques which follow the ideas of the Zienkiewicz-Zhu (ZZ) error estimator proposed by Zienkiewicz and Zhu (1987).

The literature on error estimation techniques for mesh-based PU methods is still scarce. One of the first steps in that direction was made in the context of the generalized finite element method (GFEM) by Strouboulis et al. (2001). The authors proposed a recovery-based error estimator which provides good results for h-adapted meshes. In a later work, two new *a posteriori* error estimators for GFEM were presented in Strouboulis et al. (2006). The first one was based on patch residual indicators and provided an accurate theoretical upper bound estimate, but its computed version severely underestimated the exact error. The second one was an error estimator based on a recovered displacement field and its performance was closely related to the quality of the GFEM solution. This recovery technique constructs a recovered solution on patches using a basis enriched with handbook functions.

Barros et al. (2004) presented an adaptive scheme for GFEM based on an indicator given by an error measure evaluated using the element residual method. The p-refinement strategy takes advantage of the GFEM enrichment to define the polynomial functions.

In order to obtain a recovered stress field that improves the accuracy of the stresses obtained by XFEM , Xiao and Karihaloo (2006) proposed a moving least squares (MLS) fitting adapted to the XFEM framework which considers the use of statically admissible basis functions. Nevertheless, the recovered stress field was not used to obtain an error indicator.

Pannachet et al. (2009) worked on error estimation for mesh adaptivity in XFEM for cohesive crack problems. Two error estimates were used, one based on the error in the energy norm and another that considers the error in a local quantity of interest (QoI). The error estimation was based on solving a series of local problems with prescribed homogeneous boundary conditions.

Panetier et al. (2010) presented an extension to enriched approximations of the constitutive relation error (CRE) technique already available to evaluate error bounds in FEM . This procedure has been used to obtain local error bounds on quantities of interest for XFEM problems.

Bordas and Duflot (2007) and Bordas et al. (2008) proposed a recovery-based error estimator for 2D and 3D XFEM approximations for cracks known as the extended moving least squares (XMLS). This method intrinsically enriched an MLS formulation to include information about the singular fields near the crack tip. Additionally, it used a diffraction method to introduce the discontinuity in the recovered field. This error estimator provided accurate results with effectivity indices close to unity (optimal value) for 2D and 3D fracture mechanics problems. Later, Duflot and Bordas (2008) proposed a global derivative recovery formulation extended to XFEM problems. The recovered solution was sought in a space spanned by the near-tip strain fields obtained from differentiating the Westergaard asymptotic expansion. Although the results provided by this technique were not as accurate as those in Bordas and Duflot (2007) and Bordas et al. (2008), they were deemed by the authors to require less computational power. Similarly, Prange et al. (2012) used a recovery procedure based on a least square fitting for linear elastic fracture mechanics (LEFM) problems and for inelastic materials in combination with XFEM . The technique considers enrichment functions for the projected stress field such that it is able to reproduce discontinuities along the crack faces and stress singularities at crack tips.

Ródenas et al. (2008) presented a modification of the SPR technique tailored to the XFEM framework called SPR$_{\text{XFEM}}$. This technique was based on three key features: (i) the use of a *singular+smooth* stress field decomposition procedure around the crack tip similar to that described by Ródenas et al. (2006) for FEM ; (ii) direct calculation of recovered stresses at integration points using the PU, and (iii) use of different stress interpolation polynomials at each side of the crack when a crack intersects a patch. In order to obtain an equilibrated smooth recovered stress field, a simplified version of the SPR-C technique presented by Ródenas et al. (2007) was used. This simplified SPR-C imposed the fulfillment of the boundary equilibrium equation at boundary nodes but did not impose the satisfaction of the internal equilibrium equations.

Chahine et al. (2008) and Nicaise et al. (2011) worked in *a priori* error estimates for the XFEM variants using a cut-off function and a fixed enrichment area. They studied the convergence of the error in the H_1 norm and the SIF.

Rüter and Stein (2011) derived residual-type goal-oriented error estimators for the J-integral used in LEFM problems approximated using XFEM . The procedure provides upper error bounds and is based on solving local Neumann problems for the primal and dual solutions using equilibrated tractions which rely on L_2-orthogonal functions with respect to the XFEM shape functions.

Gerasimov et al. (2012) worked on explicit residual error estimation for 2D XFEM approximations. The authors proposed an upper bound of the discretization error, as well as associated local error indicators, by the use of a specific quasi-interpolation operator of averaging type. Furthermore, alternative branch functions for the XFEM enrichment which fulfill divergence-free conditions near the singularity and traction-free along the crack faces were derived from this study.

6.2 Error Estimation in the Energy Norm. The ZZ Error Estimator

The approximate nature of the FEM and XFEM formulations implies a discretization error. The energy norm is commonly used in the context of elasticity problems solved using the FEM to quantify this error. The exact error in energy norm $\|\mathbf{e}\| = \|\mathbf{u} - \mathbf{u}^h\|$ is evaluated from:

$$\|\mathbf{e}\|^2 = \int_{\Omega} \left(\boldsymbol{\sigma} - \boldsymbol{\sigma}^h \right)^T \mathbf{D}^{-1} \left(\boldsymbol{\sigma} - \boldsymbol{\sigma}^h \right) d\Omega \tag{6.1}$$

Zienkiewicz and Zhu (1987) proposed the use of the following expression, commonly denoted as the ZZ error estimator, to obtain an estimate of $\|\mathbf{e}\|$:

$$\|\mathbf{e}\|^2 \approx \|\mathbf{e}_{es}\|^2 = \int_\Omega \left(\boldsymbol{\sigma}^* - \boldsymbol{\sigma}^h\right)^T \mathbf{D}^{-1} \left(\boldsymbol{\sigma}^* - \boldsymbol{\sigma}^h\right) d\Omega \tag{6.2}$$

or alternatively for the strains:

$$\|\mathbf{e}_{es}\|^2 = \int_\Omega \left(\varepsilon^* - \varepsilon^h\right)^T \mathbf{D} \left(\varepsilon^* - \varepsilon^h\right) d\Omega \tag{6.3}$$

where $\boldsymbol{\sigma}^h$ represents the stress field provided by the FEM, $\boldsymbol{\sigma}^*$ is the recovered stress field, which is a better approximation to the exact solution than $\boldsymbol{\sigma}^h$, and $\mathbf{D}$ is the elasticity matrix of the constitutive relation $\boldsymbol{\sigma} = \mathbf{D}\varepsilon$. As the error is quantified using an integral scalar quantity, the domain Ω can refer both to the full domain of the problem or to a local subdomain (as for example an element). Error estimators should have good behavior at a global level to obtain an accurate assessment of the overall quality of the FE analysis. However, in some cases, one could have a very accurate estimation of the exact global error which has been obtained through compensation of severe underestimations and overestimations of local errors. Therefore, the error estimators should also have good behavior at the local level as the information relative to the error at each element is a key ingredient of the automatic adaptive refinement techniques.

The recovered gradients $\boldsymbol{\sigma}^*$ are usually interpolated in each element using the shape functions $\mathbf{N}$ of the underlying FE approximation and the values of the recovered gradients calculated at the nodes $\tilde{\boldsymbol{\sigma}}^*$. The following expression is used to evaluate each of the stress components:

$$\boldsymbol{\sigma}^*(\mathbf{x}) = \mathbf{N}\tilde{\boldsymbol{\sigma}}^*, \tag{6.4}$$

The accuracy of an error estimator is usually measured using the value of the effectivity index θ evaluated in problems for which the exact solution is known. This index can be evaluated for the whole domain of the problem or, alternatively, for any element e in the mesh:

$$\theta = \frac{\|\mathbf{e}_{es}\|}{\|\mathbf{e}\|} \tag{6.5}$$

$$\theta^e = \frac{\|\mathbf{e}_{es}^e\|}{\|\mathbf{e}^e\|} \tag{6.6}$$

Under certain conditions, the ZZ error estimator is asymptotically exact, i.e., the effectivity θ tends to one as the exact error $\|\mathbf{e}\|$ tends to zero. Zienkiewicz and Zhu (1992b) showed that asymptotic exactness is obtained if the error in energy norm of $\boldsymbol{\sigma}^*$ converges at a higher rate than of the FE solution $\boldsymbol{\sigma}^h$.

One disadvantage of residual type error estimators is that they require the computation of specific values that are useless for the FE analysis (Díez et al., 1998). On the contrary, a reason for the widespread use of the recovery-type error estimators is that they require the evaluation of $\boldsymbol{\sigma}^*$ that is not only going to be used for the evaluation of the error but also as a solution that in the vast majority of practical applications is more accurate than the raw FE solution.

The literature contains different approaches to build $\boldsymbol{\sigma}^*$. The basis of two of these approaches: the *Superconvergent Patch Recovery (SPR)* technique and the *Moving Least Squares (MLS)* approach are going to be described here. Their adaptation to the XFEM framework will then be described in the following section.

6.2.1 The SPR Technique

Due to its simple implementation and robustness (Babuška et al., 1994a,b, 1997) the SPR developed by Zienkiewicz and Zhu (1992a,b) is of common use today. In the original SPR technique proposed by these authors, each of the stress components $\sigma_i^*(\mathbf{x})$ of $\sigma^*(\mathbf{x})$ is obtained using a polynomial expansion:

$$\sigma_i^*(\mathbf{x}) = \mathbf{p}(\mathbf{x})\mathbf{a}_i \tag{6.7}$$

This expansion is defined over a *patch* formed by the set of contiguous elements connected to a given vertex node, where $\mathbf{p}(\mathbf{x})$ is the polynomial basis of the degree used in the interpolation of the displacements (for example, $\mathbf{p}(\mathbf{x}) = \{1, x, y, x^2, xy, y^2\}$ for 2D quadratic elements) and $\mathbf{a}_i$ are the unknown coefficients. The values of $\mathbf{a}_i$ are obtained using a least squares fit to the values of the FE stresses σ^h evaluated at superconvergence points. To do this, the following functional is defined for each stress component i:

$$\Pi(\mathbf{a}) = \sum_{j=1}^{nsp} \left(\sigma^*(\mathbf{x}_j) - \sigma^h(\mathbf{x}_j)\right)^2 = \sum_{i=1}^{nsp} \left(\mathbf{p}(\mathbf{x}_j)\mathbf{a} - \sigma^h(\mathbf{x}_j)\right)^2 \tag{6.8}$$

where nsp is the number of stress sampling points in the patch used in the least squares fit. Minimizing Π with respect to the vector of unknown coefficients $\mathbf{a}$ leads to the system of equations $\mathbf{MA} = \mathbf{G}$ where:

$$\mathbf{M} = \sum_{j=1}^{nsp} \mathbf{p}^T(\mathbf{x}_j)\mathbf{p}(\mathbf{x}_j) \quad \text{and} \quad \mathbf{G} = \sum_{j=1}^{nsp} \mathbf{p}^T(\mathbf{x}_j)\sigma^h(\mathbf{x}_j) \tag{6.9}$$

As the recovery procedure is based on superconvergent results, the rate of convergence of the recovered solution is higher than that for the FE solution. Therefore, the ZZ error estimator combined with the SPR technique is asymptotically exact. Once $\sigma_i^*(\mathbf{x})$ has been evaluated, the only value retained is the value of the stress recovered at vertex node. The recovered stresses at nodes other than vertex nodes (for example, mid-side nodes) are obtained by evaluating (6.7) at these nodes. As the stresses at these nodes can be evaluated from different patches, the average of the different stress values is considered.

Equation (6.4) is transformed into:

$$\sigma^*(\mathbf{x}) = \mathbf{N}\tilde{\sigma}^* = \sum_{i=1}^{n_e} N_i(\mathbf{x})\sigma_i^*(\mathbf{x}_i) = \sum_{i=1}^{n_e} N_i(\mathbf{x})\mathbf{p}(\mathbf{x}_i)\mathbf{a}_i \tag{6.10}$$

where n_e is the number of nodes in the element under consideration.

Note that the stress interpolation (6.10) guarantees stress continuity in σ^*. However, the polynomial fitting is unable to provide any other desired characteristic like the satisfaction of the equilibrium or compatibility equations. After the publication of the SPR technique, a great number of articles followed (see, for example, Aalto (1997); Aalto and Åman (1999); Aalto and Isoherranen (1997); Benedetti et al. (2006); Blacker and Belytschko (1994); Kvamsdal and Okstad (1998); Lee et al. (1997); Park et al. (1999); Ramsay and Maunder (1996); Wiberg et al. (1994); Wiberg and Abdulwahab (1993)). In most of these works, to improve the performance of the technique, the authors tried to consider the physics of the problem including the equilibrium equations in one way or the other.

The recovery technique has normally been used to recover the stress field; however, it can also be used to obtain a recovered displacement field (Wiberg et al., 1992).

The SPR technique is well suited for smooth problems. Nevertheless, the technique does not produce accurate results in the vicinity of singular points (crack tips or re-entrant corners) as the polynomial expansion used to represent the stresses in the patch is unable to reproduce singular stress values. Inaccurate results derived from the use of the original SPR technique in the XFEM context have been reported by Bordas and Duflot (2007) and Ródenas et al. (2008).

6.2.2 The MLS Approach

The MLS technique, developed by mathematicians to build and fit surfaces, is an alternative to the SPR technique to obtain recovered fields. The MLS technique was used in the context of error estimation in FEM by Tabbara et al. (1994). The MLS technique is a weighted least squares formulation biased toward the test point where the value of a function ϕ is asked that provides a continuous representation of ϕ (displacements, strains, or stresses in our case). As in the SPR, a polynomial expansion is used to define the components i of the recovered field ϕ^* in the form:

$$\phi_i^*(\mathbf{x}) = \mathbf{p}(\mathbf{x})\mathbf{a}_i(\mathbf{x}) \tag{6.11}$$

where $\mathbf{p}$ represents a polynomial basis and $\mathbf{a}$ are unknown coefficients. Note that in the SPR case, the recovered field is evaluated at nodes and then interpolated in the interior of the elements using either (6.4) or (6.10). However, in the MLS approach, such an interpolation is not performed and the evaluation of the recovered field at a given point $\mathbf{x}$ always requires the evaluation of the unknown coefficients $\mathbf{a}$ for that particular point. This increases the computational cost of the recovery process with respect to the SPR technique but, on the other hand, will provide better continuity properties in the recovered field.

For 2D, the expression to evaluate the two components of the recovered displacement field reads:

$$\mathbf{u}^*(\mathbf{x}) = \begin{Bmatrix} u_x^*(\mathbf{x}) \\ u_y^*(\mathbf{x}) \end{Bmatrix} = \mathbf{P}(\mathbf{x})\mathbf{A}(\mathbf{x}) = \begin{bmatrix} \mathbf{p}(\mathbf{x}) & \mathbf{0} \\ \mathbf{0} & \mathbf{p}(\mathbf{x}) \end{bmatrix} \begin{Bmatrix} \mathbf{a}_x(\mathbf{x}) \\ \mathbf{a}_y(\mathbf{x}) \end{Bmatrix} \tag{6.12}$$

The expression to evaluate the three components of the recovered stress field would similarly be:

$$\sigma^*(\mathbf{x}) = \begin{Bmatrix} \sigma_{xx}^*(\mathbf{x}) \\ \sigma_{yy}^*(\mathbf{x}) \\ \sigma_{xy}^*(\mathbf{x}) \end{Bmatrix} = \mathbf{P}(\mathbf{x})\mathbf{A}(\mathbf{x}) = \begin{bmatrix} \mathbf{p}(\mathbf{x}) & \mathbf{0} & \mathbf{0}* \\ \mathbf{0} & \mathbf{p}(\mathbf{x}) & \mathbf{0} \\ \mathbf{0} & \mathbf{0} & \mathbf{p}(\mathbf{x}) \end{bmatrix} \begin{Bmatrix} \mathbf{a}_{xx}(\mathbf{x}) \\ \mathbf{a}_{yy}(\mathbf{x}) \\ \mathbf{a}_{xy}(\mathbf{x}) \end{Bmatrix} \tag{6.13}$$

The expressions of the type $\phi^*(\mathbf{x}) = \mathbf{P}(\mathbf{x})\mathbf{A}(\mathbf{x})$, that involve all the components of the field to be evaluated, could also be used in the case of the SPR technique. These types of expressions are useful to impose constrains between all the unknown coefficients in order to force the recovered field to satisfy the desired properties (internal and boundary equilibrium equations, etc.).

Suppose that χ is a point within $\Omega_{\mathbf{x}}$, being $\Omega_{\mathbf{x}}$ the support corresponding to a point $\mathbf{x}$ defined by a distance (radius) $R_{\Omega_{\mathbf{x}}}$. The MLS approximation for each component i of field ϕ at χ is given by:

$$\phi_i^*(\mathbf{x}, \chi) = \mathbf{p}(\chi)\mathbf{a}_i(\mathbf{x}) \quad \forall \chi \in \Omega_{\mathbf{x}} \tag{6.14}$$

To obtain the coefficients $\mathbf{a}$ we would minimize the following functional:

$$J(\mathbf{x}) = \sum_{j=1}^{n} W\left(\mathbf{x} - \chi_j\right)\left[\phi^*\left(\mathbf{x}, \chi_j\right) - \phi^h\left(\chi_j\right)\right]^2 \tag{6.15}$$

where $\phi^h\left(\chi_j\right)$ are, in this case, values of function ϕ evaluated with XFEM at each of the n points where the function is sampled within $\Omega_{\mathbf{x}}$.

Alternatively, to evaluate $J(\mathbf{x})$, we can also adopt the *Continuous Moving Least Squares Approximation* described in Liu (2003). The following functional would then be minimized:

$$J(\mathbf{x}) = \int_{\Omega_{\mathbf{x}}} W(\mathbf{x} - \chi)\left[\phi^*(\mathbf{x}, \chi) - \phi^h(\chi)\right]^2 d\chi \tag{6.16}$$

Evaluating $\partial J / \partial \mathbf{A} = 0$ from (6.15) results in the linear system $\mathbf{M}(\mathbf{x})\mathbf{A}(\mathbf{x}) = \mathbf{G}(\mathbf{x})$ used to evaluate $\mathbf{A}$, where:

$$\mathbf{M}(\mathbf{x}) = \sum_{j=1}^{n} W\left(\mathbf{x} - \chi_j\right)\mathbf{P}^T\left(\chi_j\right)\mathbf{P}\left(\chi_j\right)d\chi$$

$$\mathbf{G}(\mathbf{x}) = \sum_{j=1}^{n} W\left(\mathbf{x} - \chi_j\right)\mathbf{P}^T\left(\chi_j\right)\phi^h\left(\chi_j\right)d\chi \tag{6.17}$$

or their integral counterparts if (6.16) is considered.

In the previous equations, W is the MLS weighting function, which can be for example taken as the fourth-order spline, commonly used the MLS related literature:

$$W(\mathbf{x} - \chi) = \begin{cases} 1 - 6s^2 + 8s^3 - 3s^4 & if \ |s| \leq 1 \\ 0 & if \ |s| > 1 \end{cases} \tag{6.18}$$

where s denotes the normalized distance function given by:

$$s = \frac{\|\mathbf{x} - \chi\|}{R_{\Omega_{\mathbf{x}}}} \tag{6.19}$$

Note that the expressions for $\mathbf{M}$ and $\mathbf{G}$ in (6.9) for the SPR case are similar to those in (6.17) for the MLS with the difference of the weighting function W included in the later case.

6.3 Recovery-based Error Estimation in XFEM

The standard SPR and MLS techniques are based on the use of polynomial functions to describe the recovered fields. These types of functions are not appropriate to describe the singularity in the XFEM context. On the other hand, the standard implementations of these techniques are not well suited to describe the discontinuous solution at the crack faces. Therefore, the standard versions of these two techniques have been modified and adapted to the XFEM context for their use in error estimation in energy norm. The main adaptations of the SPR and MLS techniques to XFEM are as follows:

- *The SPR-CX technique* derived from the error estimator developed by Ródenas et al. (2008) and summarized in Section 6.3.1.
- *The XMLS technique* proposed by Bordas and Duflot (2007) and summarized in Section 6.3.2.
- *The MLS-CX technique* proposed by Ródenas et al. (2012) summarized in Section 6.3.3

These techniques provide recovered fields that are used to evaluate the estimated error in the energy norm by means of the expression shown in (6.2) or in (6.3). They can be considered as *extended recovery techniques* as, in one way or another, they enrich the result using the know characteristics of the solution, as in XFEM .

6.3.1 The SPR-CX Technique

The SPR-CX error estimator is a stress recovery technique that can be considered as an adaptation of the original SPR technique (Zienkiewicz and Zhu, 1992a) to the XFEM framework. This technique incorporates a number of enhancements Díez et al., 2007; Ródenas et al., 2007, 2008, 2010) that have proved to be very effective for an accurate evaluation of σ^* and, thus, for an accurate error estimation both at local and global levels. The key features of the technique are as follows:

- Use of a *singular + smooth* stress decomposition technique;
- Satisfaction of C^0 continuity using a PU approach;
- Local satisfaction of equilibrium and compatibility equations at each patch;
- Use of two different recovered stress fields in patches containing the crack;
- Continuous instead of discrete least squares approximation to σ^h.

These modifications are going to be described in the following sections. We will assume that, at each patch, the recovered stresses $\boldsymbol{\sigma}*$ are defined in a Cartesian coordinates system $(\bar{x}, \bar{y})$ parallel to the global coordinates system (x, y) centered at each patch assembly node. Only vertex nodes will be considered as patch assembly nodes.

We will use (6.13) to express the recovered stress field at each patch as this format is very convenient to impose the constraints required to satisfy the equilibrium and compatibility equations.

6.3.1.1 *Singular+smooth* stress field splitting

We previously commented that the SPR technique is not well suited for the recovery of singular stress fields because the polynomial representation is unable to reproduce the stresses at the singular point. The *singular+smooth* stress field splitting is a major modification of the SPR technique aimed at solving this problem by introducing known information about the solution during the recovery process. In singular problems in linear elasticity, the exact stress field $\boldsymbol{\sigma}$ can be decomposed into two stress fields, a smooth field $\boldsymbol{\sigma}_{smo}$ and a singular field $\boldsymbol{\sigma}_{sing}$:

$$\boldsymbol{\sigma} = \boldsymbol{\sigma}_{smo} + \boldsymbol{\sigma}_{sing} \tag{6.20}$$

Similarly, $\boldsymbol{\sigma}^*$ can also be expressed as the contribution of one smooth field $\boldsymbol{\sigma}^*_{smo}$ and one singular $\boldsymbol{\sigma}^*_{sing}$, each of them evaluated following a different recovery procedure:

$$\boldsymbol{\sigma}^* = \boldsymbol{\sigma}^*_{smo} + \boldsymbol{\sigma}^*_{sing} \tag{6.21}$$

In the vicinity of the singularity, the stresses are dominated by the first term of the asymptotic expansion of the solution, i.e., the singular term, which is expressed as a function of K_I and K_{II}. Let us call K_I^* and K_{II}^* to the SIFs evaluated as part of the XFEM analysis. Therefore, substituting these values into the first terms of the asymptotic expansion of the exact stress field in the vicinity of a crack tip provides us a recovered version $\boldsymbol{\sigma}^*_{sing}$ of the singular stress field:

$$
\begin{Bmatrix} \sigma^*_{sing11}(r,\phi) \\ \sigma^*_{sing22}(r,\phi) \\ \sigma^*_{sing12}(r,\phi) \end{Bmatrix} = \frac{K_I^*}{\sqrt{2\pi r}} \begin{Bmatrix} \cos\frac{\phi}{2}\left(1 - \sin\frac{\phi}{2}\sin\frac{3\phi}{2}\right) \\ \cos\frac{\phi}{2}\left(1 + \sin\frac{\phi}{2}\sin\frac{3\phi}{2}\right) \\ \sin\frac{\phi}{2}\cos\frac{\phi}{2}\cos\frac{3\phi}{2} \end{Bmatrix} + \frac{K_{II}^*}{\sqrt{2\pi r}} \begin{Bmatrix} -\sin\frac{\phi}{2}\left(2 + \cos\frac{\phi}{2}\cos\frac{3\phi}{2}\right) \\ \sin\frac{\phi}{2}\cos\frac{\phi}{2}\cos\frac{3\phi}{2} \\ \cos\frac{\phi}{2}\left(1 - \sin\frac{\phi}{2}\sin\frac{3\phi}{2}\right) \end{Bmatrix} \tag{6.22}
$$

An FE approximation to the smooth part $\boldsymbol{\sigma}^h_{smo}$ can be obtained subtracting $\boldsymbol{\sigma}^*_{sing}$ from the raw XFEM stress field $\boldsymbol{\sigma}^h$:

$$\boldsymbol{\sigma}^h_{smo} := \boldsymbol{\sigma}^h - \boldsymbol{\sigma}^*_{sing}. \tag{6.23}$$

$\boldsymbol{\sigma}^h_{smo}$ can be considered as an FE approximation to the smooth part of the solution because it is a discontinuous representation of a stress field where the singular part has been subtracted. An SPR-type technique, well suited for smooth problems, can then be applied over $\boldsymbol{\sigma}^h_{smo}$ to obtain $\boldsymbol{\sigma}^*_{smo}$. Once $\boldsymbol{\sigma}^*_{smo}$ and $\boldsymbol{\sigma}^*_{sing}$ have been evaluated, the final recovered stresses in the patch $\boldsymbol{\sigma}^*$ can be evaluated using (6.21).

The technique is computationally very efficient as it only makes use of the values of K_I and K_{II} which are a standard output of the XFEM analysis. In any case, there is no need to use this splitting technique in the whole problem domain. The use of this technique is only required in the vicinity of the singular point, as the stresses far from the singular point can be adequately recovered by SPR-type techniques. In any case, the splitting technique must always be used if part of the patch is contained within the XFEM enriched area of the domain (Ródenas et al., 2008). The case of multiple crack tips can be easily handled by simply defining splitting areas around each of the singular points.

6.3.1.2 Satisfaction of C^0 continuity using a partition of unity approach

In the original SPR technique proposed by Zienkiewicz and Zhu (1992a), although one could use (6.7) to evaluate the stresses at any point in the patch, the only values retained are the recovered stresses at the patch assembly nodes. Blacker and Belytschko (1994) proposed the use of the so-called *Conjoint Polynomials Enhancement*, which is a PU approach to enforce C^0 continuity from the local (patch-wise) definitions of the recovered stresses:

$$\boldsymbol{\sigma}^*(\mathbf{x}) = \sum_{i=1}^{n_v} N_i^v(\mathbf{x})\boldsymbol{\sigma}_i^*(\mathbf{x}), \qquad (6.24)$$

where N_i^v are the linear version of shape functions associated with the vertex nodes n_v. Note that in this equation the shape functions are used, in each element, to interpolate the different functions $\boldsymbol{\sigma}_i^*(\mathbf{x})$ evaluated from the different vertex nodes as opposite to the case of (6.10) where the nodal values of stresses $\boldsymbol{\sigma}_i^*(\mathbf{x}_i)$ are used in the interpolation. If linear elements are used, then N_i^v in (6.24) and N_i in (6.10) are the same, but they will be different in higher-order elements.

6.3.1.3 Satisfaction of equilibrium and compatibility equations

The SPR technique defines the stresses at each patch in terms of a polynomial expansion, see (6.7). The internal equilibrium and compatibility equations can be written in terms of the stresses. Therefore, one can introduce constraints relating the coefficients used to define the components of the recovered stresses at the patch so that $\boldsymbol{\sigma}^*$ satisfies the desired equations. This section is devoted to the evaluation of the constraints between these coefficients. Once the constraints have been evaluated its satisfaction will be enforced by means of the *Lagrange Multipliers Technique*. Let us assume that a quadratic 2D polynomial basis is used to define $\boldsymbol{\sigma}^*$. Then, each of its components σ_i^* ($i = xx, yy, xy$) and their first and second partial derivatives, expressed in a local Cartesian coordinate system $(\tilde{x}, \tilde{y})$, centered at the patch assembly node and parallel to the global Cartesian coordinate system (x, y), are given by:

$$
\begin{array}{rlllllll}
\sigma_i^*(\tilde{x}, \tilde{y}) = & +a_{i1} & +a_{i2}\tilde{x} & +a_{i3}\tilde{y} & +a_{i4}\tilde{x}^2 & +a_{i5}\tilde{x}\tilde{y} & +a_{i6}\tilde{y}^2 \\[2mm]
\dfrac{\partial \sigma_i^*(\tilde{x}, \tilde{y})}{\partial \tilde{x}} = & & +a_{i2} & & +2a_{i4}\tilde{x} & +a_{i5}\tilde{y} & \\[3mm]
\dfrac{\partial \sigma_i^*(\tilde{x}, \tilde{y})}{\partial \tilde{y}} = & & & +a_{i3} & & +a_{i5}\tilde{x} & +2a_{i6}\tilde{y} \\[3mm]
\dfrac{\partial^2 \sigma_i^*(\tilde{x}, \tilde{y})}{\partial \tilde{x}^2} = & & & & +2a_{i4} & & \\[3mm]
\dfrac{\partial^2 \sigma_i^*(\tilde{x}, \tilde{y})}{\partial \tilde{y}^2} = & & & & & & +2a_{i6} \\[3mm]
\dfrac{\partial^2 \sigma_i^*(\tilde{x}, \tilde{y})}{\partial \tilde{x} \partial \tilde{y}} = & & & & & +a_{i5} &
\end{array}
\qquad (6.25)
$$

The 2D *internal equilibrium equations* are given by:

$$
\begin{aligned}
\frac{\partial \sigma_{\tilde{x}\tilde{x}}}{\partial \tilde{x}} + \frac{\partial \sigma_{\tilde{x}\tilde{y}}}{\partial \tilde{y}} + b_{\tilde{x}} &= 0 \\[2mm]
\frac{\partial \sigma_{\tilde{x}\tilde{y}}}{\partial \tilde{x}} + \frac{\partial \sigma_{\tilde{y}\tilde{y}}}{\partial \tilde{y}} + b_{\tilde{y}} &= 0
\end{aligned}
\qquad (6.26)
$$

The second-order polynomial expansion of the stresses can represent linear body loads **b**. Let us assume that the body loads can be represented as linear functions:

$$
\begin{aligned}
b_{\tilde{x}}(\tilde{x}, \tilde{y}) &= b_{x1} + b_{x2}\tilde{x} + b_{x3}\tilde{y} \\
b_{\tilde{y}}(\tilde{x}, \tilde{y}) &= b_{y1} + b_{y2}\tilde{x} + b_{y3}\tilde{y}
\end{aligned}
\tag{6.27}
$$

Hence, considering the expressions given in (6.25) and (6.27), the internal equilibrium Equation (6.26) could be written for $\boldsymbol{\sigma}^*$ as:

$$
\begin{aligned}
(a_{xx2} + 2a_{xx4}\tilde{x} + a_{xx5}\tilde{y}) + (a_{xy3} + a_{xy5}\tilde{x} + 2a_{xy6}\tilde{y}) + (b_{x1} + b_{x2}\tilde{x} + b_{x3}\tilde{y}) &= 0 \\
(a_{xy2} + 2a_{xy4}\tilde{x} + a_{xy5}\tilde{y}) + (a_{yy3} + a_{yy5}\tilde{x} + 2a_{yy6}\tilde{y}) + (b_{y1} + b_{y2}\tilde{x} + b_{y3}\tilde{y}) &= 0
\end{aligned}
\tag{6.28}
$$

Therefore, the following constrains between the a coefficients should be satisfied to enforce the exact satisfaction of the equilibrium equation in the patch:

$$
\begin{aligned}
a_{xx2} + a_{xy3} = -b_{x1}, \qquad & 2a_{xx4} + a_{xy5} = -b_{x2}, \qquad && a_{xx5} + 2a_{xy6} = -b_{x3} \\
a_{xy2} + a_{yy3} = -b_{y1}, \qquad & 2a_{xy4} + a_{yy5} = -b_{y2}, \qquad && a_{xy5} + 2a_{yy6} = -b_{y3}
\end{aligned}
\tag{6.29}
$$

A similar procedure is used to impose the satisfaction of the *boundary equilibrium equation*. Let $\tilde{\mathbf{x}}_j$ be a point of the boundary of the domain, located in the patch of node i, where we want to impose the boundary equilibrium equation. Let us express the stress vector $\boldsymbol{\sigma}^*$ as $\boldsymbol{\sigma}^{*\prime}$, defined in a Cartesian reference system (x', y') such that x' is parallel to the outward normal vector at point $\tilde{\mathbf{x}}_j$, rotated an angle α with respect to $\tilde{x}$. The stress vector at $\tilde{\mathbf{x}}_j$ is given by:

$$
\boldsymbol{\sigma}^{*\prime}(\tilde{\mathbf{x}}_j) = \mathbf{R}(\alpha)\boldsymbol{\sigma}^*(\tilde{\mathbf{x}}_j)
\tag{6.30}
$$

where $\mathbf{R}$ is the stress rotation matrix:

$$
\mathbf{R} = \begin{bmatrix} \mathbf{r}_{x'x'} \\ \mathbf{r}_{y'y'} \\ \mathbf{r}_{x'y'} \end{bmatrix} = \begin{bmatrix} \cos^2 \alpha & \sin^2 \alpha & \sin(2\alpha) \\ \sin^2 \alpha & \cos^2 \alpha & -\sin(2\alpha) \\ -\sin(2\alpha)/2 & \sin(2\alpha)/2 & \cos(2\alpha) \end{bmatrix}
\tag{6.31}
$$

Let us use (6.13) to define $\mathbf{t}^*(\tilde{\mathbf{x}}_j)$ as the vector composed by the $x'x'$ and $y'y'$ components of $\boldsymbol{\sigma}^{*\prime}(\tilde{\mathbf{x}}_j)$:

$$
\mathbf{t}^*(\tilde{\mathbf{x}}_j) = \mathbf{R}_n(\alpha)\boldsymbol{\sigma}^*(\tilde{\mathbf{x}}_j) = \begin{bmatrix} \mathbf{r}_{x'x'} \\ \mathbf{r}_{x'y'} \end{bmatrix}_{2\times3} \begin{bmatrix} \mathbf{p}(\tilde{\mathbf{x}}_j) & 0 & 0 \\ 0 & \mathbf{p}(\tilde{\mathbf{x}}_j) & 0 \\ 0 & 0 & \mathbf{p}(\tilde{\mathbf{x}}_j) \end{bmatrix} \begin{Bmatrix} \mathbf{a}_{xx} \\ \mathbf{a}_{yy} \\ \mathbf{a}_{xy} \end{Bmatrix}
\tag{6.32}
$$

The satisfaction of the boundary equilibrium equation at $\mathbf{x}_j$ implies:

$$
\mathbf{t}^*(\tilde{\mathbf{x}}_j) = \mathbf{t}_n^{ex}(\tilde{\mathbf{x}}_j)
\tag{6.33}
$$

where the vector $\mathbf{t}_n^{ex}(\tilde{\mathbf{x}}_j)$ represents the exact known boundary tractions projected over the normal outward vector at $\tilde{\mathbf{x}}_j$.

The following expression defines the constraint equations that should be imposed to enforce the satisfaction of the equilibrium equation at $\tilde{\mathbf{x}}_j$.

$$\begin{bmatrix} \mathbf{r}_{x'x'} \\ \mathbf{r}_{x'y'} \end{bmatrix} \begin{bmatrix} \mathbf{p}(\tilde{\mathbf{x}}_j) & \mathbf{0} & \mathbf{0} \\ \mathbf{0} & \mathbf{p}(\tilde{\mathbf{x}}_j) & \mathbf{0} \\ \mathbf{0} & \mathbf{0} & \mathbf{p}(\tilde{\mathbf{x}}_j) \end{bmatrix} \begin{Bmatrix} \mathbf{a}_{xx} \\ \mathbf{a}_{yy} \\ \mathbf{a}_{xy} \end{Bmatrix} = \mathbf{t}_n^{ex}(\tilde{\mathbf{x}}_j) \tag{6.34}$$

Equation (6.34) defines two constraint equations (one for each component of the boundary tractions). Assuming straight contours and p^{th} order recovered stresses, we can impose the satisfaction of (6.34) at any $p + 1$ points on the boundary to enforce the exact satisfaction of the boundary equilibrium Equation (6.33) all along the boundary contained in the patch, provided the applied tractions can be represented by a pth order polynomial. Otherwise, obviously, the boundary equilibrium will not be fully satisfied[1].

Finally, the *2D compatibility equation* can be expressed in terms of stresses (see Timoshenko and Goodier, 1951) as:

$$\frac{\partial^2}{\partial \tilde{y}^2}\left(k\sigma_{\tilde{x}\tilde{x}} - vq\sigma_{\tilde{y}\tilde{y}}\right) + \frac{\partial^2}{\partial \tilde{x}^2}\left(k\sigma_{\tilde{y}\tilde{y}} - vq\sigma_{\tilde{x}\tilde{x}}\right) - 2(1+v)\frac{\partial^2\sigma_{\tilde{x}\tilde{y}}}{\partial \tilde{x}\partial \tilde{y}} = 0 \tag{6.35}$$

where

$$\begin{cases} k = 1, \ q = 1 & \text{for plane stress} \\ k = (1 - v)^2, \ q = (1 + v) & \text{for plane strain} \end{cases}$$

Considering the second partial derivatives shown in (6.25) the compatibility Equation (6.35) can be now written as:

$$k2a_{xx6} - vq2a_{yy6} + k2a_{yy4} - vq2a_{xx4} - 2(1+v)a_{xy5} = 0 \tag{6.36}$$

Equation (6.36) defines the constraint equation that must be imposed to enforce σ^* to satisfy (6.35) at any point into the patch.

6.3.1.4 Assembly of patches containing the crack

For patches intersected by the crack (see Figure 6.1), the recovery technique uses different stress interpolation polynomials on each side of the crack. The stress discontinuity along the crack can therefore be easily represented.

As the resulting sub-patches can contain a single integration subdomain, see, for example, the upper part of patch B in Figure 6.1 (and, thus, a reduced number of integration points), it must be ensured that each subdomain contains XFEM – computed stresses at least at the same number of points as the number of terms in the polynomials used to represent σ^*. If first-order polynomials (having three unknown coefficients) were used to define σ^*, then

---

1 Note that if the *singular + smooth* stress splitting technique is used in a patch that contains the Neumann boundary, then the boundary tractions must also be split. The boundary tractions to be considered for the satisfaction of the boundary equilibrium equation in the smooth problem are given by:

$$\mathbf{t}_{smo} = \mathbf{t} - \mathbf{t}_{sing}^*$$

where $\mathbf{t}_{sing}^*$ is the projection of σ_{sing}^*.

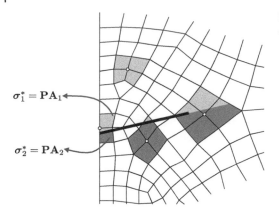

Figure 6.1 Patches intersected by the crack. Two patches are assembled, one on each side of the crack.

$\sigma_1^* = \mathbf{P}\mathbf{A}_1$

$\sigma_2^* = \mathbf{P}\mathbf{A}_2$

at least three integration points should be used in each integration subdomain. Labbe and Garon (1995) used this procedure in a FEM framework to overcome the difficulties associated with using SPR in patches containing few elements.

If the patch is contained within the *singular+smooth* stress decomposition area, then this procedure is used to obtain σ_{smo}^* at each part of the crack. Note that the σ_{sing}^* is discontinuous along the crack faces. Therefore, the resulting recovered stress field in the patch $\sigma^* = \sigma_{smo}^* + \sigma_{sing}^*$ will also be discontinuous. Once we have a different stress interpolation at each part of the crack, we can impose constrains to force the recovered stresses to satisfy equilibrium and compatibility constraints at each side. Note that crack faces are considered as any other external boundary when imposing the contour equilibrium equation. When the crack faces are not in contact, the set of equations used to evaluate the stresses at one side of the patch is not coupled with the set at the other side of the crack. However, if crack faces are in contact, one could impose constrain equations between both sets of equations to enforce stress continuity (in the stresses normal to the surface and in the shear stresses) along the crack surfaces.

6.3.1.5 Continuous least squares approximation to σ^h

In FEM, all the elements in a patch have the same number of integration points. So, we can consider that there is a uniform density of integration points whose values of σ^h will be used to obtain σ^* using the least squares approximation. However, in the XFEM context, for integration purposes, each element is subdivided into integration subdomains, each of them having its own integration points. This is especially important around the crack tip where a high density of integration points is required. To account for the different densities of integration points in the patch the discrete definition of the functional in (6.8) is substituted if redefined as a continuous functional using:

$$\Pi(\mathbf{a}) = \int_{\Omega_i} \left(\sigma^*(\mathbf{x}) - \sigma^h(\mathbf{x})\right)^2 d\Omega = \int_{\Omega_i} \left(\mathbf{p}(\mathbf{x})\mathbf{a} - \sigma^h(\mathbf{x})\right)^2 d\Omega \tag{6.37}$$

that transforms the expressions of $\mathbf{M}$ and $\mathbf{G}$ given in (6.9) into their continuous counterparts that are numerically integrated using the values of the stresses at the stress sampling points in the patch:

$$\mathbf{M} = \int_{\Omega_i} \mathbf{p}^T(\mathbf{x})\mathbf{p}(\mathbf{x})d\Omega \quad \approx \sum_{j=1}^{nsp} \mathbf{p}^T(\mathbf{x}_j)\mathbf{p}(\mathbf{x}_j)H(\mathbf{x}_j)$$

$$\mathbf{G} = \int_{\Omega_i} \mathbf{p}^T(\mathbf{x})\sigma^h(\mathbf{x})d\Omega \quad \approx \sum_{j=1}^{nsp} \mathbf{p}^T(\mathbf{x}_j)\sigma^h(\mathbf{x}_j)H(\mathbf{x}_j) \tag{6.38}$$

where nsp is the number of stress sampling points in the patch and $H(\mathbf{x}_j)$ represents the area associated to each of these sampling points. We can say that (6.38) represents an improved version of (6.9) where each of the sampling points is weighted by its corresponding area.

6.3.1.6 Nearly statically admissible recovered stress field σ^*

The SPR-CX technique provides a recovered stress field that is equilibrated at patch level and continuous. Continuity of σ^* is obtained from the local descriptions of the (equilibrated) stresses in each patch σ_i^* by means of the PU concept using Equation (6.24). Taking into account that $\sum_{i=1}^{n_v} N_i^v = 1$ and that at each patch $\nabla \cdot \sigma_i^* = -\mathbf{b}$, the divergence of σ^* is given by:

$$\nabla \cdot \sigma^* = \nabla \sum_{i=1}^{n_v} N_i^v \sigma_i^* = \sum_{i=1}^{n_v} \nabla \cdot N_i^v \sigma_i^* + \sum_{i=1}^{n_v} N_i^v \underbrace{\nabla \cdot \sigma_i^*}_{-\mathbf{b}} = -\mathbf{s} - \mathbf{b} \tag{6.39}$$

with

$$\mathbf{s} = \sum_{i=1}^{n_v} \nabla \cdot N_i^v \sigma_i^* \tag{6.40}$$

The previous expression shows that σ^* does not satisfy the internal equilibrium equation but a modified version of this equation, where the term $\mathbf{s}$ represents the small lacks of internal equilibrium introduced by the use of the PU concept to enforce the stress continuity. In elements where the *singular + smooth* stress splitting has been used, these lacks of equilibrium come only from σ_{smo}^* because σ_{sing}^* (see Equation (6.22)) satisfies the equilibrium equations as it is the exact solution of a linear elasticity problem. A polynomial basis of degree p can represent stresses of degree p but only body loads $\mathbf{b}$ of degree $p - 1$. This would represent a further source of equilibrium defaults if the actual body loads cannot be represented by the polynomial basis selected. A similar problem can appear if the polynomial basis selected to represent the stress field is unable to represent the exact boundary tractions $\mathbf{t}$. In this case, a boundary residual traction $\mathbf{r} = \sigma^*\mathbf{n} - \mathbf{t}$, being $\mathbf{n}$ the outward normal surface vector, would also be present. Therefore, the recovered stress field σ^* provided by the SPR-CX technique would satisfy the following modified versions of the equilibrium equations, where $\mathbf{s}$ and $\mathbf{r}$ are small compared respectively to $\mathbf{b}$ and $\mathbf{t}$ if the SPR-CX is used:

$$-\nabla \cdot \sigma^* = \mathbf{b} + \mathbf{s} \qquad\qquad \text{in } \Omega \tag{6.41}$$

$$\sigma^*\mathbf{n} = \mathbf{t} + \mathbf{r} \qquad\qquad \text{on } \Gamma_N \tag{6.42}$$

6.3.1.7 The Westergaard problem. Numerical model

The Westergaard problem (Gdoutos, 1993) has been selected to present the results obtained with the recovery techniques described in this chapter. This problem has been selected because it is one of the few LEFM problems in mixed mode where the analytical solution is available. Giner et al. (2005) and Ródenas et al. (2008) show the explicit expressions for the stress fields in terms of the spatial coordinates.

The Westergaard problem corresponds to an infinite plate loaded at infinity with biaxial tractions $\sigma_{x\infty} = \sigma_{y\infty} = \sigma_\infty$ and shear traction τ_∞, presenting a crack of length $2a$ as shown in Figure 6.2. Combining the externally applied loads we can obtain different loading conditions: pure mode I, II, or mixed mode.

In the numerical model, only a finite portion of the domain ($a = 1$ and $b = 4$ in Figure 6.2) is considered. The projection of the stress distribution corresponding to the analytical Westergaard solution for modes I and II, given by the expressions below, is applied to its boundary:

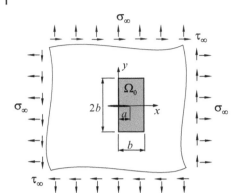

Figure 6.2 Westergaard problem. Infinite plate with a crack of length $2a$ under uniform tractions σ_∞ (biaxial) and τ_∞. Finite portion of the domain Ω_0, modeled with FE.

$$\sigma_x^I(x,y) = \frac{\sigma_\infty}{\sqrt{|t|}}\left[\left(x\cos\frac{\phi}{2} - y\sin\frac{\phi}{2}\right) + y\frac{a^2}{|t|^2}\left(m\sin\frac{\phi}{2} - n\cos\frac{\phi}{2}\right)\right]$$

$$\sigma_y^I(x,y) = \frac{\sigma_\infty}{\sqrt{|t|}}\left[\left(x\cos\frac{\phi}{2} - y\sin\frac{\phi}{2}\right) - y\frac{a^2}{|t|^2}\left(m\sin\frac{\phi}{2} - n\cos\frac{\phi}{2}\right)\right] \tag{6.43}$$

$$\tau_{xy}^I(x,y) = y\frac{a^2\sigma_\infty}{|t|^2\sqrt{|t|}}\left(m\cos\frac{\phi}{2} + n\sin\frac{\phi}{2}\right)$$

$$\sigma_x^{II}(x,y) = \frac{\tau_\infty}{\sqrt{|t|}}\left[2\left(y\cos\frac{\phi}{2} + x\sin\frac{\phi}{2}\right) - y\frac{a^2}{|t|^2}\left(m\cos\frac{\phi}{2} + n\sin\frac{\phi}{2}\right)\right]$$

$$\sigma_y^{II}(x,y) = y\frac{a^2\tau_\infty}{|t|^2\sqrt{|t|}}\left(m\cos\frac{\phi}{2} + n\sin\frac{\phi}{2}\right) \tag{6.44}$$

$$\tau_{xy}^{II}(x,y) = \frac{\tau_\infty}{\sqrt{|t|}}\left[\left(x\cos\frac{\phi}{2} - y\sin\frac{\phi}{2}\right) + y\frac{a^2}{|t|^2}\left(m\sin\frac{\phi}{2} - n\cos\frac{\phi}{2}\right)\right]$$

where the stress fields are expressed as a function of x and y, with origin at the center of the crack. The parameters t, m, n, and ϕ are defined as:

$$t = (x + iy)^2 - a^2 = (x^2 - y^2 - a^2) + i(2xy) = m + in$$

$$m = \text{Re}(t) = \text{Re}(z^2 - a^2) = x^2 - y^2 - a^2$$

$$n = \text{Im}(t) = (z^2 - a^2) = 2xy \tag{6.45}$$

$$\phi = \text{Arg}(\bar{t}) = \text{Arg}(m - in) \qquad \text{with } \phi \in [-\pi, \pi],\ i^2 = -1$$

For the problem analyzed, the exact value of the SIF is given as:

$$K_{\text{I},ex} = \sigma\sqrt{\pi a} \qquad\qquad K_{\text{II},ex} = \tau\sqrt{\pi a} \tag{6.46}$$

Three different loading configurations corresponding to the *pure mode I* ($\sigma_\infty = 100$, $\tau_\infty = 0$), *pure mode II* ($\sigma_\infty = 0$, $\tau_\infty = 100$), and *mixed mode* ($\sigma_\infty = 30$, $\tau_\infty = 90$) cases of the Westergaard problem are considered.

Sequences of uniformly refined structured (Figure 6.3) and unstructured (Figure 6.4) meshes are considered. In the first case, the mesh sequence is defined so that the crack tip always coincided with a node. For a more general scenario, this condition has not been applied for the unstructured meshes.

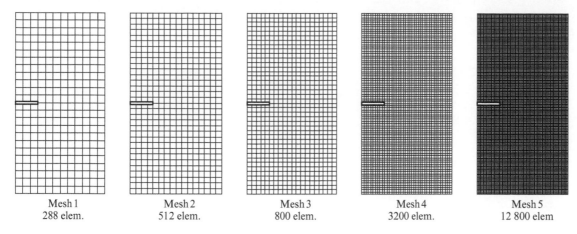

Figure 6.3 Sequence of structured meshes.

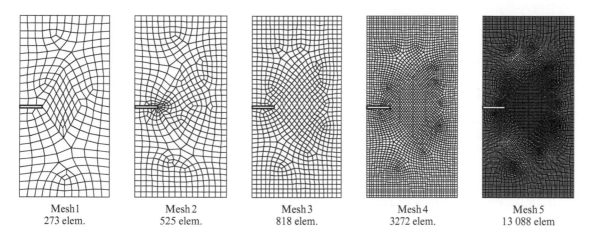

Figure 6.4 Sequence of unstructured meshes.

The XFEM model considers geometrical enrichment with a radius $r_e = 0.5$ without blending elements. Bilinear elements are used in the models. Young's modulus is $E = 10^7$ and Poisson's ratio $v = 0.333$. For the numerical integration of standard elements, a 2×2 Gaussian quadrature rule is considered. For split elements, a seven Gauss points quadrature is used in each triangular subdomain, and a 5×5 quasipolar integration (Béchet et al., 2005) in the subdomains of the element containing the crack tip.

For a fare comparison between the different recovery techniques described in this chapter, the numerical examples shown in this and the following sections will use the model of the Westergaard problem previously described.

6.3.1.8 Numerical results

In the numerical analysis, a *singular+smooth* decomposition area of radius $\rho = 0.5$ equal to the radius r_e is used for the fixed enrichment area. The radius for the Plateau function used to extract the SIF required to evaluate σ^*_{sing} is $r_q = 0.9$.

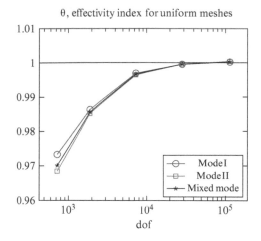
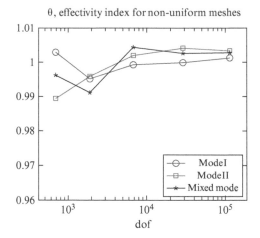

Figure 6.5 Evolution of the global effectivity index for the error estimator based on the SPR-CX technique with uniform (left) and non-uniform (right) meshes.

Figure 6.5 shows the evolution of the global effectivity index obtained for the sequences of uniform and non-uniform meshes. The behavior in all cases is quite similar, showing effectivity indexes very close to the desired value $\theta = 1$.

The behavior of the estimates is also preserved at the local level. The local effectivity is represented in Figure 6.6 by the quantity D defined at elements as:

$$D = \theta^e - 1 \qquad if \ \theta^e \geq 1$$
$$D = 1 - 1/\theta^e \qquad if \ \theta^e < 1 \tag{6.47}$$

Index D is particularly well suited for the local representation of the effectivity index because it fairly compares[2] the underestimation of the error ($D < 0$) and the overestimation ($D > 0$). The good behavior of the estimates results in values of D close to zero.

Figure 6.6 displays the distribution of D in the mesh 2 of the sequences of meshes shown in Figures 6.3 and 6.4. Note that the values of D are very close to zero. This accuracy would be useful to efficiently drive adaptive processes.

This high accuracy of the error estimator both at local and global levels is a direct consequence of the accuracy of the recovered stress field σ^*. Figure 6.7 represents a sample of the values of the stress error at the integration points obtained for each stress component and for the von Mises stresses ($\sigma_{xx}, \sigma_{yy}, \sigma_{xy}$, and σ_{vm}), for the XFEM and recovered stresses considering mixed mode and non-uniform meshes. To obtain these graphs, the value at each integration point has been represented as a constant value in a prescribed area surrounding each integration point. The figures under each graph represent the error range in each case. These figures show that the error range of the recovered stress field is about one order of magnitude smaller than that of the XFEM solution. This is a very interesting characteristic of these types of error estimation techniques, as the process for the evaluation of the error involves the evaluation of an enhanced recovered field that can be used instead of the FEM or XFEM solutions.

---

2 Note that error overestimations would be represented by values of θ_e in $(1, \infty)$ whereas underestimations would only be represented in $(0, 1)$.

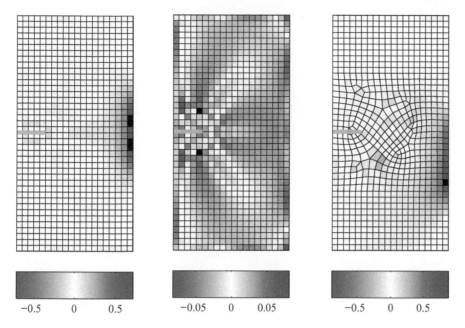

Figure 6.6 Local effectivity index obtained with the SPR-CX technique. Mesh 2 of the uniform and non-uniform sequences.

6.3.2 The XMLS Technique

The XMLS technique extends the work of Tabbara et al. (1994) for FEM to enriched approximations. The general idea of the XMLS is to use the displacement solution provided by XFEM to obtain a recovered displacement field $\mathbf{u}^*$ and then the recovered strain field ε^* that would be used in (6.3) for the error estimation in energy norm (Bordas and Duflot, 2007; Bordas et al., 2008).

The main adaptations proposed by Bordas and Duflot (2007) for the use of the MLS procedure in the XFEM framework are as follows:

- Use of a diffraction criterion to describe the crack geometry;
- Enrichment of the polynomial basis with functions that describe the first-order asymptotic expansion at the crack tip.

6.3.2.1 Crack geometry described by a diffraction criterion
To describe the discontinuity, the distance s in the weight function W, see (6.18), is redefined using the *diffraction criterion* (Belytschko et al., 1996) as depicted in Figure 6.8. The weight function is continuous except across the crack faces, since the points at the other side of the crack are not considered as part of the support. Near the crack tip, the value of $W(\chi)$ diminishes as the crack hides the point. When the point χ is hidden by the crack the following expression is used, otherwise (6.19) is used:

$$s = \frac{\left\|\mathbf{x} - \chi_\lambda\right\| + \left\|\chi - \chi_\lambda\right\|}{R_{\Omega_\mathbf{x}}} \tag{6.48}$$

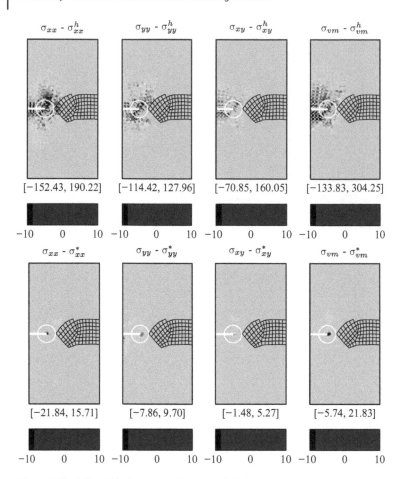

Figure 6.7 Mixed Mode, non-uniform mesh 2. Exact error in σ_{xx}, σ_{yy}, σ_{xy} and σ_{vm} obtained by XFEM and SPR-CX. Under each graph is indicated the range of the error in stresses.

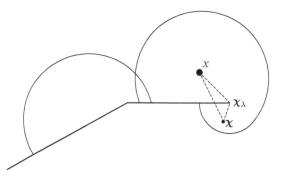

Figure 6.8 Diffraction criterion to introduce the discontinuity in the XMLS approximation.

6.3.2.2 Enrichment of the polynomial basis

The MLS basis can include not only polynomial terms, but also other terms. In order to obtain a more accurate description of the singular stress field around the singularity, Bordas and Duflot (2007) and Bordas et al. (2008) enriched the basic linear polynomial basis **p** with functions that describe the first-order asymptotic expansion at the crack tip:

$$F(r,\phi) = \left[\sqrt{r}\sin\left(\frac{\phi}{2}\right), \sqrt{r}\cos\left(\frac{\phi}{2}\right), \sqrt{r}\sin\left(\frac{\phi}{2}\right)\cos(\phi), \sqrt{r}\cos\left(\frac{\phi}{2}\right)\sin(\phi)\right] \tag{6.49}$$

such that:

$$\mathbf{p} = [1, x, y, [F_1(r,\phi), F_2(r,\phi), F_3(r,\phi), F_4(r,\phi)]] \tag{6.50}$$

Note that although the enriched basis can reproduce the singular behavior of the solution around the crack tip, the resulting recovered field not necessarily satisfies the equilibrium equations.

6.3.2.3 Numerical results

Figure 6.9 shows the evolution of the global effectivity index obtained by means of the use of the XMLS technique. The behavior in all cases is quite similar, showing effectivity indexes that asymptotically tend to the desired value $\theta = 1$ although not as close to 1 as in the SPR-CX case.

The behavior of the error estimate at the elements of the second mesh of the sequences is represented in terms of the local effectivity D in Figure 6.10. The values of D are close to zero; however, there is a noticeable overestimation close to the singular point when compared to Figure 6.15. In this case, the enrichment of the basis is not as accurate as the splitting technique when capturing the singular features of the solution.

The graphs shown in Figure 6.11 represent a sample of the values of the stress error at the integration points obtained for each stress component and for the von Mises stresses for the recovered stress field. These errors can be compared with the corresponding errors of the XFEM stresses in Figure 6.7. These figures show that the error decreases in the domain of the problem where the solution is smooth but is not particularly better in the regions influenced by the singularity. This behavior indicates that the splitting approach provides a more accurate representation of the singular field than the enrichment of the basis used in the XMLS and is in agreement with the observations made for the local effectivity index.

6.3.3 The MLS-CX Technique

The MLS-CX technique was developed by Ródenas et al. (2012) as an attempt to evaluate a statically admissible recovered stress field that would be useful to obtain upper bounds of the error in energy norm. Statical admissibility is obtained if the stress field is continuous and equilibrated. As previously explained, the SPR-CX technique warranties local (patch-wise) equilibrium. Equation (6.24) was used to enforce continuity. Nevertheless, this post-process slightly modifies the global equilibrium and, as a result, the final stress field is continuous but the

Figure 6.9 Evolution of the global effectivity index for the error estimator based on the XMLS technique with uniform and non-uniform meshes.

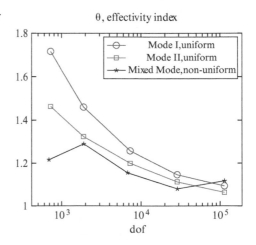

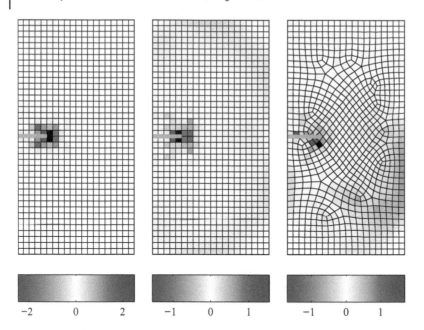

Figure 6.10 Local effectivity index obtained with the XMLS technique. Mesh 2 of the uniform and non-uniform sequences.

equilibrium condition is not exactly satisfied. The MLS technique avoids this post-process because it directly provides a continuous recovered field. Therefore, the MLS technique would be a good alternative to obtain a statically admissible recovered stress field.

The objective of this section is to show the MLS-CX technique, which makes use of the ideas previously described to enforce the satisfaction of the equilibrium equations in the SPR-CX. The MLS technique requires two main improvements:

- To enforce the satisfaction of boundary equilibrium;
- To enforce the satisfaction of internal equilibrium.

Additionally, the following improvements, taken from the SPR-CX and XMLS techniques, will be used to improve the accuracy of the resulting recovered stress field:

- Use of the *singular + smooth* stress decomposition technique;
- Use of the continuous instead of the discrete MLS approximation to σ^h;
- Use of a diffraction criterion to describe the crack geometry.

6.3.3.1 Satisfaction of the boundary equilibrium equation

The boundary equilibrium equation must be satisfied at each point along the contour. There would be two cases, one for the points whose support does not contain the boundary (the boundary constraint would not be enforced in this case) and another for the points whose support contains it.

Let $\mathbf{x}$ be a point whose support $\Omega_{\mathbf{x}}$ contains part of the Neumann boundary Γ_N. The first alternative to consider boundary equilibrium in the evaluation of $\sigma^*(\mathbf{x})$ is to enforce the equilibrium at $\Gamma_N \bigcap \Omega_{\mathbf{x}}$ using the Lagrange Multipliers technique. Note that this technique would create a discontinuity in σ^* in the interior of the domain as the use of the Lagrange Multipliers technique abruptly appears when approaching the boundary.

A *nearest point* approach, that smoothly introduces the satisfaction of the boundary equilibrium equation, can be used to avoid creating discontinuities in σ^*. For a point $\mathbf{x} \in \Omega$ whose support $\Omega_{\mathbf{x}}$ intersects the boundary Γ_N,

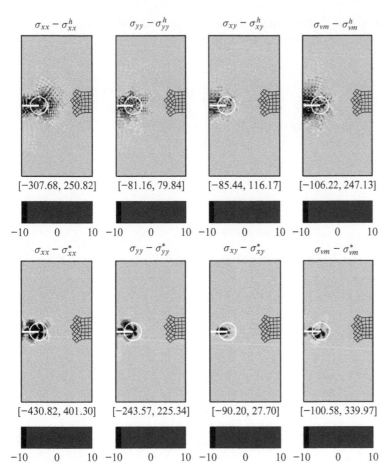

Figure 6.11 Mixed Mode, non-uniform mesh 2. Exact error in σ_{xx}, σ_{yy}, σ_{xy} and σ_{vm} obtained by the XMLS technique. Under each graph is indicated the range of the error in stresses.

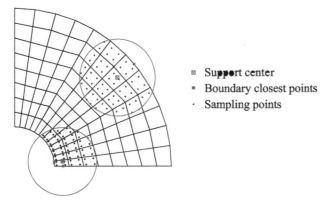

⊠ Support center
▪ Boundary closest points
· Sampling points

Figure 6.12 MLS support with boundary conditions applied on the nearest boundary points.

the equilibrium constraints are imposed only in the closest points $\chi_j \in \Gamma_N$ on the boundaries within the support to **x**, as shown in Figure 6.12.

Let $\boldsymbol{\sigma}^{*'}(\mathbf{x},\chi)$ be the stress vector $\boldsymbol{\sigma}^*(\mathbf{x},\chi)$ expressed in a coordinate system $x'y'$ aligned with Γ_N at χ_j such that x' is the outward normal vector, rotated an angle α with respect to axis x of the global coordinates system:

$$\boldsymbol{\sigma}^{*'}(\mathbf{x},\chi) = \mathbf{R}(\alpha)\boldsymbol{\sigma}^*(\mathbf{x}) = \begin{bmatrix} \mathbf{r}_{x'x'} \\ \mathbf{r}_{y'y'} \\ \mathbf{r}_{x'y'} \end{bmatrix} \boldsymbol{\sigma}^*(\mathbf{x},\chi)$$

$$= \begin{bmatrix} \cos^2\alpha & \sin^2\alpha & \sin(2\alpha) \\ \sin^2\alpha & \cos^2\alpha & -\sin(2\alpha) \\ -\sin(2\alpha)/2 & \sin(2\alpha)/2 & \cos(2\alpha) \end{bmatrix} \boldsymbol{\sigma}^*(\mathbf{x},\chi)$$

(6.51)

where $\mathbf{R}$ is the stress rotation matrix.

The MLS functional expressed in its continuous version (see (6.16)) and incorporating the boundary constraints would read:

$$J(\mathbf{x}) = \int_{\Omega_\mathbf{x}} W(\mathbf{x}-\chi)\left[\boldsymbol{\sigma}^*(\mathbf{x},\chi) - \boldsymbol{\sigma}^h(\chi)\right]^2 d\chi +$$

$$\sum_{j=1}^{nbc} W'\left(\mathbf{x}-\chi_j\right)\left[\boldsymbol{\sigma}'^*_i\left(\mathbf{x},\chi_j\right) - \boldsymbol{\sigma}'^{ex}_i\left(\chi_j\right)\right]^2$$

(6.52)

$$= \int_{\Omega_\mathbf{x}} W(\mathbf{x}-\chi)\left[\mathbf{P}(\chi)\mathbf{A}(\mathbf{x}) - \boldsymbol{\sigma}^h(\chi)\right]^2 d\chi +$$

$$\sum_{j=1}^{nbc} W'\left(\mathbf{x}-\chi_j\right)\left[\mathbf{r}_{i'}(\alpha)\mathbf{P}(\chi_j)\mathbf{A}(\mathbf{x}) - \boldsymbol{\sigma}'^{ex}_i\left(\chi_j\right)\right]^2 \quad i' = x'x', x'y'$$

where *nbc* is the number of points χ_j on the boundary used to enforce the known boundary constraints $\boldsymbol{\sigma}'^{ex}_i$ (in general, those would be the normal $\boldsymbol{\sigma}_{x'x'}$ and tangential $\boldsymbol{\sigma}_{x'y'}$ stresses). Evaluating $\partial J/\partial \mathbf{A} = 0$ results in the linear system $\mathbf{M}(\mathbf{x})\mathbf{A}(\mathbf{x}) = \mathbf{G}(\mathbf{x})$ used to evaluate $\mathbf{A}$, where, in this case:

$$\mathbf{M} = \int_{\Omega_\mathbf{x}} W(\mathbf{x}-\chi)\mathbf{P}^T(\chi)\mathbf{P}(\chi)d\chi + \sum_{j=1}^{nbc} W'(\mathbf{x}-\chi_j)\mathbf{P}^T(\chi_j)\mathbf{r}_{i'}^T\mathbf{r}_{i'}\mathbf{P}(\chi_j)$$

(6.53)

$$\mathbf{G} = \int_{\Omega_\mathbf{x}} W(\mathbf{x}-\chi)\mathbf{P}^T(\chi)\boldsymbol{\sigma}^h(\chi)d\chi + \sum_{j=1}^{nbc} W'(\mathbf{x}-\chi_j)\mathbf{P}^T(\chi_j)\mathbf{r}_{i'}^T\boldsymbol{\sigma}'^{ex}_i(\chi_j)$$

(6.54)

In the previous equations, W' is a weighting function defined as:

$$W'(\mathbf{x}-\chi_j) = \frac{W(\mathbf{x}-\chi_j)}{s} = \begin{cases} \dfrac{1}{s} - 6s + 8s^2 - 3s^3 & \text{if } |s| \leq 1 \\ 0 & \text{if } |s| > 1 \end{cases}$$

(6.55)

This function has two main characteristics:

1. Contains W (see (6.18)); therefore, the term used to take into account the information about the boundary equilibrium smoothly appears in the functional $J(\mathbf{x})$. As a result, $\boldsymbol{\sigma}^*$ will be continuous in Ω.
2. W' also includes s^{-1} such that the weight of the boundary constraint in $J(\mathbf{x})$ increases when approaching the boundary (when $\mathbf{x} \to \chi_j$ $s \to 0$); therefore, $\boldsymbol{\sigma}^*$ will tend to exactly satisfy boundary equilibrium as $\mathbf{x} \to \chi_j$ (see Figure 6.13). Note that to estimate the error using the numerical integration in (6.2), the value of $\boldsymbol{\sigma}^*$ is

never evaluated on the boundary (where $s = 0$) because the integration points considered are always inside the elements. The term s^{-1} could become a source of numerical errors as the system of equations obtained might become ill-conditioned in the vicinity of the boundary. However, as shown in (6.19), s is a distance normalized with the radius of the support $R_{\Omega_x} = 2 \times element\ size$. Then, s^{-1} will not take high values, not even when mesh refinement locates integration points $\mathbf{x}$ very close to the boundary in global coordinates.

If the recovered stresses are to be evaluated at a particular point $\mathbf{x}$ on the boundary, where $s = 0$, and in order to avoid the singular term $1/s$ into the functional (6.52), we would eliminate the boundary terms in this functional and use the Lagrange Multipliers technique to enforce $\boldsymbol{\sigma}^*$ to satisfy boundary equilibrium at $\mathbf{x}$. The resulting recovered stress field will still be continuous if the *closest point approach* is used at points not on the boundary.

6.3.3.2 Satisfaction of the internal equilibrium equation
The recovered stress field $\boldsymbol{\sigma}^*$ must also satisfy the internal equilibrium equation:

$$\nabla \cdot \boldsymbol{\sigma}^* + \mathbf{b} = \mathbf{0} \tag{6.56}$$

The required constraints are imposed via the Lagrange Multipliers technique. To do this the spatial derivatives of $\boldsymbol{\sigma}^*$ must be evaluated. Considering (6.13), $\nabla \cdot \boldsymbol{\sigma}^*$ is expressed as:

$$\nabla \cdot \boldsymbol{\sigma}^* = (\nabla \cdot \mathbf{P})\mathbf{A} + \mathbf{P}(\nabla \cdot \mathbf{A}) \tag{6.57}$$

The first terms in (6.57) can be directly evaluated by differentiating the polynomial basis, or, alternatively, using an equilibrated polynomial basis. Previous works have only considered the first term in the satisfaction of the appropriate equations, thus only providing a pseudo-satisfaction of these equations (Huerta et al., 2004). Therefore, the second term in (6.57) must also be obtained. To evaluate it, we differentiate the linear system $\mathbf{MA} = \mathbf{G}$:

$$(\nabla \cdot \mathbf{M})\mathbf{A} + \mathbf{M}(\nabla \cdot \mathbf{A}) = \nabla \cdot \mathbf{G} \tag{6.58}$$

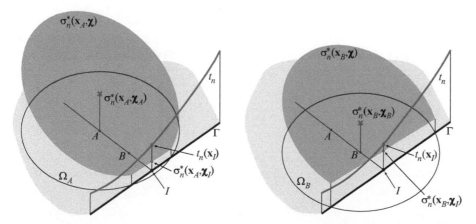

Figure 6.13 Satisfaction of boundary equilibrium. $\sigma_n^*(\mathbf{x}_A,\chi)$ and $\sigma_n^*(\mathbf{x}_B,\chi)$ are the values of $\boldsymbol{\sigma}^*(\mathbf{x},\chi)$, projected along the direction normal to boundary Γ at I, in the supports Ω_A and Ω_B of the points A and B, whose nearest point on Γ is I. t_n represents the normal tractions applied on Γ. Note that $\sigma_n^*(\mathbf{x},\chi_I) \neq t_n(\mathbf{x}_I)$ although $\sigma_n^*(\mathbf{x}_B,\chi_I)$ is more accurate than $\sigma_n^*(\mathbf{x}_B,\chi_I)$ is more accurate than $\sigma_n^*(\mathbf{x}_A,\chi_I)$. Thus, as $\mathbf{x} \to \mathbf{x}_I$, $\sigma_n^*(\mathbf{x},\chi_I) \to t_n(\mathbf{x}_I)$ and, similarly the value of the stresses evaluated at the center of the support $\sigma_n^*(\mathbf{x},\mathbf{x}) \to t_n(\mathbf{x}_I)$.

Evaluating $\nabla \cdot \mathbf{A}$ from (6.58), replacing in (6.57) and expanding leads to:

$$\frac{\partial \boldsymbol{\sigma}^*}{\partial x} = \left(\frac{\partial \mathbf{P}}{\partial x} - \mathbf{PM}^{-1}\frac{\partial \mathbf{M}}{\partial x}\right)\mathbf{A} + \mathbf{PM}^{-1}\frac{\partial \mathbf{G}}{\partial x} = \mathbf{E}_{,x}\mathbf{A} + \mathbf{f}_{,x} \tag{6.59}$$

$$\frac{\partial \boldsymbol{\sigma}^*}{\partial y} = \left(\frac{\partial \mathbf{P}}{\partial y} - \mathbf{PM}^{-1}\frac{\partial \mathbf{M}}{\partial y}\right)\mathbf{A} + \mathbf{PM}^{-1}\frac{\partial \mathbf{G}}{\partial y} = \mathbf{E}_{,y}\mathbf{A} + \mathbf{f}_{,y} \tag{6.60}$$

where the partial derivatives of $\mathbf{M}$ and $\mathbf{G}$ with respect, for example, to x are given by:

$$\frac{\partial \mathbf{M}}{\partial x} = \int_{\Omega_x} \frac{\partial W(\mathbf{x} - \boldsymbol{\chi})}{\partial x}\mathbf{P}^T(\boldsymbol{\chi})\mathbf{P}(\boldsymbol{\chi})d\boldsymbol{\chi} + \sum_{j=1}^{nbc} \frac{\partial W'(\mathbf{x} - \boldsymbol{\chi}_j)}{\partial x}\mathbf{P}^T(\boldsymbol{\chi}_j)\mathbf{r}_{i'}^T\mathbf{r}_{i'}\mathbf{P}(\boldsymbol{\chi}_j) \tag{6.61}$$

$$\frac{\partial \mathbf{G}}{\partial x} = \int_{\Omega_x} \frac{\partial W(\mathbf{x} - \boldsymbol{\chi})}{\partial x}\mathbf{P}^T(\boldsymbol{\chi})\boldsymbol{\sigma}^h(\boldsymbol{\chi})d\boldsymbol{\chi} + \sum_{j=1}^{nbc} \frac{\partial W'(\mathbf{x} - \boldsymbol{\chi}_j)}{\partial x}\mathbf{P}^T(\boldsymbol{\chi}_j)\mathbf{r}_{i'}^T\boldsymbol{\sigma}_t'^{ex}(\boldsymbol{\chi}_j) \tag{6.62}$$

Differentiating (6.18) and (6.55), we obtain:

$$\frac{\partial W(\mathbf{x} - \boldsymbol{\chi})}{\partial x} = \frac{\partial W(\mathbf{x} - \boldsymbol{\chi})}{\partial s}\frac{\partial s}{\partial x} \tag{6.63}$$

$$\frac{\partial W'(\mathbf{x} - \boldsymbol{\chi}_j)}{\partial x} = \frac{\partial W'(\mathbf{x} - \boldsymbol{\chi}_j)}{\partial s}\frac{\partial s}{\partial x} \tag{6.64}$$

where $\partial s/\partial x$ can be obtained from (6.19) or, alternatively, from (6.48). Equations (6.59) and (6.60) are expressed as a function of $\mathbf{A}$, so, we can write the two terms of the internal equilibrium equation (6.56) as a function of the vector of unknowns $\mathbf{A}$:

$$\frac{\partial \sigma_{xx}^*}{\partial x} + \frac{\partial \sigma_{xy}^*}{\partial y} + b_x = \left(\mathbf{E}_{xx,x} + \mathbf{E}_{xy,y}\right)\mathbf{A} + \left(f_{xx,x} + f_{xy,y}\right) + b_x = 0 \tag{6.65}$$

$$\frac{\partial \sigma_{xy}^*}{\partial x} + \frac{\partial \sigma_{yy}^*}{\partial y} + b_y = \left(\mathbf{E}_{xx,y} + \mathbf{E}_{yy,y}\right)\mathbf{A} + \left(f_{xy,x} + f_{yy,y}\right) + b_y = 0 \tag{6.66}$$

where $[\ ]_{i,j}$ ($i = xx, yy, xy$ and $j = x, y$) represents the row in $[\ ]_{,j}$ corresponding to the ith component of the stresses. These expressions define the constraints between the coefficients $\mathbf{A}$ required to satisfy the internal equilibrium equation at $\mathbf{x}$. Lagrange Multipliers are used to impose these constraint equations.

The use of the Lagrange Multipliers technique to impose the equilibrium constraint (6.65) and (6.66) in (6.52) leads to the following system of equations:

$$\begin{bmatrix} \mathbf{M} & \mathbf{C}^T \\ \mathbf{C} & \mathbf{0} \end{bmatrix}\begin{bmatrix} \mathbf{A} \\ \lambda \end{bmatrix} = \begin{bmatrix} \mathbf{G} \\ \mathbf{B} \end{bmatrix} \tag{6.67}$$

where $\mathbf{C}$ and $\mathbf{B}$ are the terms used to impose the constraint equations and λ is the vector of Lagrange Multipliers.

However, in (6.58) it was assumed that $\mathbf{A}$ is evaluated solving $\mathbf{MA} = \mathbf{G}$, although, operating by blocks in (6.67), the following system of equations is obtained:

$$\mathbf{MA} + \mathbf{C}^T\lambda = \mathbf{G} \tag{6.68}$$

Therefore, this formulation neglects the term $\mathbf{C}^T\lambda$ when evaluating $\nabla\mathbf{A}$. This implies that the internal equilibrium equation is not fully satisfied. In any case, this approximation represents an enhancement with respect to the pseudo satisfaction of equilibrium as indicated by Huerta et al. (2004). The use of this recovery technique normally produces overestimations of the error in energy norm although small small underestimations could be obtained as a consequence of the approximation used.

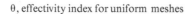

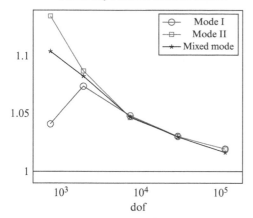

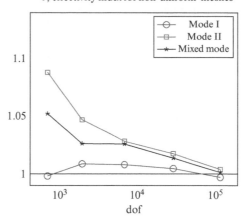

Figure 6.14 Evolution of the global effectivity index for the error estimator based on the MLS-CX technique with uniform (left) and non-uniform (right) meshes.

As in the case of the use of the SPR-CX technique (see (6.41)), the recovered stress field σ^* obtained with the MLS-CX technique satisfies a modified version of the internal equilibrium equation. Note that in this case the procedure used to consider the boundary tractions enforces the satisfaction of the boundary equilibrium equation without any modification.

In this case, the residual of the internal equilibrium equation **s** can be evaluated at any point in the domain, taking into account (6.57), from:

$$\nabla \cdot \sigma^* + \mathbf{b} = (\nabla \cdot \mathbf{P})\mathbf{A} + \mathbf{P}(\nabla \cdot \mathbf{A}) + \mathbf{b} = -\mathbf{s} \tag{6.69}$$

6.3.3.3 Numerical results

Figure 6.14 shows the evolution of the global effectivity index obtained using the MLS-CX technique. For the three loading modes, the effectivity is close to the optimal value $\theta = 1$ and improves as we increase the number of dof. In this example, for uniform meshes, the recovery technique yields values of the effectivity $\theta > 1$. For non-uniform meshes, the results are similar although for two meshes under mode I the technique underestimates the true error.

The performance of the estimate at the element level is represented in terms of the local effectivity D in Figure 6.15 for the second mesh of the sequences for mode I, mode II, and mixed mode. The local values of D indicate a better performance than the XMLS but still less accurate than the splitting approach close to the crack tip.

The graphs shown in Figure 6.16 represent a sample of the values of the stress error at the integration points obtained for each stress component and for the von Mises stresses for the recovered stress field. These errors can be compared with the corresponding errors of the XFEM stresses in Figures 6.7 and 6.11. As the splitting procedure helps to accurately describe the singular behavior of the solution, the exact error in the recovered field is approximately one order of magnitude smaller than the error of the FE solution. The results are similar to those provided by the SPR-CX technique and outperform those of the XMLS in the vicinity of the crack tip.

6.3.4 On the Roles of Enhanced Recovery and Admissibility

The numerical results presented by Bordas and Duflot (2007); Bordas et al. (2008), and Ródenas et al. (2008) showed promising accuracy for both the XMLS and the SPR$_{\text{XFEM}}$ techniques. It is apparent in those papers that

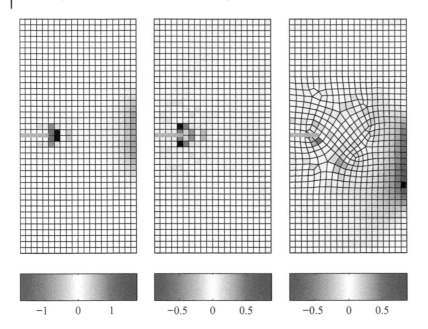

Figure 6.15 Local effectivity index obtained with the MLS-CX technique. Mesh 2 of the uniform and non-uniform sequences.

SPR_{XFEM} led to effectivity indexes remarkably close to unity. Yet, these papers do not shed any significant light on the respective roles played by two important ingredients in those error estimators, namely:

1. The statical admissibility of the recovered solution;
2. The enrichment of the recovered solution.

A systematic study of the results obtained when considering these two features in recovery-based error estimators for LEFM problems solved with XFEM techniques is available in González-Estrada et al. (2012).

The analysis of the effect of these two features can be easily done considering the SPR-CX and XMLS techniques and two additional techniques, respectively denoted as SPR-C and the SPR-X, that partially use the enhancements of SPR-CX:

- **SPR-C technique**. This partial implementation of the SPR-CX technique considers a recovery procedure that enforces both internal and boundary equilibrium, but does not include the *singular+smooth* splitting technique.
- **SPR-X technique**. The *singular+smooth* stress splitting is considered in this case without enforcing any kind of equilibration in the smooth part of the solution. Note that although the evaluation of the smooth recovered field σ^*_{smo} does not consider any kind of equilibration, the recovered singular field σ^*_{sing} is fully equilibrated because (6.22) satisfies the equilibrium equations as this equation is the solution of a LEFM problem.

Table 6.1 summarizes the main features of the different recovery procedures we considered.

An example of the performance of each of these four techniques can be observed in Figure 6.17 which represents the local effectivity D in the vicinity of the crack tip for the fourth structured mesh in the Mode I case. The results in the rest of the cases are similar. The figure shows that, although accurate error estimations are obtained along the boundaries of the model, the SPR-C technique is clearly not able to reproduce the singular fields, which is shown by large values of D inside the enriched region. This behavior of the SPR-C technique inside the enriched region is not surprising if we take into account that the SPR-C technique tries to represent the solution by means of polynomial

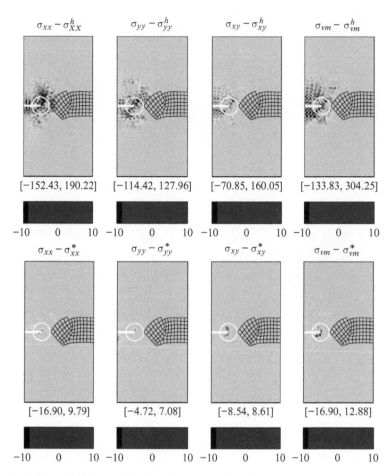

Figure 6.16 Mixed Mode, non-uniform mesh 2. Exact error in σ_{xx}, σ_{yy}, σ_{xy} and σ_{vm} obtained by the MLS-CX technique. Under each graph is indicated the range of the error in stresses.

Table 6.1 Comparison of features for the different recovery procedures.

	Singular functions considered	Equilibrated singular functions	Locally equilibrated field
SPR-X	Yes	Yes	No
SPR-C	No	Not applicable	Yes
SPR-CX	Yes	Yes	Yes
XMLS	Yes	No	No

functions in an area where the XFEM solution is already making use of enrichment functions to represent the singularity. The XMLS and SPR-X techniques make use of singular functions to represent the solution around the crack tip; however, the use of an equilibrated singular solution in the case of the SPR-X technique produces a considerable improvement with respect to the XMLS technique, which, in any case, produces very accurate results along the boundaries of the domain, including the crack faces, where the SPR-X technique shows a poor behavior.

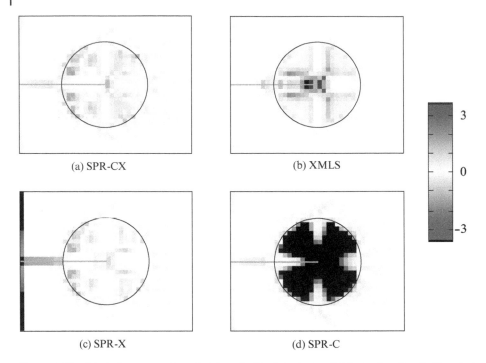

(a) SPR-CX

(b) XMLS

(c) SPR-X

(d) SPR-C

Figure 6.17 Mode I, structured mesh 4. Local effectivity index D (ideal value $D = 0$). The circles denote the enriched zones around the crack tips. Dark blue and red colors indicate values of D out of range.

This behavior is clearly improved by the SPR-CX technique which provides accurate error estimations both into the enriched area and along the boundaries of the model.

Figure 6.18 is used now to compare the global accuracy obtained with the different recovery techniques in terms of energy norm, whose theoretical error convergence rate is $\mathcal{O}(h^p)$, h being the element size and p the polynomial degree of the FE interpolation functions (1 for linear elements), or $\mathcal{O}(\mathrm{dof}^{-p/2})$ in terms of the number of dof. Figure 6.18(left) shows the convergence of the estimated errors in the energy norm $\|\mathbf{e}_{es}\|$ (see Equation (6.2)). The convergence of the exact error $\|\mathbf{e}\|$ is shown for comparison. It can be observed that the evolution of exact error and that of the estimated error obtained with the SPR-CX and SPR-X techniques are practically indistinguishable in the graph, although the numerical values show that the SPR-CX technique is slightly more accurate, leading to the best approximation to the exact error in the energy norm. The convergence rates of error estimations provided by the XMLS, SPR-X, and SPR-CX techniques are practically coincident with that of the exact error. The non-singular formulation SPR-C greatly overestimates the error with a suboptimal convergence rate. The convergence rate of the estimated error can be used as a quality measure of the error estimator in cases where the exact solution is not available as this value should be (or at least should tend to be) equal to the convergence rate of the XFEM solution. Error estimators not converging at the optimal convergence rate, as in the case of the SPR-C technique, should be disregarded.

The accuracy and convergence rate of the error in energy norm of the recovered field $\|\mathbf{e}^*\| = \|\mathbf{u} - \mathbf{u}^*\|$ is compared for the error estimators in Figure 6.18(right) in order to evaluate the quality of the recovery techniques. In terms of accuracy, SPR-CX is the best method closely followed by SPR-X. XMLS errors are half an order of magnitude larger than SPR-CX/SPR-X, but superior to the non-enriched non-enriched recovery method SPR-C. It can be seen that the XMLS, the SPR-CX, and the SPR-X techniques provide convergence rates higher than the convergence rate for the exact error $\|\mathbf{e}\|$, which implies that the error estimator is asymptotically exact (Zienkiewicz and Zhu, 1992b),

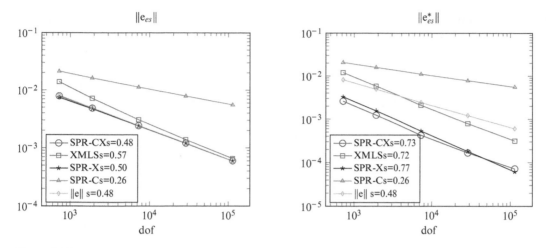

Figure 6.18 Mode I, structured meshes: convergence of the estimated error in the energy norm $\|\mathbf{e}_{es}\|$ (left) and convergence of the exact error of the recovered solutions. The evolution of the exact error of the xfem solution $\|\mathbf{e}\|$ is shown for comparison.

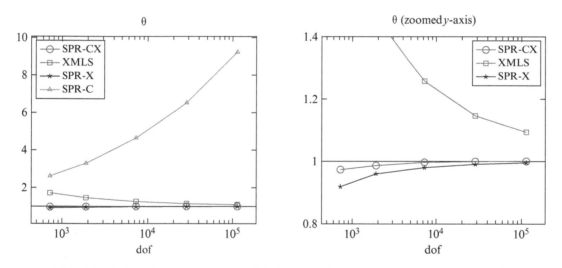

Figure 6.19 Global effectivity index θ for mode I and structured meshes.

i.e., the estimated error tends to the exact error when the mesh is refined. Again, these results indicate that the SPR-C technique cannot be used to estimate the error of the XFEM solution.

The accuracy of the error estimator obtained with the different recovery techniques can be clearly analyzed in Figure 6.19, which shows the evolution of the effectivity index θ of the error estimator with respect to the number of dof. The SPR-C technique does not consider the use of near-tip fields to evaluate the recovered solution; therefore, although it includes the satisfaction of the equilibrium equations, the effectivity of the error estimator does not converge to the optimal value ($\theta = 1$). In the other three cases where near crack tips functions have been considered, the error estimator can be considered as asymptotically exact as the effectivity index will tend to the optimal value. This shows the importance of introducing a description of the singular field in the recovery process, i.e., of using *extended recovery techniques* in XFEM .

Regarding the enforcement of the equilibrium equations, we can see that an equilibrated formulation (SPR-CX) leads to better effectivities compared to other non-equilibrated configurations (SPR-X, XMLS), which is an indication of the advantages associated with equilibrated recoveries in this context.

Still, in Figure 6.19, the results for the SPR-CX and XMLS show that the values of the global effectivity index are close to (and less than) unity for the SPR-CX estimator, while the XMLS is a lot less effective but shows effectivities larger than 1.

6.4 Recovery Techniques in Error Bounding. Practical Error Bounds.

Error estimators can normally overestimate or underestimate the exact error. For this reason, the use of bounding techniques that guarantee certain levels of accuracy is preferred. Dual analysis techniques have been used in Pereira et al. (1999) and de Almeida and Pereira (2006) to obtain upper error bounds from an equilibrated and a compatible FE solutions. The residual-type error estimators are considered to have a sounder mathematical basis and are the most commonly used technique to obtain error bounds. These methods build up a statically admissible stress field, naturally providing upper bounds of the energy norm of the error (Ainsworth and Oden, 2000; Bangerth and Rannacher, 2003; Díez et al., 1998, 2003). This has been a traditional advantage of the residual type estimators over recovery-based error estimators which were unable to produce guaranteed upper bounds of the error in energy norm. However, Díez et al. (2007) proposed an error bounding technique based on a nearly equilibrated recovered stress field. They obtained upper error bounds using a recovered stress field that almost satisfies the equilibrium equations, and then corrected the estimate evaluated with this field with a term that takes into account the lack of equilibrium. Then, Ródenas et al. (2010) made use of the SPR-CX technique to extend this approach to the XFEM context.

Díez et al. (2007) showed that if σ^* is continuous and equilibrated, then the ZZ error estimator is an upper bound of the error in energy norm. Note that σ^h is the stress field derived from the kinematically admissible displacement field provided by (X)FEM. Then, if σ^* is continuous and equilibrated and thus statically admissible, the integral in (6.2) would represent the square of the CRE $\mathbf{e}_{cre}$ associated with the pair $(\sigma^*, \mathbf{u}^h)$. Using the Práger-Synge theorem (Práger and Synge, 1947), Ladevèze et al. (1999) and Panetier et al. (2010, 2009) have shown that $\mathbf{e}_{cre}$ satisfies $||\mathbf{e}|| \leq \mathbf{e}_{cre}$.

Díez et al. (2007) and Ródenas et al. (2010) showed that an upper bound of the error in energy norm can be obtained with the following expression, that corrects the expression of the ZZ error estimator in Equation (6.2) using two terms to account for possible internal and boundary equilibrium residuals, $\mathbf{s}$ and $\mathbf{r}$:

$$||\mathbf{e}||^2 \leq \int_{\Omega} \left(\sigma^* - \sigma^h \right)^T \mathbf{D}^{-1} \left(\sigma^* - \sigma^h \right) d\Omega - 2 \int_{\Omega} \mathbf{e} \cdot \mathbf{s} d\Omega - 2 \int_{\Gamma_N} \mathbf{e} \cdot \mathbf{r} d\Gamma \tag{6.70}$$

where $\mathbf{e}$ is the exact error in the displacements field, $\mathbf{e} = \mathbf{u} - \mathbf{u}^h$. In the case of the SPR-CX technique, the residual of the internal equilibrium equation $\mathbf{s}$ is evaluated using Equation (6.40) whereas the residual of the boundary equilibrium equation $\mathbf{r}$ can be evaluated using:

$$\mathbf{r} = \sigma^* \cdot \mathbf{n} - \mathbf{t} = \sum_{i=1}^{n_v} \left(N_i^v \sigma_i^* \right) \cdot \mathbf{n} - \mathbf{t} = \sum_{i=1}^{n_v} N_i^v \left(\sigma_i^* \cdot \mathbf{n} - \mathbf{t} \right) \tag{6.71}$$

In the case of the MLS-CX technique, $\mathbf{s}$ is evaluated using (6.69)

The computer implementation of the upper error bound in Equation (6.70) requires the estimation of $\mathbf{e}$. Ródenas et al. (2010) proposed a procedure to estimate $\mathbf{e}$ based on the use of a sequence of increasingly refined meshes used for the analysis. Let us consider a sequence of N meshes. The displacement field $\mathbf{u}^h_{(N)}$ obtained as the solution from mesh N can be considered as an approximation to the exact solution of the problem, $\mathbf{u} \approx \mathbf{u}^h_{(N)}$. This allows the evaluation of an approximation of $\mathbf{e}$ for the first $N - 1$ meshes of the sequence:

$$\mathbf{e}_{(i)} = \mathbf{u} - \mathbf{u}^h_{(i)} \approx \mathbf{u}^h_{(N)} - \mathbf{u}^h_{(i)} = \mathbf{e}_{es(i)} \tag{6.72}$$

The use of (6.72) in the evaluation of $\mathbf{e}_{es(i)}$, $i = 1, ..., N - 1$ requires the projection of the displacement field of mesh N on mesh i. This is a standard data interpolation technique, especially simple and efficient if nested meshes are used. Finally, it is possible to obtain an estimate of $-2 \int_{\Omega} \mathbf{e} \cdot \mathbf{s} d\Omega$ and $-2 \int_{\Gamma_N} \mathbf{e} \cdot \mathbf{r} d\Gamma$ for mesh N considering the values of these integrals evaluated for the previous meshes and the use of extrapolation from mesh $N-1$ taking into account the convergence rate of these integrals numerically evaluated from the mesh sequence. As the evaluation of $\mathbf{e}_{es}$ introduces an approximation, the implementation of Equation (6.70) could produce underestimations of the exact error in energy norm. Figure 6.20 shows the evolution of the exact and estimated values of the integrals used in (6.70) as correction terms for the error estimator. These graphs show that the technique used to estimate $\mathbf{e}$ at each mesh provides very accurate approximations of the correction terms in (6.70); thus, in the vast majority of cases the error is correctly bounded, although small underestimations of the error could be obtained in some cases.

As Figure 6.20 shows, in the case of the SPR-CX technique, the value of $-2 \int_{\Omega} \mathbf{e} \cdot \mathbf{s} d\Omega$ is considerably higher than $-2 \int_{\Gamma_N} \mathbf{e} \cdot \mathbf{r} d\Gamma$ because the technique used to impose boundary equilibrium is very accurate (see Section 6.3.1); therefore, in practice, this last term can be neglected. Note that, in fact, this term is not present when the MLS-CX technique is considered as in the implementation of the MLS-CX technique σ^* will satisfy equilibrium at any point along the Neumann boundary. The figure shows that the technique used to estimate the values of the correction terms is very accurate.

Figure 6.21 shows the effectivity index obtained with the SPR-CX technique in the examples analyzed. The effectivity of the error estimator without any correction terms (E_{SPR-CX} curve) is compared with the effectivity index of the upper bound obtained when the correction term corresponding to the residuals of the internal equilibrium equation ($-2 \int_{\Omega} \mathbf{e} \mathbf{s} d\Omega$) is considered. Two curves will represent the results obtained when the residual $\mathbf{s}$ is taken into account. Curve $E_{UB_{ex}}$ represents the effectivity of the upper bound when the exact error $\mathbf{e}$ in the displacement field is used. On the other hand, the E_{UB} curve represents the effectivity of the upper bound obtained using the estimation of $\mathbf{e}$ evaluated using (6.72) and the extrapolation of $-2 \int_{\Omega} \mathbf{e} \mathbf{s} d\Omega$ from mesh $N-1$ to mesh N previously described. This figure shows that although the error estimator (E_{SPR-CX} curve) can underestimate or overestimate the exact error in energy norm, the technique proposed to evaluate upper bounds has always generated sharp overestimations of the exact error.

The technique described for the evaluation on an upper bound of the error in energy norm based on the use of the SPR-CX technique neglects the effect of residuals of boundary equilibrium and uses an estimation of the correction term used to account for the residual of the internal equilibrium equation. Therefore, this technique can only be considered as a computed version of the upper bound which could produce underestimations of the exact error in some cases. However, although the error bounds obtained are not mathematically guaranteed, the

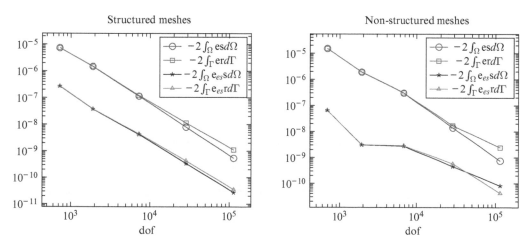

Figure 6.20 SPR-CX technique. Evolution of estimated and exact values of the correction terms in (6.70) form for mode I in the cases of structured(left) and un-structured(right) meshes.

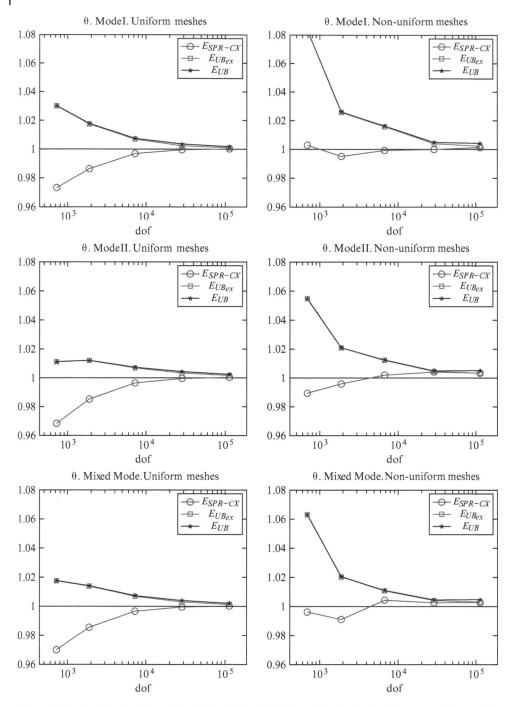

Figure 6.21 SPR-CX technique. Evolution of the global effectivity index for the error estimator (E_{SPR-CX} curve) and the upper bound obtained including both the exact ($E_{UB_{ex}}$ curve) and the estimated (E_{UB} curve) correction terms for uniform (left) and non-uniform (right) meshes.

numerical results show that it can be considered as a valid technique to obtain practical sharp upper bounds of the error in energy norm in the XFEM framework.

The equilibrium defaults evaluated using (6.40) in the case of the SPR-CX technique are due to the use of (6.24) to enforce stress continuity. However, in the case of the MLS-CX technique, continuity in σ^* is directly obtained. Considering $\mathbf{MA} = \mathbf{G}$ during the MLS-CX recovery process instead of (6.68) is an accurate approximation that also leads to small equilibrium defaults $\mathbf{s}$ that can be evaluated using (6.69). As seen in Figure 6.14, the accuracy of this approximation has always lead to effectivities greater than 1, except for Mode I in the case of non-uniform meshes, where the exact error was slightly underestimated in meshes 1 and 5, leading to effectivity indexes of $\theta = 0.9981$ and $\theta = 0.9970$ using the error estimator in (6.2) without any correction terms. Very small underestimations of the exact error in energy norm have also been reported in some cases in the FEM context.

Therefore, practical upper bounds of the error in energy norm in XFEM can be obtained with the ZZ error estimation using the SPR-CX technique and the correction term to consider the internal equilibrium residual $\mathbf{s}$ or, alternatively, using the MLS-CX technique without any correction terms. In the first case, the displacement projection technique (which implies the evaluation of a sequence of meshes) used to obtain an estimate of the error of the displacement field $\mathbf{e}_{es}$ could represent a drawback of the technique that could be partially alleviated if nested meshes are used. Alternatively, a displacement recovery technique that would produce an approximation $\mathbf{u}^*$ to the exact solution $\mathbf{u}$ could also be considered. In the MLS-CX case, no correction terms are required to obtain practical upper bounds at the expense of a higher computational cost. Figure 6.22 compares the results previously displayed for the upper error bounds based on the use of the SPR-CX and MLS-CX. As previously mentioned, in both cases, the proposed computer implementation of the upper bound does not mathematically guarantee that the error values obtained are always an upper bound of the exact error, but, because of their accuracy and their ability to obtain upper bounds in the XFEM context in numerical experiments with benchmark problems, these two techniques represent a practical alternative to the use of other error bounding techniques. In any case, the graphs in this figure show that the upper error bounds obtained with the SPR-CX technique are usually more accurate than those obtained with the MLS-CX technique. Each of these two techniques to obtain practical upper error bounds have their own pros and cons which will be exposed below to help the reader who wants to use them to make a decision about which one of them should be implemented.

SPR-CX technique:

- In the examples analyzed, the SPR-CX technique was, in general, more accurate than the MLS-CX technique.
- It is an element-based technique; therefore, it is useful for FEM, XFEM, or other numerical techniques based on an element structure. However, it cannot be directly applied if there is not such an element structure, like in the meshless methods.
- It requires the use of a sequence of meshes, at least 3, to be able to project the results evaluated in the last mesh to the rest of the meshes in the sequence and to extrapolate the value of the error correction term from the last-but-one mesh to the last mesh. This projection process can be time consuming, although its cost could be alleviated if mesh splitting based on a hierarchical data structure is used for mesh refinement.
- Enforcing-continuous recovered stress field is cumbersome along element edges with non-conforming nodes where C^0 continuity must be enforced.
- The number of systems of equations to be solved is equal to the number of vertex nodes in the mesh. Values at integration points into the elements are obtained by interpolation.
- In SPR-based techniques, the boundary equilibrium equation is only exactly satisfied if the boundary tractions can be represented by the polynomials used to evaluate σ^*. For example, SPR-based techniques will not be able to exactly reproduce the boundary tractions given by the solution of the Westergaard problem along the boundary of the domain used to model the problem as this solution involves non-polynomial terms. Curved contours can also be a source of non-satisfaction of the boundary equilibrium equation. For example, a constant pressure applied on a circular contour cannot be represented by the available SPR-based techniques.

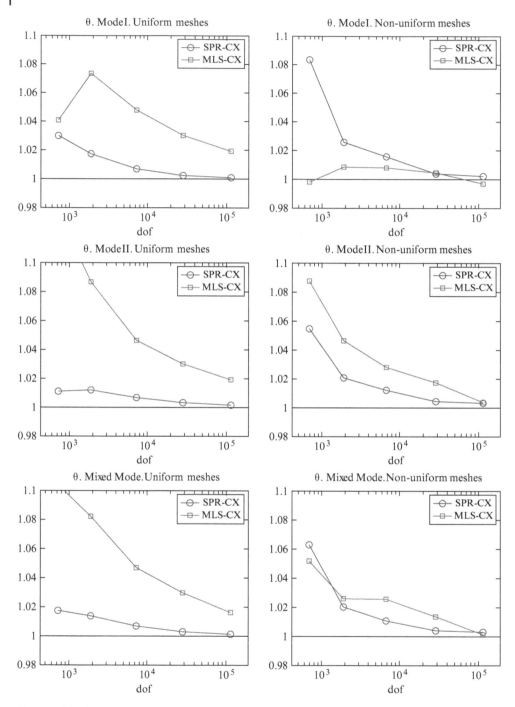

Figure 6.22 Evolution of effectivity index of upper error bounds. Comparison between results obtained with SPR-CX and MLS-CX techniques.

MLS-CX technique:

- The error estimation procedure can be carried out by means of the use of one single mesh. The use of a sequence of meshes is not required in this case.
- The recovery process does not need an element mesh and would therefore be useful for meshless methods. It does not require any special procedure to enforce C^0 continuity; therefore, adapting the MLS-CX recovery technique to other numerical methods is easier than adapting the SPR-CX recovery technique.
- One system of equations must be solved at each point where the recovered stress field is required. Reducing the number of integrations points will thus reduce the computational cost of this technique at the expense of reducing the accuracy of the integration.
- The closest point approach considered in the MLS-CX technique provides exact satisfaction of the boundary equilibrium equation at any point on the boundaries with imposed tractions.

6.5 Error Estimation in Quantities of Interest

The energy norm has been traditionally considered as the magnitude used to improve the quality of FE discretizations via adaptive analysis where the error in energy norm is progressively reduced. However, the goal of numerical simulations is the determination of a particular QoI which is needed for taking decisions during the design process, and not the accurate evaluation of the energy norm. Therefore, it is important to guarantee the quality of the analyses by controlling the error of the approximation in terms of the QoI rather than the global energy norm. Significant advances in the late 1990s introduced a new approach focused on the evaluation of error estimates of local quantities which are crucial in applications (Ainsworth and Oden, 2000; Ladevèze et al., 1999; Paraschivoiu et al., 1997). In order to obtain an error estimate for the QoI, two different problems are solved: the primal problem which is the FE approximation under consideration and the dual or adjoint problem which is related to the (non)linear functional that describes the QoI. Goal-oriented error estimators have been usually developed from the basis of residual formulations and the widely used strategy of solving a dual problem (Ainsworth and Oden, 2000; Oden and Prudhomme, 2001). Although limited, use of recovery techniques to evaluate error bounds in quantities of interest can be found in, e.g., Cirak and Ramm (1998) and Rüter and Stein (2006). Rüter and Stein (2006) verified that although the use of a standard SPR technique for the recovery of the primal and dual solutions did not provide an error estimate which guarantees the upper bound property, the SPR-based error estimators were sharper than those based on residuals.

This section presents a short introduction to the basics of error estimation in quantities of interest. We will show how the ideas behind the error estimators measured in energy norm may be utilized to estimate the error in a particular QoI (Ainsworth and Oden, 2000). The extended strategy consists of solving a primal problem, which is the problem at hand, and a dual problem useful to extract information on the QoI. From the FE solutions of both problems, it is possible to estimate the contribution of each of the elements to the error in the QoI. This error measure allows to adapt the mesh using procedures similar to traditional techniques based on the estimated error in energy norm. The adaptation of recovery-based error estimation techniques to error estimation in QoI in the context of XFEM will be described in this section, focusing our attention on the case where the QoI are the values of the SIFs.

6.5.1 Recovery-based Estimates for the Error in Quantities of Interest

Consider the linear elasticity problem related to the original or primary problem to be solved, which henceforth will be called the *primal problem*. Denote, in vectorial form, σ and ε as the stresses and strains, $\mathbf{D}$ as the elasticity matrix of the constitutive relation $\sigma = \mathbf{D}\varepsilon$, and the unknown displacement field $\mathbf{u}$, which values in $\Omega \subset \mathbb{R}^2$. $\mathbf{u}$ is the solution of the boundary value problem given by:

$$-\nabla \cdot \boldsymbol{\sigma}(\mathbf{u}) = \mathbf{b} \qquad \text{in } \Omega \qquad (6.73)$$

$$\boldsymbol{\sigma}(\mathbf{u}) \cdot \mathbf{n} = \mathbf{t} \qquad \text{on } \Gamma_N \qquad (6.74)$$

$$\mathbf{u} = \bar{\mathbf{u}} \qquad \text{on } \Gamma_D, \qquad (6.75)$$

where Γ_N and Γ_D denote the Neumann and Dirichlet boundaries with $\partial\Omega = \overline{\Gamma_N \cup \Gamma_D}$ and $\Gamma_N \cap \Gamma_D = \emptyset$, $\mathbf{b}$ are body loads, and $\mathbf{t}$ are the tractions imposed along Γ_N.

Consider the initial stresses $\boldsymbol{\sigma}_0$ and strains ε_0, the symmetric bilinear form $a : (\mathcal{V} + \bar{\mathbf{u}}) \times \mathcal{V} \to \mathbb{R}$ and the continuous linear form $\ell : \mathcal{V} \to \mathbb{R}$ are defined by:

$$a(\mathbf{u}, \mathbf{v}) := \int_\Omega \boldsymbol{\sigma}^T(\mathbf{u})\varepsilon(\mathbf{v})\mathrm{d}\Omega = \int_\Omega \boldsymbol{\sigma}^T(\mathbf{u})\mathbf{D}^{-1}\boldsymbol{\sigma}(\mathbf{v})\mathrm{d}\Omega \qquad (6.76)$$

$$\ell(\mathbf{v}) := \int_\Omega \mathbf{v}^T \mathbf{b}\mathrm{d}\Omega + \int_{\Gamma_N} \mathbf{v}^T \mathbf{t}\mathrm{d}\Gamma + \int_\Omega \boldsymbol{\sigma}^T(\mathbf{v})\varepsilon_0\mathrm{d}\Omega - \int_\Omega \varepsilon^T(\mathbf{v})\boldsymbol{\sigma}_0\mathrm{d}\Omega. \qquad (6.77)$$

With these notations, the variational form of the problem reads (Verfürth, 1999):

$$\text{Find } \mathbf{u} \in (\mathcal{V} + \bar{\mathbf{u}}) : \forall \mathbf{v} \in \mathcal{V} \quad a(\mathbf{u}, \mathbf{v}) = \ell(\mathbf{v}) \qquad (6.78)$$

where $\mathcal{V}$ is the standard test space for the elasticity problem such that $\mathcal{V} = \{\mathbf{v} \mid \mathbf{v} \in [H^1(\Omega)]^2, \mathbf{v}|_{\Gamma_D}(\mathbf{x}) = \mathbf{0}\}$.

Let $\mathbf{u}^h$ be an FE approximation of $\mathbf{u}$. The solution for the discrete counterpart of the variational problem in (6.78) lies in a subspace $(\mathcal{V}^h + \bar{\mathbf{u}}) \subset (\mathcal{V} + \bar{\mathbf{u}})$ associated with a mesh of FEs of characteristic size h, and it is such that:

$$\forall \mathbf{v}^h \in \mathcal{V}^h \subset \mathcal{V} \quad a(\mathbf{u}^h, \mathbf{v}^h) = \ell(\mathbf{v}^h) \qquad (6.79)$$

Let us define $Q : \mathcal{V} \to \mathbb{R}$ as a bounded linear functional representing some QoI, acting on the space $\mathcal{V}$ of admissible functions for the problem at hand. The goal is to estimate the error in $Q(\mathbf{u})$ when calculated using the value of the approximate solution $\mathbf{u}^h$:

$$Q(\mathbf{u}) - Q(\mathbf{u}^h) = Q(\mathbf{u} - \mathbf{u}^h) = Q(\mathbf{e}) \qquad (6.80)$$

As it will be shown later, $Q(\mathbf{v})$ may be interpreted as the work associated with a displacement field $\mathbf{v}$ and a distribution of forces specific to each type of QoI. So if we particularize $Q(\mathbf{v})$ for $\mathbf{v} = \mathbf{u}$, this force distribution will allow to extract information concerning the magnitude of interest associated with the solution of the problem in (6.78).

A standard procedure to evaluate $Q(\mathbf{e})$ consists of solving the auxiliary dual problem (also called *adjoint* or *extraction* problem) defined as:

$$\text{Find } \mathbf{w}_Q \in \mathcal{V} : a(\mathbf{v}, \mathbf{w}_Q) = Q(\mathbf{v}) \quad \forall \mathbf{v} \in \mathcal{V}. \qquad (6.81)$$

An exact representation for the measure error $Q(\mathbf{e})$ in terms of the solution of the dual problem can be simply obtained by substituting $\mathbf{v} = \mathbf{e}$ in (6.81) and considering that, due to the Galerkin orthogonality, $a(\mathbf{e}, \mathbf{w}_Q^h) = 0$ such that:

$$Q(\mathbf{e}) = a(\mathbf{e}, \mathbf{w}_Q) = a(\mathbf{e}, \mathbf{w}_Q) - a(\mathbf{e}, \mathbf{w}_Q^h) = a(\mathbf{e}, \mathbf{w}_Q - \mathbf{w}_Q^h) = a(\mathbf{e}, \mathbf{e}_Q) \qquad (6.82)$$

Note that in (6.82) the error in the QoI is related to the errors in the FE approximations $\mathbf{u}^h$ and $\mathbf{w}_Q^h$. On that account, the available procedures to estimate the error in energy norm may be considered to obtain estimates of the error in the QoI. We will therefore consider (6.82) and a ZZ-type approach to describe an error estimator for QoI.

Equation (6.82) can be rewritten as:

$$Q(\mathbf{e}) = a(\mathbf{e}, \mathbf{e}_Q) = \int_\Omega \left(\boldsymbol{\sigma}_p - \boldsymbol{\sigma}_p^h \right) \mathbf{D}^{-1} \left(\boldsymbol{\sigma}_d - \boldsymbol{\sigma}_d^h \right) d\Omega \tag{6.83}$$

where $\boldsymbol{\sigma}_p$ is the stress field associated with the primal solution and $\boldsymbol{\sigma}_d$ is the one associated with the dual solution.

As the exact values of $\boldsymbol{\sigma}_p$ and $\boldsymbol{\sigma}_d$ are unknown, we can follow Zienkiewiz and Zhu's idea to derive an estimator for the error in the QoI substituting these field by their recovered versions $\boldsymbol{\sigma}_p^*$ and $\boldsymbol{\sigma}_d^*$:

$$Q(\mathbf{e}) \approx Q(\mathbf{e}_{es}) = \int_\Omega \left(\boldsymbol{\sigma}_p^* - \boldsymbol{\sigma}_p^h \right) \mathbf{D}^{-1} \left(\boldsymbol{\sigma}_d^* - \boldsymbol{\sigma}_d^h \right) d\Omega \tag{6.84}$$

6.5.2 The Stress Intensity Factor as QoI: Error Estimation

The accurate evaluation of the SIFs K_I and K_{II} in LEFM is of key importance as these factors are characterizing parameters in fracture mechanics problems. For that reason, it results interesting to evaluate error estimates defined as a function of the SIF, considering this parameter as the QoI. The procedure to evaluate the SIF will be described in this section.

To evaluate the SIF in XFEM approximations it is a common practice to use the interaction integral in its equivalent domain integral (EDI) form. There are different expressions already available to evaluate EDI integrals for singular problems. For the sake of generality, we are going to consider the expression shown in Szabó and Babuška (1991), which is a generalization of the interaction integral for the singular problem of a V-notch plate.

Let us consider the general singular problem of a V-notch domain subjected to loads in the infinite and with a singularity as shown in Figure 6.23.

The analytic solution for this singular elasticity problem can be found in Szabó and Babuška (1991) where, in accordance with the polar reference system in Figure 6.23, the displacement and stress fields at points sufficiently close to the corner can be described as:

$$\mathbf{u}(r, \phi) = K_I r^{\lambda_I} \boldsymbol{\Psi}_I(\lambda_I, \phi) + K_{II} r^{\lambda_{II}} \boldsymbol{\Psi}_{II}(\lambda_{II}, \phi) \tag{6.85}$$

$$\boldsymbol{\sigma}(r, \phi) = K_I \lambda_I r^{\lambda_I - 1} \boldsymbol{\Phi}_I(\lambda_I, \phi) + K_{II} \lambda_{II} r^{\lambda_{II} - 1} \boldsymbol{\Phi}_{II}(\lambda_{II}, \phi) \tag{6.86}$$

where r is the radial distance to the corner, λ_m (with $m = $ I,II) are the eigenvalues that determine the order of the singularity, $\boldsymbol{\Psi}_m$, $\boldsymbol{\Phi}_m$ are sets of trigonometric functions that depend on the angular position ϕ, and K_m are

Figure 6.23 Elastic solid with a V-notch.

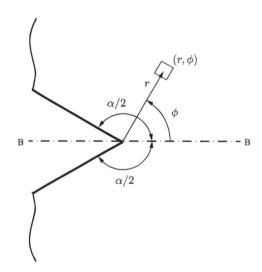

the so-called generalized stress intensity factors (GSIFs). Therefore, the eigenvalues λ and the GSIFs K define the singular field.

In practice, the eigenvalue λ is easily known in advance because it depends solely on the corner angle α and can be obtained as the smallest positive root of the following characteristic equations:

$$\sin \lambda_I \alpha + \lambda_I \sin \alpha = 0$$

$$\sin \lambda_{II} \alpha - \lambda_{II} \sin \alpha = 0 \tag{6.87}$$

The set of trigonometric functions for the displacement and stress fields under model I are given by (Szabó and Babuška, 1991):

$$\boldsymbol{\Psi}_I(\lambda_I, \phi) = \begin{Bmatrix} \boldsymbol{\Psi}_{I,x}(\lambda_I, \phi) \\ \boldsymbol{\Psi}_{I,y}(\lambda_I, \phi) \end{Bmatrix}$$

$$= \frac{1}{2\mu} \begin{Bmatrix} (\kappa - Q_I(\lambda_I + 1)) \cos \lambda_I \phi - \lambda_I \cos(\lambda_I - 2)\phi \\ (\kappa + Q_I(\lambda_I + 1)) \sin \lambda_I \phi + \lambda_I \sin(\lambda_I - 2)\phi \end{Bmatrix} \tag{6.88}$$

$$\boldsymbol{\Phi}_I(\lambda_I, \phi) = \begin{Bmatrix} \boldsymbol{\Phi}_{I,xx}(\lambda_I, \phi) \\ \boldsymbol{\Phi}_{I,yy}(\lambda_I, \phi) \\ \boldsymbol{\Phi}_{I,xy}(\lambda_I, \phi) \end{Bmatrix}$$

$$= \begin{Bmatrix} (2 - Q_I(\lambda_I + 1)) \cos(\lambda_I - 1)\phi - (\lambda_I - 1) \cos(\lambda_I - 3)\phi \\ (2 + Q_I(\lambda_I + 1)) \cos(\lambda_I - 1)\phi + (\lambda_I - 1) \cos(\lambda_I - 3)\phi \\ Q_I(\lambda_I + 1) \sin(\lambda_I - 1)\phi + (\lambda_I - 1) \sin(\lambda_I - 3)\phi \end{Bmatrix} \tag{6.89}$$

where κ is the Kolosov's constant, μ is the shear modulus, and Q is a constant for a given notch angle:

$$Q_I = -\frac{\left(\dfrac{\lambda_I - 1}{\lambda_I + 1}\right) \cdot \sin\left((\lambda_I - 1)\dfrac{\alpha}{2}\right)}{\sin\left((\lambda_I + 1)\dfrac{\alpha}{2}\right)} \tag{6.90}$$

$$\tag{6.91}$$

For mode II we have:

$$\boldsymbol{\Psi}_{II}(\lambda_{II}, \phi) = \begin{Bmatrix} \boldsymbol{\Psi}_{II,x}(\lambda_{II}, \phi) \\ \boldsymbol{\Psi}_{II,y}(\lambda_{II}, \phi) \end{Bmatrix}$$

$$= \frac{1}{2\mu} \begin{Bmatrix} (\kappa - Q_{II}(\lambda_{II} + 1)) \sin \lambda_{II} \phi - \lambda_{II} \sin(\lambda_{II} - 2)\phi \\ -(\kappa + Q_{II}(\lambda_{II} + 1)) \cos \lambda_{II} \phi - \lambda_{II} \cos(\lambda_{II} - 2)\phi \end{Bmatrix} \tag{6.92}$$

$$\boldsymbol{\Phi}_{II}(\lambda_{II}, \phi) = \begin{Bmatrix} \boldsymbol{\Phi}_{II,xx}(\lambda_{II}, \phi) \\ \boldsymbol{\Phi}_{II,yy}(\lambda_{II}, \phi) \\ \boldsymbol{\Phi}_{II,xy}(\lambda_{II}, \phi) \end{Bmatrix}$$

$$= \begin{Bmatrix} (2 - Q_{II}(\lambda_{II} + 1)) \sin(\lambda_{II} - 1)\phi - (\lambda_{II} - 1) \sin(\lambda_{II} - 3)\phi \\ (2 + Q_{II}(\lambda_{II} + 1)) \sin(\lambda_{II} - 1)\phi + (\lambda_{II} - 1) \sin(\lambda_{II} - 3)\phi \\ Q_{II}(\lambda_{II} + 1) \cos(\lambda_{II} - 1)\phi + (\lambda_{II} - 1) \cos(\lambda_{II} - 3)\phi \end{Bmatrix} \tag{6.93}$$

$$Q_{II} = -\frac{\sin\left((\lambda_{II} - 1)\dfrac{\alpha}{2}\right)}{\sin\left((\lambda_{II} + 1)\dfrac{\alpha}{2}\right)} \tag{6.94}$$

For the particular case of $\alpha/2 = \pi$ the problem corresponds to a crack as considered in the context of Fracture Mechanics.

In Szabó and Babuška (1991), the following expression to evaluate the integral is presented:

$$K^{(1,2)} = -\frac{1}{C} \int_{\Omega^*} \left(\sigma_{jk}^{(1)} u_k^{(2)} - \sigma_{jk}^{(2)} u_k^{(1)} \right) \frac{\partial q}{\partial x_j} d\Omega \tag{6.95}$$

where $\mathbf{u}^{(1)}$, $\sigma^{(1)}$ are the displacement and stress fields from the XFEM solution, and $\mathbf{u}^{(2)}$, $\sigma^{(2)}$ are the auxiliary fields used to extract the SIFs in mode I or mode II. q is a function used to define the extraction zone Ω^* which must take the value of 1 at the singular point and 0 at the boundary Γ. This equation clearly shows that the GSIF is a linear functional of the solution.

In (6.95), the auxiliary fields for the problem in mode I are defined as:

$$\mathbf{u}^{(2)}(r, \phi) = r^{-\lambda_I} \mathbf{\Psi}_I(-\lambda_I, \phi) \tag{6.96}$$

$$\sigma^{(2)}(r, \phi) = -\lambda_I r^{-\lambda_I - 1} \mathbf{\Phi}_I(-\lambda_I, \phi) \tag{6.97}$$

and C is a constant that can be evaluated for example for mode I using the expression:

$$C_I = \int_{-\alpha/2}^{\alpha/2} [\lambda_I \Xi_I(\lambda_I, \phi) \cdot \mathbf{\Psi}_I(-\lambda_I, \phi) - (-\lambda_I) \Xi_I(-\lambda_I, \phi) \cdot \mathbf{\Psi}_I(\lambda_I, \phi)] \, d\phi \tag{6.98}$$

where $\mathbf{\Psi}_I$ are the displacement trigonometric functions given in (6.88) and Ξ_I are the trigonometric functions associated with the traction vector related to $\mathbf{\Phi}_I$, i.e.:

$$\Xi_I = \begin{Bmatrix} \Phi_{I,xx}(\lambda, \phi) \cos \phi + \Phi_{I,xy}(\lambda, \phi) \sin \phi \\ \Phi_{I,xy}(\lambda, \phi) \cos \phi + \Phi_{I,yy}(\lambda, \phi) \sin \phi \end{Bmatrix} \tag{6.99}$$

Equation (6.95) allows us to evaluate the magnitudes of interest K_I and K_{II} and will be used to define the dual problem used to evaluate the error in these magnitudes. Expanding this equation for $j = x, y$ and $k = x, y$ results in:

$$
\begin{aligned}
K^{(1,2)} = -\frac{1}{C} \int_{\Omega^*} \Bigg[&\left(\sigma_{xx}^{(1)} u_x^{(2)} + \sigma_{xy}^{(1)} u_y^{(2)} \right) \frac{\partial q}{\partial x} + \left(\sigma_{xy}^{(1)} u_x^{(2)} + \sigma_{yy}^{(1)} u_y^{(2)} \right) \frac{\partial q}{\partial y} \\
&- \left(\sigma_{xx}^{(2)} u_x^{(1)} + \sigma_{xy}^{(2)} u_y^{(1)} \right) \frac{\partial q}{\partial x} - \left(\sigma_{xy}^{(2)} u_x^{(1)} + \sigma_{yy}^{(2)} u_y^{(1)} \right) \frac{\partial q}{\partial y} \Bigg] d\Omega
\end{aligned}
\tag{6.100}
$$

Rearranging terms we can write:

$$
\begin{aligned}
K^{(1,2)} = -\frac{1}{C} \int_{\Omega^*} \Bigg[&\sigma_{xx}^{(1)} \left(u_x^{(2)} \frac{\partial q}{\partial x} \right) + \sigma_{yy}^{(1)} \left(u_y^{(2)} \frac{\partial q}{\partial y} \right) + \sigma_{xy}^{(1)} \left(u_x^{(2)} \frac{\partial q}{\partial y} + u_y^{(2)} \frac{\partial q}{\partial x} \right) \\
&- u_x^{(1)} \left(\sigma_{xx}^{(2)} \frac{\partial q}{\partial x} + \sigma_{xy}^{(2)} \frac{\partial q}{\partial y} \right) - u_y^{(1)} \left(\sigma_{xy}^{(2)} \frac{\partial q}{\partial x} + \sigma_{yy}^{(2)} \frac{\partial q}{\partial y} \right) \Bigg] d\Omega
\end{aligned}
\tag{6.101}
$$

$$
K^{(1,2)} = -\frac{1}{C} \int_{\Omega^*} \left[\sigma_{xx}^{(1)} \quad \sigma_{yy}^{(1)} \quad \sigma_{xy}^{(1)} \right] \underbrace{\begin{bmatrix} u_x^{(2)} \frac{\partial q}{\partial x} \\ u_y^{(2)} \frac{\partial q}{\partial y} \\ u_x^{(2)} \frac{\partial q}{\partial y} + u_y^{(2)} \frac{\partial q}{\partial x} \end{bmatrix}}_{:= \vartheta_u} - \left[u_x^{(1)} \quad u_y^{(1)} \right] \underbrace{\begin{bmatrix} \sigma_{xx}^{(2)} \frac{\partial q}{\partial x} + \sigma_{xy}^{(2)} \frac{\partial q}{\partial y} \\ \sigma_{xy}^{(2)} \frac{\partial q}{\partial x} + \sigma_{yy}^{(2)} \frac{\partial q}{\partial y} \end{bmatrix}}_{:= \vartheta_\sigma} d\Omega \tag{6.102}
$$

$$
K^{(1,2)} = -\frac{1}{C} \int_{\Omega^*} \sigma^{h^T} \vartheta_u - \mathbf{u}^{h^T} \vartheta_\sigma d\Omega \approx \sum_{e=1}^{Ne} \mathbf{u}_e^{h^T} \left(-\frac{1}{C} \right) \int_{\Omega^* \cap \Omega_e} \mathbf{B}^T \mathbf{D} \vartheta_u - \mathbf{N} \vartheta_\sigma d\Omega = \sum_{e=1}^{Ne} \mathbf{u}_e^{h^T} \mathbf{f}_e \tag{6.103}
$$

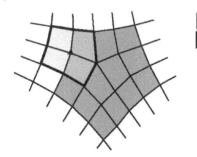

▨	DoI
☐	Patch

Figure 6.24 Representation of a patch partially overlapping Ω_i. Different system of equations are assembled at each side of the interface Γ_i.

where Ne is the number of elements in the mesh and $\mathbf{f}_e$ represents the vector of equivalent forces at nodes of element e corresponding to the dual problem. This will allow us to assemble and solve the dual problem that would be used to evaluate the error in K.

Once the primal and dual problem have been solved with XFEM , we have $\mathbf{u}_p^h$ $\mathbf{u}_d^h$ and we are able to evaluate the estimation of the error in K using the recovery-based error estimator in Equation (6.84).

As in the case of the error estimation in energy norm, the quality of the error estimator will depend on the quality of the recovered stress fields. The SPR-CX technique previously described will be considered in the evaluation of the recovered stress fields. This technique can be used in the evaluation of the primal and the dual recovered stress field.

One must take into account that applying the SPR-CX technique requires the analytical expressions of the applied body loads and boundary tractions. As body forces $\mathbf{b}_p$ and boundary tractions $\mathbf{t}_p$ are known, patch-wise internal and boundary equilibrium can be forced by the SPR-CX technique for the primal problem. However, their counterparts in the dual problem, $\mathbf{b}_d$ and $\mathbf{t}_d$, in Ω^* are unknown; therefore, the SPR-CX technique cannot force internal and boundary equilibrium in the extraction area Ω^* defined by function q. This simplification is quite accurate as the biggest impact on the accuracy of the error estimator comes from the use of the *singular+smooth* stress splitting technique which will be used also for the dual problem[3]. Full equilibrium can be imposed at patches outside Ω^*, where $\mathbf{b}_d = \mathbf{0}$ and $\mathbf{t}_d = \mathbf{0}$. Note that the compatibility equation can be forced at any patch as it does not depend neither on $\mathbf{b}_d$ nor on $\mathbf{t}_d$.

Patches partially overlapping Ω_i represent a special case. Different functions must be used at each side of Γ_i to describe the recovered stresses. Note that the procedure to do this could be the same procedure used to represent different stress fields at each side of the crack.

Let us consider the Westergaard problem shown in Section 6.3.1. The model is defined by a square domain $(-5, -5) \times (5, 5)$ which exhibits a crack defined by the segment with the endpoints $(-5, 0)$ and $(0, 0)$. In the numerical analyses, we use a geometrical enrichment defined by a circular fixed enrichment area $B(x_0, r_e)$ with radius $r_e = 2$, with its center at the crack tip x_0. Bilinear elements are considered in the models, using a sequence of uniformly refined meshes. For the numerical integration of standard elements we use a 2×2 Gaussian quadrature rule. The elements intersected by the crack are split into triangular integration subdomains that do not contain the crack. We use seven Gauss points in each triangular subdomain, and a 5×5 quasipolar integration in the subdomains of the element containing the crack tip. We do not consider correction for blending elements.

To evaluate the SIF K we use an EDI. For the primal problem we consider a square plateau function q centered at the crack tip, as shown in Figure 6.25. $q = 1$ for the domain defined by an inner square with side length 6 and $q = 0$ for the part of the domain outside the outer square with side length 8, q is interpolated in-between the two squares. Thus, this plateau function defines the subdomain Ω_i for the extraction of the QoI. As the dual problem is also a singular problem we have to evaluate a second SIF for the recovery of the dual solution. In that case, we use a plateau function such that $q = 1$ for all nodes inside a square with side length 4.9 and $q = 0$ otherwise.

---

3 The interaction integral (6.95) can be considered for the evaluation of the SIFs in both the primal and the dual problems.

Figure 6.25 Domain of interest for the extraction of the stress intensity factor.

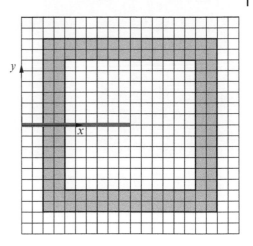

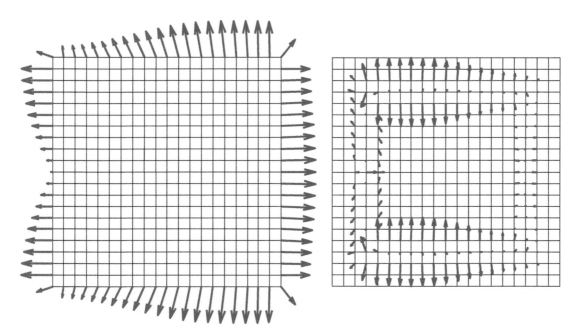

Figure 6.26 Equivalent forces at nodes for the primal (left) and dual (right) problems.

In Figure 6.26, we represent the equivalent nodal forces used to solve the primal and dual problems. The Dirichlet boundary constraints are the same for both models. For the dual problem, we can see that the forces are distributed along the nodes located in the domain of interest.

The yy-component of the stress field for the raw FE and the recovered solutions for the dual problem are represented in Figure 6.27. The enrichment area is indicated with a circle. Notice how the recovery procedure smoothes the stresses along the interface of the domain of interest. As the dual problem is also characterized by the crack, we have to evaluate the corresponding SIF and perform the *singular+smooth* decomposition of the stress field.

Figure 6.28 shows the evolution of the effectivity index θ as we increase the number of dof. We consider as quantities of interest the two SIFs characterizing two different loading conditions, i.e., mode I and mode II. We can see that for both quantities the error estimator yields effectivities close to the optimal value $\theta = 1$.

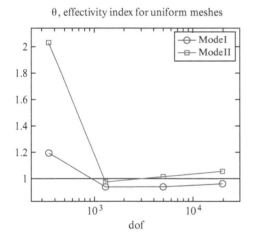

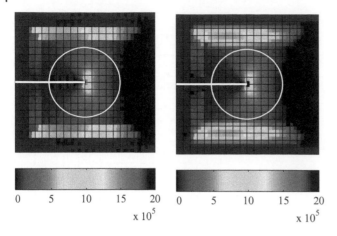

Figure 6.27 FE (left) and recovered fields (right) σ_{yy} for the dual problem.

θ, effectivity index for uniform meshes

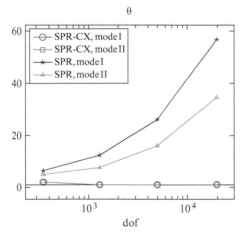

Figure 6.28 Evolution of the effectivity index θ considering the SIF as quantity of interest under mode I and mode II loading conditions.

Figure 6.29 Evolution of the effectivity index θ for the SPR-CX and SPR. Mode I and mode II loading conditions.

Figure 6.29 compares the results of the proposed SPR-CX recovery with the standard SPR technique. In particular, the SPR cannot properly recover singular fields; thus, the error estimate provided by the technique does not converge to the exact error. This behavior is similar for the two loading modes.

References

Aalto, J. (1997). Built-in field equations for recovery procedures. *Computers & Structures* 64 (1–4): 157–176.

Aalto, J. and Åman, M. (1999). Polynomial representations for patch recovery procedures. *Computers & Structures* 73 (1–5): 119–146.

Aalto, J. and Isoherranen, H. (1997). An element by element recovery method with built-in field equations. *Computers & Structures* 64 (1–4): 177–196.

ABAQUS. (2009). *ABAQUS/Standard User's Manual, v.6.9*. Pawtucket, Rhode Island: Hibbitt, Karlsson & Sorensen, Inc.

Ainsworth, M. and Oden, J. (1993). A unified approach to a posteriori error estimation using element residual methods. *Numerische Mathematik* 50: 23–50.

Ainsworth, M. and Oden, J.T. (2000). *A posteriori Error Estimation in Finite Element Analysis*. John Wiley & Sons.

Babuška, I. and Miller, A. (1987). A feedback finite element method with a posteriori error estimation: Part I. The finite element method and some basic properties of the a posteriori error estimator. *Computer Methods in Applied Mechanics and Engineering* 61 (1): 1–40.

Babuška, I. and Rheinboldt, W.C. (1978). A-posteriori error estimates for the finite element method. *International Journal for Numerical Methods in Engineering* 12 (10): 1597–1615.

Babuška, I. and Rheinboldt, W.C. (1979). Analysis of optimal finite-element meshes in R1. *Mathematics of Computation* 33: 435–463.

Babuška, I., Strouboulis, T., and Upadhyay, C.S. (1994a) A model study of the quality of a posteriori error estimators for linear elliptic problems. Error estimation in the interior of patchwise uniform grids of triangles. *Computer Methods in Applied Mechanics and Engineering* 114 (3–4): 307–378.

Babuška, I., Strouboulis, T., Upadhyay, C.S., et al. (1994b). Validation of a posteriori error estimators by numerical approach. *International Journal for Numerical Methods in Engineering* 37 (7): 1073–1123.

Babuška, I., Strouboulis, T., and Upadhyay, C.S. (1997). A model study of the quality of a posteriori error estimators for finite element solutions of linear elliptic problems, with particular reference to the behaviour near the boundary. *International Journal for Numerical Methods in Engineering* 40 (14): 2521–2577.

Bangerth, W. and Rannacher, R. (2003). *Adaptive Finite Element Methods for Differential Equations*. Basel: ETH, Zürich, Birkhäuser.

Bank, R.E. and Weiser, A. (1985). Some A posteriori error estimators for elliptic partial differential equations. *Mathematics of Computation* 44: 283–301.

Barros, F.B., Proenca, S.P.B., and de Barcellos, C.S. (2004). On error estimator and p-adaptivity in the generalized finite element method. *International Journal for Numerical Methods in Engineering* 60 (14): 2373–2398.

Béchet, E., Minnebo, H., Moës, N., and Burgardt, B. (2005). Improved implementation and robustness study of the X-FEM for stress analysis around cracks. *International Journal for Numerical Methods in Engineering* 64 (8): 1033–1056.

Belytschko, T., Krongauz, Y., Fleming, M., Organ, D. and Liu, W.K. (1996). Smoothing and accelerated computations in the element free Galerkin method. *Journal of Computational and Applied Mathematics* 74 (1–2): 111–126.

Benedetti, A., de Miranda, S., and Ubertini, F. (2006). A posteriori error estimation based on the superconvergent recovery by compatibility in patches. *International Journal for Numerical Methods in Engineering* 67 (1): 108–131.

Blacker, T. and Belytschko, T. (1994). Superconvergent patch recovery with equilibrium and conjoint interpolant enhancements. *International Journal for Numerical Methods in Engineering* 37 (3): 517–536.

Bordas, S.P.A. and Duflot, M. (2007). Derivative recovery and a posteriori error estimate for extended finite elements. *Computer Methods in Applied Mechanics and Engineering* 196 (35–36): 3381–3399.

Bordas, S.P.A. and Moran, B. (2006). Enriched finite elements and level sets for damage tolerance assessment of complex structures. *Engineering Fracture Mechanics* 73 (9): 1176–1201.

Bordas, S.P.A., Duflot, M., and Le, P. (2008). A simple error estimator for extended finite elements. *Communications in Numerical Methods in Engineering* 24 (11): 961–971.

CENAERO (2011). *Morfeo. Gosselies*. Belgium: CENAERO.

Chahine, E., Laborde, P., and Renard, Y. (2008). Crack tip enrichment in the XFEM using a cutoff function. *International Journal for Numerical Methods in Engineering* 75 (6): 629–646.

Cirak, F. and Ramm, E. (1998). A posteriori error estimation and adaptivity for linear elasticity using the reciprocal theorem. *Computer Methods in Applied Mechanics and Engineering* 156 (1–4): 351–362.

de Almeida, J.P.M. and Pereira, O.J.B.A. (2006). Upper bounds of the error in local quantities using equilibrated and compatible finite element solutions for linear elastic problems. *Computer Methods in Applied Mechanics and Engineering* 195 (4–6): 279–296.

Díez, P., Egozcue, J.J. and Huerta, A. (1998). A posteriori error estimation for standard finite element analysis. *Computer Methods in Applied Mechanics and Engineering* 163 (1–4): 141–157.

Díez, P., Parés, N., and Huerta, A. (2003). Recovering lower bounds of the error by postprocessing implicit residual a posteriori error estimates. *International Journal for Numerical Methods in Engineering* 56 (10): 1465–1488.

Díez, P., Ródenas, J.J., and Zienkiewicz, O.C. (2007). Equilibrated patch recovery error estimates: Simple and accurate upper bounds of the error. *International Journal for Numerical Methods in Engineering* 69 (10): 2075–2098.

Duflot, M. (2007). A study of the representation of cracks with level sets. *International Journal for Numerical Methods in Engineering* 70 (11): 1261–1302.

Duflot, M. and Bordas, S.P.A. (2008). A posteriori error estimation for extended finite elements by an extended global recovery. *International Journal for Numerical Methods in Engineering* 76: 1123–1138.

Gdoutos, E.E. (1993). *Fracture Mechanics and Introduction: Solid Mechanics and its Applications*. Dordrecht: Kluwer Academic Publishers.

Gerasimov, T., Rüter, M. and Stein, E. (2012). An explicit residual-type error estimator for Q 1 -quadrilateral extended finite element method in two-dimensional linear elastic fracture mechanics. *International Journal for Numerical Methods in Engineering* 90 (April): 1118–1155.

Giner, E., Fuenmayor, F.J., Baeza, L., and Tarancón, J.E. (2005). Error estimation for the finite element evaluation of GI and GII in mixed-mode linear elastic fracture mechanics. *Finite Elements in Analysis and Design* 41 (11–12): 1079–1104.

González-Estrada, O.A., Ródenas, J.J., Bordas, S.P.A., et al. (2012). On the role of enrichment and statical admissibility of recovered fields in a-posteriori error estimation for enriched finite element methods. *Engineering Computations*, 29 (8), 814–841.

Huerta, A., Vidal, Y., and Villon, P. (2004). Pseudo-divergence-free element free Galerkin method for incompressible fluid flow. *Computer Methods in Applied Mechanics and Engineering* 193 (12–14): 1119–1136.

Kvamsdal, T. and Okstad, K.M. (1998). Error estimation based on superconvergent patch recovery using statically admissible stress fields. *International Journal for Numerical Methods in Engineering* 42 (3): 443–472.

Labbe, P. and Garon, A. (1995). A robust implementation of Zienkiewicz and Zhu's local patch recovery method. *Communications in Numerical Methods in Engineering* 11 (5): 427–434.

Ladevèze, P. and Leguillon, D. (1983). Error estimate procedure in the finite element method and applications. *SIAM Journal on Numerical Analysis* 20 (3): 485–509.

Ladevèze, P., Rougeot, P., Blanchard, P., and Moreau, J.P. (1999). Local error estimators for finite element linear analysis. *Computer Methods in Applied Mechanics and Engineering* 176 (1–4): 231–246.

Lee, T., Park, H.C., and Lee, S.W. (1997). A superconvergenet stress recovery technique with equilibrium constraint. *International Journal for Numerical Methods in Engineering* 40 (6): 1139–1160.

Liu, G.R. (2003). MFree shape function construction mesh free methods. In: *Moving beyond the Finite Element Method*, Chapter 5, 693. Florida: CRC Press Boca Ratón.

Melenk, J.M. and Babuška, I. (1996). The partition of unity finite element method: Basic theory and applications. *Computer Methods in Applied Mechanics and Engineering* 139 (1–4): 289–314.

Moës, N., Dolbow, J., and Belytschko, T. (1999). A finite element method for crack growth without remeshing. *International Journal for Numerical Methods in Engineering* 46 (1): 131–150.

Nicaise, S., Renard, Y., and Chahine, E. (2011). Optimal convergence analysis for the extended finite element method. *International Journal for Numerical Methods in Engineering* 86: 528–548.

Oden, J.T. and Prudhomme, S. (2001). Goal-oriented error estimation and adaptivity for the finite element method. *Computers & Mathematics with Applications* 41 (5–6): 735–756.

Panetier, J., Ladevèze, P., and Chamoin, L. (2010). Strict and effective bounds in goal-oriented error estimation applied to fracture mechanics problems solved with XFEM. *International Journal for Numerical Methods in Engineering* 81 (6): 671–700.

Panetier, J., Ladevèze, P., and Louf, F. (2009). Strict bounds for computed stress intensity factors. *Computers & Structures* 87 (15–16): 1015–1021.

Pannachet, T., Sluys, L.J., and Askes, H. (2009). Error estimation and adaptivity for discontinuous failure. *International Journal for Numerical Methods in Engineering* 78 (5): 528–563.

Paraschivoiu, M., Peraire, J., and Patera, A.T. (1997). A posteriori finite element bounds for linear-functional outputs of elliptic partial differential equations. *Computer Methods in Applied Mechanics and Engineering* 150 (1–4): 289–312.

Park, H.C., Shin, S.H., and Lee, S.W. (1999). A superconvergent stress recovery technique for accurate boundary stress extraction. *International Journal for Numerical Methods in Engineering* 45 (9): 1227–1242.

Pereira, O.J.B.A., de Almeida, J.P.M., and Maunder, E.A.W. (1999). Adaptive methods for hybrid equilibrium finite element models. *Computer Methods in Applied Mechanics and Engineering* 176 (1–4): 19–39.

Práger, W. and Synge, J.L. (1947). Approximation in elasticity based on the concept of function space. *Quarterly of Applied Mathematics* 5: 241–269.

Prange, C., Loehnert, S., and Wriggers, P. (2012). Error estimation for crack simulations using the XFEM. *International Journal for Numerical Methods in Engineering* 91: 1459–1474.

Ramsay, A.C.A. and Maunder, E.A.W. (1996). Effective error estimation from continuous, boundary admissible estimated stress fields. *Computers & Structures* 61 (2): 331–343.

Ródenas, J.J., Giner, E., Tarancón, J.E., and González-Estrada, O.A. (2006). A recovery error estimator for singular problems using singular+smooth field splitting. In: *Fifth International Conference on Engineering Computational Technology* (ed. B.H.V. Topping, G. Montero, and R. Montenegro). Scotland: Civil-Comp Press, Stirling.

Ródenas, J.J., González-Estrada, O.A., Díez, P. and Fuenmayor, F.J. (2010). Accurate recovery-based upper error bounds for the extended finite element framework. *Computer Methods in Applied Mechanics and Engineering* 199 (37–40): 2607–2621.

Ródenas, J.J., González-Estrada, O.A., Fuenmayor, F.J. and Chinesta, F. (2012). Enhanced error estimator based on a nearly equilibrated moving least squares recovery technique for FEM and XFEM. *Computational Mechanics* 52, 321–344.

Ródenas, J.J., González-Estrada, O.A., Tarancón, J.E., and Fuenmayor, F.J. (2008). A recovery-type error estimator for the extended finite element method based on singular+smooth stress field splitting. *International Journal for Numerical Methods in Engineering* 76 (4): 545–571.

Ródenas, J.J., Tur, M., Fuenmayor, F.J., and Vercher, A. (2007). Improvement of the superconvergent patch recovery technique by the use of constraint equations: The SPR-C technique. *International Journal for Numerical Methods in Engineering* 70 (6): 705–727.

Rüter, M. and Stein, E. (2006). Goal-oriented a posteriori error estimates in linear elastic fracture mechanics. *Computer Methods in Applied Mechanics and Engineering* 195 (4–6): 251–278.

Rüter, M. and Stein, E. (2011). Goal-oriented residual error estimats for XFEM approximations in LEFM. In: *Recent Developments and Innovative Applications in Computational Mechanics* (ed. D. Mueller-Hoeppe, P. Wriggers, S. Loehnert, and S. Reese), Chapter 26, 231–238. Berlin: Springer.

Strouboulis, T., Copps, K., and Babuška, I. (2001). The generalized finite element method. *Computer Methods in Applied Mechanics and Engineering* 190 (32–33): 4081–4193.

Strouboulis, T., Zhang, L., Wang, D., and Babuška, I. (2006). A posteriori error estimation for generalized finite element methods. *Computer Methods in Applied Mechanics and Engineering* 195 (9–12): 852–879.

Szabó, B.A. and Babuška, I. (1991). *Finite Element Analysis*. New York: John Wiley & Sons.

Tabbara, M., Blacker, T., and Belytschko, T. (1994). Finite element derivative recovery by moving least square interpolants. *Computer Methods in Applied Mechanics and Engineering* 117 (1–2): 211–223.

Timoshenko, S.P. and Goodier, J.N. (1951). *Theory of Elasticity*, 2e. New York: McGraw-Hill.

Verfürth, R. (1999). A review of a posteriori error estimation techniques for elasticity problems. *Computer Methods in Applied Mechanics and Engineering* 176: 419–440.

Wiberg, N.E., Abdulwahab, F., and Ziukas, S. (1994). Enhanced superconvergent patch recovery incorporating equilibrium and boundary conditions. *International Journal for Numerical Methods in Engineering* 37 (20): 3417–3440.

Wiberg, N.E. and Abdulwahab, F. (1993). Patch recovery based on superconvergent derivatives and equilibrium. *International Journal for Numerical Methods in Engineering* 36 (16): 2703–2724.

Wiberg, N.E., Zeng, L., and Li, X. (1992). Error estimation and adaptivity in elastodynamics. *Computer Methods in Applied Mechanics and Engineering* 101 (1–3): 369–395.

Wyart, E., Coulon, D., Duflot, M., et al. (2007). A substructured FE-shell/XFE-3D method for crack analysis in thin-walled structures. *International Journal for Numerical Methods in Engineering* 72 (7): 757–779.

Xiao, Q.Z. and Karihaloo, B.L. (2006). Improving the accuracy of XFEM crack tip fields using higher order quadrature and statically admissible stress recovery. *International Journal for Numerical Methods in Engineering* 66 (9): 1378–1410.

Zienkiewicz, O.C. and Zhu, J.Z. (1987). A simple error estimator and adaptive procedure for practical engineering analysis. *International Journal for Numerical Methods in Engineering* 24 (2): 337–357.

Zienkiewicz, O.C. and Zhu, J.Z. (1992a). The superconvergent patch recovery and a posteriori error estimates. Part 1: The recovery technique. *International Journal for Numerical Methods in Engineering* 33 (7): 1331–1364.

Zienkiewicz, O.C. and Zhu, J.Z. (1992b). The superconvergent patch recovery and a posteriori error estimates. Part 2: Error estimates and adaptivity. *International Journal for Numerical Methods in Engineering* 33 (7): 1365–1382.

7

φ-FEM: An Efficient Simulation Tool Using Simple Meshes for Problems in Structure Mechanics and Heat Transfer

Stéphane Cotin[1], Michel Duprez[1], Vanessa Lleras[2], Alexei Lozinski[3,], and Killian Vuillemot[1]*

[1] *MIMESIS team, Inria Nancy - Grand Est, MLMS team, Université de Strasbourg, France*
[2] *IMAG, Univ Montpellier, CNRS, Montpellier, France*
[3] *Université de Franche-Comté, CNRS, LmB, Besançon, France*
* *Corresponding author*

7.1 Introduction

Taking the geometrical complexity into account is one of the major issues in the computational mechanics. Although some spectacular advances in mesh generation have been achieved in recent years, constructing and using the meshes fitting the geometry of, for example, human organs may still be prohibitively expensive in realistic three-dimensional (3D) configurations. Moreover, when the geometry is changing in time or on iterations of an optimization algorithm, the mesh should be frequently adapted, either by complete remeshing (expensive) or by moving the nodes (may lead to a degradation of the mesh quality, impacting the accuracy and the stability of computations).

Geometrically unfitted methods, i.e., the numerical methods using the computational meshes that do not fit the boundary of the domain, and/or the internal interfaces, have been widely investigated in the computational mechanics for decades. Their variants come under the name of immersed boundary (Mittal and Iaccarino, 2005) or fictitious domain (Glowinski et al., 1994) methods. However, these classical approaches suffer from poor accuracy because of their rudimentary (but easy to implement) treatment of the boundary conditions, c.f. (Girault and Glowinski, 1995). For example, in the case of the linear elasticity equations, these methods start by extending the displacement $\mathbf{u}$, from the physical domain Ω to a fictitious domain (typically a rectangular box) $\mathcal{O} \supset \Omega$ assuming that $\mathbf{u}$ still solves the same governing equations on $\mathcal{O}$ as on Ω. This creates an artificial singularity on the boundary of Ω (a jump in the normal derivative) so that the resulting numerical approximation is, at best, $\sqrt{h}$-accurate in the energy norm with whatever finite elements (FEs) (from now on, h denotes the mesh size).

The last two decades have seen the arrival of more accurate geometrically unfitted methods such as extended finite element method (XFEM) (Haslinger and Renard, 2009; Moës et al., 1999), Cut Finite Element Method (CutFEM) (Burman et al., 2015; Burman and Hansbo, 2010, 2012; Hansbo et al., 2017) and Shifted Boundary Method (SBM) (Atallah et al., 2021; Main and Scovazzi, 2018). We are citing here only the methods based on the FE approach; the list would be much longer if the methods based on finite differences were included. In the case of XFEM/CutFEM, the optimal accuracy, i.e., the same convergence rates as those of the standard finite element method (FEM) on a geometrically fitted mesh, is achieved at the price of a considerable sophistication in the implementation of boundary conditions. The idea is to introduce the unfitted mesh (known as the *active mesh*) starting

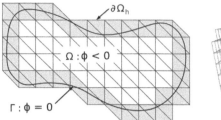
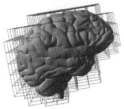

Figure 7.1 Left: Meshes and notations for a 2D domain $\Omega = \{\phi < 0\}$; the computational mesh $\mathcal{T}_h$ is obtained from a structured background mesh and is represented by both white and gray triangles forming the domain Ω_h; the gray triangles constitute the sub-mesh $\mathcal{T}_h^\Gamma$ occupying the domain Ω_h^Γ. Right: a more involved example of an active mesh in 3D that can be used in ϕ-FEM; a hexahedral mesh covering a brain geometry.

from the simple background mesh and getting rid of the cells lying entirely outside the physical domain, as illustrated at Figure 7.1. The FEs are then set up on the active mesh, the variational formulation is imposed on the *physical* domain, and an appropriate stabilization is added. In practice, one should thus compute the integrals on the actual boundary and on the parts of the active mesh cells cut by the boundary (the cut cells). To this end, one should typically construct a boundary fitted mesh, now only locally near the boundary and only for the numerical integration purposes, but the generation of a non-trivial mesh is still not completely avoided.

On the other hand, the non-trivial integration is completely absent from SBM. This method introduces again an active mesh as a sub-mesh of the background mesh (unlike CutFEM, the active mesh here contains only the cells inside Ω) and then imposes the approximate boundary conditions on the boundary of the active mesh by a Taylor expansion around the actual boundary. The absence of non-standard numerical integration is an important practical advantage of SBM over XFEM/CutFEM. We note however that, to the best of our knowledge, SBM is readily available only for the lowest order FE. Moreover, in the case of Neumann boundary conditions, the original version of SBM (Main and Scovazzi, 2018) needs an extrapolation of the second derivatives of the solution that makes its implementation rather tricky. This difficulty can be alleviated if the problem is recast in a mixed form introducing the secondary variables for the gradient (Nouveau et al., 2019).

In this chapter, we present yet another unfitted FE-based method, first introduced in Duprez and Lozinski (2020) and Duprez et al. (2021) and baptized ϕ-FEM to emphasize the prominent role played in it by the level set (LS) function, traditionally denoted by ϕ. From now on, we suppose that the physical domain is characterized by a given LS function[1]:

$$\Omega = \{\phi < 0\}. \tag{7.1}$$

Similarly to CutFEM/XFEM/SBM, we suppose that Ω is embedded into a simple background mesh and we introduce the *active* computational mesh $\mathcal{T}_h$ as in CutFEM, c.f. Figure 7.1. However, unlike CutFEM, we abandon the variational formulation on Ω. We rather introduce a non-standard formulation on the extended domain Ω_h (slightly larger than Ω) occupied by the active mesh $\mathcal{T}_h$. The general procedure is as follows:

- Extend the governing equations from Ω to Ω_h and write down a formal variational formulation on Ω_h without taking into account the boundary conditions on $\partial\Omega$.
- Impose the boundary conditions using appropriate ansatz or additional variables, explicitly involving the LS ϕ which provides the link to the actual boundary. For instance, the homogeneous Dirichlet boundary conditions ($\boldsymbol{u} = 0$ on $\partial\Omega$) can be imposed by the ansatz $\boldsymbol{u} = \phi\boldsymbol{w}$ thus reformulating the problem in terms of the new

---

1 In some settings presented further, the level set ϕ will describe an interior interface inside Ω rather than the geometry of Ω itself.

unknown w (modifications for non-homogeneous conditions, mixed boundary conditions, and other settings are introduced further in the text).

- Add appropriate stabilization, including the ghost penalty (Burman, 2010) as in CutFEM plus a least square imposition of the governing equation on the mesh cells near the boundary, to guarantee coerciveness/stability on the discrete level.

This approach allows us to achieve the optimal accuracy using classical FE spaces of any order and the usual numerical integration: all the integrals in ϕ-FEM can be computed by standard quadrature rules on entire mesh cells and on entire boundary facets; no integration on cut cells or on the actual boundary is needed. This is the principal advantage of ϕ-FEM over CutFEM/XFEM. Moreover, we can cite the following features of ϕ-FEM which distinguish it from both CutFEM/XFEM and SBM:

- FE of any order can be straightforwardly used in ϕ-FEM. The geometry is naturally taken into account with the needed optimal accuracy: it suffices to approximate the LS function ϕ by piecewise polynomials of the same degree as that used for the primal unknown. This should be contrasted to CutFEM where a special additional treatment is needed if one uses FEM of order ≥ 2. Indeed, a piecewise linear representation of the boundary is not sufficient in this case. One needs either a special implementation of the isoparametric method (Lehrenfeld, 2016) or a local correction by Taylor expansions (Boiveau et al., 2018). The extension to higher order FE is not trivial for SBM either.
- Contrary to SBM, ϕ-FEM is based on a purely variational formulation so that the existing standard FEM libraries suffice to implement it. The geometry of the domain comes into the formulation only through the LS ϕ. We emphasize that ϕ is not necessarily the signed distance to the boundary of Ω. It is sufficient to give to the method any ϕ satisfying (7.1) which is the minimal imaginable geometrical input. This can be contrasted with SBM which assumes that the distance to the actual boundary in the normal direction is known on all the boundary facets of the active mesh.

Moreover, ϕ-FEM is designed so that the matrices of the problems on the discrete level are reasonably conditioned, i.e., their condition numbers are of the same order as those of a standard fitting FEM on a mesh of comparable size. ϕ-FEM shares this feature with both CutFEM/XFEM and SBM.

Up to now, the ϕ-FEM approach has been proposed, tested, and substantiated mathematically only in some simplest settings: Poisson equation with Dirichlet boundary conditions (Duprez and Lozinski, 2020), or with Neumann/Robin boundary conditions (Duprez et al., 2021). The goal of the present chapter is to demonstrate its applicability to some more sophisticated governing equations arising in the computational mechanics. In Section 7.2, we adapt ϕ-FEM to the linear elasticity equations accompanied by either pure Dirichlet boundary conditions, or with mixed conditions (both Dirichlet and Neumann on parts of the boundary). In Section 7.3, we consider the interface problem (elasticity with material coefficients abruptly changing over an internal interface). Section 7.4 is devoted to the treatment of internal cracks. Finally, our method is adapted to the heat equation in Section 7.5. In all these settings, we start by deriving an appropriate variant of ϕ-FEM and then illustrate it by numerical tests on manufactured solutions. We also compare the accuracy and efficiency of ϕ-FEM with those of the standard fitted FEM on the meshes of similar size, revealing the substantial gains that can be achieved by ϕ-FEM in both the accuracy and the computational time.

All the codes used in the present work have been implemented thanks to the open libraries `fenics` (Alnæs et al., 2015) and `multiphenics` (Ballarin and Rozza, 2020). They are available at the link
https://github.com/michelduprez/phi-FEM-an-efficient-simulation-tool-using-simple-meshes-for-problems-in-structure-mechanics.git

7.2 Linear Elasticity

In this section, we consider the static linear elasticity for homogeneous and isotropic materials. The governing equation for the displacement $\boldsymbol{u}$ is thus:

$$\text{div } \sigma(\boldsymbol{u}) + \boldsymbol{f} = 0, \tag{7.2}$$

where the stress $\sigma(\boldsymbol{u})$ is given by:

$$\sigma(\boldsymbol{u}) = 2\mu\varepsilon(\boldsymbol{u}) + \lambda(\text{div } \boldsymbol{u})I,$$

$\varepsilon(\boldsymbol{u}) = \frac{1}{2}(\nabla\boldsymbol{u} + \nabla\boldsymbol{u}^T)$ is the strain tensor, and Lamé parameters λ, μ are defined via the Young modulus E and the Poisson coefficient ν by:

$$\mu = \frac{E}{2(1+\nu)} \text{ and } \lambda = \frac{E\nu}{(1+\nu)(1-2\nu)}. \tag{7.3}$$

Equation (7.2) is posed in a domain Ω, which can be 2D or 3D, and should be accompanied with Dirichlet and Neumann boundary conditions on $\Gamma = \partial\Omega$. We assume that Γ is decomposed into two disjoint parts, $\Gamma = \Gamma_D \cup \Gamma_N$ with $\Gamma_D \neq \varnothing$, and

$$\boldsymbol{u} = \boldsymbol{u}^g \text{ on } \Gamma_D, \tag{7.4}$$

$$\sigma(\boldsymbol{u})\boldsymbol{n} = \boldsymbol{g} \text{ on } \Gamma_N, \tag{7.5}$$

with the given displacement $\boldsymbol{u}^g$ on Γ_D and the given force $\boldsymbol{g}$ on Γ_N.

Let us first recall the weak formulation of this problem (to be compared with forthcoming ϕ-FEM formulations): find the vector field $\boldsymbol{u}$ on Ω s.t. $\boldsymbol{u}|_{\Gamma_D} = \boldsymbol{u}^g$ and

$$\int_{\Omega} \sigma(\boldsymbol{u}) : \nabla\boldsymbol{v} = \int_{\Omega} \boldsymbol{f} \cdot \boldsymbol{v} + \int_{\Gamma_N} \boldsymbol{g} \cdot \boldsymbol{v}, \quad \forall\boldsymbol{v} \text{ on } \Omega \text{ such that } \boldsymbol{v}|_{\Gamma_D} = 0. \tag{7.6}$$

This is obtained by multiplying the equation by a test function $\boldsymbol{v}$, integrating over Ω and taking into account the boundary conditions. Formulation (7.6) is routinely used to construct conforming FE methods, which necessitate a mesh that fits the domain Ω in order to approximate the integrals on Ω and Γ_N and to impose $\boldsymbol{u} = \boldsymbol{u}^g$ on Γ_D.

We now consider the situation where a fitting mesh of Ω is not available. We rather assume that Ω is inscribed in a box $\mathcal{O}$ which is covered by a simple background mesh $\mathcal{T}_h^{\mathcal{O}}$. We further introduce the computational mesh $\mathcal{T}_h$ (also referred to as the active mesh) by getting rid of cells lying entirely outside Ω. In practice, Ω is given by the LS function ϕ: $\Omega = \{\phi < 0\}$. Usually, the LS is known only approximately. Accordingly, we assume that we are given an FE function ϕ_h, i.e., a piecewise polynomial function on mesh $\mathcal{T}_h^{\mathcal{O}}$, which approximates sufficiently well ϕ. The selection of the mesh cells forming the active mesh is done on the basis of ϕ_h rather than ϕ:

$$\mathcal{T}_h := \{T \in \mathcal{T}_h^{\mathcal{O}} : T \cap \{\phi_h < 0\} \neq \varnothing\}. \tag{7.7}$$

The domain occupied by $\mathcal{T}_h$ is denoted by Ω_h, i.e., $\Omega_h = (\cup_{T \in \mathcal{T}_h} T)^o$. In some of our methods, we shall also need a sub-mesh of $\mathcal{T}_h$, referred to as $\mathcal{T}_h^{\Gamma}$, consisting of the cells intersected with the curve (surface) $\{\phi_h = 0\}$, approximating Γ:

$$\mathcal{T}_h^{\Gamma} := \{T \in \mathcal{T}_h^{\mathcal{O}} : T \cap \{\phi_h = 0\} \neq \varnothing\}. \tag{7.8}$$

The domain covered by mesh $\mathcal{T}_h^{\Gamma}$ will be denoted by Ω_h^{Γ}, c.f. Figure 7.1.

The starting point of all variants of ϕ-FEM is a variational formulation of problem (7.2) extended to Ω_h, in which we do not impose any boundary conditions since they are lacking on $\partial\Omega_h$. We thus assume that the right-hand side $\boldsymbol{f}$ is given on the whole Ω_h rather than on Ω alone, and suppose moreover that $\boldsymbol{u}$ can be extended from Ω to Ω_h as the solution to the governing Equation (7.2), now posed on Ω_h instead of Ω. In a usual manner, we take then any test function $\boldsymbol{v}$ on Ω_h, multiply the governing equation by $\boldsymbol{v}$ and integrate it over Ω_h. This gives the following formulation: find a vector field $\boldsymbol{u}$ on Ω_h such that:

$$\int_{\Omega_h} \boldsymbol{\sigma}(\boldsymbol{u}) : \nabla \boldsymbol{v} - \int_{\partial \Omega_h} \boldsymbol{\sigma}(\boldsymbol{u})\boldsymbol{n} \cdot \boldsymbol{v} = \int_{\Omega_h} \boldsymbol{f} \cdot \boldsymbol{v}, \quad \forall \boldsymbol{v} \text{ on } \Omega_h \tag{7.9}$$

We emphasize that this formulation is fundamentally different from the standard formulation (7.6). First of all, no boundary conditions are incorporated in (7.9) so that we cannot expect it to admit a unique solution. Furthermore, if we add somehow the boundary conditions on $\partial \Omega$ to (7.9), which we shall do indeed when constructing our ϕ-FEM variants, the resulting formulation will still be ill posed, meaning that its solution (on the continuous level) either does not exist, or is not unique. However, we shall be able to turn these problems into well-defined numerical schemes by adding an appropriate stabilization on the discrete level.

7.2.1 Dirichlet Conditions

Let us first consider the case of pure Dirichlet conditions: $\Gamma = \Gamma_D$. On the continuous level, we thus want to impose $\boldsymbol{u} = \boldsymbol{u}^g$ on $\Gamma = \Gamma_D = \{\phi = 0\}$ on top of the general formulation (7.9) of the problem on Ω_h. We consider here two options to achieve this: (i) **direct** Dirichlet ϕ-FEM, as proposed in (Duprez and Lozinski, 2020), introducing a new unknown $\boldsymbol{w}$ and redefining $\boldsymbol{u}$ through the product $\phi \boldsymbol{w}$ which automatically vanishes on Γ; (ii) **dual** Dirichlet ϕ-FEM, inspired by (Duprez et al., 2021), keeping the original unknown $\boldsymbol{u}$ and imposing $\boldsymbol{u} = \boldsymbol{u}^g$ on Γ with the aid of an auxiliary variable $\boldsymbol{p}$ in a least square manner. In more details, our two approaches can be described as follows:

- **Direct Dirichlet ϕ-FEM** *(on continuous level).* Supposing that $\boldsymbol{u}^g$ is actually given on the whole Ω_h rather than on Γ alone, we make the ansatz as follows:

$$\boldsymbol{u} = \boldsymbol{u}^g + \phi \boldsymbol{w}, \text{ on } \Omega_h \tag{7.10}$$

and substitute it into (7.9). To make the formulation more symmetric we also replace the test functions $\boldsymbol{v}$ by $\phi \boldsymbol{z}$. This yields: find a vector field $\boldsymbol{w}$ on Ω_h such that:

$$\int_{\Omega_h} \boldsymbol{\sigma}(\phi \boldsymbol{w}) : \nabla(\phi \boldsymbol{z}) - \int_{\partial \Omega_h} \boldsymbol{\sigma}(\phi \boldsymbol{w})\boldsymbol{n} \cdot \phi \boldsymbol{z} = \int_{\Omega_h} \boldsymbol{f} \cdot \phi \boldsymbol{z}$$

$$- \int_{\Omega_h} \boldsymbol{\sigma}(\boldsymbol{u}^g) : \nabla(\phi \boldsymbol{z}) + \int_{\partial \Omega_h} \boldsymbol{\sigma}(\boldsymbol{u}^g)\boldsymbol{n} \cdot \phi \boldsymbol{z}, \quad \forall \boldsymbol{z} \text{ on } \Omega_h. \tag{7.11}$$

The idea is thus to work with the new unknown $\boldsymbol{w}$ on Ω_h, discretize it by FEM starting from the variational formulation above, and to reconstitute the approximation to $\boldsymbol{u}$ by the ansatz (7.10).

- **Dual Dirichlet ϕ-FEM** *(on continuous level).* We now suppose that $\boldsymbol{u}^g$ is defined on Ω_h^Γ, c.f. (7.8), rather than on the whole of Ω_h. We keep the primal unknown $\boldsymbol{u}$ in (7.9) and we want to impose

$$\boldsymbol{u} = \boldsymbol{u}^g + \phi \boldsymbol{p}, \text{ on } \Omega_h^\Gamma \tag{7.12}$$

on top of it, with a new auxiliary unknown $\boldsymbol{p}$ on Ω_h^Γ. The new variable $\boldsymbol{p}$ lives beside $\boldsymbol{u}$ inside a variational formulation that combines (7.9) with (7.12): find vector fields $\boldsymbol{u}$ on Ω_h and $\boldsymbol{p}$ on Ω_h^Γ such that:

$$\int_{\Omega_h} \boldsymbol{\sigma}(\boldsymbol{u}) : \nabla \boldsymbol{v} - \int_{\partial \Omega_h} \boldsymbol{\sigma}(\boldsymbol{u})\boldsymbol{n} \cdot \boldsymbol{v} + \gamma \int_{\Omega_h^\Gamma} (\boldsymbol{u} - \phi \boldsymbol{p}) \cdot (\boldsymbol{v} - \phi \boldsymbol{q})$$

$$= \int_{\Omega_h} \boldsymbol{f} \cdot \boldsymbol{v} + \gamma \int_{\Omega_h^\Gamma} \boldsymbol{u}^g \cdot (\boldsymbol{v} - \phi \boldsymbol{q}), \quad \forall \boldsymbol{v} \text{ on } \Omega_h, \boldsymbol{q} \text{ on } \Omega_h^\Gamma \tag{7.13}$$

with a positive parameter γ. Comparing the direct and dual variants, we observe that the expressions (7.10) and (7.12) are of course pretty similar, but their roles are quite different in the corresponding methods. The variable $\boldsymbol{w}$ replaces $\boldsymbol{u}$ in (7.11), while $\boldsymbol{p}$ lives alongside $\boldsymbol{u}$ in (7.13). The introduction of the additional variable $\boldsymbol{p}$ makes the dual method only slightly more expensive than the direct one, since this new variable is introduced only on a narrow strip around Γ. On the other hand, a certain advantage of the dual variant over the direct one lies in the fact that both ϕ and $\boldsymbol{u}^g$ should be here known only locally around Γ since they enter into Equation (7.13) only on Ω_h^Γ. This can facilitate the construction of ϕ and $\boldsymbol{u}^g$ in practice. More importantly, it is the dual method that we shall be able to adapt to various, more and more complicated settings below.

As mentioned above, both variational problems (7.11) and (7.13) are derived on a very formal level. They are not valid in any mathematically rigorous way: we cannot expect to have a meaningful boundary value problems on a domain Ω_h with no boundary conditions on $\partial\Omega_h$, while prescribing some conditions on a curve (surface) Γ which is inside Ω_h. However, both formulations can serve as starting problems to write down FE problems which become well posed once an appropriate stabilization is added.

We start by introducing the FE spaces: fix an integer $k \geq 1$ and let:

$$V_h := \left\{ \boldsymbol{v}_h : \Omega_h \to \mathbb{R}^d : \boldsymbol{v}_{h|T} \in \mathbb{P}^k(T)^d \;\; \forall T \in \mathcal{T}_h, \; \boldsymbol{v}_h \text{ continuous on } \Omega_h \right\}. \tag{7.14}$$

For future reference, we introduce the local version of this space for any sub-mesh $\mathcal{M}_h$ of $\mathcal{T}_h$ and polynomial degree $l \geq 0$

$$Q_h^l(\mathcal{M}_h) := \left\{ \boldsymbol{q}_h : \mathcal{M}_h \to \mathbb{R}^d : \boldsymbol{q}_{h|T} \in \mathbb{P}^l(T)^d \;\; \forall T \in \mathcal{M}_h, \; \boldsymbol{q}_h \text{ continuous on } \mathcal{M}_h \text{ if } l \geq 0 \right\}. \tag{7.15}$$

In particular, we shall need the space $Q_h^k(\Omega_h^\Gamma)$ on the sub-mesh Ω_h^Γ in the dual version of Dirichlet ϕ-FEM.

The two variants of ϕ-FEM introduced above can now be written on the fully discrete level as:

- **Direct Dirichlet ϕ-FEM**: find $\boldsymbol{w}_h \in V_h$ such that:

$$\int_{\Omega_h} \sigma(\phi_h \boldsymbol{w}_h) : \nabla(\phi_h \boldsymbol{z}_h) - \int_{\partial\Omega_h} \sigma(\phi_h \boldsymbol{w}_h)\boldsymbol{n} \cdot \phi_h \boldsymbol{z}_h + G_h(\phi_h \boldsymbol{w}_h, \phi_h \boldsymbol{z}_h) + J_h^{lhs}(\phi_h \boldsymbol{w}_h, \phi_h \boldsymbol{z}_h)$$

$$= \int_{\Omega_h} \boldsymbol{f} \cdot \phi_h \boldsymbol{z}_h - \int_{\Omega_h} \sigma(\boldsymbol{u}_h^g) : \nabla(\phi_h \boldsymbol{z}_h) + \int_{\partial\Omega_h} \sigma(\boldsymbol{u}_h^g)\boldsymbol{n} \cdot \phi_h \boldsymbol{z}_h,$$

$$+ J_h^{rhs}(\phi_h \boldsymbol{z}_h), \quad \forall \boldsymbol{z}_h \in V_h \tag{7.16}$$

and set $\boldsymbol{u}_h = \boldsymbol{u}_h^g + \phi_h \boldsymbol{w}_h$. Here, $\phi_h, \boldsymbol{u}_h^g$ are FE approximations for $\phi, \boldsymbol{u}^g$ on the whole Ω_h, and $G_h, J_h^{lhs}, J_h^{rhs}$ stand for the stabilization terms:

$$G_h(\boldsymbol{u}, \boldsymbol{v}) := \sigma_D h \sum_{E \in \mathcal{F}_h^\Gamma} \int_E [\sigma(\boldsymbol{u})\boldsymbol{n}] \cdot [\sigma(\boldsymbol{v})\boldsymbol{n}], \tag{7.17}$$

$$J_h^{lhs}(\boldsymbol{u}, \boldsymbol{v}) := \sigma_D h^2 \sum_{T \in \mathcal{T}_h^\Gamma} \int_T \text{div } \sigma(\boldsymbol{u}) \cdot \text{div } \sigma(\boldsymbol{v}), \quad J_h^{rhs}(\boldsymbol{v}) := -\sigma_D h^2 \sum_{T \in \mathcal{T}_h^\Gamma} \int_T \boldsymbol{f} \cdot \text{div } \sigma(\boldsymbol{v}). \tag{7.18}$$

The stabilization G_h (7.17) is known as the ghost penalty. σ_D in (7.17) is a positive stabilization parameter which should be chosen sufficiently big (in a mesh independent manner). $\mathcal{F}_h^\Gamma$ stands for the set of internal facets of mesh $\mathcal{T}_h$ which are also the facets of $\mathcal{T}_h^\Gamma$ (these are the facets either intersected by Γ, or belonging to the cells intersected by Γ). Stabilization (7.17) was first introduced in Burman (2010) in the form of penalization of jumps in the normal derivatives of the FE solution. Here, we prefer to penalize the jumps of internal elastic forces, following Claus and Kerfriden (2018), thus controlling appropriate combinations of the derivatives, rather than the normal derivatives themselves. We emphasize however that the original ghost penalty in Hansbo et al. (2017) also involved the jumps of higher order derivatives of $\boldsymbol{u}$ (up to the highest order of polynomials present in the FE formulation), while our variant affects the first-order derivatives only. We can allow ourselves to reduce the order of stabilized derivatives thanks to the presence of additional stabilization terms J_h^{lhs} (7.18), as first suggested in Duprez and Lozinski (2020) (a similar idea can also be found in Elfverson et al., 2019). The combination of G_h and J_h^{lhs} allows us indeed to get rid of possible spurious oscillations of the approximate solution on "badly cut" cells near Γ and to guarantee the coerciveness of the bilinear form in our FE formulation. Note that the terms J_h^{lhs} are not consistent by themselves but they are consistently compensated by their right-hand side counterpart J_h^{rhs}. Indeed, the exact solution satisfies div $\boldsymbol{\sigma}(\boldsymbol{u}) = -\boldsymbol{f}$ so that $J_h^{lhs}(\boldsymbol{u}, \boldsymbol{v}) = J_h^{rhs}(\boldsymbol{v})$ if $\boldsymbol{u}$ is the exact solution.

- Dual ϕ-FEM-Dirichlet: find $\boldsymbol{u}_h \in V_h$, $\boldsymbol{p}_h \in Q_h^k(\Omega_h^\Gamma)$ such that:

$$\int_{\Omega_h} \boldsymbol{\sigma}(\boldsymbol{u}_h) : \nabla \boldsymbol{v}_h - \int_{\partial\Omega_h} \boldsymbol{\sigma}(\boldsymbol{u}_h)\boldsymbol{n} \cdot \boldsymbol{v}_h + \frac{\gamma}{h^2} \int_{\Omega_h^\Gamma} \left(\boldsymbol{u}_h - \frac{1}{h}\phi_h \boldsymbol{p}_h\right) \cdot \left(\boldsymbol{v}_h - \frac{1}{h}\phi_h \boldsymbol{q}_h\right)$$

$$+ G_h(\boldsymbol{u}_h, \boldsymbol{v}_h) + J_h^{lhs}(\boldsymbol{u}_h, \boldsymbol{v}_h)$$

$$= \int_{\Omega_h} \boldsymbol{f} \cdot \boldsymbol{v}_h + \frac{\gamma}{h^2} \int_{\Omega_h^\Gamma} \boldsymbol{u}_h^g \cdot \left(\boldsymbol{v}_h - \frac{1}{h}\phi_h \boldsymbol{q}_h\right) + J_h^{rhs}(\boldsymbol{v}_h), \quad \forall \boldsymbol{v}_h \in V_h, \, \boldsymbol{q}_h \in Q_h^k(\Omega_h^\Gamma). \quad (7.19)$$

With respect to (7.13, we have added here the factors $\frac{1}{h}$, $\frac{1}{h^2}$. They serve to control the condition numbers, c.f. Duprez et al. (2021). The stabilizations $G_h, J_h^{lhs}, J_h^{rhs}$ are again defined by (7.17) and (7.18).

Test case: Let $\mathcal{O}$ be the square $(0, 1)^2$ and $\mathcal{T}_h^{\mathcal{O}}$ a uniform mesh on $\mathcal{O}$. Let Ω be the circle centered at the point $(0.5, 0.5)$ of radius $\frac{\sqrt{2}}{4}$. The LS function ϕ is thus given by:

$$\phi(x, y) = -\frac{1}{8} + (x - 0.5)^2 + (y - 0.5)^2. \quad (7.20)$$

We take the elasticity parameters $E = 2$ and $\nu = 0.3$, and the scheme parameters $\gamma = \sigma_D = 20.0$. We use $\mathbb{P}^2$-Lagrange polynomials for both FE spaces V_h and Q_h, i.e., we set $k = 2$ in (7.14) and (7.15). We finally choose a manufactured exact solution:

$$\boldsymbol{u} = \boldsymbol{u}_{ex} := (\sin(x)\exp(y), \sin(y)\exp(x)) \quad (7.21)$$

giving the right-hand side $\boldsymbol{f}$ by substitution to (7.2) and the boundary conditions $\boldsymbol{u}^g = \boldsymbol{u}_{ex}$ on Γ. In order to set up both ϕ-FEM schemes above, we should extend $\boldsymbol{u}^g$ from Γ to Ω_h (in the case of the direct method) or to Ω_h^Γ (in the case of the dual method). To mimic the realistic situation where $\boldsymbol{u}^g$ is known on Γ only, we prefer not to extend $\boldsymbol{u}^g$ by $\boldsymbol{u}_{ex}$ everywhere. We rather set

$$\boldsymbol{u}^g = \boldsymbol{u}_{ex}(1 + \phi), \quad \text{on } \Omega_h \text{ or on } \Omega_h^\Gamma$$

adding to $\boldsymbol{u}_{ex}$ a perturbation which vanishes on the boundary.

The typical active meshes $\mathcal{T}_h$ and $\mathcal{T}_h^\Gamma$ for ϕ-FEM are illustrated in Figure 7.2 (left). Besides the direct ϕ-FEM (7.16) and the dual ϕ-FEM (7.19), we shall present the numerical results obtained by the standard FEM with $\mathbb{P}^2$-Lagrange polynomials on fitted meshes for approximately the same values of h, as illustrated in Figure 7.2 (right). The results obtained by both variants of ϕ-FEM and by the standard FEM are reported in Figures 7.3 and 7.4.

We first illustrate the numerical convergences order for the relative errors in L^2 and H^1 norms in Figure 7.3. We observe that both variants of ϕ-FEM demonstrate indeed the expected optimal convergence orders: h^2 is the H^1-semi-norm and h^3 in the L^2-norm, and the direct variant performs significantly better than the dual one. This

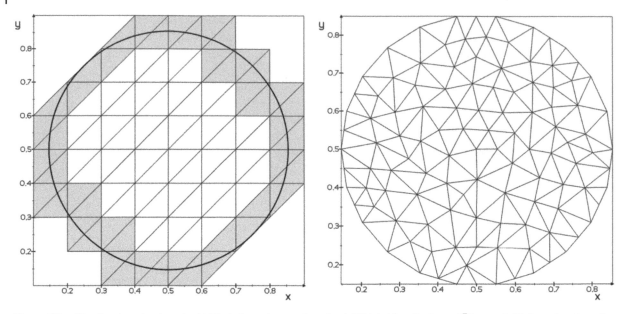

Figure 7.2 Circular domain given by (7.20). Left: active meshes for ϕ-FEM (with cells from $\mathcal{T}_h^{\Gamma}$ in gray). Right: a fitted mesh for the standard FEM.

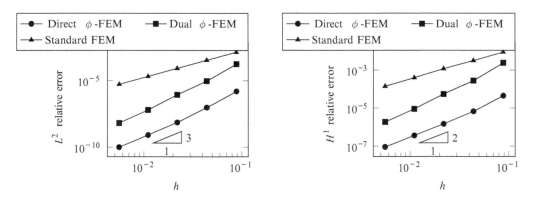

Figure 7.3 Test case with pure Dirichlet conditions. L^2 relative errors on the left, H^1 relative errors on the right.

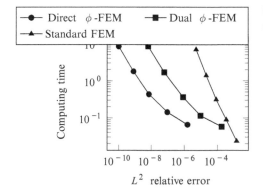

Figure 7.4 Test case with pure Dirichlet conditions. Computing time (in seconds) versus the L^2 relative errors.

can be attributed to a better representation of the solution near the boundary in the direct variant: indeed, it is effectively approximated there by fourth-order polynomials ($\mathbb{P}^2$ for $\boldsymbol{w}_h$ times $\mathbb{P}^2$ for ϕ_h). Moreover, both ϕ-FEMs, even the dual one, significantly outperform the standard FEM (the latter is even of a sub-optimal order in the L^2-norm). This can be partially attributed to a coarse geometry approximation. Indeed, we use triangular meshes so that the curved boundary of Ω is actually approximated by a collection of straight segments, i.e., the boundary facets of the fitted mesh, c.f. Figure 7.2 (right). The superior efficiency of ϕ-FEM with respect to the standard FEM is further confirmed by Figure 7.4. We report there the computing times on different meshes for the three methods and set them against the relative L^2 error. These computing times include assembling of the FE matrices and resolution of the resulting linear systems. For a given relative error, the calculations are always much faster with ϕ-FEM than with the standard FEM. The advantage would be even more significant if the mesh generation times were included, since the construction of active meshes in ϕ-FEM only involves choosing a subset of cells according to a simple criterion, and some renumbering of the degrees of freedom. We do not dispose however of an efficient implementation of cell selection algorithm at the moment. All our computations are performed using the Python interface for the popular FEniCS computing platform, and the selection of active cells is done by a simple, non-optimized Python script.

7.2.2 Mixed Boundary Conditions

We now consider the much more complicated case of mixed conditions (7.4)–(7.5) on the boundary $\Gamma = \Gamma_N \cup \Gamma_D$ with $\Gamma_D \neq \emptyset$ and $\Gamma_N \neq \emptyset$. This setting is challenging for any geometrically unfitted method since the junction between the Dirichlet and Neumann boundary parts can occur inside a mesh cell, so that approximating polynomials in this cell should account simultaneously for both boundary conditions. In Lehrenfeld (2016), it is demonstrated that the linear elasticity with mixed boundary conditions can be successfully treated by Cut-FEM. A rigorous mathematical substantiation allowing of the low regularity of the solution is available in Burman et al. (2020). Here, we shall adapt ϕ-FEM (in the dual form) to the mixed boundary conditions by adopting a "lazy" approach: we choose to do not impose any boundary conditions on a mesh cell if the Dirichlet/Neumann junction happens to be inside it.

To set up the geometry of the problem, we recall that the domain Ω is given by the LS function ϕ, $\Omega = \{\phi < 0\}$, and assume furthermore that the boundary partition into the Dirichlet and Neumann parts is governed by a secondary LS ψ:

$$\Gamma_D = \Gamma \cap \{\psi < 0\}, \quad \Gamma_N = \Gamma \cap \{\psi > 0\}.$$

Introducing the active meshes $\mathcal{T}_h$ and $\mathcal{T}_h^\Gamma$ as above, c.f. (7.7), (7.8) and Figure 7.1, we want now further partition the sub-mesh $\mathcal{T}_h^\Gamma$ into two parts: $\mathcal{T}_h^{\Gamma_D}$ around Γ_D, serving to impose the Dirichlet boundary conditions, and $\mathcal{T}_h^{\Gamma_N}$ around Γ_N for the Neumann ones. The natural choice for these is given by:

$$\mathcal{T}_h^{\Gamma_D} := \{T \in \mathcal{T}_h^\Gamma : \psi \leqslant 0 \text{ on } T\} \qquad \text{and} \qquad \mathcal{T}_h^{\Gamma_N} := \{T \in \mathcal{T}_h^\Gamma : \psi \geqslant 0 \text{ on } T\}. \tag{7.22}$$

As before, we denote the domains occupied by meshes $\mathcal{T}_h, \mathcal{T}_h^\Gamma, \mathcal{T}_h^{\Gamma_D}, \mathcal{T}_h^{\Gamma_N}$ by $\Omega_h, \Omega_h^\Gamma, \Omega_h^{\Gamma_D}, \Omega_h^{\Gamma_N}$, respectively. Note that these definitions may leave a small number of cells of $\mathcal{T}_h^\Gamma$ out of both $\mathcal{T}_h^{\Gamma_D}$ and $\mathcal{T}_h^{\Gamma_N}$. Indeed, there may be mesh cells, near the junction of Dirichlet and Neumann parts, where ψ changes sign inside the cell, so that ψ is neither everywhere positive not everywhere negative on such a cell. This is illustrated in Figure 7.8 (left) where the Dirichlet/Neumann junction is supposed at $x = 0.5$, i.e., the secondary LS is $\psi(x, y) = 0.5 - y$, c.f. Figure 7.5. The active mesh cells intersected by Γ in Figure 7.8 are either on the Dirichlet side (they thus form $\mathcal{T}_h^{\Gamma_D}$ and are colored

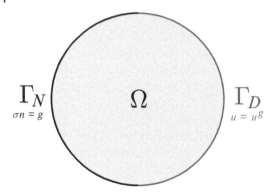

Figure 7.5 Test case with mixed boundary conditions: the geometry of Dirichlet and Neumann boundary parts.

in medium-dark gray), or on the Neumann side (they thus form $\mathcal{T}_h^{\Gamma_N}$ and are colored in dark gray), or in between (they are then in $\mathcal{T}_h^{\Gamma}$ but not in $\mathcal{T}_h^{\Gamma_D}$ or $\mathcal{T}_h^{\Gamma_N}$, and are colored in light gray).

Assuming once more that $\boldsymbol{u}$, the solution to (7.2), (7.4), and (7.5), can be extended from Ω to Ω_h as the solution to the same governing Equation (7.2), we introduce a ϕ-FEM scheme, combining the dual ϕ-FEM Dirichlet approach, as introduced in (7.13) and (7.19), with the indirect imposition of Neumann boundary condition as proposed in Duprez et al. (2021). We thus keep $\boldsymbol{u}$ as the primary unknown on Ω_h and recall that it satisfies the variational formulation (7.9). The Dirichlet boundary condition affects the solution on $\Omega_h^{\Gamma_D}$ through the introduction of the auxiliary variable $\boldsymbol{p}_D$ there. We thus adapt (7.12) from the pure Dirichlet case as:

$$\boldsymbol{u} = \boldsymbol{u}^g + \phi \boldsymbol{p}_D, \quad \text{on } \Omega_h^{\Gamma_D}. \tag{7.23}$$

We have assumed here that $\boldsymbol{u}^g$ is extended from Γ_D to $\Omega_h^{\Gamma_D}$.

The Neumann boundary condition will affect $\boldsymbol{u}$ on $\Omega_h^{\Gamma_N}$ through the introduction of two auxiliary variables there. We first introduce a tensor-valued variable $\boldsymbol{y}$ on $\Omega_h^{\Gamma_N}$ setting $\boldsymbol{y} = -\sigma(\boldsymbol{u})$. It remains to impose $\boldsymbol{y}\boldsymbol{n} = -\boldsymbol{g}$ on Γ_N. To this end, we note that the outward-looking unit normal $\boldsymbol{n}$ is given on Γ by $\boldsymbol{n} = \frac{1}{|\nabla\phi|}\nabla\phi$ so that the Neumann boundary condition is satisfied by setting $\boldsymbol{y}\nabla\phi + \boldsymbol{g}|\nabla\phi| = -\boldsymbol{p}_N\phi$ on $\Omega_h^{\Gamma_N}$ where $\boldsymbol{p}_N$ is yet another (vector-valued) auxiliary variable on $\Omega_h^{\Gamma_N}$. This can be summarized as:

$$\boldsymbol{y} + \sigma(\boldsymbol{u}) = 0, \quad \text{on } \Omega_h^{\Gamma_N}, \tag{7.24a}$$

$$\boldsymbol{y}\nabla\phi + \boldsymbol{p}\phi = -\boldsymbol{g}|\nabla\phi|, \quad \text{on } \Omega_h^{\Gamma_N}. \tag{7.24b}$$

Note that the combination of (7.23) with (7.24a) and (7.24b) does not impose the mixed Dirichlet/Neumann conditions on the whole of Γ since the latter may be not completely covered by $\Omega_h^{\Gamma_D} \cup \Omega_h^{\Gamma_N}$. Fortunately, this defect of the formulation on the continuous level can be repaired on the discrete level by adding the appropriate stabilization to the FE discretization.

To describe the resulting FE scheme, we start by introducing the FE spaces. As before, we fix an integer $k \geq 1$ and keep the space V_h, as defined in (7.14), for the approximation $\boldsymbol{u}_h$ of the primary variable $\boldsymbol{u}$. We also need the spaces for the approximation of the auxiliary variables $\boldsymbol{p}_{h,D}$ and $\boldsymbol{p}_{h,N}$, respectively $Q_h^k(\Omega_h^{\Gamma_D})$ and $Q_h^{k-1}(\Omega_h^{\Gamma_N})$ as defined in (7.15), as well as the space $Z_h(\Omega_h^{\Gamma_N})$ to approximate $\boldsymbol{y}$, where for each sub-mesh $\mathcal{M}_h$ of $\mathcal{T}_h$, $Z_h(\mathcal{M}_h)$ is defined by:

$$Z_h(\mathcal{M}_h) := \left\{ \boldsymbol{z}_h : \mathcal{M}_h \to \mathbb{R}^{(d \times d)} : \boldsymbol{z}_{h|T} \in \mathbb{P}^k(T)^{(d \times d)} \ \forall T \in \mathcal{M}_h, \ \boldsymbol{z}_h \text{ continuous on } \mathcal{M}_h \right\}. \tag{7.25}$$

Now, combining the variational formulation (7.9) with (7.23) and (7.24a–7.24b) imposed in a least squares manner, we get the following scheme: find $\boldsymbol{u}_h \in V_h$, $\boldsymbol{p}_{h,D} \in Q_h^k(\Omega_h^{\Gamma_D})$, $\boldsymbol{y}_h \in Z_h(\Omega_h^{\Gamma_N})$ and $\boldsymbol{p}_{h,N} \in Q_h^{k-1}(\Omega_h^{\Gamma_N})$ such that:

$$\int_{\Omega_h} \boldsymbol{\sigma}(\boldsymbol{u}_h) : \nabla \boldsymbol{v}_h - \int_{\partial\Omega_h \setminus \partial\Omega_{h,N}} \boldsymbol{\sigma}(\boldsymbol{u}_h)\boldsymbol{n} \cdot \boldsymbol{v}_h + \int_{\partial\Omega_{h,N}} \boldsymbol{y}_h \boldsymbol{n} \cdot \boldsymbol{v}_h$$

$$+ \gamma_u \int_{\Omega_h^{\Gamma_N}} (\boldsymbol{y}_h + \boldsymbol{\sigma}(\boldsymbol{u}_h)) : (\boldsymbol{z}_h + \boldsymbol{\sigma}(\boldsymbol{v}_h)) + \frac{\gamma_p}{h^2} \int_{\Omega_h^{\Gamma_N}} \left(\boldsymbol{y}_h \nabla\phi_h + \frac{1}{h}\boldsymbol{p}_{h,N}\phi_h \right) \cdot \left(\boldsymbol{z}_h \nabla\phi_h + \frac{1}{h}\boldsymbol{q}_{h,N}\phi_h \right)$$

$$+ \frac{\gamma}{h^2} \int_{\Omega_h^{\Gamma_D}} (\boldsymbol{u}_h - \frac{1}{h}\phi_h \boldsymbol{p}_{h,D}) \cdot (\boldsymbol{v}_h - \frac{1}{h}\phi_h \boldsymbol{q}_{h,D}) + G_h(\boldsymbol{u}_h, \boldsymbol{v}_h) + J_h^{lhs,D}(\boldsymbol{u}_h, \boldsymbol{v}_h) + J_h^{lhs,N}(\boldsymbol{y}_h, \boldsymbol{z}_h)$$

$$= \int_{\Omega_h} \boldsymbol{f} \cdot \boldsymbol{v}_h + \frac{\gamma}{h^2} \int_{\Omega_h^D} \boldsymbol{u}_h^g \cdot (\boldsymbol{v}_h - \frac{1}{h}\phi_h \boldsymbol{q}_{h,D}) - \frac{\gamma_p}{h^2} \int_{\Omega_h^{\Gamma_N}} \boldsymbol{g} \cdot |\nabla\phi_h|(\boldsymbol{z}_h \cdot \nabla\phi_h + \frac{1}{h}\boldsymbol{q}_{h,N}\phi_h)$$

$$+ J_h^{rhs,D}(\boldsymbol{v}_h) + J_h^{rhs,N}(\boldsymbol{z}_h),$$

$$\forall \boldsymbol{v}_h \in V_h, \boldsymbol{q}_{h,D} \in Q_h^k(\Omega_h^{\Gamma_D}), \boldsymbol{z}_h \in Z_h(\Omega_h^{\Gamma_N}), \boldsymbol{q}_{h,N} \in Q_h^{k-1}(\Omega_h^{\Gamma_N}). \quad (7.26)$$

We have added here the ghost stabilization G_h defined by (7.17) as in the pure Dirichlet case. The additional stabilizations terms J_h^{lhs}, J_h^{rhs} are now adapted from (7.18) and separated into the terms acting on $\boldsymbol{u}_h$ on the Dirichlet cells of $\mathcal{T}_h^\Gamma$ (and also those not marked) and the terms acting on $\boldsymbol{y}_h$ on the Neumann cells:

$$J_h^{lhs,D}(\boldsymbol{u}, \boldsymbol{v}) := \sigma_D h^2 \sum_{T \in \mathcal{T}_h^\Gamma \setminus \mathcal{T}_h^{\Gamma_N}} \int_T \text{div }\boldsymbol{\sigma}(\boldsymbol{u}) \cdot \text{div }\boldsymbol{\sigma}(\boldsymbol{v}),$$

$$J_h^{rhs,D}(\boldsymbol{v}) := -\sigma_D h^2 \sum_{T \in \mathcal{T}_h^\Gamma \setminus \mathcal{T}_h^{\Gamma_N}} \int_T \boldsymbol{f} \cdot \text{div }\boldsymbol{\sigma}(\boldsymbol{v}),$$

$$J_h^{lhs,N}(\boldsymbol{y}, \boldsymbol{z}) = \gamma_{div} \int_{\Omega_h^{\Gamma_N}} \text{div }\boldsymbol{y} \cdot \text{div }\boldsymbol{z}, \qquad\qquad J_h^{rhs,N}(\boldsymbol{z}) = \gamma_{div} \int_{\Omega_h^{\Gamma_N}} \boldsymbol{f} \cdot \text{div }\boldsymbol{z}. \quad (7.27)$$

These stabilizations are consistent with the governing equations $\text{div }\boldsymbol{\sigma}(\boldsymbol{u}) = -\boldsymbol{f}$, rewritten as $\text{div }\boldsymbol{y} = \boldsymbol{f}$, using (7.24a), wherever possible, i.e., on $\Omega_h^{\Gamma_N}$. Note that a similar treatment is applied to the boundary integral terms on $\partial\Omega$ in (7.9). In (7.26), they are rewritten in terms of $\boldsymbol{y}$, using (7.24a) and (7.24b), wherever possible. We thus introduce a part of the boundary $\partial\Omega_h$, referred to as $\partial\Omega_{h,N}$, formed by the boundary facets of $\mathcal{T}_h$ belonging to the cells in $\mathcal{T}_h^{\Gamma_N}$. We replace $\boldsymbol{\sigma}(\boldsymbol{u}_h)$ by $-\boldsymbol{y}_h$ on $\partial\Omega_{h,N}$, while keeping the boundary term as is on the remaining part of the boundary. All this contributes to the coerciveness of the bilinear form in (7.26) and good conditioning of the matrix as can be proven following the ideas of Duprez et al. (2021). We emphasize again that neither Dirichlet nor Neumann boundary conditions are imposed in any way in scheme (7.26) on the cells in $\mathcal{T}_h^\Gamma \setminus (\mathcal{T}_h^{\Gamma_D} \cup \mathcal{T}_h^{\Gamma_N})$ (the cells in light gray in Figure 7.8). On the other hand, both stabilizations G_h and J_h are active on the whole $\mathcal{T}_h^\Gamma$, comprising these cells not marked as Dirichlet or Neumann.

Test case: We are now going to present some numerical results with method (7.26) highlighting the optimal convergence of ϕ-FEM and comparing it with a standard FEM. We use the same geometry (7.20), elasticity parameters and the exact solution (7.21) as for the case of pure Dirichlet conditions on page 197. We set furthermore the Dirichlet boundary conditions (7.4) for $x > 0.5$ and the Neumann boundary conditions (7.5) for $x < 0.5$, c.f. Figure 7.5, i.e., we choose the secondary LS as $\psi = 0.5 - x$. The data $\boldsymbol{u}^g$ and $\boldsymbol{g}$ are computed from the exact solution. In ϕ-FEM, they should be extended from Γ to appropriate portion of the strip Ω_h^Γ. We choose these extensions as:

$$\begin{cases} \boldsymbol{u}^g = \boldsymbol{u}_{ex}(1 + \phi), & \text{on } \Omega_h^\Gamma \cap \{x \geqslant 0.5\}, \\ \boldsymbol{g} = \boldsymbol{\sigma}(\boldsymbol{u}_{ex})\frac{\nabla\phi}{\|\nabla\phi\|} + \boldsymbol{u}_{ex}\phi, & \text{on } \Omega_h^\Gamma \cap \{x < 0.5\}. \end{cases}$$

Again, both expressions are perturbed away from Γ to mimic the real-life situation where the data are available only on Γ. The stabilization parameters are set to $\gamma_{div} = \gamma_u = \gamma_p = 1.0$, $\sigma = 0.01$, and $\gamma = \sigma_D = 20.0$.

We start by studying mesh configurations where the Dirichlet-Neumann junction line $\{x = 0.5\}$ happens to be covered by the mesh facets both in the background mesh used by ϕ-FEM, and in the fitted mesh used by FEM, as illustrated in Figure 7.6. All the boundary cells in $\mathcal{T}_h^\Gamma$ are marked in this case either as Dirichlet or as Neumann ones, according to the criterion (7.22), giving, respectively, medium-dark gray and dark gray cells in Figure 7.6 (left). There is no ambiguity for the standard FEM fitted meshes: all the boundary facets are straightforwardly marked either as Dirichlet or as Neumann, c.f. Figure 7.6 (right) with the same color code as for the unfitted mesh. The results obtained by both ϕ-FEM (7.26) and the standard FEM, using $\mathbb{P}^2$-Lagrange polynomials for $\boldsymbol{u}_h$ in both cases, are reported in Figure 7.7. On the left, the relative errors are plotted with respect to the mesh step. We observe again the optimal convergence orders for ϕ-FEM, while the convergence of the standard FEM is sub-optimal in the L^2-norm. The ϕ-FEM approach is again systematically more precise in both norms. On the right side of the same figure, we plot the computing times and notice again that ϕ-FEM is less expensive than the standard FEM.

Let us now turn to a less artificial mesh configuration where the Dirichlet/Neumann junction point can turn up inside a mesh cell of the background mesh, or inside a boundary facet of the fitted mesh. We study these situations on a series of meshes, as illustrated in Figure 7.8. In the case of the background meshes used for ϕ-FEM, we ensure in particular that there are no vertical grid line with the abscissa $x = 0.5$ so that there are exactly four cells in $\mathcal{T}_h^\Gamma$ that are neither in $\mathcal{T}_h^{\Gamma_D}$ nor in $\mathcal{T}_h^{\Gamma_N}$ (yellow cells on the left side of Figure 7.8). We recall that scheme (7.26) does not impose any boundary conditions on these cells, but retains the stabilization there (in particular, the governing equation is still re-enforced on these cells in the least squares manner). Note that the fitted FEM is not straightforward to implement in this case either, since the Dirichlet boundary conditions cannot be strongly imposed on the boundary facets which turn up only partially on the Dirichlet side. We bypass this difficulty by treating the Dirichlet conditions by penalization, so that the "standard" FEM is now defined as: find $\boldsymbol{u}_h$ in the $\mathbb{P}^k$

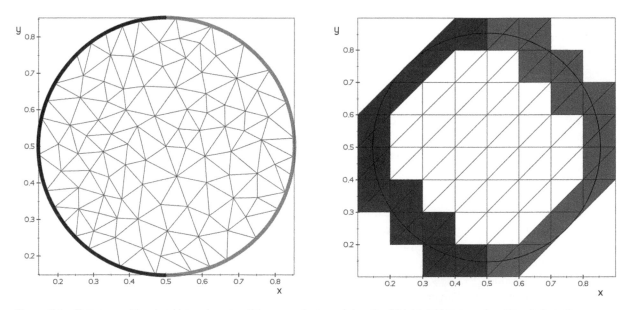

Figure 7.6 Test case with mixed boundary conditions, meshes resolving the Dirichlet/Neumann junction. Left: active meshes for ϕ-FEM, medium-dark gray for $\mathcal{T}_h^{\Gamma_D}$, dark gray for $\mathcal{T}_h^{\Gamma_N}$. Right: a mesh for standard FEM, medium-dark gray boundary facets on Γ_D, dark gray boundary facets on Γ_N.

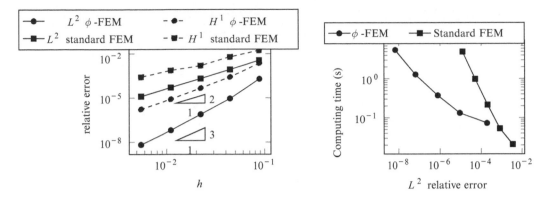

Figure 7.7 Test case with mixed boundary conditions, results on meshes as in Figure 7.6. Left: L^2 and H^1 relative errors under the mesh refinement. Right: computing time versus the L^2 relative error.

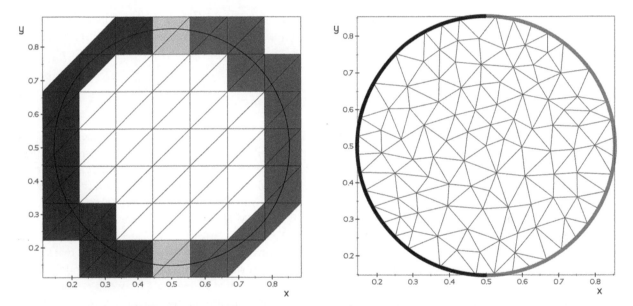

Figure 7.8 Test case with mixed boundary conditions, meshes not resolving the Dirichlet/Neumann junction. Left: active meshes for ϕ-FEM, medium-dark gray for $\mathcal{T}_h^{\Gamma_D}$, dark gray for $\mathcal{T}_h^{\Gamma_N}$, light gray for $\mathcal{T}_h^{\Gamma}$ otherwise unmarked. Right: a mesh for standard FEM, medium-dark gray boundary facets on Γ_D, dark gray boundary facets on Γ_N, note that some boundary facets contain both Dirichlet and Neumann parts.

FE space (without any restrictions on the boundary) such that:

$$\int_\Omega \boldsymbol{\sigma}(\boldsymbol{u}_h) : \nabla \boldsymbol{v}_h + \frac{1}{\varepsilon} \int_{\Gamma_D} \boldsymbol{u}_h \cdot \boldsymbol{v}_h = \int_\Omega \boldsymbol{f} \cdot \boldsymbol{v}_h + \int_{\Gamma_N} \boldsymbol{g} \cdot \boldsymbol{v}_h + \frac{1}{\varepsilon} \int_{\Gamma_D} \boldsymbol{u}^g \cdot \boldsymbol{v}_h \tag{7.28}$$

for all $\boldsymbol{v}_h$ in the same FE space as $\boldsymbol{u}_h$, with a small parameter $\varepsilon > 0$.

The mesh refinement study in this case is reported in Figure 7.9. Comparing the results with those of Figure 7.7 (obtained on idealized unrealistic meshes without any unmarked cells), we observe that the behavior of ϕ-FEM (7.26) is almost unaffected by the presence (or not) of the unmarked "light gray" cells, although the convergence

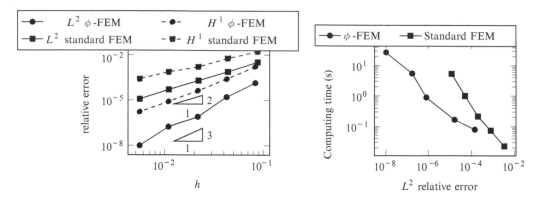

Figure 7.9 Test case with mixed boundary conditions, results on meshes as in Figure 7.8. Left: L^2 and H^1 relative errors under the mesh refinement. Right: computing time versus the L^2 relative error.

curve for the L^2 relative error is now slightly less regular. In particular, the conclusions about the relative merits of φ-FEM and the fitted FEM, now in version (7.28), remain unchanged: φ-FEM is more precise on comparable meshes and less expensive in terms of the computing times for a given error tolerance.

7.3 Linear Elasticity with Multiple Materials

We now consider the case of interfaces problems, i.e., partial differential equations with coefficients jumping across an interface, which can cut the computational mesh in an arbitrary manner. The simplest meaningful example in the realm of linear elasticity is given by structures consisting of multiple materials having different elasticity parameters. This situation has already been treated in XFEM (Annavarapu et al., 2012; Carraro and Wetterauer, 2015; Xiao et al., 2020; Yuanming et al., 2020), CutFEM (Burman et al., 2019; Hansbo and Hansbo, 2002; Hansbo, 2005; Lehrenfeld and Reusken, 2018), and SBM (Li et al., 2020) paradigms. We are now going to demonstrate the applicability of φ-FEM in this context.

Let us assume that the structure occupies a domain Ω and it consists of two materials that occupy two subdomains Ω_1 and Ω_2 separated by the interface Γ. To fix the ideas, we further assume that Ω_1 is surrounded by Ω_2, so that the interface Γ can actually be described as $\Gamma = \partial\Omega_1$, as illustrated in Figure 7.10. We also assume that the displacement $\boldsymbol{u}$ is given on the external boundary (these assumptions are not restrictive and the forthcoming method can be easily adapted to other situations, e.g., with Γ touching $\partial\Omega$ or with Neumann boundary conditions on the external boundary). We then consider the problem for the displacement $\boldsymbol{u}$ on Ω:

$$\begin{cases} -\operatorname{div}\boldsymbol{\sigma}(\boldsymbol{u}) & = \boldsymbol{f}\,,\ \text{on}\ \Omega\backslash\Gamma\,, \\ \boldsymbol{u} & = \boldsymbol{u}^g\,,\ \text{on}\ \partial\Omega\,, \\ [\boldsymbol{u}] & = 0\,,\ \text{on}\ \Gamma\,, \\ [\boldsymbol{\sigma}(\boldsymbol{u})\boldsymbol{n}] & = 0\,,\ \text{on}\ \Gamma\,, \end{cases} \tag{7.29}$$

where $\boldsymbol{n}$ is the unit normal pointing from Ω_1 to Ω_2, and the brackets $[\cdot]$ stand for the jump across Γ. The elasticity parameters are assumed constant on each subdomain, but different from each other. The stress tensor is thus given by:

$$\boldsymbol{\sigma}(\boldsymbol{u}) = \begin{cases} \boldsymbol{\sigma}_1(\boldsymbol{u}) = 2\mu_1\varepsilon(\boldsymbol{u}) + \lambda_1(\operatorname{div}\boldsymbol{u})I\,,\ \text{on}\ \Omega_1\,, \\ \boldsymbol{\sigma}_2(\boldsymbol{u}) = 2\mu_2\varepsilon(\boldsymbol{u}) + \lambda_2(\operatorname{div}\boldsymbol{u})I\,,\ \text{on}\ \Omega_2\,, \end{cases}$$

Figure 7.10   Geometry with the interface Γ: elasticity with multiple materials.

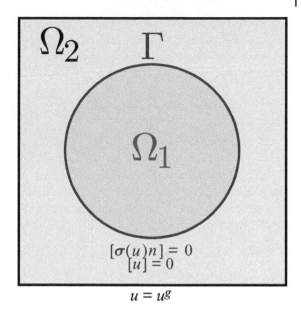

$$[\sigma(u)n] = 0$$
$$[u] = 0$$

$$u = u^g$$

with the Lamé parameters λ_i and μ_i defined via the formulas (7.3) with given $E_i, \nu_i, i = 1, 2$. Introducing the displacements $\boldsymbol{u}_i = \boldsymbol{u}|_{\Omega_i}, i = 1, 2$ on Ω_1 and Ω_2 separately, problem (7.29) can be rewritten as the system of two coupled sub-problems:

$$\begin{cases} -\operatorname{div} \boldsymbol{\sigma}_i(\boldsymbol{u}_i) & = \boldsymbol{f}, \text{ on } \Omega_i, \, i = 1, 2, \\ \boldsymbol{u}_2 & = \boldsymbol{u}^g, \text{ on } \partial\Omega, \\ \boldsymbol{u}_1 & = \boldsymbol{u}_2, \text{ on } \Gamma, \\ \boldsymbol{\sigma}_1(\boldsymbol{u}_1)\boldsymbol{n} & = \boldsymbol{\sigma}_2(\boldsymbol{u}_2)\boldsymbol{n}, \text{ on } \Gamma. \end{cases} \tag{7.30}$$

We suppose that Ω is sufficiently simple-shaped so that a matching mesh T_h on Ω is easily available (again, this assumption is not restrictive; we have seen that a complex-shape domain Ω can be also treated by ϕ-FEM). On the contrary, the mesh $\mathcal{T}_h$ is not supposed to match the internal interface Γ and we are going to adapt ϕ-FEM to this situation. The starting point is the reformulation (7.30). We are thus going to discretize separately $\boldsymbol{u}_1$ on Ω_1 and $\boldsymbol{u}_2$ on Ω_2. To this end, we introduce two active meshes $\mathcal{T}_{h,1}$ and $\mathcal{T}_{h,2}$, sub-meshes of $\mathcal{T}_h$, constructed by retaining in $\mathcal{T}_{h,i}$ the cells of $\mathcal{T}_h$ having a non-empty intersection with Ω_i. In practice, the subdomains are defined through an LS ϕ:

$$\Omega_1 = \{\phi > 0\} \cap \Omega, \qquad \Omega_2 = \{\phi < 0\}, \qquad \Gamma = \{\phi = 0\} \cap \Omega.$$

The sub-meshes $\mathcal{T}_{h,i}$ are defined using a piecewise-polynomial approximation ϕ_h of ϕ, rather than ϕ itself:

$$\mathcal{T}_{h,1} := \{T \in \mathcal{T}_h : T \cap \{\phi_h > 0\} \neq \varnothing\} \text{ and } \mathcal{T}_{h,2} := \{T \in \mathcal{T}_h : T \cap \{\phi_h < 0\} \neq \varnothing\}. \tag{7.31}$$

We also introduce the sub-mesh $\mathcal{T}_h^\Gamma$ as the intersection $\mathcal{T}_{h,1} \cap \mathcal{T}_{h,2}$ and denote by $\Omega_{h,1}, \Omega_{h,2}, \Omega_h^\Gamma$ the domains covered by meshes $\mathcal{T}_{h,1}, \mathcal{T}_{h,2}, \mathcal{T}_h^\Gamma$, respectively. Similarly to the simpler settings considered above, the unknowns $\boldsymbol{u}_1$ and $\boldsymbol{u}_2$, living physically on Ω_1 and Ω_2, will be discretized on larger domains $\Omega_{h,1}$ and $\Omega_{h,2}$, introducing artificial extensions on narrow fictitious strips near Γ. On the discrete level, the unknowns will be thus redoubled on the joint sub-mesh

$\mathcal{T}_h^\Gamma$. Several auxiliary unknowns will be introduced on Ω_h^Γ similar to the case of mixed boundary conditions above (indeed, we have to discretize both Dirichlet and Neumann conditions on the interface Γ in the current setting).

We now put the program above into the equations, first on the continuous level. Similarly to (7.9), the unknowns $\boldsymbol{u}_i$ extended to larger domains Ω_h^i satisfy formally the variational formulations, c.f. the first equation in (7.30):

$$\int_{\Omega_{h,i}} \boldsymbol{\sigma}_i(\boldsymbol{u}_i) : \nabla\boldsymbol{v}_i - \int_{\partial\Omega_{h,i}} \boldsymbol{\sigma}_i(\boldsymbol{u}_i)\boldsymbol{n}_i \cdot \boldsymbol{v}_i = \int_{\Omega_{h,i}} \boldsymbol{f} \cdot \boldsymbol{v}_i, \quad \forall \boldsymbol{v}_i \text{ on } \Omega_{h_i} \text{ s.t.} \boldsymbol{v}_i = 0 \text{ on } \partial\Omega. \tag{7.32}$$

Here, with a slight abuse of notations, $\partial\Omega_{h,i}$ denotes the component of the boundary of $\Omega_{h,i}$ other than $\partial\Omega$, and $\boldsymbol{n}_i$ denotes the unit normal vector on $\partial\Omega_{h,i}$ pointing outside $\Omega_{h,i}$. The boundary conditions on the external boundary $\partial\Omega$, i.e., the second equation in (7.30), will be imposed strongly. The remaining equations in (7.30), i.e., the interface conditions on Γ, will be imposed by introduction of auxiliary variables on Ω_h^Γ: the vector-valued $\boldsymbol{p}$ (similar to the dual version of φ-FEM for the Dirichlet boundary conditions above) and matrix-valued $\boldsymbol{y}_1, \boldsymbol{y}_2$ (similar to φ-FEM for the Neumann boundary conditions). This gives, c.f. the last two equations in (7.30):

$$\boldsymbol{u}_1 - \boldsymbol{u}_2 + \boldsymbol{p}\phi = 0, \quad \text{on } \Omega_h^\Gamma, \tag{7.33}$$

$$\boldsymbol{y}_i + \boldsymbol{\sigma}_i(\boldsymbol{u}_i) = 0, \quad \text{on } \Omega_h^\Gamma, i = 1, 2, \tag{7.34}$$

$$\boldsymbol{y}_1\nabla\phi - \boldsymbol{y}_2\nabla\phi = 0, \quad \text{on } \Omega_h^\Gamma. \tag{7.35}$$

Equation (7.35) above extends the last equation in (7.30) from Γ to Ω_h^Γ since the normal on Γ is collinear with the vector $\nabla\phi$ there.

We are now going to discretize Equations (7.32)–(7.35). We fix an integer $k \geq 1$ and introduce the FE spaces for the primary variables $\boldsymbol{u}_i$:

$$V_{h,i} := \{\boldsymbol{v}_h : \Omega_{h,i} \to \mathbb{R}^d : \boldsymbol{v}_{h|T} \in \mathbb{P}^k(T)^d \ \forall T \in \mathcal{T}_h, \ \boldsymbol{v}_h \text{ continuous on } \Omega_{h,i},$$

$$\text{and } \boldsymbol{v}_h = I_h\boldsymbol{u}^g \text{ on } \partial\Omega\} \tag{7.36}$$

with the standard FE interpolation I_h, and their homogeneous counterparts $V_{h,i}^0$ with the constraint $\boldsymbol{v}_h = 0$ on $\partial\Omega$, to be used for the test functions. We recall moreover the spaces $Q_h(\Omega_h^\Gamma)$ and $Z_h(\Omega_h^\Gamma)$ defined respectively by (7.15) and (7.25). Combining (7.32) with (7.33)–(7.35) taken in the least square sense gives the following scheme: find $\boldsymbol{u}_{h,1} \in V_{h,1}, \boldsymbol{u}_{h,2} \in V_{h,2}, \boldsymbol{p}_h \in Q_h^k(\Omega_h^\Gamma), \boldsymbol{y}_{h,1}, \boldsymbol{y}_{h,2} \in Z_h(\Omega_h^\Gamma)$ such that:

$$\sum_{i=1}^2 \int_{\Omega_{h,i}} \boldsymbol{\sigma}_i(\boldsymbol{u}_{h,i}) : \nabla\boldsymbol{v}_{h,i} + \sum_{i=1}^2 \int_{\partial\Omega_{h,i}} \boldsymbol{y}_{h,i}\boldsymbol{n} \cdot \boldsymbol{v}_h$$

$$+ \frac{\gamma_p}{h^2} \int_{\Omega_h^\Gamma} (\boldsymbol{u}_{h,1} - \boldsymbol{u}_{h,2} + \frac{1}{h}\boldsymbol{p}_h\phi_h) \cdot (\boldsymbol{v}_{h,1} - \boldsymbol{v}_{h,2} + \frac{1}{h}\boldsymbol{q}_h\phi_h)$$

$$+ \gamma_u \sum_{i=1}^2 \int_{\Omega_h^\Gamma} (\boldsymbol{y}_{h,i} + \boldsymbol{\sigma}_i(\boldsymbol{u}_{h,i})) : (\boldsymbol{z}_{h,i} + \boldsymbol{\sigma}_i(\boldsymbol{v}_{h,i}))$$

$$+ \frac{\gamma_y}{h^2} \int_{\Omega_h^\Gamma} (\boldsymbol{y}_{h,1}\nabla\phi_h - \boldsymbol{y}_{h,2}\nabla\phi_h) \cdot (\boldsymbol{z}_{h,1}\nabla\phi_h - \boldsymbol{z}_{h,2}\nabla\phi_h)$$

$$+ \sum_{i=1}^2 \left(G_h(\boldsymbol{u}_{h,i}, \boldsymbol{v}_{h,i}) + J_h^{lhs,N}(\boldsymbol{y}_{h,i}, \boldsymbol{z}_{h,i})\right) = \sum_{i=1}^2 \int_{\Omega_{h,i}} \boldsymbol{f} \cdot \boldsymbol{v}_{h,i} \quad + \sum_{i=1}^2 J_h^{rhs,N}(\boldsymbol{z}_{h,i}),$$

$$\forall \boldsymbol{v}_{h,1} \in V_{h,1}^0, \boldsymbol{v}_{h,2} \in V_{h,2}^0, \boldsymbol{q}_h \in Q_h^k(\Omega_h^\Gamma), \boldsymbol{z}_{h,1}, \boldsymbol{z}_{h,2} \in Z_h(\Omega_h^\Gamma). \tag{7.37}$$

Similarly to the previous settings, we have added here the ghost stabilization G_h defined by (7.17) and the additional stabilization $J_h^{rhs,N}$ defined by (7.27) with $\Omega_h^{\Gamma_N}$ replaced by Ω_h^{Γ} and imposing div $\mathbf{y}_i = \mathbf{f}$ on Ω_h^{Γ} in the least squares sense.

Test case: Consider $\Omega = (0, 1)^2$ and Ω_1, Ω_2 defined by the LS ϕ:

$$\phi(x, y) = -R^2 + (x - 0.5)^2 + (y - 0.5)^2\,,$$

with $R = 0.3$ as illustrated in Figure 7.10. We want to solve (7.29) with the manufactured radial solution:

$$\mathbf{u} = \mathbf{u}_{ex} = \begin{cases} \dfrac{1}{E_1}(\cos(r) - \cos(R))(1, 1)^T & \text{if } r < R, \\ \dfrac{1}{E_2}(\cos(r) - \cos(R))(1, 1)^T & \text{else,} \end{cases}$$

where $r = \sqrt{(x - 0.5)^2 + (y - 0.5)^2}$. Thus:

$$\mathbf{f} = -\operatorname{div}(\sigma_1((\cos(r) - \cos(R))(1, 1)^T)/E_1$$

and $\mathbf{u}_g = \mathbf{u}_{ex}$.

The material parameters are given by $E_1 = 7$, $E_2 = 2.28$, and $\nu_1 = \nu_2 = 0.3$. The meshes used for ϕ-FEM and for the standard FEM are illustrated in Figure 7.11. In the latter case, the mesh should resolve the interface $r = R$ so that the solution $\mathbf{u}_h \in V_h$ is obtained by the straight forward scheme:

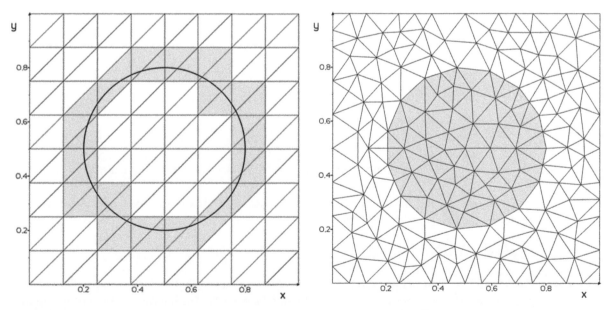

Figure 7.11 Linear elasticity with multiple materials. Left: a mesh used for ϕ-FEM (Ω_h^{Γ} painted in gray). Right: a mesh matching the interface for standard FEM (gray and white represent the two materials).

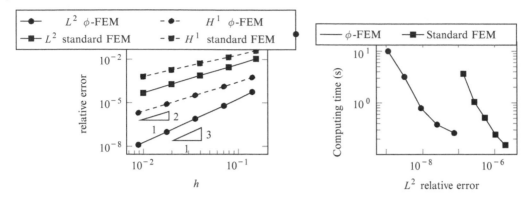

Figure 7.12 Test case with multiple materials. Left: H^1 and L^2 relative error obtained with φ-FEM and the standard FEM. Right: computing times for φ-FEM and the standard FEM.

$$\sum_{i=1}^{2} \int_{\Omega_{h,i}} \sigma_i(\boldsymbol{u}_h) : \nabla \boldsymbol{v}_h = \int_{\Omega} \boldsymbol{f} \cdot \boldsymbol{v}_h, \ \forall \, \boldsymbol{v}_h \in V_h^0, \tag{7.38}$$

where V_h is the conforming $\mathbb{P}^k$ FE space approximating $\boldsymbol{u}_g$ on $\partial\Omega$ and V_h^0 is its homogeneous analog. The results obtained with φ-FEM (7.37) and FEM (7.38) using $\mathbb{P}^2$ piecewise polynomials ($k = 2$) are reported in Figure 7.12. The conclusions remain the same as in the previous setting: φ-FEM is more precise on comparable meshes and less expensive in terms of the computing times for a given error tolerance.

7.4 Linear Elasticity with Cracks

We now want to consider the linear elasticity problem posed on a cracked domain $\Omega \setminus \Gamma_f$ with Γ_f being a line (a surface) inside Ω:

$$\begin{cases} -\operatorname{div} \boldsymbol{\sigma}(\boldsymbol{u}) & = \boldsymbol{f}, \text{ on } \Omega \setminus \Gamma_f, \\ \boldsymbol{u} & = \boldsymbol{u}^g, \text{ on } \partial\Omega, \\ \sigma(\boldsymbol{u})\boldsymbol{n} & = \boldsymbol{g}, \text{ on } \Gamma_f. \end{cases} \tag{7.39}$$

This problem is actually what XFEM was originally designed for, c.f. Moës et al. (1999). We are now going to adapt φ-FEM to it.

In practice, the crack geometry is given by the primary LS ϕ (to locate the line or surface of the crack) and the secondary LS ψ (to locate the tip or the front of the crack):

$$\Gamma_f := \Omega \cap \{\phi = 0\} \cap \{\psi < 0\}.$$

To fix the ideas, let us suppose that the line (surface) $\Gamma := \{\phi = 0\}$ splits Ω into two subdomains Ω_1 and Ω_2, characterized by $\{\phi < 0\}$ and $\{\phi > 0\}$, respectively, as illustrated in Figure 7.13. The interface Γ thus consists of the fracture location Γ_f and the remaining (fictitious) part Γ_{int}:

$$\Gamma_{int} := \Omega \cap \{\phi = 0\} \cap \{\psi > 0\}.$$

Figure 7.13 Geometry notations to represent the crack. Γ_{int} and Γ_f represent the fictitious interface and the actual crack, respectively.

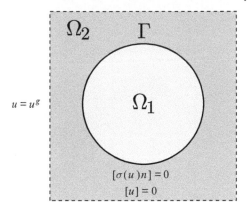

In order to reuse the ϕ-FEM scheme (7.37) introduced for the interface problem above, we reformulate problem (7.39) in terms of two separate unknowns $\boldsymbol{u}_i = \boldsymbol{u}|_{\Omega_i}, i = 1, 2$:

$$
\begin{cases}
-\operatorname{div}\sigma(\boldsymbol{u}_i) &= \boldsymbol{f}, \text{ on } \Omega_i, \\
\boldsymbol{u}_i &= \boldsymbol{u}^g, \text{ on } \partial\Omega, \\
[\boldsymbol{u}] &= 0, \text{ on } \Gamma_{int}, \\
[\sigma(\boldsymbol{u})\boldsymbol{n}] &= 0, \text{ on } \Gamma_{int}, \\
\sigma(\boldsymbol{u})\boldsymbol{n} &= \boldsymbol{g}, \text{ on } \Gamma_f.
\end{cases}
\tag{7.40}
$$

We are interested again in a situation where Ω is sufficiently simple-shaped so that a matching mesh T_h on Ω is easily available, but this mesh does not match the internal interface Γ. As in the preceding section, we are thus going to discretize separately $\boldsymbol{u}_1$ on Ω_1 and $\boldsymbol{u}_2$ on Ω_2 starting from the reformulation (7.40). To this end, we introduce two active sub-meshes $\mathcal{T}_{h,1}, \mathcal{T}_{h,2}$ as in (7.31), based on the piecewise polynomial approximation ϕ_h of ϕ. We also introduce the interface mesh $\mathcal{T}_h^\Gamma = \mathcal{T}_{h,1} \cap \mathcal{T}_{h,2}$, which we further split into two sub-meshes with respect to the secondary LS ψ, similarly to our treatment of the mixed boundary conditions, c.f. (7.22):

$$
\mathcal{T}_h^{\Gamma_f} := \{T \in \mathcal{T}_h^\Gamma : \psi \leqslant 0 \text{ on } T\} \qquad \text{and} \qquad \mathcal{T}_h^{\Gamma_{int}} := \{T \in \mathcal{T}_h^\Gamma : \psi \geqslant 0 \text{ on } T\}.
$$

Note that there may be some cells in $\mathcal{T}_h^\Gamma$ that are not marked as either $\mathcal{T}_h^{\Gamma_f}$ or $\mathcal{T}_h^{\Gamma_{int}}$. This is illustrated by the mesh example on the right of Figure 7.14, where the cells in $\mathcal{T}_h^{\Gamma_f}$ and $\mathcal{T}_h^{\Gamma_{int}}$ are painted in medium-dark gray and dark gray, respectively, but there remain some cells $\mathcal{T}_h^\Gamma$ that are in neither of these categories. These are painted in light gray on the picture. These are the cells intersected by the line $\{\psi = 0\}$. The crack tip happens to be thus inside one of the light gray cells.

Everything is now set up to adapt the ϕ-FEM approaches of the two preceding sections to the equations in (7.40). We choose an integer $k \geq 1$ and introduce first the FE spaces $V_{h,1}, V_{h,2}$ together with their homogeneous counterparts $V_{h,1}^0, V_{h,2}^0$ as in (7.36) to approximate $\boldsymbol{u}_1$ and $\boldsymbol{u}_2$. These will be used in the discretization of the variational formulation of the first equation in (7.40) together with the boundary conditions on $\partial\Omega$. The remaining equations in (7.40), i.e., the relations on Γ_{int} and Γ_f, will be treated by the introduction of auxiliary variables on the appropriate parts of Ω_h^Γ (the domain of the mesh $\mathcal{T}_h^\Gamma$):

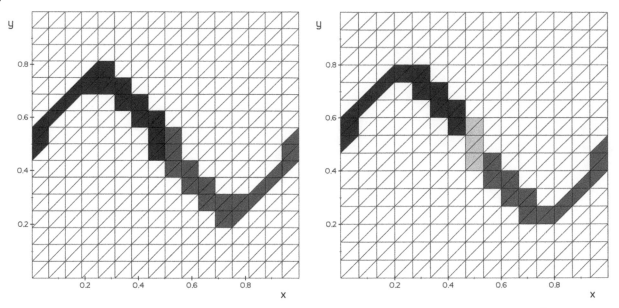

Figure 7.14 Test case with a crack, meshes used for ϕ-FEM. Left: a mesh resolving the crack tip; the cells in $\mathcal{T}_h^{\Gamma_{int}}$ in dark gray; the cells in $\mathcal{T}_h^{\Gamma_f}$ in medium-dark gray. Right: a mesh not resolving the crack tip; in addition to dark gray and medium-dark gray cells, there are light gray cells not belonging to $\mathcal{T}_h^{\Gamma_{int}}$ or $\mathcal{T}_h^{\Gamma_f}$.

- The vector-valued unknown $\boldsymbol{p}$ and the matrix-valued unknowns $\boldsymbol{y}_1, \boldsymbol{y}_2$ on $\Omega_h^{\Gamma_{int}}$ (the domain of the mesh $\mathcal{T}_h^{\Gamma_{int}}$). These will serve to impose the continuity of both the displacement and the normal force on Γ_{int} thorough the equations:

$$\boldsymbol{u}_1 - \boldsymbol{u}_2 + \boldsymbol{p}\phi = 0, \quad \text{on } \Omega_h^{\Gamma_{int}},$$

$$\boldsymbol{y}_i = -\sigma(\boldsymbol{u}_i), \quad \text{on } \Omega_h^{\Gamma_{int}},$$

$$\boldsymbol{y}_1 \cdot \nabla\phi - \boldsymbol{y}_2 \cdot \nabla\phi = 0, \quad \text{on } \Omega_h^{\Gamma_{int}},$$

which are exactly the same as (7.33)–(7.35) with the only exception that they are posed on the appropriate portion of Ω_h^{Γ} rather than on entire Ω_h^{Γ}. These variables will be discretized in FE spaces $Q_h^k(\Omega_h^{\Gamma_{int}})$ for $\boldsymbol{p}$ and $Z_h(\Omega_h^{\Gamma_{int}})$ for $\boldsymbol{y}_1, \boldsymbol{y}_2$, defined by (7.15) and (7.25), respectively.

- The vector-valued unknowns $\boldsymbol{p}_i^N$ and the matrix-valued unknown $\boldsymbol{y}_i^N$, $i = 1, 2$ on $\Omega_h^{\Gamma_f}$ (the domain of the mesh $\mathcal{T}_h^{\Gamma_f}$). These will serve to impose the Neumann boundary conditions on both sides of Γ_f thorough the equations:

$$\boldsymbol{y}_i^N = -\sigma(\boldsymbol{u}_i), \quad \text{on } \Omega_h^{\Gamma_f},$$

$$\boldsymbol{y}_i^N \nabla\phi + \boldsymbol{p}_i^N \phi + \boldsymbol{g}|\nabla\phi| = 0, \quad \text{on } \Omega_h^{\Gamma_f},$$

which are exactly the same as (7.24a-7.24b) with the only exception that the domain is renamed to $\Omega_h^{\Gamma_f}$ from $\Omega_h^{\Gamma_N}$. These variables will be discretized in FE spaces $Q_h^{k-1}(\Omega_h^{\Gamma_f})$ for $\boldsymbol{p}_i^N$ and $Z_h(\Omega_h^{\Gamma_f})$ for $\boldsymbol{y}_i^N$, defined again by (7.15) and (7.25), respectively.

Note that the combination of equations above does not impose the appropriate interface conditions on the whole of Γ since the latter may be not completely covered by $\Omega_h^{\Gamma_f} \cup \Omega_h^{\Gamma_{int}}$. Fortunately, this defect of the formulation on the continuous level can be repaired on the discrete level by adding the appropriate stabilization to the FE discretization, similarly to what we have already seen in the setting with mixed boundary conditions.

All this results in the following FE scheme: find $\boldsymbol{u}_{h,1} \in V_{h,1}$, $\boldsymbol{u}_{h,2} \in V_{h,2}$, $\boldsymbol{p}_h \in Q_h^k(\Omega_h^{\Gamma_D})$, $\boldsymbol{y}_{h,1}, \boldsymbol{y}_{h,2} \in Z_h(\Omega_h^{\Gamma_N})$, $\boldsymbol{p}_{h,1}^N, \boldsymbol{p}_{h,2}^N \in Q_h^{k-1}(\Omega_h^{\Gamma_N})$, $\boldsymbol{y}_{h,1}^N, \boldsymbol{y}_{h,2}^N \in Z_h(\Omega_h^{\Gamma_N})$ such that:

$$
\sum_{i=1}^{2} \Big(\int_{\Omega_{h,i}} \boldsymbol{\sigma}(\boldsymbol{u}_{h,i}) : \nabla \boldsymbol{v}_{h,i} + \int_{\partial \Omega_{h,i,int}} \boldsymbol{y}_{h,i} \boldsymbol{n} \cdot \boldsymbol{v}_{h,i} + \int_{\partial \Omega_{h,i,f}} \boldsymbol{y}_{h,i}^N \boldsymbol{n} \cdot \boldsymbol{v}_{h,i}
$$

$$
- \int_{\partial \Omega_{h,i} \backslash (\partial \Omega_{h,i,int} \cup \partial \Omega_{h,i,f})} \boldsymbol{\sigma}(\boldsymbol{u}_{h,i}) \boldsymbol{n} \cdot \boldsymbol{v}_{h,i} \Big)
$$

$$
+ \frac{\gamma_p}{h^2} \int_{\Omega_h^{\Gamma_{int}}} \big(\boldsymbol{u}_{h,1} - \boldsymbol{u}_{h,2} + \tfrac{1}{h} \boldsymbol{p}_h \phi_h\big) \cdot \big(\boldsymbol{v}_{h,1} - \boldsymbol{v}_{h,2} + \tfrac{1}{h} \boldsymbol{q}_h \phi_h\big)
$$

$$
+ \gamma_u \sum_{i=1}^{2} \int_{\Omega_h^{\Gamma_{int}}} \big(\boldsymbol{y}_{h,i} + \boldsymbol{\sigma}(\boldsymbol{u}_{h,i})\big) : \big(\boldsymbol{z}_{h,i} + \boldsymbol{\sigma}(\boldsymbol{v}_{h,i})\big)
$$

$$
+ \frac{\gamma_y}{h^2} \int_{\Omega_h^{\Gamma_{int}}} \big(\boldsymbol{y}_{h,1} \nabla \phi_h - \boldsymbol{y}_{h,2} \nabla \phi_h\big) \cdot \big(\boldsymbol{z}_{h,1} \nabla \phi_h - \boldsymbol{z}_{h,2} \nabla \phi_h\big)
$$

$$
+ \gamma_{u,N} \sum_{i=1}^{2} \int_{\Omega_h^{\Gamma_f}} \big(\boldsymbol{y}_{h,i}^N + \boldsymbol{\sigma}(\boldsymbol{u}_{h,i})\big) : \big(\boldsymbol{z}_{h,i}^N + \boldsymbol{\sigma}(\boldsymbol{v}_{h,i})\big)
$$

$$
+ \frac{\gamma_{p,N}}{h^2} \sum_{i=1}^{2} \int_{\Omega_h^{\Gamma_f}} \big(\boldsymbol{y}_{h,i}^N \nabla \phi_h + \tfrac{1}{h} \boldsymbol{p}_{h,i}^N \phi_h\big) \cdot \big(\boldsymbol{z}_{h,i}^N \nabla \phi_h + \tfrac{1}{h} \boldsymbol{q}_{h,i}^N \phi_h\big)
$$

$$
+ \sum_{i=1}^{2} \Big(G_h \big(\boldsymbol{u}_{h,i}, \boldsymbol{v}_{h,i}\big) + J_h^{lhs,int} \big(\boldsymbol{y}_{h,i}, \boldsymbol{z}_{h,i}\big) + J_h^{lhs,f} \big(\boldsymbol{y}_{h,i}^N, \boldsymbol{z}_{h,i}^N\big) \Big)
$$

$$
= \sum_{i=1}^{2} \int_{\Omega_{h,i}} \boldsymbol{f} \cdot \boldsymbol{v}_{h,i} - \frac{\gamma_{p,N}}{h^2} \sum_{i=1}^{2} \int_{\Omega_h^{\Gamma_f}} \boldsymbol{g} |\nabla \phi_h| \big(\boldsymbol{z}_{h,i}^N \nabla \phi_h + \tfrac{1}{h} \boldsymbol{q}_{h,i}^N \phi_h\big)
$$

$$
+ \sum_{i=1}^{2} \Big(J_h^{rhs,int} \big(\boldsymbol{z}_{h,i}\big) + J_h^{rhs,f} \big(\boldsymbol{z}_{h,i}^N\big) \Big),
$$

$\forall \boldsymbol{v}_{h,1} \in V_{h,1}^0, \boldsymbol{v}_{h,2} \in V_{h,2}^0, \boldsymbol{q}_h \in Q_h^k(\Omega_h^{\Gamma_D}), \boldsymbol{z}_{h,1}, \boldsymbol{z}_{h,2} \in Z_h, \boldsymbol{q}_{h,1}^N, \boldsymbol{q}_{h,2}^N \in Q_h^{k-1}(\Omega_h^{\Gamma_N})$,

$$
\boldsymbol{z}_{h,1}^N, \boldsymbol{z}_{h,2}^N \in Z_h(\Omega_h^{\Gamma_N}). \quad (7.41)
$$

As usual, we have added here the ghost stabilization G_h (7.17) and the additional stabilizations $J_h^{lhs,int}, J_h^{lhs,f}$ (accompanied by the their counterparts on the right-hand side for the consistency) that are copied from the $J_h^{lhs,N}$ in (7.27) but adjusted to the corresponding sub-meshes:

$$
J_h^{lhs,int}(\boldsymbol{y}, \boldsymbol{z}) = \gamma_{div} \int_{\Omega_h^{\Gamma_{int}}} \operatorname{div} \boldsymbol{y} \cdot \operatorname{div} \boldsymbol{z}, \quad J_h^{lhs,f}(\boldsymbol{y}, \boldsymbol{z}) = \gamma_{div} \int_{\Omega_h^{\Gamma_f}} \operatorname{div} \boldsymbol{y} \cdot \operatorname{div} \boldsymbol{z}.
$$

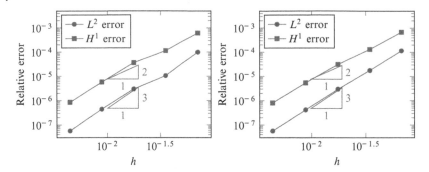

Figure 7.15 Test case with a crack, H^1 and L^2 relative errors. Left: on meshes resolving the crack tip. Right: on meshes not resolving the crack tip.

The boundary integrals are rewritten in terms of $\boldsymbol{y}_i, \boldsymbol{y}_i^N$ wherever possible. We have here denoted by $\partial\Omega_{h,i}$ the part of the boundary of $\Omega_{h,i}$ other than $\partial\Omega$ and introduced $\partial\Omega_{h,i,int}$ as the part of $\partial\Omega_{h,i}$ formed by the boundary facets of $\mathcal{T}_{h,i}$ belonging to the cells in $\mathcal{T}_h^{\Gamma_{int}}$. The same for $\partial\Omega_{h,i,f}$.

Test case: Let $\Omega = (0,1)^2$ and the interface Γ be given by the LS:

$$\phi(x,y) = y - \frac{1}{4}\sin(2\pi x) - \frac{1}{2}.$$

We choose the crack tip to be at $x = 0.5$ so that:

$$\Gamma_{int} := \{\phi = 0\} \cap \{x < 0.5\} \quad \text{and} \quad \Gamma_f := \{\phi = 0\} \cap \{x > 0.5\}.$$

This is the setting represented in Figure 7.13.

We use the φ-FEM (7.41) to solve (7.39) with the manufactured solution:

$$\boldsymbol{u} = \boldsymbol{u}_{ex} = (\sin(x) \times \exp(y), \sin(y) \times \exp(x))^T$$

which gives $\boldsymbol{f}$, $\boldsymbol{g}$, and $\boldsymbol{u}^g$ by substitution. The force on the crack $\boldsymbol{g}$ should be extended to a vicinity of Γ_f and we implement it by:

$$\boldsymbol{g} = \sigma(\boldsymbol{u}_{ex})\frac{\nabla\phi}{\|\nabla\phi\|} + \phi\boldsymbol{u}_{ex}.$$

We choose $\gamma_u = \gamma_p = \gamma_{div} = \gamma_{u,N} = \gamma_{p,N} = \gamma_{div,N} = 1.0$, $\sigma_p = 1.0$, and $\sigma_D = 20.0$.

We have conducted two series of numerical experiments using φ-FEM (7.41) with $\mathbb{P}^2$ Lagrange polynomials ($k = 2$) on families of meshes presented in Figure 7.14, either resolving the crack tip (the mesh on the left) or not (the mesh on the right). The results are reported in Figure 7.15. We see that φ-FEM converges optimally, giving very similar results on both types of meshes.

7.5 Heat Equation

We finally demonstrate the applicability of the φ-FEM approach to time-dependent problems. We take the example of the heat equation with Dirichlet boundary conditions: given a bounded domain $\Omega \in \mathbb{R}^d$, the initial conditions

u^0 on Ω, and the final time $T > 0$, find the scalar field $u = u(x, t)$ such that:

$$\begin{cases} u_t - \Delta u = f & \text{in } \Omega \times (0, T), \\ u = 0 & \text{on } \Gamma \times (0, T), \\ u(., 0) = u^0 & \text{in } \Omega. \end{cases} \tag{7.42}$$

We are interested again in the situation where a fitting mesh of Ω is not available. We rather assume that Ω is inscribed in a box $\mathcal{O}$ which is covered by a simple background mesh $\mathcal{T}_h^{\mathcal{O}}$, and introduce the active mesh $\mathcal{T}_h$ as in (7.7). We then follow the Direct Dirichlet ϕ-FEM approach (7.11), (7.16) with the following modifications:

- We introduce the uniform partition of the time interval $I = [0, T]$ into time steps of length Δt by the nodes $t_i = i\Delta t$. We discretize then (7.42) in time using implicit Euler scheme. On the continuous level this is formally written as: find u^n (the approximation to u at time t_n) in the form $u^n = \phi w^n$ successively for $n = 1, 2, \ldots$ solving

$$\frac{\phi w^n - \phi w^{n-1}}{\Delta t} - \Delta(\phi w^n) = f^n \tag{7.43}$$

where $f^n(\cdot) = f(t_n, \cdot)$.
- We extend (7.43) to Ω_h, integrate by parts on Ω_h, and discretize the resulting variational formulation using an FE space and adding appropriate stabilizations.

The ϕ-FEM for (7.42) thus reads as: find $w_h^n \in V_h$ for $n = 1, 2, \ldots$ with V_h defined by (7.14) such that:

$$\int_{\Omega_h} \frac{\phi_h w_h^n}{\Delta t} \phi_h v_h + \int_{\Omega_h} \nabla(\phi_h w_h^n) \cdot \nabla(\phi_h v_h) - \int_{\partial\Omega_h} \frac{\partial}{\partial n}(\phi_h w_h^n)\phi_h v_h$$

$$+ \sigma_D h \sum_{E \in \mathcal{F}_h^\Gamma} \int_E [\partial_n(\phi_h w_h^n)] \cdot [\partial_n(\phi_h v_h)] - \sigma h^2 \sum_{T \in \mathcal{T}_h^\Gamma} \int_T \left(\frac{\phi_h w_h^n}{\Delta t} - \Delta(\phi_h w_h^n) \right) \Delta(\phi_h v_h)$$

$$= \int_{\Omega_h} \left(\frac{\phi_h w_h^{n-1}}{\Delta t} + f^n \right) \phi_h v_h - \sigma h^2 \sum_{T \in \mathcal{T}_h^\Gamma} \int_T \left(\frac{\phi_h w_h^{n-1}}{\Delta t} + f^n \right) \Delta(\phi_h v_h). \tag{7.44}$$

We have added here the ghost stabilization, similar to (7.17) but in simpler scalar setting, and additional stabilization inspired by (7.18). The idea for the latter is to take the governing equation in the strong form, which is now (7.43), and to impose it in a least squares manner cell by cell.

Test case: We consider again the geometry of Ω and of the surrounding box $\mathcal{O}$ as in our first test case on page 197. In particular, the LS is given by (7.20) so that Ω is the circle centered at $(0.5, 0.5)$. Examples of meshes used both by ϕ-FEM and by the standard FEM are given in Figure 7.2. We want to solve (7.42) with the manufactured solution:

$$u = u_{ex} = \exp(x) \sin(2\pi y) \sin(t)$$

and extrapolated boundary conditions

$$u^g = u_{ex}(1 + \phi).$$

We are going to compare the convergence of the ϕ-FEM (7.44) with that of the standard FEM using $\mathbb{P}^1$ Lagrange polynomials in space and the implicit Euler scheme in time in both cases. The ϕ-FEM stabilization parameter is taken as $\sigma = 20$. The results are reported in Figures 7.16 and 7.17, for $\Delta t = h$ and $\Delta t = 10h^2$, respectively. Once again, ϕ-FEM converges faster than standard FEM. In the test considered here, the predominant source of error seems to be in the time discretization. In particular, we observe only $O(h)$ convergence in the L^2-norm in space in

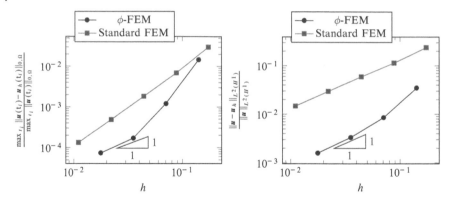

Figure 7.16 Test case for the heat equation; $\Delta t = h$. Left: $L^\infty(0, T; L^2(\Omega))$ relative errors. Right: $L^2(0, T; H^1(\Omega))$ relative errors.

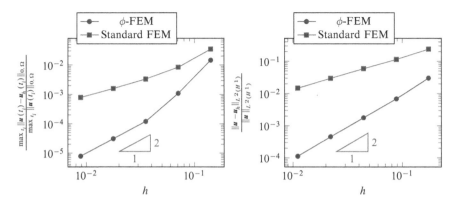

Figure 7.17 Test case for the heat equation; $\Delta t = 10h^2$. Left: $L^\infty(0, T; L^2(\Omega))$ relative errors. Right: $L^2(0, T; H^1(\Omega))$ relative errors.

the regime $\Delta t = h$ in Figure 7.16. A cleaner second order in time should be possible to achieve using the BDF2 marching scheme, but this remains out of the scope of the present paper.

7.6   Conclusions and Perspectives

φ-FEM is a relative newcomer to the field of unfitted FE methods. Up to now, it was only applied to scalar second-order elliptic equations with pure Dirichlet or pure Neumann/Robin boundary conditions in Duprez and Lozinski (2020) and Duprez et al. (2021). The purpose of the present contribution is to demonstrate its applicability to more sophisticated settings including the linear elasticity with mixed boundary conditions and material properties jumping across the internal interfaces, elasticity with cracks, and the heat transfer. In all the cases considered here, the numerical tests confirm the optimal accuracy on manufactured smooth solutions. φ-FEM is easily implementable in standard FEM packages (we have chosen FEniCS for the numerical illustration in this chapter). In particular, φ-FEM uses classical FE spaces and avoids the mesh generation and any non-trivial numerical integration.

Interestingly, our methods systematically outperform the standard FEM on comparable meshes. This can be attributed to a better representation of the boundary and of the solution near the boundary, as opposed to the approximation of the domain by a polyhedron/polygon in standard FEM. We recall that the computing times, reported in some of our tests with ϕ-FEM and favorably compared with those of the standard FEM, only include assembling of the matrices and the resolution of the linear systems. It would be interesting to add the mesh generation time to the comparison, which should be even more in favor of ϕ-FEM (when efficiently implemented).

Admittedly, the test cases presented in this contribution do not comprise all the complexity of the real-life problems. We have restricted ourselves to simple geometries in two dimensions only. Even more importantly, we have tested the methods only on smooth solutions, which is not supposed to happen in practice in problems with cracks, for example. Taking accurately into account the singularity at the crack tip remains an important challenge for the future ϕ-FEM developments. A relatively easily implementable approach would be to combine ϕ-FEM with a local mesh refinement by quadtree/octree structures near the crack tip (front). We emphasize that such a refinement should be necessary only in the vicinity of the front, since the discontinuous solution along the crack should be efficiently approximated by ϕ-FEM on a reasonably coarse unfitted mesh.

The mathematical analysis of the schemes presented in this chapter is in progress. We also plan to adapt ϕ-FEM to fluid-structure simulations starting by the creeping flow around of a Newtonian fluid (Stokes equations) in the presence of rigid particles.

References

Alnæs, M., Blechta, J., Hake, J. et al. (2015). The FEniCS Project Version 1.5. *Archive of Numerical Software*. University Library Heidelberg.

Annavarapu, C., Hautefeuille, M., and Dolbow, J. (2012). A robust Nitsche's formulation for interface problems. *Computer Methods in Applied Mechanics and Engineering* 225–228: 44–54.

Atallah, N., Canuto, C., and Scovazzi, G. (2021). The shifted boundary method for solid mechanics. *International Journal for Numerical Methods in Engineering* 1–36.

Ballarin, F. and Rozza, G. (2020). *Multiphenics*. https://mathlab.sissa.it/multiphenics.

Boiveau, T., Burman, E., Claus, S., and Larson, M. (2018). Fictitious domain method with boundary value correction using penalty-free Nitsche method. *Journal of Numerical Mathematics* 26 (2): 77–95.

Burman, E. (2010). Ghost penalty. *Comptes Rendus. Mathématique* 348 (21–22): 1217–1220.

Burman, E., Claus, S., Hansbo, P., et al. (2015). CutFEM: Discretizing geometry and partial differential equations. *International Journal for Numerical Methods in Engineering* 104 (7): 472–501.

Burman, E., Elfverson, D., Hansbo, P. et al. (2019). Hybridized CutFEM for elliptic interface problems. *SIAM Journal on Scientific Computing* 41 (5): A3354–A3380.

Burman, E. and Hansbo, P. (2010). Fictitious domain finite element methods using cut elements: I. A stabilized Lagrange multiplier method. *Computer Methods in Applied Mechanics and Engineering* 199 (41): 2680–2686.

Burman, E. and Hansbo, P. (2012). Fictitious domain finite element methods using cut elements: II. A stabilized Nitsche method. *Applied Numerical Mathematics* 62 (4): 328–341.

Burman, E., Hansbo, P., and Larson, M. (2020). Low regularity estimates for CutFEM approximations of an elliptic problem with mixed boundary conditions.

Carraro, T. and Wetterauer, S. (2015). On the implementation of the eXtended finite element method (XFEM) for interface problems.

Claus, S. and Kerfriden, P. (2018). A stable and optimally convergent LaTIn-CutFEM algorithm for multiple unilateral contact problems. *International Journal for Numerical Methods in Engineering* 113 (6): 938–966.

Duprez, M., Lleras, V., and Lozinski, A. (2021). A new φ-FEM approach for problems with natural boundary conditions. *Numerical Methods Partial Differential Equations to Appear* 39 (1): 281–303.

Duprez, M. and Lozinski, A. (2020). φ-FEM: A finite element method on domains defined by level-sets. *SIAM Journal on Numerical Analysis* 58 (2): 1008–1028.

Elfverson, D., Larson, M., and Larsson, K. (2019). A new-least squares stabilized Nitsche method for cut isogeometric analysis. *Computer Methods in Applied Mechanics and Engineering* 349: 1–16.

Girault, V. and Glowinski, R. (1995). Error analysis of a fictitious domain method applied to a Dirichlet problem. *Japan Journal of Industrial and Applied Mathematics* 12 (3): 487.

Glowinski, R., Pan, T., and Periaux, J. (1994). A fictitious domain method for Dirichlet problem and applications. *Computer Methods in Applied Mechanics and Engineering* 111 (3–4): 283–303.

Hansbo, A. and Hansbo, P. (2002). An unfitted finite element method, based on Nitsche's method, for elliptic interface problems. *Computer Methods in Applied Mechanics and Engineering* 191 (47): 5537–5552.

Hansbo, P. (2005). Nitsche's method for interface problems in computational mechanics. *Gammmitteilungen* 28: 183–206.

Hansbo, P., Larson, M., and Larsson, K. (2017). Cut finite element methods for linear elasticity problems. In: *Geometrically Unfitted Finite Element Methods and Applications, Volume 121 of Lecture. Notes Computer Science Engineering,* 25–63. Cham: Springer.

Haslinger, J. and Renard, Y. (2009). A new fictitious domain approach inspired by the extended finite element method. *SIAM Journal on Numerical Analysis* 47 (2): 1474–1499.

Lehrenfeld, C. (2016). High order unfitted finite element methods on level set domains using isoparametric mappings. *Computer Methods in Applied Mechanics and Engineering* 300: 716–733.

Lehrenfeld, C. and Reusken, A. (2018). Analysis of a high-order unfitted finite element method for elliptic interface problems. *IMA Journal of Numerical Analysis* 38 (3): 1351–1387.

Li, K., Atallah, N., Main, G., and Scovazzi, G. (2020). The shifted interface method: A flexible approach to embedded interface computations. *International Journal for Numerical Methods in Engineering* 121 (3): 492–518.

Main, A. and Scovazzi, G. (2018). The shifted boundary method for embedded domain computations. Part I: Poisson and Stokes problems. *Journal of Computational Physics* 372: 972–995.

Mittal, R. and Iaccarino, G. (2005). Immersed boundary methods. *Annual Review of Fluid Mechanics* 37: 239–261.

Moës, N., Dolbow, J., and Belytschko, T. (1999). A finite element method for crack growth without remeshing. *International Journal for Numerical Methods in Engineering* 46 (1): 131–150.

Nouveau, L., Ricchiuto, M., and Scovazzi, G. (2019). High-order gradients with the shifted boundary method: An embedded enriched mixed formulation for elliptic PDEs. *Journal of Computational Physics* 398: 108898, 28.

Xiao, Y., Zhai, F., Zhang, L., and Zheng, W. (2020). High-order finite element methods for interface problems: theory and implementations. In: *Spectral and High Order Methods for Partial Differential Equations ICOSAHOM 2018* (ed. S. Sherwin, D. Moxey, J. Peiró, P. Vincent, and C. Schwab). *Lecture Notes in Computational Science and Engineering, Volume 134,* 167–177. Cham: Springer.

Yuanming, X., Jinchao, X., and Fei, W. (2020). High-order extended finite element methods for solving interface problems. *Computer Methods in Applied Mechanics and Engineering* 364: 112964.

8

eXtended Boundary Element Method (XBEM) for Fracture Mechanics and Wave Problems

Jon Trevelyan

Durham University, UK

8.1 Introduction

The boundary element method (BEM) is rooted in the classical theories of potential flow and integral equations, so that its development can be traced back at least as far as finite element methods (FEMs). Indeed, it can be viewed as a discretized form of the integral equations of Fredholm. As the name of the method suggests, the BEM requires the discretization of only the boundary of the object or domain to be analyzed, and not its volume as required in the FEM. Thus, a BEM model of a two-dimensional (2D) object will consist of elements that are line segments describing portions of the perimeter contour, while boundary elements for a 3D object will describe portions of the object's surface area.

This reduction of dimensionality is a major benefit over the FEM, for a number of reasons: the reduction in data preparation effort and the improved robustness of automatically meshed models are two fairly obvious benefits. However, on further reflection we find other advantages, such as the fact that we can reverse all the element normals so that the analysis domain comprises the infinite region surrounding an internal boundary. This is used to great effect for problems of wave scattering, for example, in which the analysis domain might be the region surrounding a scattering obstacle. Such problems can be very efficiently solved using just a few elements. A further benefit of boundary-only meshing is the more straightforward conversion of computer-aided design (CAD) models into analysis models (or, taking this further, the integration of analysis within CAD systems). This is becoming yet more simplified with the emergence of isogeometric analysis formulations.

However, to present only the advantages of BEM over FEM would be to ignore the disadvantages. The primary drawback is the limited range of application to problems for which a Green's function is available. This is a "fundamental solution" to the governing differential equation which describes the effect, over an infinite domain, of a point source (or point force) at some location $\mathbf{p}$. This generally complicates the use of the method for problems involving material nonlinearity. There are also disadvantages in implementation of BEM, since at its heart it requires the evaluation of singular integrals, i.e., integrals of functions that go to infinity at a point. While these can be evaluated using a variety of methods, they do present intellectual challenges that are not faced in finite element coding.

Partition of Unity Methods, First Edition. Stéphane P. A. Bordas, Alexander Menk, and Sundararajan Natarajan.
© 2024 John Wiley & Sons Ltd. Published 2024 by John Wiley & Sons Ltd.

There are different flavors of the BEM:

1. The **direct BEM**, in which we solve directly for the nodal unknowns (e.g., displacements and tractions).
2. The **indirect BEM**, in which we solve for a distributed source density around the boundary that yields the correct solution.

A further classification of BEM approaches is as follows:

1. The **collocation BEM**, in which the linear system of equations is derived by applying the boundary integral equation (BIE) at a discrete set of points, i.e., enforcing the condition that the BIE is satisfied at this set of points.
2. The **Galerkin BEM**, in which the BIE is integrated in a weighted residual sense, i.e., enforcing the condition that the BIE is satisfied weakly when weighted by a set of test functions.

In this text, we consider only the direct, collocation form of the BEM. First, we describe the conventional BEM formulation (though a detailed derivation is not included since this is standard and available in a great many texts). This is then extended by introducing a "partition of unity (PU)" enrichment, and afterward enriched formulations are presented for fracture mechanics and wave scattering. As for extended finite element method (XFEM), the idea behind PU enriched BEM schemes is that the solution is sought in a space of functions that exhibits better approximation properties than the conventional piecewise polynomial basis of shape functions. In the fracture mechanics section, we use the asymptotic behavior of displacements in the vicinity of a crack tip, and in the wave scattering section, we use an enrichment using plane waves. These are fundamentally different forms of enrichment since:

- in the fracture mechanics case the enrichment is injected locally in the elements at crack tips, while the majority of elements are conventional;
- in the wave scattering case the enrichment is applied to every node in the model.

We will also confine ourselves to 2D problems in elasticity/fracture mechanics, though present the formulation for both two dimensions and three dimensions for wave scattering.

Although some might argue about correctness of naming, for the purposes of this chapter in a book on XFEM we will call all the enriched boundary element formulations the *eXtended boundary element method* (XBEM).

8.2 Conventional BEM Formulation

The BEM is based on the development of a BIE in which the value of the primary variable (e.g., displacement or wave potential) at a point is written in terms of integrals (taken over the domain boundary) containing Green's functions and their derivatives. The continuous BIEs are discretized in a similar fashion to the discretization of the weak form integral equation in finite element formulations.

For the successful implementation of BEMs it is very important to treat the integration with care. There are many regular integrals that can be evaluated successfully using Gauss-Legendre quadrature with a sufficiently large number of integration points. However, there are a number of singular integrals that arise, i.e., in which the integrand becomes infinite at some location in the interval of integration. In spite of this, the integrals are finite and it is possible to evaluate them. In broad terms, the difficulty of evaluating them increases with the increasing order of the singularity. We now classify these integrands, as functions of the distance r from the singular point, and suggest methods of evaluating them:

- Weakly singular integrals are those for which the integrand behavior is $\mathcal{O}(\ln r)$ for 2D problems or $\mathcal{O}(r^{-1})$ for 3D problems. There are logarithmic Gauss quadrature schemes that can be used, but the cubic transformation of Telles (1987) is a simple and popular method for treating these integrals.

- Strongly singular integrals are those for which the integrand behavior is $\mathcal{O}(r^{-1})$ for 2D problems or $\mathcal{O}(r^{-2})$ for 3D problems. For some classes of problem it is possible to use a rigid body motion load case to evaluate these terms indirectly. For other problems, this is not available, and other techniques such as the method of Guiggiani et al. (1992) are useful.
- Hypersingular integrals are those for which the integrand behavior is $\mathcal{O}(r^{-2})$ for 2D problems or $\mathcal{O}(r^{-3})$ for 3D problems. The method of Guiggiani et al. (1992) provides a general technique for evaluating boundary integrals of arbitrarily high-order singularity.

8.2.1 Elasticity

We consider a domain $\Omega \subset \mathbb{R}^2$ having boundary $\partial\Omega = \Gamma$. We solve the boundary value problem in which the equations of static equilibrium, i.e.:

$$\sigma_{ij,j}(\mathbf{x}) + b_i(\mathbf{x}) = 0, \qquad \mathbf{x} \in \Omega \tag{8.1}$$

are solved subject to boundary conditions:

$$u_i(\mathbf{q}) = \bar{u}, \quad \mathbf{q} \in \Gamma_u \tag{8.2}$$

$$t_i(\mathbf{q}) = \bar{t}, \quad \mathbf{q} \in \Gamma_t \tag{8.3}$$

In the above expressions, σ is the stress tensor and b is the body force per unit volume, indices $i, j = x, y$ in the usual tensor notation, u is the displacement vector, and t the traction vector, defined by:

$$t_i = \sigma_{ij} n_j \tag{8.4}$$

where n_j is the j component of the outward pointing normal vector. The quantities $\bar{u}, \bar{t}$ are prescribed displacement and traction values, respectively, and the boundary conditions are described over Dirichlet and Neumann boundaries Γ_u and Γ_t, respectively.

The BIE may be derived readily from the reciprocal theorem, and ultimately (for cases in which $b = 0$ for simplicity) we arrive at:

$$c(\mathbf{p})u_i(\mathbf{p}) + \int_\Gamma T_{ij}(\mathbf{p},\mathbf{q})u_j(\mathbf{q})\mathrm{d}\Gamma(\mathbf{q}) = \int_\Gamma U_{ij}(\mathbf{p},\mathbf{q})t_j(\mathbf{q})\mathrm{d}\Gamma(\mathbf{q}), \qquad \mathbf{p},\mathbf{q} \in \Gamma \tag{8.5}$$

In this equation, which is sometimes known as the Somigliana identity, U_{ij} and T_{ij} are the displacement and traction fundamental solutions, respectively (i.e., the Kelvin solutions for the elastic problem), and are given by:

$$U_{ij} = \frac{1}{8\pi\mu(1-\nu)}\left[(3-4\nu)\ln\frac{1}{r}\delta_{ij} + r_{,i}\,r_{,j}\right] \tag{8.6}$$

$$T_{ij} = \frac{-1}{4\pi(1-\nu)r}\left\{r_{,n}\left[(1-2\nu)\delta_{ij} + 2r_{,i}\,r_{,j}\right] - (1-2\nu)\left[r_{,j}\,n_i - r_{,i}\,n_j\right]\right\} \tag{8.7}$$

where μ is the shear modulus, ν the Poisson's ratio, δ the Kronecker delta, n the outward normal to the boundary, the coordinate $r = |\mathbf{q}-\mathbf{p}|$, and the comma implies differentiation with respect to the subsequent indicial direction. The term c arises as a result of the strongly singular nature of the first integral in (8.5); it is dependent on the local boundary geometry at $\mathbf{p}$, and is given by:

$$c(\mathbf{p}) = \frac{\theta(\mathbf{p})}{2\pi} \tag{8.8}$$

where $\theta(\mathbf{p})$ is the angle subtended by the domain Ω at $\mathbf{p}$, so that, for example, $c(\mathbf{p}) = 0.5$ where $\mathbf{p}$ lies on a smooth boundary.

We proceed by discretizing Γ into a set of non-overlapping boundary elements, i.e.:

$$\Gamma = \bigcup_{e=1}^{E} \Gamma_e, \qquad \Gamma_a \cap \Gamma_b = \emptyset, \quad a \neq b \tag{8.9}$$

and this allows us to write the discretized form of (8.5) as:

$$c(\mathbf{p})u_i(\mathbf{p}) + \sum_{e=1}^{E} \int_{\Gamma_e} T_{ij}(\mathbf{p}, \mathbf{q})u_j(\mathbf{q})d\Gamma(\mathbf{q}) = \sum_{e=1}^{E} \int_{\Gamma_e} U_{ij}(\mathbf{p}, \mathbf{q})t_j(\mathbf{q})d\Gamma(\mathbf{q}) \tag{8.10}$$

Now we can interpolate the displacement and traction fields over each element using shape functions N_s:

$$u_j = \sum_{s=1}^{S} N_s(\xi_e)u_j^{es} \tag{8.11}$$

$$t_j = \sum_{s=1}^{S} N_s(\xi_e)t_j^{es} \tag{8.12}$$

where u_j^{es}, u_j^{es} are, respectively, the displacement and traction components in the j direction at node s of element e, and the location on the element is given in terms of its local element parametric coordinate $\xi_e \in (-1, 1)$. This allows (8.10) to be rewritten:

$$c(\mathbf{p})u_i(\mathbf{p}) + \sum_{e=1}^{E} \int_{-1}^{1} T_{ij}(\mathbf{p}, \mathbf{q}(\xi_e))N_s(\xi_e)J^e d\xi_e u_j^{es} = \sum_{e=1}^{E} \int_{-1}^{1} U_{ij}(\mathbf{p}, \mathbf{q}(\xi_e))N_s(\xi_e)J^e d\xi_e t_j^{es}. \tag{8.13}$$

where J^e is the Jacobian of coordinate transformation on element e. The shape functions, N_s, are identical in construction to finite element shape functions, so that for a three-noded line element shown in Figure 8.1, the shape functions are identical to a three-noded bar finite element, i.e.:

$$N_1(\xi_e) = \frac{-1}{2}\xi_e(1 - \xi_e) \tag{8.14}$$

$$N_2(\xi_e) = 1 - \xi_e^2 \tag{8.15}$$

$$N_3(\xi_e) = \frac{1}{2}\xi_e(1 + \xi_e) \tag{8.16}$$

In moving forward to enriched formulations such as those described in this chapter, it is important to recognize that these shape functions do not only interpolate displacements, tractions, and geometry, but they also form the basis of functions in which we seek the solution. That is to say, the displacements will be found as a linear combination of these shape functions. We can say that the set of shape functions forms our *approximation space*.

The discretized integral Equation (8.13) can be simplified as:

$$c(\mathbf{p})u_i(\mathbf{p}) + \sum_{e=1}^{E}\sum_{s=1}^{S} P_{ij}^{es}u_j^{es} = \sum_{e=1}^{E}\sum_{s=1}^{S} Q_{ij}^{es}t_j^{es} \tag{8.17}$$

where S is the number of nodes on each element, and the influence coefficients P_{ij}^{es}, Q_{ij}^{es} are the results of evaluating the boundary integrals:

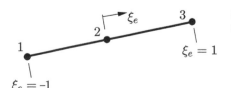

Figure 8.1 Three-noded, quadratic boundary element showing local coordinate ξ_e.

$$P_{ij}^{es} = \int_{-1}^{1} T_{ij}(\mathbf{p}, \mathbf{q}(\xi_e)) N_s(\xi_e) J^e d\xi_e \tag{8.18}$$

$$Q_{ij}^{es} = \int_{-1}^{1} U_{ij}(\mathbf{p}, \mathbf{q}(\xi_e)) N_s(\xi_e) J^e d\xi_e \tag{8.19}$$

The majority of these integrals are regular and can be evaluated using simple schemes like Gauss-Legendre quadrature. However, for cases in which the integral is taken over an element e containing point $\mathbf{p}$, care needs to be taken since the integral is singular. The details of methods for evaluating these singular boundary integrals are beyond the scope of this work; however, references are given in Section 8.2.

There now comes the collocation. We take point $\mathbf{p}$ to lie at the first node and take i to be the x-direction. Evaluation of the required boundary integrals yields an equation of the form:

$$h_1 u_x^1 + h_2 u_y^1 + h_3 u_x^2 + \dots + h_{2NN} u_y^{NN} = g_1 t_x^1 + g_2 t_y^1 + g_3 t_x^2 + \dots + g_{2NN} t_y^{NN} \tag{8.20}$$

where coefficients h_i, g_i are the results of evaluating the boundary integrals, u_x^k, t_x^k are the displacement and traction components in the x-direction at node k (with similar notation for the y-direction) and NN is the number of nodes.

Taking all nodes in turn, and letting i be the x and y directions separately at each, we arrive at a system of equations most conveniently expressed in the matrix form:

$$\mathbf{Hu} = \mathbf{Gt} \tag{8.21}$$

where $\mathbf{H}, \mathbf{G}$ are matrices containing the h and g coefficients, and $\mathbf{u}$ and $\mathbf{t}$ are the vectors of nodal displacements and tractions. Finally, by prescribing exactly half of the terms in the combination of vectors $\mathbf{u}$ and $\mathbf{t}$ as boundary conditions, the equations reduce to a linear system:

$$\mathbf{Ax} = \mathbf{b} \tag{8.22}$$

in which only vector $\mathbf{x}$ is unknown. This linear system can be solved to find the remaining unknown boundary displacements and tractions in $\mathbf{x}$.

8.2.1.1 Extensions for fracture mechanics problems

If we are to model crack problems using the collocation BEM, we run into some difficulties. Consider a crack modeled as shown in Figure 8.2a, in which figure the crack is shown with a finite crack opening displacement for clarity; we consider problems in which the crack surfaces are coincident. We can consider the boundary Γ to comprise the external boundary Γ_{ext}, the upper crack surface Γ_c^+, and the lower crack surface Γ_c^-. If the meshes on the upper and lower crack surfaces correspond, as is common, the nodes on those elements will coincide, so that in the example presented in the figure, nodes 18 and 24 will share the same coordinates. The result of this is that when these two nodes are used as the collocation point $\mathbf{p}$, they will yield identical equations. The system matrix, $\mathbf{A}$, will therefore be singular and no solution can be obtained.

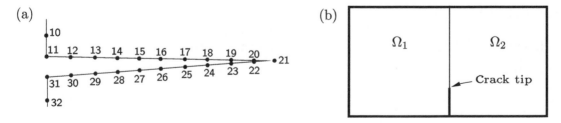

Figure 8.2 (a) Detail of nodes on crack surfaces; (b) domain decomposition.

In order to overcome this difficulty, it is possible to use a domain decomposition approach as illustrated in Figure 8.2b. The two domains, Ω_1 and Ω_2, are considered separately and we can derive linear systems of the form (8.22) for both independently. Then these two systems can be solved simultaneously, coupled together by equations of continuity in displacement and traction over the degrees of freedom located on the interface. While this works, it is undesirable, particularly for crack growth modeling as it can be difficult to design a robust algorithm to automate the generation of the two-zone model.

An alternative is the dual BEM proposed by Portela et al. (1992), in which the problem of identical equations being formed on the surfaces Γ_c^+ and Γ_c^- is overcome by using different integral equations when collocating over these two surfaces. The usual displacement boundary integral equation (DBIE) (8.5) is used when $\mathbf{p} \in \Gamma_c^+$, while the traction boundary integral equation (TBIE) is used when $\mathbf{p} \in \Gamma_c^-$. In order to derive the TBIE, the DBIE is differentiated with respect to the coordinate directions, providing expressions for strain components, and these are combined using Hooke's law with the material properties to give expressions for stress. Finally, the stresses are multiplied by the normal vector at $\mathbf{p}$ to give the tractions, yielding the TBIE, being:

$$c(\mathbf{p})t_i(\mathbf{p}) + n_j(\mathbf{p}) \int_\Gamma S_{kji}(\mathbf{p},\mathbf{q})u_k(\mathbf{q})\mathrm{d}\Gamma(\mathbf{q}) = n_j(\mathbf{p}) \int_\Gamma D_{kji}(\mathbf{p},\mathbf{q})t_k(\mathbf{q})\mathrm{d}\Gamma(\mathbf{q}), \qquad \mathbf{p},\mathbf{q} \in \Gamma \tag{8.23}$$

where S and D are derivative kernels found from the process described above. This approach obviates the need for domain decomposition since the duplicate equations are no longer formed. It does so at the expense of increasing the order of the singularity, so that hypersingular integrals are required to be evaluated. Methods for doing this are beyond the scope of this chapter, but it is mentioned here that the integrals can be found analytically for flat elements, which are most commonly used for crack modeling.

The boundary displacements and tractions are only moderately useful for fracture modeling. A measure of more engineering significance is the stress intensity factor (SIF) or, more specifically the SIFs K_I and K_II. These essentially express the amplitude of the singular stress field, and can be used to provide a ready assessment of fracture safety by comparison with material fracture toughness properties. They are also useful in determining both the rate of crack growth and direction of crack propagation in fatigue. More detail will be given later in this chapter on the nature of the SIFs and, in particular, their relation to the stress-displacement field surrounding a crack tip. However, for the moment we mention that they can be calculated in various ways, the most popular being the J-integral approach of Rice (1968).

8.2.2 Helmholtz Wave Problems

We first define our acoustic domain $\Omega \subset \mathbb{R}^d$, where $d = 2, 3$. This may be the region enclosed by a boundary $\Gamma \equiv \partial\Omega$, or alternatively the unbounded region surrounding the internal boundary (or boundaries) $\Gamma \equiv \partial\Omega$. We start with the wave equation, describing the propagation of a wave through the medium, as:

$$\nabla^2 u(\mathbf{q}, t) - \frac{1}{c^2} \frac{\partial^2 u(\mathbf{q}, t)}{\partial t^2} = 0, \qquad \mathbf{q} \in \Omega \tag{8.24}$$

where u can be thought of as a wave height above the mean water level, or as the acoustic pressure relative to the mean atmospheric pressure, c is the wave speed in the transmitting medium, and t is time. The frequency domain approaches, which we shall focus upon in this section, make the assumption that the wave field is sinusoidally varying in time with angular frequency ω, i.e., we allow the separation of variables:

$$u(\mathbf{q}, t) = \phi(\mathbf{q})e^{-i\omega t} \tag{8.25}$$

where $i = \sqrt{-1}$ and $\phi : \Omega \to \mathbb{C}$ is the acoustic potential. Once this is substituted into the wave equation (8.24), the problem reduces to:

$$\nabla^2 \phi(\mathbf{q}) + k^2 \phi(\mathbf{q}) = 0 \tag{8.26}$$

This governing differential equation is known as the Helmholtz equation. We have introduced the wave number k which may be written as:

$$k = \frac{2\pi}{\lambda} = \frac{\omega}{c} \tag{8.27}$$

and is therefore a measure of frequency. Our problem, then, is to solve for the acoustic potential field ϕ subject to boundary conditions:

$$\phi(\mathbf{q}) = \bar{\phi}, \qquad\qquad \mathbf{q} \in \Gamma_D \tag{8.28}$$

$$\frac{\partial \phi(\mathbf{q})}{\partial \mathbf{n}(\mathbf{q})} = \bar{g}, \qquad\qquad \mathbf{q} \in \Gamma_N \tag{8.29}$$

where $\bar{\phi}, \bar{g}$ are prescribed potentials and their normal derivatives, respectively, and Γ_D, Γ_N are Dirichlet and Neumann boundaries, respectively. $\mathbf{n}$ is the unit normal vector directed outward from Ω.

The use of the BEM in wave problems has become established (Brebbia and Ciskowski, 1998), and it is well understood how the Helmholtz equation may be written in an equivalent integral equation form. To do so we introduce the fundamental solution, G, to the Helmholtz equation:

$$G(\mathbf{p}, \mathbf{q}) = \begin{cases} \dfrac{i}{4} H_0^1(kr) & d = 2 \\[2mm] \dfrac{e^{ikr}}{4\pi r} & d = 3 \end{cases} \tag{8.30}$$

where $H_0^1(\cdot)$ is the Hankel function of the first kind and of order 0, and $r := |\mathbf{q} - \mathbf{p}|$ is the usual radial coordinate used in BEMs. One can think of G as being the effect at a point $\mathbf{q}$ of a singular oscillatory source at point $\mathbf{p}$. Knowledge of the fundamental solution allows us to write our conventional BIE:

$$c(\mathbf{p})\phi(\mathbf{p}) + \int_\Gamma \frac{\partial G(\mathbf{p}, \mathbf{q})}{\partial \mathbf{n}(\mathbf{q})} \phi(\mathbf{q}) d\Gamma(\mathbf{q}) = \int_\Gamma G(\mathbf{p}, \mathbf{q}) \frac{\partial \phi(\mathbf{q})}{\partial \mathbf{n}(\mathbf{q})} d\Gamma(\mathbf{q}), \qquad \mathbf{p} \in \Gamma \tag{8.31}$$

As for the BEM formulation described earlier for elasticity, a key feature of BIEs such as this is the singularity of the kernel functions. In (8.31), both points $\mathbf{p}, \mathbf{q}$ lie on the boundary Γ and, since the integrals are taken over the entire boundary Γ, the region of integration includes the point where $\mathbf{p}$ and $\mathbf{q}$ coincide. Here, there is a singularity in the fundamental solution, G, and care needs to be taken over implementation of integration schemes. As for elasticity problems, the term c arises out of the limiting process of taking $\mathbf{p}$ to the boundary, and is dependent on the local geometry at $\mathbf{p}$; it is given by (8.8), and for the smooth (C^1 continuous) boundaries we will consider in this section it can be taken as 1/2.

For exterior scattering problems, the BIE is modified by the addition of a term describing an incident wave ϕ^I, i.e., a wave that impinges upon a scattering obstacle:

$$c(\mathbf{p})\phi(\mathbf{p}) + \int_\Gamma \frac{\partial G(\mathbf{p}, \mathbf{q})}{\partial \mathbf{n}(\mathbf{q})} \phi(\mathbf{q}) d\Gamma(\mathbf{q}) = \int_\Gamma G(\mathbf{p}, \mathbf{q}) \frac{\partial \phi(\mathbf{q})}{\partial \mathbf{n}(\mathbf{q})} d\Gamma(\mathbf{q}) + \phi^I(\mathbf{p}) \tag{8.32}$$

The incident wave can take any form we choose, but engineers commonly consider it to be a plane wave propagating in some given direction, i.e.:

$$\phi^I(\mathbf{p}) = A^I e^{ik\mathbf{d}^I \cdot \mathbf{p}} \tag{8.33}$$

where A^I is the amplitude of the incident wave and $\mathbf{d}^I$ is the unit vector describing its direction of propagation.

The solution of (8.32) proceeds by discretizing the boundary Γ into E non-overlapping elements, i.e.:

$$\Gamma = \bigcup_{e=1}^{E} \Gamma_e, \qquad \Gamma_a \cap \Gamma_b = \emptyset, a \neq b \tag{8.34}$$

Each element is paramaterized in terms of a local coordinate in the usual finite element fashion, allowing us to use shape functions to interpolate quantities over elements. For 2D applications ($d = 2$) we will use $\xi_e \in [-1, 1]$ and for 3D ($d = 3$) we will use $(\xi_e, \eta_e) \in [-1, 1]^2$. Thus, we can interpolate ϕ from its values at the nodes of an S-noded element using:

$$\phi = \begin{cases} \displaystyle\sum_{s=1}^{S} N_s(\xi_e)\phi^{es}, & d = 2 \\ \displaystyle\sum_{s=1}^{S} N_s(\xi_e, \eta_e)\phi^{es}, & d = 3 \end{cases} \tag{8.35}$$

where N_s is the shape function associated with the sth node on the element and ϕ^{es} is the value of the potential at the node s on element e. In isoparametric element formulations, we can interpolate geometry in the same manner.

The discretization and interpolation allow the BIE to be restated in discrete form:

$$c(\mathbf{p})\phi(\mathbf{p}) + \sum_{e=1}^{E}\sum_{s=1}^{S} P^{es}(\mathbf{p})\phi^{es} = \sum_{e=1}^{E}\sum_{s=1}^{S} Q^{es}(\mathbf{p})\frac{\partial \phi^{es}}{\partial \mathbf{n}} + \phi^I(\mathbf{p}) \tag{8.36}$$

where the influence coefficients P^{es}, Q^{es} are the results of evaluating boundary integrals:

$$P^{es}(\mathbf{p}) = \int_{-1}^{1} \frac{\partial G(\mathbf{p}, \mathbf{q}(\xi_e))}{\partial \mathbf{n}(\mathbf{q}(\xi_e))} N_s J^e \mathrm{d}\xi_e \tag{8.37}$$

$$Q^{es}(\mathbf{p}) = \int_{-1}^{1} G(\mathbf{p}, \mathbf{q}(\xi_e)) N_s J^e \mathrm{d}\xi_e \tag{8.38}$$

where $\mathbf{q}(\xi_e)$ is the location in Cartesian space of the point corresponding to local coordinate ξ_e on the element, and J^e is the Jacobian of this coordinate transformation. In these expressions, the 2D form is shown. Corresponding expressions involving double integrals can easily be developed for three dimensions.

Placing the boundary point $\mathbf{p}$ at a node, and evaluating the boundary integrals, will reduce Equation (8.36) to a simple expression:

$$\frac{1}{2}\phi(\mathbf{p}) + h_{p1}\phi_1 + h_{p2}\phi_2 + \cdots + h_{pNN}\phi_{NN} = g_{p1}\frac{\partial \phi_1}{\partial \mathbf{n}} + g_{p2}\frac{\partial \phi_2}{\partial \mathbf{n}} + \cdots + g_{pNN}\frac{\partial \phi_{NN}}{\partial \mathbf{n}} + \phi^I(\mathbf{p}) \tag{8.39}$$

where ϕ_i is the potential at node i and NN is the total number of nodes. Placing $\mathbf{p}$ at each node in turn, the equations that arise may be assembled into a matrix system:

$$\mathbf{Hu} = \mathbf{Gv} + \mathbf{w} \tag{8.40}$$

where the matrices $\mathbf{H}$ and $\mathbf{G}$ are fully populated and unsymmetric, and multiply vectors $\mathbf{u}$ and $\mathbf{v}$ which contain the nodal potentials and their normal derivatives, respectively. The vector $\mathbf{w}$ contains the known values of ϕ^I at the nodes.

The system has NN equations but $2NN$ unknowns, but this situation is resolved by the application of the boundary conditions. For each node, we prescribe either the potential or its normal derivative according to (8.28) or (8.29). This allows our matrix system to be reduced to:

$$\mathbf{Ax = b} \tag{8.41}$$

where the matrix $\mathbf{A}$ contains the columns from $\mathbf{H}$ and $\mathbf{G}$ that multiply the nodal values still remaining unknown, all those unknowns (some potentials and some normal derivatives) lie in the vector $\mathbf{x}$ and the vector $\mathbf{b}$ contains a mixture of boundary conditions multiplied by influence coefficients and of terms from $\mathbf{w}$. This linear system can be solved for the unknowns.

Once the boundary solution has been obtained in this fashion, we can consider the problem of finding results internally in the acoustic domain. To do this, we place our point $\mathbf{p}$ at an arbitrary point in Ω and return to the BIE (8.32), in which, since the point $\mathbf{p}$ now lies in the domain and not on the boundary, the term $c(\mathbf{p})$ may be taken as 1. All terms in this expression are known with the exception of $\phi(\mathbf{p})$, allowing this internal acoustic potential to be found.

The above approach is the "collocation" form of the BEM. We have collocated the integral equation at specific points (the nodes) and solved the resulting system. We can recall that there is also a Galerkin form of the BEM, in which the BIE is weighted by a set of functions and integrated over the boundary. For a Bubnov Galerkin approach one would take the weighting functions to be the same as the basis functions defining the space in which the potential solution is sought. These are the shape functions. It is common in the mathematics literature to see the Galerkin form being favored, as it brings benefits in the mathematical numerical analysis of the integral equation formulation. However, in this section, we use the collocation form, which is common in the engineering literature as it is simpler to implement, robust, and more computationally efficient than the Galerkin form which requires an extra level of integration.

The BEM system relies on global operators; since every node has an influence on every other node, the system matrix is fully populated. For larger systems, this becomes a significant problem, and restricts practical usage of the method to smaller problems more quickly than finite element approximations, in which the system matrix is strongly banded. In an attempt to overcome this difficulty, the fast multipole method (FMM) has emerged as a powerful extension to the BEM. Here, the inter-nodal influence functions are not established through boundary integrals (8.37) and (8.38), but instead through a hierarchy of multi-level, multipole expansions through a tree structure of nodal connectivities. This effectively decouples the source and field points, $\mathbf{p}$ and $\mathbf{q}$, so that the assembly of the system is no longer a problem that scales with NN^2. The method is aimed principally at large problems, where the run time scales considerably better than the conventional BEM. The large problem size and the use of the multipole expansions themselves mean that FMM codes typically make use of iterative solvers. Only some initial attempts have been made to couple XBEM type enriched BEM simulations to the FMM. Further attempts may yet be made, and sound attractive, but care will be required to handle the ill-conditioning of the system. This aspect of enriched approximations will be discussed further in the following sections.

8.2.2.1 The nonuniqueness problem

The formulation of the BEM as presented here has been successfully used for several decades to perform acoustic simulations. However, for robustness of application, it is necessary to account for a well-known problem of nonuniqueness. This applies only to simulations of exterior problems such as scattering and radiation problems (the problem does not arise for the analysis of bounded acoustic cavities such as automobile passenger compartments), and is an artifact of using a boundary integral representation. The nonunique solution arises because the resulting matrix becomes singular at a set of frequencies. These are the eigenfrequencies of the corresponding interior acoustic Dirichlet and Neumann problems. Just to clarify, imagine we are solving in two dimensions the problem of scattering of an incident plane wave by a cylinder, so our boundary elements describe the circular boundary and the acoustic domain lies outside this circle. There is a corresponding interior problem in which the acoustic domain lies inside (not outside) the circle, and this corresponding problem exhibits some eigenfrequencies associated with the acoustic natural modes. A boundary element analysis of the exterior problem suffers from nonuniqueness of solution at this set of eigenfrequencies.

There have been various methods proposed to handle the nonuniqueness problem, of which the two most well known are the CHIEF method of Schenck (1968) and the combined integral equation approach of Burton and Miller (1971).

The CHIEF approach requires us to consider collocating at additional points inside the scatterer. These points are often called CHIEF points. Here, we can take the term $c(\mathbf{p})$ to be zero. The integration is straightforward because there are no singular cases; all integrals are regular. In theory, only a single CHIEF point is required to render the system non-singular and allow a solution to be obtained. The restriction is that, to be effective, a CHIEF point must not lie at a node of the mode shape of the interior natural acoustic mode at the frequency in question. In general, since these mode shapes are not known *a priori*, it is common to scatter several CHIEF points so that it becomes a strong probability that at least one of them will lie in an acceptable location. While this offers a simple remedy to the nonuniqueness problem, it suffers from two disadvantages: (i) there is no guarantee that the CHIEF points lie in acceptable locations, and this may be expected to be more of a problem at high frequency, since the higher modes will have more complicated mode shapes, and (ii) the extra equations cause the system to become overdetermined, requiring us to find a suitable solver.

The approach of Burton and Miller is appealing since it does not suffer from either of the two disadvantages of CHIEF. The system remains square, and it is guaranteed to result in a non-singular system from which a solution may be obtained. The method proceeds as follows. First, we differentiate the BIE (8.32) with respect to the unit outward normal at the collocation point $\mathbf{p}$:

$$c(\mathbf{p})\frac{\partial\phi(\mathbf{p})}{\partial\mathbf{n}(\mathbf{p})} + \int_{\Gamma}\frac{\partial^2 G(\mathbf{p},\mathbf{q})}{\partial\mathbf{n}(\mathbf{p})\partial\mathbf{n}(\mathbf{q})}\phi(\mathbf{q})\mathrm{d}\Gamma(\mathbf{q}) = \int_{\Gamma}\frac{\partial G(\mathbf{p},\mathbf{q})}{\partial\mathbf{n}(\mathbf{p})}\frac{\partial\phi(\mathbf{q})}{\partial\mathbf{n}(\mathbf{q})}\mathrm{d}\Gamma(\mathbf{q}) + \frac{\partial\phi^I(\mathbf{p})}{\partial\mathbf{n}(\mathbf{p})} \tag{8.42}$$

The equation we solve is a linear combination of Equations (8.32) and (8.42), i.e.:

$$c(\mathbf{p})\left(\phi(\mathbf{p}) + \eta\frac{\partial\phi(\mathbf{p})}{\partial\mathbf{n}(\mathbf{p})}\right) + \int_{\Gamma}\left(\frac{\partial G(\mathbf{p},\mathbf{q})}{\partial\mathbf{n}(\mathbf{q})} + \eta\frac{\partial^2 G(\mathbf{p},\mathbf{q})}{\partial\mathbf{n}(\mathbf{p})\partial\mathbf{n}(\mathbf{q})}\right)\phi(\mathbf{q})\mathrm{d}\Gamma(\mathbf{q}) =$$
$$\int_{\Gamma}\left(G(\mathbf{p},\mathbf{q}) + \eta\frac{\partial G(\mathbf{p},\mathbf{q})}{\partial\mathbf{n}(\mathbf{p})}\right)\frac{\partial\phi(\mathbf{q})}{\partial\mathbf{n}(\mathbf{q})}\mathrm{d}\Gamma(\mathbf{q}) + \phi^I(\mathbf{p}) + \eta\frac{\partial\phi^I(\mathbf{p})}{\partial\mathbf{n}(\mathbf{p})} \tag{8.43}$$

Here, η is a coupling parameter by which Equation (8.42) is multiplied in forming the linear combination with (8.32); it is normally taken as i/k.

The disadvantage of the Burton and Miller approach is that it has introduced a further differentiation of the fundamental solutions, and this increases the order of singularity of the integral. The equation becomes a hypersingular BIE which presents some challenges for implementation. Various researchers have proposed different regularized forms that are at worst weakly singular.

8.3 Shortcomings of the Conventional Formulations

Boundary element analysis of both types of problem presented, i.e., fracture mechanics and wave modeling, becomes inefficient in that very fine meshes may be needed in order to obtain the required accuracy. It is the aim of the XBEM formulation to achieve accurate results on much coarser meshes by using basis functions that have better approximation properties than conventional polynomials.

For the fracture mechanics problem, the principal difficulty is the stress singularity at the crack tip(s). If conventional piecewise polynomial models are used, very fine discretizations are required to achieve accurate solutions. The mesh generation burden is somewhat less severe than in the FEM case since in BEM we have boundary-only meshing, but a large number of elements can still be required and this is detrimental to computational efficiency. As a result, there has been considerable activity in developing formulations that attempt, in one way or another, to overcome the requirement for such fine meshes. One classical approach, also used in the FEM, is to make use of quarter-point boundary elements (Henshell and Shaw, 1975), in which the "mid-node" of a quadratic element is located at the quarter point of the element, i.e., 25% of the way from the crack tip to the other end node of the element. The idea is that the geometric mapping from the real Cartesian space (x, y) to the element's parametric space, ξ, becomes nonlinear, even for a flat element, and causes the usual shape functions to become an approximation space that includes the leading order terms of the asymptotic crack tip displacement field. However, Ingraffea and Manu (1980) demonstrated the size dependence of quarter-point elements preventing a general strategy for their use being formulated. Furthermore, the extent of the singular region created by the crack tip is restricted to the size of the quarter-point element when in reality it may extend further over a larger region. Further complications arise in the use of quarter-point elements for curved crack geometries.

Other attempts to improve the FEM for fracture include a hybrid-element approach, first introduced by Tong and Pian (1973) and more recently extended by Karihaloo and Xiao (2001), while a more recent approach, known as the fractal finite element method (FFEM) (Leung and Su, 1995) has been developed. The first uses a complex variable approach in which a special "hybrid" element incorporates the correct crack tip behavior. The latter technique models the singular region surrounding the crack tip as a self-similar mesh in which several layers, progressively decreasing in size, are used. The large number of unknowns created are transformed into a small number of global unknowns using appropriate interpolation functions. Both methods exhibit accurate results for relatively coarse meshes but do exhibit certain disadvantages. In particular, in the case of multiple cracks with tips in close proximity to one another, problems will occur in the formation of the "hybrid" element (hybrid-element method) and singular region (FFEM). This would also be the case for any geometrical feature that lay near the crack tip. In these cases, BEMs, in which only discretization of the boundary is required, present a distinct advantage in the context of linear elastic fracture mechanics (LEFM).

We now turn to the shortcomings of conventional BEM approaches for wave modeling. We consider one fundamental point. Unlike other problems, e.g., those in elasticity and potential flow, the solution we seek is oscillatory in nature and this requires a minimum level of discretization in order to capture successfully this behavior. A rule of thumb that is commonly applied is that the nodal spacing should be no more than about $\lambda/10$, where λ is the wavelength under consideration (for problems in the time-domain we would have to design a mesh that is capable of capturing the shortest wavelength we desire to include). The rule applies throughout the analysis domain, and not only in the regions of most interest. This simple heuristic means that problems in which the domain size is much greater than the wavelength present a severe challenge; we refer to these problems as "short wave" problems.

As a simple example, imagine an electromagnetics simulation of a radar wave of $\lambda = 10$ mm impinging upon an aircraft in free space. The wavelength implies a nodal spacing of 1 mm is required. A finite element model of only 100 m cube of air surrounding the aircraft would require a staggering 10^{15} unknowns. Furthermore, Ihlenburg and Babuška (1995) have shown how finite element approximations suffer also from *pollution error*, which is a progressive deterioration of the accuracy in capturing the required wavelength over multiple elements. The outcome of this is that, even if the nodal spacing satisfies the $\lambda/10$ heuristic, for high-frequency problems the mesh may still be inadequately refined to produce results of the required accuracy.

These difficulties led Zienkiewicz (2000) to state, in his millennial review of the state of the art in computational mechanics,

...there are some areas where our present techniques of formulation and solution prove inadequate and where important developments are yet to take place. I shall mention two of these here.
(i) The problem of short waves in acoustics, electromagnetics or surface water wave applications.
(ii) The problem of boundary layer and turbulence development in fluid mechanics.

It is the first of these two types of problems that the XBEM addresses.

8.4 Partition of Unity BEM Formulation

The XBEM presented in this chapter is an example of the use of the PU method introduced by Melenk and Babuška (1996). This is a general approach to enrichment of finite element solutions. The essential idea is that the usual piecewise polynomial basis functions are enriched by other functions that have better approximation properties. The name of the method arises from the well-known PU property of the conventional shape functions, i.e.:

$$\forall \xi \in [-1, 1], \qquad \sum_{s=1}^{S} N_s(\xi) = 1 \tag{8.44}$$

This significance is that we can inject an arbitrary function $\psi(\xi)$, writing:

$$\forall \xi \in [-1, 1], \qquad \sum_{s=1}^{S} N_s(\xi)\psi(\xi) = \psi(\xi) \tag{8.45}$$

and see that the function ψ is undistorted by the shape function interpolation. This means that *a priori* knowledge of features of the solution can be usefully injected (in the place of the function ψ) into the approximation of the solution variable over an element (or elements).

For a particular partial differential equation (PDE), we may have some information about the form of the solution, u. The PU method allows us to approximate the solution over an element using the expression:

$$u(\xi) \approx u_h(\xi) = \sum_{s=1}^{S} \sum_{m=1}^{M} A^{sm} N_s(\xi)\psi_m(\xi) \tag{8.46}$$

This recasts the problem so that the solution is sought in terms of the amplitudes A^{sm} of a family of enrichment functions ψ at each node s, and no longer in terms of nodal values of u. For example, if we know that the solution to the PDE at hand behaves like a certain function F, then F can be included in the enrichment functions. This has certain negative implications, notably on the conditioning of the system, which is known to deteriorate when enrichment is used. However, the resulting decrease in the required size of the system is an advantage that generally outweighs this drawback, and very accurate solutions can be obtained on very coarse meshes.

8.5 XBEM for Accurate Fracture Analysis

8.5.1 Williams Expansions

We define first a local polar coordinate system (ρ, θ) with its origin at the crack tip, as shown in Figure 8.3. Note that $\theta = 0$ describes the direction of the crack. Williams (1952) showed that the displacement field, asymptotically close to the crack tip, may be given in LEFM by:

Figure 8.3 Definition of coordinate system (ρ, θ) local to crack tip.

$$u_j(\rho, \theta) = K_{\mathrm{I}} B_{\mathrm{I}j}(\rho, \theta) + K_{\mathrm{II}} B_{\mathrm{II}j}(\rho, \theta) + \ldots \tag{8.47}$$

where $K_{\mathrm{I}}, K_{\mathrm{II}}$ are the mode I and mode II SIFs respectively, and:

$$B_{\mathrm{I}x}(\rho, \theta) = \frac{1}{2\mu} \sqrt{\frac{\rho}{2\pi}} \cos \frac{\theta}{2} \left[\kappa - 1 + 2 \sin^2 \frac{\theta}{2} \right] \tag{8.48}$$

$$B_{\mathrm{II}x}(\rho, \theta) = \frac{1}{2\mu} \sqrt{\frac{\rho}{2\pi}} \sin \frac{\theta}{2} \left[\kappa + 1 + 2 \cos^2 \frac{\theta}{2} \right] \tag{8.49}$$

$$B_{\mathrm{I}y}(\rho, \theta) = \frac{1}{2\mu} \sqrt{\frac{\rho}{2\pi}} \sin \frac{\theta}{2} \left[\kappa + 1 - 2 \cos^2 \frac{\theta}{2} \right] \tag{8.50}$$

$$B_{\mathrm{II}y}(\rho, \theta) = \frac{1}{2\mu} \sqrt{\frac{\rho}{2\pi}} \cos \frac{\theta}{2} \left[\kappa - 1 - 2 \sin^2 \frac{\theta}{2} \right] \tag{8.51}$$

where μ is the shear modulus, and the constant κ is given by:

$$\kappa = \begin{cases} 3 - 4\nu & \text{plane strain} \\ \dfrac{3 - \nu}{1 + \nu} & \text{plane stress} \end{cases} \tag{8.52}$$

in which ν is Poisson's ratio. It is required for the validity of (8.47) that $\rho << a$, where a is the crack length.

8.5.2 Local XBEM Enrichment at Crack Tips

Taking our lead from the XFEM developments at crack tips (e.g., Moës et al., 1999) we can manipulate the Williams expansion (8.47) to give a set of enrichment functions that include the asymptotic displacement behavior at crack tips. Thus, the displacement on element e, locally to the crack tip, might be written:

$$u_j^e(\xi_e) = \sum_{s=1}^{S} \sum_{l=1}^{L} N_s(\xi_e) \psi_l(\xi_e) A_{jl}^{es} \tag{8.53}$$

where the enrichment functions ψ_l are given by:

$$\psi(\rho, \theta) = \left\{ 1, \sqrt{\rho} \cos \frac{\theta}{2}, \sqrt{\rho} \sin \frac{\theta}{2}, \sqrt{\rho} \sin \frac{\theta}{2} \sin \theta, \sqrt{\rho} \cos \frac{\theta}{2} \sin \theta \right\} \tag{8.54}$$

where S is the number of nodes on element e, $\xi_e \in [-1, 1]$ is the usual parametric element coordinate, and L is the number of enrichment functions. In this case, $L = 5$. Note that the first enrichment function, $\psi_1 = 1$, is there to allow the conventional polynomial shape function to remain in the basis. This is required because the other enrichment functions all contain the term $\sqrt{\rho}$, and therefore ψ_1 is needed in order to allow a non-zero displacement at the crack tip, i.e., at $\rho = 0$. Since the opposing crack faces are explicitly modeled in the dual BEM by elements that can undergo different displacements, there is no need to include a Heaviside function term in the displacement expansion. The problem is reformulated so that the solution is sought in terms of the unknown coefficients A_{jl}^{es}, which multiply the enrichment function l, at the node s on element e to give the displacement contribution in direction j.

One important point should be made here. The derivative of the displacement expression (8.53) yields $\partial u_j / \partial \rho = f(\rho^{-1/2})$, and since this is a strain component we find that the strain becomes infinite at $\rho = 0$. Thus, although we enrich only displacements (with non-singular functions), the enrichment automatically includes the singular strain and stress fields at the crack tip. In the current work, we consider traction-free cracks, i.e., $t_x = t_y = 0$, so it is unnecessary to inject a singular enrichment of the traction field, but for other crack conditions it would be possible also to enrich tractions in this way.

One other feature also quickly becomes apparent for *flat cracks*. Here, $\theta = \pm\pi$ on the upper and lower crack surfaces, so the enrichment functions $\psi_2, \psi_3, \psi_4, \psi_5$ are all identical, over each element adjacent to the crack tip, apart from a scaling constant. For this reason, the set is reduced to $L = 2$ and:

$$\psi(\rho) = \left\{ 1, \sqrt{\vec{\rho}} \right\} \tag{8.55}$$

The revised displacement expansion causes the discretized integral equation (8.17) to become:

$$c(\mathbf{p})u_i(\mathbf{p}) + \sum_{e=1}^{E}\sum_{s=1}^{S}\sum_{l=1}^{L} \tilde{P}_{ijl}^{es} A_{jl}^{es} = \sum_{e=1}^{E}\sum_{s=1}^{S} Q_{ij}^{es} t_j^{es} \tag{8.56}$$

where S is the number of nodes on each element, and the influence coefficients $\tilde{P}_{ij}^{es}, Q_{ij}^{es}$ are the results of evaluating the boundary integrals:

$$\tilde{P}_{ij}^{es} = \int_{-1}^{1} T_{ij}(\mathbf{p}, \mathbf{q}(\xi_e)) N_s(\xi_e) \psi_l(\xi_e) J^e \mathrm{d}\xi_e \tag{8.57}$$

$$Q_{ij}^{es} = \int_{-1}^{1} U_{ij}(\mathbf{p}, \mathbf{q}(\xi_e)) N_s(\xi_e) J^e \mathrm{d}\xi_e \tag{8.58}$$

For reasons of control of conditioning, it is most effective to restrict the enrichment to the two elements immediately adjacent to each crack tip. The vast majority of elements are unenriched, and for these elements we can simply let $L = 1$.

The traction integral equation (8.23) can be similarly expressed in enriched form. The analysis proceeds by collocation as in the conventional BEM fashion, and the linear system solved to yield the coefficients A_{jl}^{es}, allowing the displacement field to be quickly recovered from (8.53), all of which is now known. The SIFs can then be found using a J-integral approach.

The above approach was first applied in a BEM context by (Simpson and Trevelyan, 2011).

An **alternative enrichment strategy** (Simpson and Trevelyan, 2011) is inspired by the early finite element work of Benzley (1974) and in a mesh-free context by Duflot and Nguyen-Dang (2004) and Fleming et al. (1997). Here, the enrichment is based more directly on the SIFs, K_I, K_{II}, at the crack tip and their role in the Williams expansion (8.47). We can write, for the elements adjacent to each crack tip:

$$u_j^e(\xi_e) = \sum_{s=1}^{S} N_s(\xi_e) u_j^{es} + \tilde{K}_I B_{Ij} + \tilde{K}_{II} B_{IIj} \tag{8.59}$$

and we seek as unknowns the terms u_j^{es} for all nodes s on all elements e, as well as single values of $\tilde{K}_I, \tilde{K}_{II}$ relating to each crack tip. The terms $\tilde{K}_I, \tilde{K}_{II}$ can be seen to act as aliases for the SIFs themselves. The traditional shape functions are included in the expression, as before, to allow for non-zero displacement of the crack tip. When this enrichment expansion is substituted for displacement in the BIE, we arrive at:

$$c(\mathbf{p})u_i(\mathbf{p}) + \sum_{e=1}^{E}\sum_{s=1}^{S} P_{ij}^{es}u_j^{es} + \sum_{\bar{e}=1}^{\bar{E}}\sum_{s=1}^{S} \tilde{P}_{ijl}^{\bar{e}s}\tilde{K}_{lj} = \sum_{e=1}^{E}\sum_{s=1}^{S} Q_{ij}^{es}t_j^{es}, \qquad l = \mathrm{I}, \mathrm{II} \tag{8.60}$$

Here, $\bar{e}$ denotes an element adjacent to a crack tip and $\bar{E}$ is the total number of these elements, i.e., $\bar{E} = 2$ for cases containing a single crack tip. The influence functions P_{ij}^{es}, Q_{ij}^{es} are as in (8.18) and (8.19), respectively, and the new influence function for the enrichment term is given by:

$$\tilde{P}_{ijl}^{\bar{e}s} = \int_{-1}^{1} T_{ij}(\mathbf{p}, \mathbf{q}(\xi_{\bar{e}}))N_s(\xi_{\bar{e}})B_{lj}(\xi_{\bar{e}})J^{\bar{e}}\mathrm{d}\xi_{\bar{e}}, \qquad l = \mathrm{I}, \mathrm{II} \tag{8.61}$$

8.5.3 Results

We start by considering a mode I edge crack in a finite sheet under uniaxial tension as shown in Figure 8.4. We consider the case $a = h = w/2$, start with the mesh as shown and systematically refine the mesh to observe convergence properties. Here, we apply enrichment according to (8.53) and compare the results against the unenriched dual BEM in Figure 8.5. It is clear that the enrichment gives us a benefit of approximately one order of magnitude reduction in error for the cost of a few extra degrees of freedom, and the convergence rate is somewhat improved also. In Figure 8.6, we compare the convergence properties also against quarter points elements toward the reference solution of Civelek and Erdogan (1982). In a second example, we consider a curved crack describing a circular arc in an infinite sheet under biaxial stress, as shown in Figure 8.7. For curved cracks, care needs to be taken in evaluating the J-integral in computing the SIFs; here, the approach of Chang and Wu (2007) is used[1]. In Figures 8.8 and 8.9, a comparison is presented for the J_1 and J_2 integrals (required to establish $K_\mathrm{I}, K_\mathrm{II}$, respectively). Here, the enriched dual BEM results are computed using the enrichment (8.59), and these are presented for curved elements and piecewise flat elements. Notice the improved performance of the enriched curved element formulation over the alternative modeling strategies.

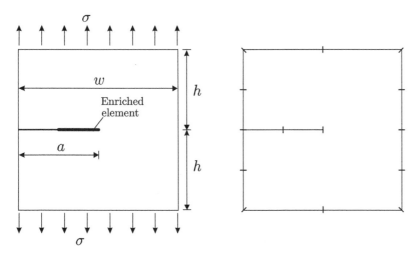

Figure 8.4 Example of a mode I edge crack in a finite sheet.

---

1 The difficulty is that there remains an integral over the portion of the crack surface that lies within the J-integral contour. This term vanishes when the crack is flat. In order to evaluate this integral, which is taken over a region that includes the singularity, we have to determine a strain energy jump across the crack tip. Chang and Wu showed how this could be evaluated indirectly.

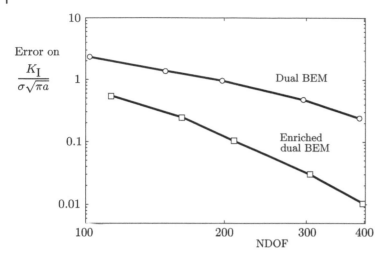

Figure 8.5 Comparison of unenriched and enriched dual BEM for mode I edge crack example.

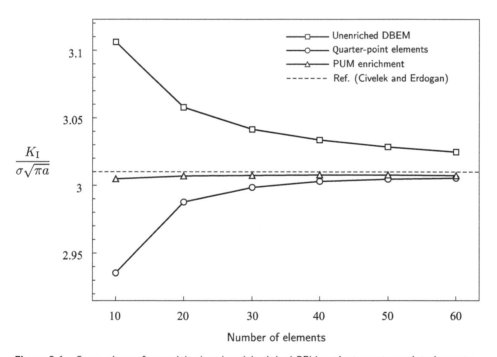

Figure 8.6 Comparison of unenriched and enriched dual BEM against quarter-point elements.

8.5.4 Auxiliary Equations and Direct Evaluation of Stress Intensity Factors

The addition of extra terms into the displacement expression, i.e., the replacement of (8.11) by either (8.53) or (8.59), increases the total number of unknowns in the problem. For this reason, collocating only at the nodes will provide an insufficient number of equations to solve the problem. An auxiliary set of equations is therefore required. The choice of how to form these auxiliary equations has significant implications in the efficient evaluation of the SIFs K_I, K_{II}.

Figure 8.7 Curved crack in infinite sheet under biaxial tension.

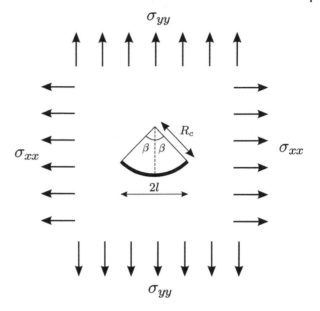

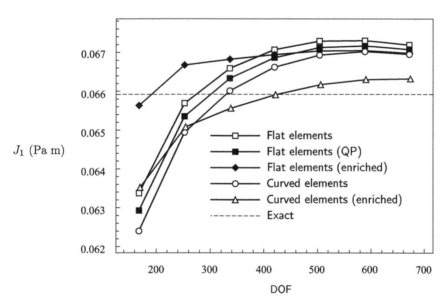

Figure 8.8 Comparison of J_1 solutions.

These equations may come from collocating at extra, non-nodal, points on the boundary. It has been found effective to locate these extra points on the enriched elements.

A key development in Simpson and Trevelyan (2011) was to expand the displacement in the vicinity of the crack tip in the form (8.59), so that the unknowns $\check{K}_{\mathrm{I}}, \check{K}_{\mathrm{II}}$ are explicitly included in the approximation space. These terms, which may be expected to approximate the SIFs, are therefore part of the solution vector and become available without the need for post-processing procedures such as displacement extrapolation or the J-integral. However, Simpson found that the accuracy was insufficient when the auxiliary equations were derived from collocation of the BIE at additional collocation points. A J-integral evaluation based on those same results, however, allows for very good accuracy.

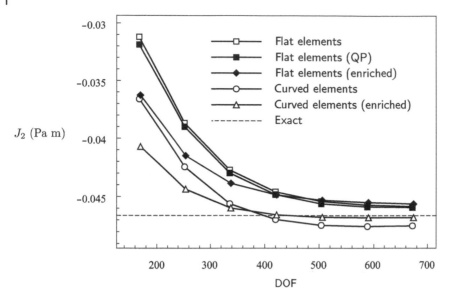

Figure 8.9 Comparison of J_2 solutions.

A revised method of forming the auxiliary equations was developed by Alatawi and Trevelyan (2015), by enforcing continuity of displacement at each crack tip; this is not automatically enforced when using discontinuous boundary elements. Because the enrichment functions go to zero at the crack tip, the displacement at the crack tip reduces to a form completely determined by the polynomial shape functions. A simple extrapolation procedure can then be used to equate the displacement at the crack tip as calculated from the upper crack surface to that calculated from the lower crack surface, and this provides the required extra equations.

The approach of Alatawi allows the terms $\tilde{K}_I, \tilde{K}_{II}$ found directly in the solution vector to approximate the SIFs with good accuracy so that for practical engineering usage a J-integral is not required. This has significant implications in computational efficiency, and these will be more fully realized when the XBEM is extended to 3D situations. It is notable to mention that Alatawi also found that the use of the J-integral allows further accuracy to be achieved for cases in which that might be required by engineers.

8.5.5 Fracture in Anisotropic Materials

A formulation was presented by Hattori et al. (2017) to allow the XBEM to provide estimates for the SIFs in anisotropic media. The Stroh formalism, which is one of the complex variable representations of elasticity theory, is used to generate expressions for the fundamental solutions and also provides a framework to write the required crack tip enrichment functions. The fundamental solutions for the isotropic problem, stated in (8.6) and (8.7), are replaced by:

$$U_{ij} = -\frac{1}{\pi} \text{Re}\left\{A_{jm}Q_{mi}\ln(z_m^{\mathbf{q}} - z_m^{\mathbf{p}})\right\} \tag{8.62}$$

$$T_{ij} = \frac{1}{\pi} \text{Re}\left\{B_{jm}Q_{mi}\frac{\mu_m n_1 - n_2}{z_m^{\mathbf{q}} - z_m^{\mathbf{p}}}\right\} \tag{8.63}$$

where $z_m = x + \mu_m y$ and A, B, μ are solutions to the eigenproblem:

$$\begin{pmatrix} -\mathbf{C}_{2ij2}^{-1}\mathbf{C}_{2ij1} & -\mathbf{C}_{2ij2}^{-1} \\ \mathbf{C}_{1ij1} - \mathbf{C}_{2ij1}^{T}\mathbf{C}_{2ij2}^{-1}\mathbf{C}_{2ij1} & -\mathbf{C}_{2ij1}^{T}\mathbf{C}_{2ij2}^{-1} \end{pmatrix}\begin{pmatrix} \mathbf{A}_m \\ \mathbf{B}_m \end{pmatrix} = \mu_m\begin{pmatrix} \mathbf{A}_m \\ \mathbf{B}_m \end{pmatrix} \tag{8.64}$$

and

$$Q = A^{-1}(L^{-1} + \bar{L}^{-1})^{-1} \tag{8.65}$$

with

$$L = iAB^{-1} \tag{8.66}$$

In the above, C is the fourth-order elasticity tensor describing the stress-strain behavior of the anisotropic material. Expressions for the derivative fundamental solutions, required for the traction BIE in the dual BEM, are given in Hattori et al. (2017).

The asymptotic displacement field at the crack tip can also be based on the Stroh formalism:

$$u_i(r, \theta) = \sqrt{\frac{2}{\pi}} \mathrm{Re} \left(K_\alpha A_{im} B_{m\alpha}^{-1} \sqrt{\rho(\cos\theta + \mu_m \sin\theta)} \right) \tag{8.67}$$

where $i, m = 1, 2, \alpha = I, II$ so we use an enrichment strategy:

$$u_j = \sum_{a=1}^{M} N^a u_j^a + \tilde{K}_I F_{Ij} + \tilde{K}_{II} F_{IIj} \tag{8.68}$$

$$F_{lj}(\rho, \theta) = \sqrt{\frac{2r}{\pi}} \begin{pmatrix} A_{11} B_{11}^{-1} \beta_1 + A_{12} B_{21}^{-1} \beta_2 & A_{11} B_{12}^{-1} \beta_1 + A_{12} B_{22}^{-1} \beta_2 \\ A_{21} B_{11}^{-1} \beta_1 + A_{22} B_{21}^{-1} \beta_2 & A_{21} B_{12}^{-1} \beta_1 + A_{22} B_{22}^{-1} \beta_2 \end{pmatrix} \tag{8.69}$$

and in which

$$\beta_i = \sqrt{\cos\theta + \mu_i \sin\theta} \tag{8.70}$$

As for the isotropic problem, the XBEM has been shown using the above formulation to provide accurate results from coarse meshes without the requirement for post-processing such as the J-integral.

8.5.6 Conclusions

In this section, it has been shown how a dual BEM formulation for the solution of problems in LEFM can be enriched locally by the injection of functions taken from the asymptotic displacement field at a crack tip. This increases the number of degrees of freedom only marginally, yet it has been shown that this enriched, XBEM formulation outperforms the conventional, piecewise polynomial dual BEM. Errors in SIFs for simple mode I problems are reduced by around one order of magnitude. An improvement of approximately the same magnitude is seen in comparison against quarter-point elements. The technique has been applied successfully to curved cracks and cracks in anisotropic media. Use of auxiliary equations that enforce continuity of displacement at the crack tip allows the SIFs to emerge directly in the solution vector without the requirement to evaluate a J-integral; however, use of the J-integral based on these same results allows enhanced accuracy to be achieved.

8.6 XBEM for Short Wave Simulation

8.6.1 Background to the Development of Plane Wave Enrichment

We can now recall the nodal spacing heuristic, which states that in order to obtain a meaningful solution with the FEM or BEM, we should use a mesh that is sufficiently dense in order to capture the sinusoidal nature of the solution. In the frequency domain, we know *a priori* the wavelength that is present in the solution by turning to Equation (8.27). Although one might argue that as few as, say, six points are sufficient to represent one wavelength of a sinusoidal function, a generally accepted rule of thumb is to use a nodal spacing of $\lambda/10$.

At this point we will define a parameter τ, which we might call a degree of freedom (DOF) density parameter, which is equal to the number of degrees of freedom used per wavelength in the discretization. So a conventional BEM simulation, in which each node has one DOF and we adhere to the heuristic of 10 nodes per wavelength, is characterized by a DOF density $\tau = 10$. Since the time taken by a direct solver will scale with the cube of the number of degrees of freedom, there is a great incentive to find ways to reduce the required number of degrees of freedom, i.e., reduce the required DOF density τ. This has been the subject of active research over the last two decades.

Although much of the work in enriched formulations (FEM and BEM) can be traced to the PU methods first proposed by Melenk and Babuška (1996), it is interesting to see some work in enriched boundary integral formulations appearing a little earlier. Guided by the high-frequency behavior which sees the potential field on the surface of a convex scatterer behaving as the incident wave itself, Abboud et al. (1995) wrote integral equations derived from the expansion of the potential as $\phi(\mathbf{q}) = A(\mathbf{q})\phi^I(\mathbf{q})$, and solved for the slowly varying function A instead of the rapidly oscillating potential ϕ. This approach can be particularly effective for convex scatterers of size $>> \lambda$. Bruno et al. (2004) have presented a Nystrom approach based on this expansion, showing results for scatterers of size up to $10^6\lambda$, in which the run time is independent of k. Similar wavenumber independent complexity has been shown for scattering from polygons by Langdon and Chandler-Wilde (2006) using an enriched boundary integral approach. de la Bourdonnaye (1994) extended the idea to a basis containing multiple plane waves, under the name "microlocal discretization."

Following the development of the PU, an increasing volume of literature appeared in which *families* of plane waves and Bessel functions were used to enrich both BEM and FEM approximations for wave problems. Space here does not permit a full review, but interested readers may see examples in the PUFEM of Laghrouche et al. (2002); the PU-BEM papers of Perrey-Debain et al. (2002, 2003, 2004), Bériot et al. (2010), and (2013a, 2013b); the variational theory of complex rays of Riou et al. (2008); the ultraweak variational formulation of Cessenat and Després (1998) and Huttunen et al. (2002); and the discontinuous enrichment method of Farhat et al. (2001). For a detailed exposition of numerical methods in high-frequency wave scattering, readers are directed to Chandler-Wilde et al. (2012). In this section, we focus on the methods developed in Perrey-Debain et al. (2004).

As a final point to make before presenting the technical development, it is worth pointing out that this is a different form of enrichment from that used in XFEM and XBEM for fracture mechanics, in the following ways:

- In XFEM and XBEM for fracture mechanics, the enrichment is confined to the elements in the vicinity of a crack. Elements used to model the bulk, uncracked material are no different from conventional piecewise polynomial FEM and BEM. For wave boundary elements, the enrichment is applied to every element in the model, since the solution will be oscillatory over all elements.
- In XFEM and XBEM for fracture mechanics, the enrichment is based on the leading order term of an analytical series expansion for the displacement asymptotically close to the crack tip. For wave boundary elements, we draw upon the understanding that any acoustic field can be expressed as a linear combination of (a suitably large number of) plane waves. In other words, the set of plane waves propagating in different directions form a complete set for representing the solutions of the Helmholtz equation.

8.6.2 Plane Wave Enrichment

Let us now move forward to the injection of enrichment in the BEM approximation. In this presentation, we restrict ourselves for simplicity to the collocation form of the BEM, and assume we will use the CHIEF approach to overcoming the nonuniqueness problem. We follow the idea of the PU to write the acoustic potential, ϕ, over element e in a plane wave expansion, i.e., we replace (8.35) by a new expression:

$$\phi(\mathbf{q}) = \begin{cases} \displaystyle\sum_{s=1}^{S} N_s(\xi) \sum_{m=1}^{M} A^{esm} e^{ik\mathbf{d}_{sm}\cdot\mathbf{q}}, & d = 2 \\ \displaystyle\sum_{s=1}^{S} N_s(\xi,\eta) \sum_{m=1}^{M} A^{esm} e^{ik\mathbf{d}_{sm}\cdot\mathbf{q}}, & d = 3 \end{cases} \tag{8.71}$$

where A^{esm} is a (complex) amplitude associated with each exponential term, which is a plane wave of unit amplitude, propagating in a direction given by the unit vector $\mathbf{d}_{sm}$. By comparing this expansion with the classical interpolation (8.35), we can see that one interpretation of this use of the PU is that the nodal potential is replaced by a linear combination of plane waves. So the unknowns are no longer the nodal values of potential, but have become the amplitudes A^{esm}.

This means that each element now has associated with it possibly a large number of degrees of freedom. Take, for example, a three-noded element in a 2D simulation (i.e., $S = 3$). If we choose to use eight plane waves at each node (i.e., $M = 8$), then there will be 24 degrees of freedom associated with this element. This is the reason we have introduced earlier our parameter τ as a measure of degrees of freedom per wavelength instead of nodes per wavelength.

We see in this application of enrichment another difference from the XFEM and XBEM in fracture mechanics, so will add another bullet point to the list in the previous section:

- It is no longer necessary to include a term in the enriched form (8.71) representing the unenriched piecewise polynomial shape functions. This is allowable because the solution for ϕ is entirely oscillatory and there is no non-oscillatory part of the solution to reproduce. Another way of looking at this is that in the formulation for fracture mechanics (XFEM and XBEM), the conventional polynomials are required to reproduce a non-zero displacement at the crack tip, and there is no analogy to this in the solution space for the Helmholtz problem.

In order to facilitate further our enriched formulation, we write our boundary conditions not in the Dirichlet and Neumann forms (8.28) and (8.29), but in the more general Robin form:

$$\frac{\partial \phi(\mathbf{q})}{\partial \mathbf{n}(\mathbf{q})} = \alpha \phi(\mathbf{q}) + \beta \tag{8.72}$$

where α and β are known, complex constants, and this allows us to rewrite the BIE (8.32):

$$c(\mathbf{p})\phi(\mathbf{p}) + \int_{\Gamma} \left(\frac{\partial G(\mathbf{p},\mathbf{q})}{\partial \mathbf{n}(\mathbf{q})} - \alpha G(\mathbf{p},\mathbf{q}) \right) \phi(\mathbf{q})\mathrm{d}\Gamma(\mathbf{q}) = \int_{\Gamma} \beta G(\mathbf{p},\mathbf{q})\mathrm{d}\Gamma(\mathbf{q}) + \phi^I(\mathbf{p}) \tag{8.73}$$

Now, for purposes of clarity, we will drop the terms showing explicit dependence of G and $\mathbf{n}$ on the points $\mathbf{p}$ and $\mathbf{q}$:

$$c(\mathbf{p})\phi(\mathbf{p}) + \int_{\Gamma} \left(\frac{\partial G}{\partial \mathbf{n}} - \alpha G \right) \phi(\mathbf{q})\mathrm{d}\Gamma(\mathbf{q}) = \int_{\Gamma} \beta G\mathrm{d}\Gamma(\mathbf{q}) + \phi^I(\mathbf{p}) \tag{8.74}$$

and we will discretize the equation by first taking the integrals element by element (2D form is shown):

$$c(\mathbf{p})\phi(\mathbf{p}) + \sum_{e=1}^{E} \int_{-1}^{1} \left(\frac{\partial G}{\partial \mathbf{n}} - \alpha G \right) \phi(\mathbf{q}(\xi_e)) J^e \mathrm{d}\xi_e = \sum_{e=1}^{E} \int_{-1}^{1} \beta G J^e \mathrm{d}\xi_e + \phi^I(\mathbf{p}) \tag{8.75}$$

and now inserting the plane wave enrichment (8.71):

$$c(\mathbf{p})\phi(\mathbf{p}) + \sum_{e=1}^{E}\sum_{s=1}^{S}\sum_{m=1}^{M} \int_{-1}^{1} \left(\frac{\partial G}{\partial \mathbf{n}} - \alpha G \right) N_s e^{ik\mathbf{d}_{sm}\cdot\mathbf{q}(\xi_e)} J^e \mathrm{d}\xi_e A^{esm} = \sum_{e=1}^{E} \int_{-1}^{1} \beta G J^e \mathrm{d}\xi_e + \phi^I(\mathbf{p}) \tag{8.76}$$

where the unknown A^{esm} denotes the amplitude associated with the plane wave number m at the node number s on element e.

8.6.3 Evaluation of Boundary Integrals

Equation (8.76) is the form of BIE that we will solve for the XBEM. Like the conventional BEM, we are required to evaluate boundary integrals to provide the influence coefficients that will populate a matrix. However, this is not as straightforward once the plane wave enrichment has been included.

The difference lies in the fact that:

1. the purpose of enriching in this way is to obtain good results from coarse meshes, being unrestricted by a $\lambda/10$ nodal spacing guideline;
2. each element will span multiple wavelengths;
3. the integrand becomes highly oscillatory over the element.

As we shall see in the next section, the XBEM has the ability to provide results of (simultaneously) considerably lower error and considerably lower number of degrees of freedom than conventional BEM. But to some extent these gains have come at the cost of the difficulty of the integral evaluations, and an increased cost per DOF.

While there are classical schemes for evaluating oscillatory integrals, such as Filon, Levin, asymptotic methods, and the method of stationary phase, their application is complicated by the awkward phase function and the presence of stationary points in (or, worse still, near to) the element. New methods of evaluating these integrals have become a subject for some research over the last few years. There are the asymptotic method of Iserles and Nørsett (2004), numerical steepest descent of Honnor and Trevelyan (2010), use of polynomial approximations of Harris and Chen (2009), Filon-Clenshaw-Curtis of Kim et al. (2009), and an element subdivision approach using the curved coordinate in which the integrand oscillates (Trevelyan and Honnor, 2009). No clear conclusion has yet emerged about a preferred scheme, and the subject remains an open one for research. In the meantime, a Gauss-Legendre scheme is robust though suboptimal in runtime terms.

8.6.4 Collocation Strategy and Solution

In our discussion of the conventional BEM, we developed a BIE (8.36) and then chose a series of points at which to place our source point **p**. These points all give different equations, in the same unknowns (the nodal potentials). Given enough equations these equations generated are assembled to form a square system of equations that can be solved. Since the number of degrees of freedom is equal to the number of nodes, it is convenient to place the source point **p** at each node in turn to ensure a sufficient number of equations.

Now we move to the XBEM with the plane wave enrichment, and each of the N nodes has M degrees of freedom associated with it. It is clear, then, that placing the source point (i.e., collocating) only at the nodes will not provide us with enough equations. In fact, we will need to collocate at a total of MN points. One way to do this is to distribute these points approximately uniformly around the boundary.

The implication of this comes in the first term of our integral equation (8.76); this term contains $\phi(\mathbf{p})$, the potential at the collocation point. In conventional BEM, collocating only at nodes, this is straightforward to handle because $\phi(\mathbf{p})$ is one of the unknowns. This first term, then, causes us to add the term $c(\mathbf{p})$ to the diagonal terms in the matrix **H**. Remember, that on a smooth boundary $c(\mathbf{p})$ may be taken as $1/2$.

For collocation at non-nodal locations, we need to write the term $\phi(\mathbf{p})$ in the plane wave expansion (8.71) over the element on which it lies. The result is that the factor of $1/2$ is distributed among the degrees of freedom associated with the element, so that multiple matrix entries will be modified in the process of building in the first term of (8.76). This term can then be expressed by:

$$c(\mathbf{p}) \sum_{s=1}^{S} N_s(\bar{\xi}) \sum_{m=1}^{M} \bar{A}_{sm} e^{ik\mathbf{d}_{sm} \cdot \mathbf{p}}$$

where the overbar denotes the fact that we are considering the element on which the collocation point **p** lies.

Having gone through this process for all collocation points, we arrive at a system of equations that is expressed in matrix form as:

$$\mathbf{Ax} = \mathbf{b} \tag{8.77}$$

where the matrix $\mathbf{A}$ contains the results of evaluating the boundary integrals on the left-hand side of the BIE along with the distributed $c(\mathbf{p})$ terms, the vector $\mathbf{b}$ contains the terms on the right-hand side of the BIE, and the vector $\mathbf{x}$ contains the unknown plane wave amplitudes A^{esm}. Solving this set of equations yields the amplitudes, and the potential solution can quickly be recovered from (8.71).

8.6.5 Results

The first example is of the scattering of an incident plane wave by a hard cylinder of radius a. The analysis is performed in two dimensions, so the domain Ω is infinite in the exterior and is bounded in the interior by a circular boundary Γ. The wavenumber is k and we investigate errors as a function of the non-dimensionalized measure of frequency ka. Note that ka can be conveniently thought of as the number of wavelengths required to describe the perimeter of the cylinder. The error measure used is defined as the relative error norm, i.e.:

$$\varepsilon_2 = \frac{||\phi - \phi^{\text{ex}}||_{L^2(\Gamma)}}{||\phi^{\text{ex}}||_{L^2(\Gamma)}} \tag{8.78}$$

where ϕ is the potential solution obtained by the numerical simulation and ϕ^{ex} is the exact solution provided by Morse and Feshbach (1981).

Results of a large number of simulations of the problem of wave scattering from the hard cylinder are presented in Figure 8.10. The thick, solid line presents results of a conventional, i.e., unenriched, piecewise quadratic boundary element simulation using quadratic elements. A total of 192 nodes (and therefore 192 degrees of freedom) are used. It is noted that the simulation is satisfactory for engineering accuracy, defined as $\varepsilon_2 < 1\%$, for ka smaller than about 20, as indicated by the dashed lines superimposed on the graph. Note that, as ka increases beyond this point, the 192 degrees of freedom are no longer sufficient to provide the required accuracy. Taking the limiting value of ka as 23, this corresponds to $\tau = 8.3$. The reader is reminded that this means the BEM requires 8.3 degrees of freedom for every wavelength needed to describe the perimeter.

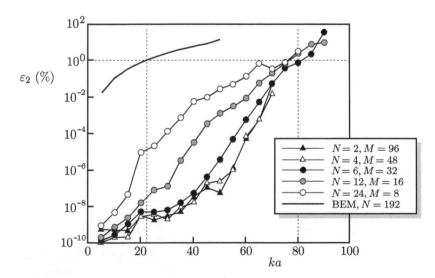

Figure 8.10 Errors, ε_2, for the solution to the wave scattering problem from a hard cylinder. N: number of nodes, M: number of plane waves per node.

The remaining simulations are conducted using the plane wave enriched XBEM formulation. All simulations are designed to use the same number of degrees of freedom, i.e., NDOF = 192, but they are accumulated using different meshes and different levels of plane wave enrichment. The degrees of freedom correspond to the amplitudes of the M plane waves located at the N nodes in the model. Three points are apparent from the graphs for different XBEM simulations:

1. The plane wave enrichment has markedly improved the accuracy over the piecewise polynomial formulation. This improvement can give as much as *eight orders of magnitude* reduction in error for the same number of degrees of freedom.
2. The improved accuracy means that the method can provide acceptable engineering accuracy for this model up to a frequency measure $ka = 80$, i.e., the enrichment effectively extends the supported frequency range for BEM simulations by a factor of nearly 4. At $ka = 80$, engineering accuracy is being obtained using only $\tau = 2.4$ degrees of freedom per wavelength.
3. The best accuracy is to be obtained by using a small number of nodes and large number of plane waves at each. The extreme case is a model containing only two nodes ($N = 2$) with 96 plane waves located at each.

The second example is taken from the study of adaptivity presented in Trevelyan and Coates (2010). This is motivated by the aim to optimize the plane wave functions in the XBEM basis so that we use the most efficient distribution of plane waves for the solution of the particular problem at hand. The example also presents an opportunity to investigate the performance of the plane wave enrichment for problems involving multiple reflections. Thus, unlike the scattering by the single convex cylinder, above, we consider the scattering of an incident plane wave propagating in direction (1,0) by a system of three perfectly reflecting cylinders, of boundaries $\Gamma_1, \Gamma_2, \Gamma_3$, respectively, these being:

- scatterer 1: a circle of boundary Γ_1, having center (0,0), radius 1, and modeled using 8 elements;
- scatterer 2: a circle of boundary Γ_2, having center (2,3), radius 2, and modeled using 16 elements;
- scatterer 3: a circle of boundary Γ_3, having center (4,−2), radius 3, and modeled using 24 elements.

We consider $\lambda = 0.25$, and start with a model using $M = 4$, i.e., four plane waves at each node, giving an overall $\tau = 2.54$. The adaptive scheme progresses by using:

- an error indicator denoted $||R||_1$. The full definition of $||R||_1$ is not provided in this chapter for reasons of brevity, but it is based on the integral over $\Gamma = \Gamma_1 \cup \Gamma_2 \cup \Gamma_3$ of the residual of the BIE. Interested readers are referred to Trevelyan and Coates (2010) for details;
- a model improvement strategy in which the number of plane waves applied at nodes in regions of large $||R||_1$ is incremented by 1 in each iteration. The new plane waves are inserted such that they bisect the angle between two pre-existing waves at that node in the iteration just completed.

The convergence of the global error norm $||R||_1$, from $||R||_1 = 0.1485$ for the initial analysis ($M = 4, \tau = 2.54$) to achieve convergence in the fourth iteration at $||R||_1 = 0.00264$, is shown in Figure 8.11. Convergence is achieved using 631 degrees of freedom at $\tau = 4.18$. Contours of the converged solution $Re(\phi)$ are shown in Figure 8.12, and show reflection from the illuminated surfaces, a clear shadow region to the right, diffraction around the sides of the scatterers and a complicated region of multiple reflections between the three cylinders. This complication is emphasized by plotting $|\phi|$ on Γ_2, as shown in Figure 8.13.

8.6.6 Choice of Basis Functions

When using the XBEM formulation for wave problems, we have some freedom in how the degrees of freedom are accumulated. It has been seen in Section 8.6.5 that it is desirable to design the discretization of the problem so that we maintain $\tau \approx 3$, i.e., we use approximately three DOF per wavelength. For example, in the 2D analysis of a

Figure 8.11 Convergence of the adaptive scheme for scattering from three perfectly reflecting cylinders.

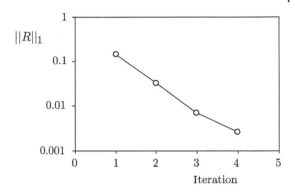

Figure 8.12 Converged solution: contours of the real part of the solution for the potential ϕ.

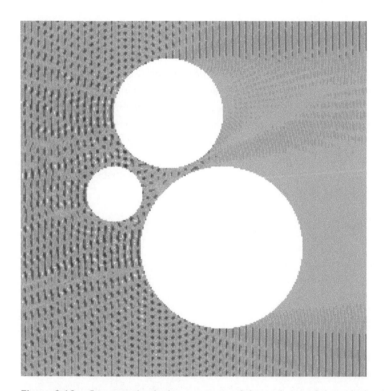

scatterer of perimeter $P = 3$ m impinged by a wave of wavelength $\lambda = 25$ mm, the perimeter describes $P/\lambda = 120$ wavelengths, and so we design a model to have $120\tau = 360$ degrees of freedom. However, we can design different models that provide those 360 degrees of freedom:

- use 10 nodes, with $M = 36$, i.e., 36 plane waves at each node;
- use 18 nodes, with $M = 20$;
- use 20 nodes, with $M = 18$;
- use 24 nodes, with $M = 15$;
- use 30 nodes, with $M = 12$.

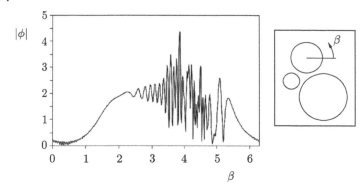

Figure 8.13 Converged solution: real part of the potential over Γ_2.

In general, as shown in Figure 8.10, the best convergence properties are found when the number of nodes is minimized and M maximized. In one sense this should come as no surprise, since it is the oscillatory plane waves and not the slowly varying shape functions that play the most important role in capturing the solution. However, it will be seen in Section 8.7 that the conditioning of the system degrades with an increasing level of enrichment. Thus, using a small number of nodes and large M is advised only until the ill-conditioning becomes too severe for the solver to handle. This can still be quite extreme ill-conditioning if an appropriate solver is used, such as the singular value decomposition (SVD), giving the rather counter-intuitive result that the most accurate solutions are to be derived from the most ill-conditioned systems.

Most works in enriched wave boundary element problems are conducted using uniformly spaced wave directions $\mathbf{d}_{sm}$. This is straightforward to design in 2D simulations, i.e.:

$$\mathbf{d}_{sm} = \left\{ \begin{array}{c} \cos\theta_{sm} \\ \sin\theta_{sm} \end{array} \right\}, \qquad \theta_{sm} = \frac{2\pi(m-1)}{M}, \qquad m = 1, 2, ..., M \qquad (8.79)$$

Further, since the high-frequency solution on the illuminated surface of a convex scatterer can be written in terms only of the incident wave, we can ensure that the incident wave direction $\mathbf{d}^I = (\cos\theta^I, \sin\theta^I)^{\mathrm{T}}$ is included in the basis by designing the plane wave directions to be:

$$\theta_{sm} = \theta^I + \frac{2\pi(m-1)}{M} \qquad (8.80)$$

In 3D simulations, it is not so straightforward to determine a set of uniformly spaced wave directions. In a few special cases, i.e., $M = 4, 6, 8, 12, 20$, an analytical set of directions can be found by analogy with the vertex locations of the platonic solids. Other distributions are available using various models, e.g., discretized cubes, latitude/longitude models, the Kurihara grid, but these all place restrictions on particular values of M. For an arbitrary M, other methods are available using an analogy with an equilibrium state of a set of charged particles on a spherical surface.

Use of a constant M at each node in the model may not be appropriate for cases in which nodes are far from being uniformly spaced. In these cases, a uniform M will give rise to some regions of the boundary being over defined while others remain described using an insufficient number of degrees of freedom. This may result in deterioration in accuracy of solution. One remedy was proposed by Bériot et al. (2010), making use of a genetic algorithm to optimize the distribution of plane waves between the nodes.

8.6.7 Scattering from Sharp Corners

An elegant use of XBEM enrichment is the extension to the scattering of waves from sharp corners. Such geometries cause a singularity to arise in the solution, the asymptotic behavior locally to the corner being (for sound-hard Neumann conditions):

$$u \approx \sum_{n=1}^{N} J_{n\alpha}(kr_b) \cos n\alpha\theta_b \tag{8.81}$$

where (r_b, θ_b) are the polar coordinates centered on the corner in question and the exterior angle at the corner is π/α. $J_{n\alpha}(\cdot)$ is the Bessel function of the first kind. The Fourier-Bessel function series is shown here truncated to the first N terms.

There is considerable literature on the scattering of waves by polygonal obstacles, and we can note a highly efficient approach of Chandler-Wilde et al. (2012), which later evolved into the hybrid numerical/asymptotic approach Groth et al. (2015). These approaches offer an attractive combination of accuracy and computational efficiency for scattering by, and transmission through, convex polygons. In order to capture the singular behavior, enrichment schemes have been successfully explored in the ultraweak variational formulation (Luostari et al., 2012) and in a least squares method in combination with the method of fundamental solutions (Barnett and Betcke, 2010).

Gilvey et al. (2020) has recently included this local enrichment in an XBEM scheme to complement the plane wave enrichment for problems of scattering by polygonal obstacles, and demonstrated that it provides accurate solutions both for convex scatterers and problems involving multiple reflections. Noting that it is only the leading order term ($n = 1$) from the Fourier-Bessel function series (8.81) that contains the singularity, this term is adopted as an enrichment function over elements in the vicinity of each corner, adding only a very small number of degrees of freedom to the simulation. The terms $n > 1$ are adequately captured by the other basis functions (polynomial or plane wave).

Figure 8.14 shows the convergence of three schemes in simulating wave scattering by a unit square obstacle with wavenumber $k = 20$. The three schemes are as follows:

- BEM: classical BEM with piecewise quadratic basis;
- XBEM: as "BEM" (above) complemented by Bessel function corner enrichment;
- PUXBEM: full plane wave basis complemented by Bessel function corner enrichment.

The plots show the reduction in the relative error norm ε, defined as (8.78) but with the use of a reference solution in the absence of an exact solution to this problem, with increasing number of degrees of freedom. It is clear that the addition of the Bessel enrichment at the corners significantly improves the approximation over the classical BEM, but the best solutions are obtained from a combination of the plane wave basis and the Bessel enrichment.

8.7 Conditioning and its Control

Researchers investigating PU methods regularly report conditioning problems. An ill-conditioned system of equations is nearly singular, and as a result the solutions obtained can be very sensitive to errors in individual matrix terms. Such a system is also likely to cause an iterative solver to exhibit poor convergence and possible divergence.

One way of looking at ill-conditioning is that it is a state that arises when a basis function either:

- closely resembles another basis function, or
- is closely approximated by a linear combination of other basis functions.

Taking first the XBEM formulation for fracture mechanics, in which the displacement enrichments in (8.53) and (8.59) are used to give better approximation properties than piecewise polynomials. Both forms provide displacement as a function of $\sqrt{\rho}$, i.e., the square root of the distance from the crack tip. For elements remote from the crack tip, the square root function is rather flat, and is approximately constant over the element. Bearing in mind the PU property of the shape functions $N_s(\xi)$, it is clear that the enrichment basis function is close to a linear combination of other basis functions.

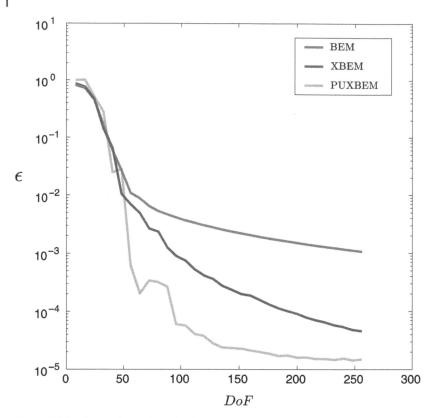

Figure 8.14 Comparison of results for scattering by a square obstacle.

In this case, the remedy is a simple one; we can restrict the enrichment to the elements in the close vicinity of the crack tip, where the $\sqrt{\rho}$ enrichment has a sufficiently large gradient. This is most effectively achieved by restricting the enrichment to the elements immediately adjacent to the crack tip.

The XBEM formulation for wave problems presents a different case. Here, we cannot apply geometric restrictions on the enrichment since we need a sufficient number of degrees of freedom per wavelength, τ, everywhere in the model, and not only at a few key locations. Moreover, if M is large, i.e., we are using a large number of plane waves to provide a PU approximation to the potential, then these waves will become closely spaced in an angular sense. In these cases, the basis functions formed by the product of shape functions with adjacent plane waves will start to resemble each other, resulting in ill-conditioning. But this is not the sole cause of ill-conditioning, which is an inherent feature of function approximation using an oscillatory basis set. As a result of the above arguments, it is found that the conditioning typically degrades with increasing levels of enrichment. It is noted that the conditioning is improved somewhat in 3D simulations; this benefit arises from the fact that the fundamental solution involves decay of the potential with r in three dimensions, and the decay is more rapid, going with r^{-1} instead of the Hankel function decay found in two dimensions, which goes with $\ln r$ for small r. This naturally makes the system matrix more diagonally dominant, which improves conditioning.

Numerical tests have shown that, although the conditioning of these systems can be poor, it can be controlled by tuning the basis set appropriately. In particular, the use of a sufficiently low τ is recommended. It has been seen that the use of $\tau < 2$ is not usually recommended, apart from some special cases, and that τ slightly larger than 2 is appropriate for high frequencies. For moderate frequencies, $\tau \approx 3$ is recommended. In addition to providing the solutions with a minimum number of degrees of freedom, it also provides for the best conditioning. In other words, the best conditioning is to be found by being careful not to overdefine the system by using more degrees of freedom than are required.

In spite of this guideline, use of a greater τ, giving more degrees of freedom, can provide for improved results as long as a solver is used that can cope with such ill-conditioned systems. Use of the truncated singular value decomposition (TSVD) can be particularly effective. This gives us the rather counter-intuitive result that the most accurate solutions are to be gained from the most ill-conditioned systems.

8.8   Conclusions

In this chapter, it has been shown that XFEM type of enrichment can be applied also in a boundary element context. Two types of problem have been considered, fracture mechanics and short wave scattering. In the former, enrichment is local to crack tips, and is based on the analytical form of the asymptotic displacement behavior as expressed in the Williams expansions. In the latter, the enrichment is applied over every element, and comprises a set of plane waves propagating in different directions that may be complemented by Fourier-Bessel functions to improve approximations at corners.

The results of both types of problem demonstrate clearly the benefits than can accrue from using this strategy of enrichment, providing for highly accurate solutions from coarse meshes.

References

Abboud, T., Nédélec, J., and Zhou, B. (1995). Improvement of the integral equation method for high frequency problems. In: *3rd International Conference on Mathematical Aspects of Wave Propagation Phenomena*, 178–187. Philadelphia: SIAM.

Alatawi, I. and Trevelyan, J. (2015). A direct evaluation of stress intensity factors using the extended dual boundary element method. *Engineering Analysis with Boundary Elements* 52: 56–63.

Barnett, A. and Betcke, T. (2010). An exponentially convergent nonpolynomial finite element method for time-harmonic scattering from polygons. *SIAM Journal on Scientific Computing* 32 (3): 1417–1441.

Benzley, S. (1974). Representation of singularities with isoparametric finite elements. *International Journal for Numerical Methods in Engineering* 8: 537–545.

Bériot, H., Perrey-Debain, E., Tahar, M.B., and Vayssade, C. (2010). Plane wave basis in Galerkin BEM for bidimensional wave scattering. *Engineering Analysis with Boundary Elements* 34: 130–143.

Brebbia, C. and Ciskowski, R. (1998). *Boundary Element Methods in Acoustics*. Southampton: CMP and Elsevier Applied Science.

Bruno, O., Geuzaine, C., Munro, J., and Reitich, F. (2004). Prescribed error tolerances within fixed computational times for scattering problems of arbitrarily high frequency: The convex case. *Philosophical Transactions Royal Society London A* 362 (1916): 629–645.

Burton, A. and Miller, G. (1971). The application of integral equation methods to the numerical solution of some exterior boundary value problems. *Philosophical Transactions Royal Society London A* 323: 201–210.

Cessenat, O. and Després, B. (1998). Application of an ultra weak variational formulation of elliptic PDES to the two dimensional Helmholtz problem. *SIAM Journal on Numerical Analysis* 35: 255–299.

Chandler-Wilde, S., Graham, I., Langdon, S., and Spence, E. (2012). Numerical-asymptotic boundary integral methods in high-frequency acoustic scattering. *Acta Numerica* 21: 89–305.

Chang, J. and Wu, D. (2007). Computation of mixed-mode stress intensity factors for curved cracks in anisotropic elastic solids. *Engineering Fracture Mechanics* 74: 1360–1372.

Civelek, M. and Erdogan, F. (1982). Crack problems for a rectangular sheet and an infinite strip. *International Journal of Fracture* 19: 139–159.

de la Bourdonnaye, A. (1994). A microlocal discretization method and its utilization for a scattering problem. *Comptes Rendus de l'Académie des Sciences* 38: 385–388.

Duflot, M. and Nguyen-Dang, H. (2004). A meshless method with enriched weight functions for fatigue crack growth. *International Journal for Numerical Methods in Engineering* 59: 1945–1961.

Farhat, C., Harari, I., and Franca, L. (2001). The discontinuous enrichment method. *Computer Methods in Applied Mechanics and Engineering* 190: 6455–6479.

Fleming, M., Chu, Y., Moran, B., and Belytschko, T. (1997). Enriched element-free Galerkin methods for crack tip fields. *International Journal for Numerical Methods in Engineering* 40: 1483–1504.

Gilvey, B., Trevelyan, J., and Hattori, G. (2020). Singular enrichment functions for Helmholtz scattering at corner locations using the boundary element method. *International Journal for Numerical Methods in Engineering* 121 (3): 519–533.

Groth, S., Hewett, D., and Langdon, S. (2015). Hybrid numerical/asymptotic approximation for high-frequency scattering by penetrable convex polygons. *IMA Journal of Applied Mathematics* 80 (2): 324–353.

Guiggiani, M., Krishnasamy, G., and Rudolphi, T. (1992). A general algorithm for the numerical solution of hypersingular boundary integral equations. *ASME Journal of Applied Mechanics* 59: 604–614.

Harris, P. and Chen, K. (2009). An efficient method for evaluating the integral of a class of highly oscillatory functions. *Journal of Computing and Applied Mathematics* 230: 433–442.

Hattori, G., Alatawi, I., and Trevelyan, J. (2017). An extended boundary element method formulation for the direct calculation of the stress intensity factors in fully anisotropic materials. *International Journal for Numerical Methods in Engineering* 109: 965–981.

Henshell, R. and Shaw, K. (1975). Crack-tip finite elements are unnecessary. *International Journal for Numerical Methods in Engineering* 9: 495–507.

Honnor, M. and Trevelyan, J. (2010). Numerical evaluation of two-dimensional partition of unity boundary integrals for Helmholtz problems. *Journal of Computational and Applied Mathematics* 234: 1556–1562.

Huttunen, T., Monk, P., and Kaipio, J. (2002). Computation aspects of the ultra weak variational formulation. *Journal of Computational Physics* 182: 27–46.

Ihlenburg, F. and Babuška, I. (1995). Dispersion analysis and error estimation of Galerkin finite element methods for Helmholtz equation. *International Journal for Numerical Methods in Engineering* 38: 3745–3774.

Ingraffea, A. and Manu, C. (1980). Stress intensity factor computation in three dimensions with quarter point elements. *International Journal for Numerical Methods in Engineering* 15: 1427–1445.

Iserles, A. and Nørsett, S. (2004). On quadrature methods for highly oscillatory integrals and their implementation. *BIT Numerical Mathematics* 44: 755–772.

Karihaloo, B. and Xiao, Q. (2001). Accurate determination of the coefficients of elastic crack tip asymptotic field by a hybrid crack element with *p*-adaptivity. *Engineering Fracture Mechanics* 68: 1609–1630.

Kim, T., Dominguez, V., Graham, I., and Smyshlyaev, V. (2009). Recent progress on hybrid numerical-asymptotic methods for high-frequency scattering problems. *7th UK Conference on Boundary Integral Methods,* University, Nottingham, 15–23.

Laghrouche, O., Bettess, P. and Astley, R. (2002). Modelling of short wave diffraction problems using approximating systems of plane waves. *International Journal for Numerical Methods in Engineering* 54: 1501–1533.

Langdon, S. and Chandler-Wilde, S. (2006). A wavenumber independent boundary element method for as acoustic scattering problem. *SIAM Journal of Numerical Analysis* 43: 2450–2477.

Leung, A. and Su, R. (1995). Mixed-mode two-dimensional crack problem by fractal two level finite element method. *Engineering Fracture Mechanics* 51 (6): 889–895.

Luostari, T., Huttunen, T., and Monk, P. (2012). The ultra weak variational formulation using Bessel basis functions. *Communications* in *Computational Physics* 11 (2): 400–414.

Melenk, J. and Babuška, I. (1996). The partition of unity finite element method. *Computer Methods in Applied Mechanics and Engineering* 139: 289–314.

Moës, N., Dolbow, J., and Belytschko, T. (1999). A finite element method for crack growth without remeshing. *International Journal for Numerical Methods in Engineering* 46: 131–150.

Morse, P. and Feshbach, H. (1981). *Methods of Theoretical Physics*. McGraw-Hill Education.

Peake, M., Trevelyan, J., and Coates, G. (2013a). Extended isogeometric boundary element method (XIBEM) for two-dimensional Helmholtz problems. *Computer Methods in Applied Mechanics and Engineering* 259: 93–102.

Peake, M., Trevelyan, J., and Coates, G. (2013b). Novel basis functions for the partition of unity boundary element method for Helmholtz problems. *International Journal for Numerical Methods in Engineering* 93 (9): 905–918.

Perrey-Debain, E., Trevelyan, J., and Bettess, P. (2002). New special wave boundary elements for short wave problems. *Communications in Numerical Methods in Engineering* 18 (4): 259–268.

Perrey-Debain, E., Trevelyan, J., and Bettess, P. (2003). Plane wave approximation in the direct collocation boundary element method for radiation and wave scattering: Numerical aspects and applications. *Journal of Sound and Vibration* 261 (5): 839–858.

Perrey-Debain, E., Trevelyan, J., and Bettess, P. (2004). Wave boundary elements: A theoretical overview presenting applications in scattering of short waves. *Engineering Analysis with Boundary Elements* 28: 131–141.

Portela, A., Aliabadi, M., and Rooke, D. (1992). The dual boundary element method – effective implementation for crack problems. *International Journal for Numerical Methods Engineering* 33: 1569–1287.

Rice, J. (1968). A path independent integral and approximate analysis of strain concentration by notches and cracks. *Journal of Applied Mechanics* 35: 379–386.

Riou, H., Ladevèze, P., and Sourcis, B. (2008). The multiscale VTCR approach applied to acoustics problems. *Journal of Computational Acoustics* 16 (4): 487–505.

Schenck, H. (1968). Improved integral formulation for acoustic radiation problems. *Journal of the Acoustical Society of America* 44: 41–58.

Simpson, R. and Trevelyan, J. (2011). Evaluation of J1 and J2 integrals for curved cracks using an enriched boundary element method. *Engineering Fracture Mechanics* 78: 623–637.

Simpson, R. and Trevelyan, J. (2011). A partition of unity enriched dual boundary element method for accurate computations in fracture mechanics. *Computer Methods in Applied Mechanics and Engineering* 200: 1–10.

Telles, J. (1987). A self-adaptive coordinate transformation for efficient numerical evaluation of general boundary element integrals. *International Journal for Numerical Methods in Engineering* 24: 959–973.

Tong, P. and Pian, T. (1973). On the convergence of the finite element method for problems with singularity. *International Journal for Numerical Methods in Engineering* 9: 313–321.

Trevelyan, J. and Coates, G. (2010). On adaptive definition of the plane wave basis for wave boundary elements in acoustic scattering: The 2d case. *Computer Modelling in Engineering & Sciences* 55 (2): 147–168.

Trevelyan, J. and Honnor, M. (2009). A numerical coordinate transformation for efficient evaluation of oscillatory integrals over wave boundary elements. *Journal of Integral Equations and Applications* 21: 447–468.

Willians, M. (1952). Stress singularities resulting from various boundary conditions in angular corners of plates in extension. *Journal of Applied Mechanics* 19: 526–528.

Zienkiewicz, O. (2000). Achievements and some unsolved problems of the finite element method. *International Journal for Numerical Methods in Engineering* 47: 9–28.

9

Combined Extended Finite Element and Level Set Method (XFE-LSM) for Free Boundary Problems

Ravindra Duddu

Department of Civil and Environmental Engineering, Vanderbilt University, USA

As described in Chapter 3, it is convenient to represent embedded interfaces using a scalar level set function ϕ; however, if the interface is moving with some velocity $\vec{v}$ one needs to solve a first-order nonlinear partial differential equation (PDE) given by the level set evolution to determine its new location. The interface velocity $\vec{v}$ can be obtained by numerically solving the governing elliptic or parabolic PDEs associated with the physics of the problem; by using the extended finite element method (XFEM) to solve these PDEs, one can eliminate the need for the finite element mesh to coincide with the embedded interface. The level set method proposed in Osher and Sethian (1988) and Sethian (1999b) enables us to capture the motion of interface in a consistent and efficient way through fast marching techniques. An important point is that the interface velocity $\vec{v}$ is only defined for points that lie on the interface, but to solve the level set equation, $\vec{v}$ needs to be defined at all points in the computational domain, which necessitates the use of velocity extensions (Adalsteinsson and Sethian, 1999). By coupling the XFEM with the level set method (Chopp and Sukumar, 2003; Duddu et al., 2008), the evolution of embedded interfaces due to complex physical phenomena can be simulated efficiently and accurately without requiring complex meshing or remeshing schemes.

In this chapter, a robust numerical scheme to couple the XFEM with the level set method for simulating interface evolution in two dimensions is described along with the key concepts of the level set method such as the derivation of the basic level set equation, the procedure to extend the interface velocity, the discretization of the evolution equation using upwind finite difference in space and the explicit forward Euler method in time, and the implementation of the fast marching technique. In order to demonstrate the efficacy of the scheme, an example problem related to environmental microbiology involving the evolution of a bacterial biofilm interface is chosen (Duddu et al., 2009), and the convergence and first-order accuracy of the method are numerically demonstrated. This combined methodology was applied to study precipitate evolution in binary nickel-aluminum alloys (Duddu et al., 2011; Zhao et al., 2013) and localized corrosion pit evolution in stainless steel alloys (Duddu, 2014); however, they are not discussed herein. Before proceeding any further, it is recommended that the reader get acquainted with the concepts of Chapter 5.

9.1 Motivation

Many problems in engineering and physical sciences require us to solve PDEs in a domain Ω, a segment of whose boundary is unknown in advance. This unknown interface segment Γ_{int} of the domain boundary Γ is a free

Partition of Unity Methods, First Edition. Stéphane P. A. Bordas, Alexander Menk, and Sundararajan Natarajan.

boundary and so these problems are referred to as free boundary problems (FBPs), which is now regarded as a separate branch of mathematics (Friedman, 1982). Perhaps, the most classic FBP is the melting of ice (or solidification of water) wherein there is a fluid-solid interface that is evolving depending on the temperature field in the domain. Other examples of FBPs encountered in engineering sciences include: growth of bacterial biofilms (microbiology); metallic melt solidification and dendritic growth, precipitate coarsening or Ostwald ripening in alloys, localized corrosion of alloys (material science); growth of solid tumor and of scar tissue during wound healing (biomedicine); fracture in solids (mechanics); melting of ice sheets, mountain building process (geosciences). These problems contain irregular domains, embedded interfaces, discontinuities (in material properties/phases, or in the unknown field or its gradients), singularities, and boundary layers; therefore, specialized numerical methods are required to solve them. Thus, the motivation for developing the combined extended finite element and level set method (XFE-LSM) stems from the need for robust, accurate, and efficient numerical methods to investigate FBPs or moving interface problems encountered in engineering and physical sciences.

The main challenge with solving FBPs using the standard finite element method (FEM) is that the presence of an arbitrarily shaped moving interface requires us to generate the finite element mesh at each time step. This remeshing step can be computationally expensive and cumbersome in two dimensions and even more so in three dimensions. Alternatively, a moving mesh method may be employed, which relies on updating the mesh so that it conforms to the interface (Beckett et al., 2001; Huang and Russell, 1999; Li and Petzold, 1997); however, it is not computationally attractive when the free boundary evolves into complicated shapes in two or three dimensions or when multiple free boundaries coalesce (merge) and/or collapse (vanish). A number of finite difference based approaches for solving elliptic equations in conjunction with moving interfaces exist in the literature, including the immersed interface method (LeVeque and Li, 1994), the immersed boundary method (Peskin, 1977), and the ghost fluid method (Fedkiw et al., 1999). While these methods work well for fluid mechanics, for problems in solid mechanics the quality of the finite difference solution is not sufficient to accurately compute deformations and stresses. In contrast, the combined XFE-LSM, described herein, is attractive due to its accuracy and efficiency for solving FPBs involving solid mechanical formulations and evolving interfaces.

9.2 The Level Set Method

In this section, the finite difference implementation of the level set method is briefly reviewed; for a more detailed account the interested reader is referred to Sethian (1999b) and Chopp (2012).

9.2.1 The Level Set Representation of the Embedded Interface

The level set and fast marching methods were first introduced in Osher and Sethian (1988), Sethian (1996), Adalsteinsson and Sethian (1999), and Sethian (1999a). The level set method is based on representing an interface as the zero level set of some higher dimensional function, ϕ. Thus, the interface location at any time t is given by the zero contour of the scalar level set function, $\phi(\vec{x}, t)$, that is:

$$\Gamma_{\text{int}} = \{\vec{x} \in \Omega \mid \phi(\vec{x}, t) = 0\}. \tag{9.1}$$

such that $\phi > 0$ on one side of the interface and $\phi < 0$ on the other side. For stability purposes, it is preferable to set $\phi(\vec{x}, 0)$ equal to the signed distance to the initial interface location and maintain $\phi(\vec{x}, t)$ as a signed distance function (Chopp, 2012). For simple interfaces the signed distance function can easily be constructed using analytical expressions. Although for many applications, this type of approach is quite useful, it is not suited for general use. For complex interfaces one needs to use numerical initialization techniques detailed in Sethian (1999b) to set ϕ equal to the signed distance function as:

$$\phi(\vec{x}, 0) = \pm \min_{\vec{x}_0 \in \Gamma_{\text{int}}} ||\vec{x} - \vec{x}_0|| \text{ in } \Omega. \tag{9.2}$$

Therefore, to numerically construct the initial level set function, $\phi(\vec{x}, 0)$, one needs to calculate the sign of ϕ ($\pm$) and the perpendicular (minimum) distance between each point $\vec{x}$ in the domain Ω and its projection $\vec{x}_0$ on the interface Γ_{int}.

The level set method is formulated in an Eulerian framework and is an "interface capturing method" because the moving interface is captured, rather than tracked, by the zero contour of the level set function at any time. The method has the following advantages:

1. The interface is implicitly defined by the level set function and no parametrization (explicit description) of the interface is needed.
2. Unlike the interface tracking methods (e.g., marker particle method; Du et al., 2006) there are no moving grid points. Therefore, fixed grid finite difference approximations can be used.
3. Complex geometries and topological changes such as interface merging, vanishing, etc., can be captured without any difficulty.
4. The method can be naturally extended to three dimensions (Sussman and Puckett, 2000).
5. The level set function could be used to calculate the interface curvature and the unit normal at any given point on the interface.

On the other hand, there are some disadvantages:

1. It is computationally more expensive compared to interface tracking methods as the interface is described by a function which is one dimension higher than the dimension of the interface. In other words, whereas $N \times N$ grid points are required to represent the interface using the level set method, only N points are required to represent the interface using the marker particle method. However, the computational expense involved with the level set method can be reduced by using narrow band methods but the implementation is more complicated (Adalsteinsson and Sethian, 1995, 1999; Chopp, 1993).
2. The level set method originally proposed in Osher and Sethian (1988) is only first-order accurate and is not very suitable for certain class of problems (e.g., impact problems) where higher order accuracy is needed to capture the interface motion. Recently, second-order accurate numerical implementations of the level set method were proposed; however, they are computationally complex and expensive (Min and Gibou, 2007; Shepel et al., 2005).

9.2.2 The Basic Level Set Evolution Equation

Now that a level set function, ϕ, is defined such that the interface, $\Gamma_{\text{int}} = \phi^{-1}(0)$, the next task is to simulate interface motion with a corresponding evolution equation for $\phi(\vec{x}, t)$. For the sake of the derivation of the evolution equation, let us consider a point denoted by its spatial location $\vec{x}_0(t)$ on the interface that is moving with velocity:

$$\vec{v}(\vec{x}_0, t) = \frac{d\vec{x}_0(t)}{dt}. \tag{9.3}$$

Since $\vec{x}_0(t)$ is on the interface at all times, then it must be that $\phi(\vec{x}_0, t)$ is identically equal to zero for all t, that is, $\phi(\vec{x}_0, t) \equiv 0$. Therefore, the material time derivative of $\phi(\vec{x}_0, t)$ must be zero at all times, that is:

$$0 = \frac{d}{dt}(\phi(\vec{x}_0(t), t)) = \frac{\partial\phi}{\partial t} + \nabla\phi \cdot \frac{d\vec{x}_0(t)}{dt} = \frac{\partial\phi}{\partial t} + \nabla\phi \cdot F\vec{n}, \tag{9.4}$$

where $\nabla\phi$ denotes the spatial gradient of the level set function in an Eulerian description. Note that only the normal component of the velocity $F = \vec{v} \cdot \vec{n}$ changes the topology of the interface, while the tangential component does not, so only the normal speed F appears in the above evolution equation. The normal to the interface is also the normal to the zero level set and is given by:

$$\vec{n} = \frac{\nabla \phi}{\|\nabla \phi\|}. \tag{9.5}$$

Plugging the above equation into Equation (9.4) gives:

$$\frac{\partial \phi}{\partial t} + \nabla \phi \cdot F \frac{\nabla \phi}{\|\nabla \phi\|} = 0.$$

$$\frac{\partial \phi}{\partial t} + F\|\nabla \phi\| = 0, \tag{9.6}$$

where $F(\vec{x}, t)$ is the scalar normal speed function.

Equation (9.6) is the basic evolution equation for the level set method. Note that the normal speed function, F, is only defined on the interface, but Equation (9.6) assumes that F is known everywhere in the computational domain. Moreover, since the interface seldom passes directly through the grid points, F will not generally be evaluated anywhere on the interface. Therefore, the normal interface speed F needs to be extended from the interface to the whole domain and the numerical procedure for velocity (interface speed) extension is described in Section 9.2.3. Thus, the basic evolution equation reads:

$$\frac{\partial \phi}{\partial t} + F^{\text{ext}}\|\nabla \phi\| = 0. \tag{9.7}$$

where F^{ext} is the extended normal speed function. The above equation is subject to periodic or zero flux boundary conditions (BCs) on the external domain boundary Γ and is imposed using the ghost point method (Chopp, 2012). In order to evolve the interface, the above Equation (9.7) is solved using an appropriate upwind finite difference scheme and the fast marching method for finite difference based level set method (Barth and Sethian, 1998; Osher and Sethian, 1988; Sethian, 1999b; Sussman et al., 1999; Sussman and Fatemi, 1999; Sussman et al., 1997, 1994), which is described in Section 9.2.4. Alternatively, Equation (9.7) may be solved within the framework of the FEM (Tornberg and Engquist, 2000), SUPG FEM (Shepel et al., 2005), or the discontinuous Galerkin method (Yan and Osher, 2011). Although the finite element based level set schemes avoid the need for interpolation between level set finite difference grid and the finite element mesh in the combined XFE-LSM, these implementations are computationally more expensive than the finite difference based level set method that utilizes the fast marching technique.

9.2.3 Velocity Extension

As discussed in the previous section, the velocity extension procedure is necessary because the normal interface speed F is known only at points that lie on the interface; however, the level set method requires that F be defined at all nodes in the finite difference grid. For stability, it is preferable that $\phi(\vec{x}, t)$ is maintained as a signed distance function, consequently, its gradient must satisfy the eikonal equation, $\|\nabla \phi\| = 1$ (Evans, 1998). A suitable extended speed function F^{ext} that preserves the signed distance property of ϕ at all times can be derived as given below (Adalsteinsson and Sethian, 1999):

$$\|\nabla \phi\| = \nabla \phi \cdot \nabla \phi = 1,$$

$$\frac{\mathrm{d}}{\mathrm{d}t}(\nabla \phi \cdot \nabla \phi) = 0,$$

$$2\,\nabla\left(\frac{\mathrm{d}\phi}{\mathrm{d}t}\right) \cdot \nabla \phi = 0.$$

Using Equation (9.7) in the above equation gives:

$$2 \, \nabla(-F^{\text{ext}}\|\nabla\phi\|) \cdot \nabla\phi = 0,$$
$$\nabla F^{\text{ext}} \cdot \nabla\phi = 0. \tag{9.8}$$

The above equation also has a geometric interpretation as remarked in Chopp (2012). Noticing that $\nabla\phi$ is in the direction of the normal to the interface, Equation (9.8) states that $\nabla F^{\text{ext}} \cdot \vec{n} = 0$ implying that F^{ext} is constant along lines normal to the interface. In order to construct F^{ext} from the interface speed function F, Equation (9.8) is solved using the fast marching method and its numerical implementation is given in detail in Sethian (1999b) and Chopp (2001). Accordingly, to determine $F^{\text{ext}}_{i,j}$ at a particular grid point $\vec{x}_{i,j} = (x_{i,j}, y_{i,j})$, Equation (9.8) is discretized using the finite difference approximation as:

$$\left(\frac{F^{\text{ext}}_{i+1,j} - F^{\text{ext}}_{i,j}}{\Delta x}\right)\left(\frac{\phi_{i+1,j} - \phi_{i,j}}{\Delta x}\right) + \left(\frac{F^{\text{ext}}_{i,j} - F^{\text{ext}}_{i,j-1}}{\Delta y}\right)\left(\frac{\phi_{i,j} - \phi_{i,j-1}}{\Delta y}\right) = 0. \tag{9.9}$$

where Δx and Δy are the grid spacings in the x and y coordinate directions, respectively, and i, j are the grid point indices. Note that F and ϕ are scalar functions and the comma is only used to separate the two indices denoting the grid point. For a more concise notation, the directional derivatives $D_{-x}, D_{+x}, D_{-y}, D_{+y}$ along x, y directions are defined as:

$$D_{-x}\phi_{i,j} = \frac{\phi_{i,j} - \phi_{i-1,j}}{\Delta x},$$

$$D_{+x}\phi_{i,j} = \frac{\phi_{i+1,j} - \phi_{i,j}}{\Delta x},$$

$$D_{-y}\phi_{i,j} = \frac{\phi_{i,j} - \phi_{i,j-1}}{\Delta y},$$

$$D_{+y}\phi_{i,j} = \frac{\phi_{i,j+1} - \phi_{i,j}}{\Delta y}. \tag{9.10}$$

Solving Equation (9.9) for $F^{\text{ext}}_{i,j}$ gives:

$$F^{\text{ext}}_{i,j} = \frac{F^{\text{ext}}_{i+1,j}\left(\frac{-D_{+x}\phi_{i,j}}{\Delta x}\right) + F^{\text{ext}}_{i,j-1}\left(\frac{D_{-y}\phi_{i,j}}{\Delta y}\right)}{\left(\frac{-D_{+x}\phi_{i,j}}{\Delta x}\right) + \left(\frac{D_{-y}\phi_{i,j}}{\Delta y}\right)}. \tag{9.11}$$

Note that the above defined notation is used only in this chapter for conciseness and is not to be confused with the indicial notation for tensors or the comma notation for spatial differentiation used in Chapter 5.

It can be shown that the above expression in Equation (9.11) is a directional weighted average, as explained in Chopp (2012). To illustrate this, let us consider the interface, $\phi^{-1}(0)$, to be evolving outward as shown in Figure 9.1. The level set function is defined such that $\phi < 0$ on the inward side of the interface; consequently, ϕ is necessarily a decreasing function of time because as the interface grows outward ϕ values become more negative at the grid points. In that sense, the value of ϕ at a point is also a measure of the time t at which the interface crosses that point. Thus, due to the outward propagation of the interface, the information is traveling from low values of ϕ to high values of ϕ; therefore, $\phi_{i,j} > \phi_{i+1,j}$, $\phi_{i,j-1}$ (see Figure 9.1); so, we have $D_{+x}\phi_{i,j} < 0$ and $D_{-y}\phi_{i,j} > 0$, and consequently, the terms in the parentheses in Equation (9.11) are all positive, resulting in a weighted average. Now,

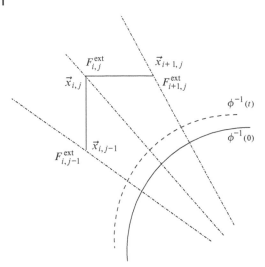

Figure 9.1 Illustration of weighted average for *F*.

F^{ext} could be defined everywhere in the domain as given below:

$$
\begin{cases}
\nabla F^{\text{ext}} \cdot \nabla \phi &= 0 \quad \text{in } \Omega^h, \\
F^{\text{ext}} &= F \quad \text{on } \Gamma_{\text{int}},
\end{cases}
\tag{9.12}
$$

where Ω^h denotes the discretized computational domain. The above equation essentially states that at the interface Γ_{int} the extended speed function F^{ext} is equal to the normal interface speed $F = \vec{v} \cdot \vec{n}$, which is obtained by post-processing the numerical solution to the underlying physical problem, as discussed in Section 9.3.8 within the context of biofilm growth. Note that the above-described velocity extension procedure is derived such that it preserves the signed distance property of the level set function as it is evolved (Adalsteinsson and Sethian, 1999); therefore, reinitialization of the level set function is seldom required. However, if the interface is evolved for long times over many time steps, an intermediate reinitialization step may be necessary.

9.2.4 Level Set Function Update

The level set evolution Equation (9.7) is of the general form:

$$
\frac{\partial \phi}{\partial t} + H(\nabla \phi) = 0.
\tag{9.13}
$$

This is a type of Hamilton-Jacobi equation (Barth and Sethian, 1998; Osher and Sethian, 1988; Sethian, 1999b) where H is called the Hamiltonian. In the level set method, the theory from hyperbolic conservation laws is used to approximate the above equation by a numerical Hamiltonian (flux function) as:

$$
H(\nabla \phi) \approx g(D_{-x}\phi, D_{+x}\phi, D_{-y}\phi, D_{+y}\phi).
\tag{9.14}
$$

In the case of the level set evolution equation, $H(\nabla \phi) = F^{\text{ext}}\|\nabla \phi\|$. It is important to recall from the theory of numerical methods, "that the finite difference implementation must respect the directions of the characteristics"; so, at any given grid point, "the difference operator must look in the direction from which the information is coming" (Chopp, 2012). For example, consider the front shown in Figure 9.2, advancing with speed $F = 1$ and the point A on the interface $\phi = 0$. Noticing the direction from which the information is traveling at A, we can infer that the finite differences for approximating $H(\nabla \phi)$ should look to the right and down, as indicated in Figure 9.2.

Figure 9.2 Example flow with speed $F = 1$.

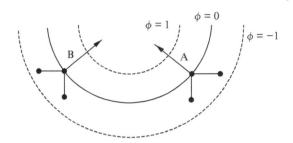

Therefore, the directional finite differences $D_{+x}\phi$ and $D_{-y}\phi$ should be used for approximating $\|\nabla\phi\|$. Using the same argument for the point B in Figure 9.2, we can infer that the finite differences should look left and down instead. If F^{ext} changes sign, that is, if the interface is propagating downward rather than upward in Figure 9.2, then our choice of directional derivatives should also change accordingly. Hence, $\text{sign}(F^{\text{ext}})$ needs to be considered when choosing the correct finite difference approximation. Thus, the general approximation for $H(\nabla\phi) = F^{\text{ext}}\|\nabla\phi\|$ is given by:

$$
\begin{aligned}
F_{i,j}^{\text{ext}}\|\nabla\phi\| = F_{i,j}^{\text{ext}}(&\max(\text{sign}(F_{i,j}^{\text{ext}})D_{-x}\phi_{i,j}, -\text{sign}(F_{i,j}^{\text{ext}})D_{+x}\phi_{i,j}, 0)^2 \\
&+ \max(\text{sign}(F_{i,j}^{\text{ext}})D_{-y}\phi_{i,j}, -\text{sign}(F_{i,j}^{\text{ext}})D_{+y}\phi_{i,j}, 0)^2),
\end{aligned}
\tag{9.15}
$$

where $F_{i,j}^{\text{ext}}$ is the value of normal speed at the grid point $\vec{x}_{i,j}$ on the finite difference grid. Finally, the level set function is updated using the explicit forward Euler method in time as given by:

$$
{}^{n+1}\phi_{i,j} = {}^{n}\phi_{i,j} - F_{i,j}^{\text{ext}}\|\nabla\phi\|\Delta t,
\tag{9.16}
$$

where n denotes the time step, Δt is the time step size, and $F_{i,j}^{\text{ext}}\|\nabla\phi\|$ is computed at the nth time step according to Equation (9.15). For the stability of the explicit update scheme, Δt should be less than the maximum time step allowed $\Delta t_{\max}$, as given by the Courant-Friedrichs-Lewy (CFL) condition (Courant et al., 1928):

$$
\Delta t \leq \Delta t_{\max} = \frac{h_{\min}}{F_{\max}};
\tag{9.17}
$$

however, for accuracy we choose $\Delta t = 0.5\Delta t_{\max}$. For a regular square finite difference grid, the minimum grid size is simply the grid size, that is, $h_{\min} = \Delta x = \Delta y$ and the maximum interface speed $F_{\max} = \max\{F(\vec{x}_0, t)\}$ for all points $\vec{x}_0$ lie on the interface $\Gamma_{\text{int}}(t)$.

9.2.5 Coupling the Level Set Method with the XFEM

In this section, the overview of the strategy used for coupling the level set method with the XFEM is presented (see Figure 9.3), along with a concise pseudo-code for the level set method. Given the initial interface location, ${}^{n}\phi$, at time $t = t_n$, the XFEM is used to discretize the governing equations of the physical system on a finite element mesh using triangular or bilinear quadrilateral elements in two dimensions. The resulting system of equations is solved using a Newton-Raphson scheme (only for nonlinear systems) and a sparse linear solver (e.g., PARDISO (Schenk et al., 2001) or PETSc (Balay et al., 2012)). The obtained numerical solution is then post-processed to determine the normal interface speed function ${}^{n}F$ at time step t_n. The numerical implementation of the XFEM is illustrated in the next Section 9.3 for the biofilm growth problem. To evolve the interface in time, the level set method is used to solve the basic evolution Equation (9.7) and determine the new geometry of the interface given by ${}^{n+1}\phi$. The level set method is implemented on a two-dimensional rectangular finite difference grid using upwind finite differences and the fast marching scheme. Finally, ${}^{n+1}\phi$ is interpolated onto the finite element mesh using an element-wise bicubic interpolation scheme (Press et al., 1992). In the case when the finite element nodes and the finite difference

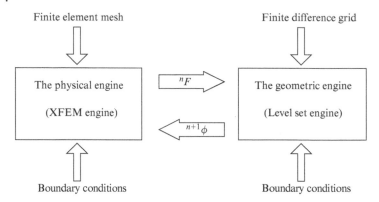

Figure 9.3 Coupling of the XFEM with the level set method.

grid points coincide, then the above interpolation step can be avoided. The coupling of the level set method with the XFEM to evolve the embedded interface is schematically shown in Figure 9.3.

Below we summarize the implementation of the level set method, but for a more detailed description the reader is referred to Chopp (2001). Note that in the notation used herein, the left superscript indicates the time step n or $n + 1$. Assuming that we have defined a suitable finite difference grid and given the current level set function $^n\phi$ and the interface speed ^nF from the XFEM engine (see Figure 9.3):

1. Segregate the finite difference grid points into three adjoint sets: Accepted, Tentative, and Distant, based on how close or distant they are to the interface. For example, the closest grid points belonging to the rectangular grid elements cut by the interface are added to the "Accepted" set, the points which possibly are to be accepted next are added to the "Tentative" set, and the grids points too far away are added to the "Distant" set;
2. Find the projections of the "Accepted" points on the interface using an element-wise bicubic polynomial interpolation to generate a local subgrid representation of $^n\phi$. Determine the perpendicular distance of the "Accepted" points, which gives the exact signed distance at these points;
3. Determine the normal interface speed ^nF at the interface projections of the "Accepted" points using the XFEM solution (see Section 9.3.8). Now set these values equal to extended speed $^nF^{\text{ext}}$ at all grid points belonging to the initial "Accepted" set;
4. Then use the fast marching method to construct the extended speed function $^nF^{\text{ext}}$ at all the remaining "Tentative" and "Distant" points (one-by-one) using the Equation (9.11);
5. Update the level set function in time to obtain $^{n+1}\phi$ according to Equations (9.15) and (9.16) using a stable time step calculated according to the CFL condition (9.17);
6. Apply BCs (zero flux or periodic) to the new level set function $^{n+1}\phi$ on the domain boundary Γ using the ghost point method, as discussed in Chopp (2012).

9.3 Biofilm Evolution

In this section, we shall demonstrate the viability of the combined method for FBPs by considering an example problem involving the evolution of bacterial biofilms. For a more detailed account on the evolution of the water-biofilm (fluid-solid) interface during biofilm growth the interested reader is referred to Duddu et al. (2008) and Duddu et al. (2009).

9.3.1 Biofilms

Biofilms are complex aggregations of microorganisms usually found attached to solid surfaces in fluid media. These three-dimensional heterogeneous biofilm structures provide bacteria with a favorable environment to grow. One of the important reasons why bacteria tend to aggregate in the form of biofilms is that the extracellular polymer matrix holding the bacterial cells together protects them from the environment (Rittmann et al., 2000). Biofilms have both positive and negative impacts, for example, beneficial biofilms are used for water purification in activated sludge reactors (Bryers, 2000; Rittmann and McCarty, 2001), and detrimental biofilms can make people sick by containing and spreading pathogens (Bryers, 2000) or foul heat exchangers (Bott, 1995). The shape, size, and composition of the biofilm can have a profound influence on its behavior and potential treatments, including susceptibility to sloughing (i.e., the tearing of biofilm due to surface stresses), and antibiotic resistance (Stewart, 1997).

Understanding biofilm growth behaviors and how to favor or control this growth could help biologists and engineers learn how to better avoid the harmful effects associated with biofilms or use their beneficial features. Modeling and simulation of biofilm growth in fluid media, in concert with laboratory experiments, has arisen as a means of producing quantitative tools for scientists to better understand biofilms, their geometric features, their growth, means of protection, and interaction with their environment. However, we need accurate multi-dimensional models and specialized numerical methods in order to make reliable predictions of biofilm growth and mechanical behavior. A detailed literature on the existing models and methods used to investigate biofilm growth can be found in Duddu et al. (2008) and Duddu et al. (2009).

9.3.2 Biofilm Modeling

For many years, biofilm structures have been represented by one-dimensional diffusive growth models (Rittmann and Manem, 1992; Wanner and Gujer, 1986). Usually, in the biofilm modeling literature, the solid surface to which the biofilm is attached is referred to as "substratum" and the growth rate-limiting nutrient that is diffused in the fluid medium (e.g., oxygen or nitrogen) is referred to as "substrate"; so, we will use the same terminology herein. In a one-dimensional model, the biofilm is considered to be a flat uniform layer growing on the top of the substratum in stagnant fluid (aqueous) media and so the diffusion of substrate molecules is only in the direction perpendicular to the substratum. However, experimental investigations show that biofilms do not grow in a one-dimensional pattern, but rather grow in the form of microbial clusters taking many different shapes. In most cases, biofilms grow in flowing fluid media as mushroom-shaped clusters, so it is important to consider the three-dimensional flow field around the biofilm. Moreover, biofilms are highly porous structures with connected channels in which there can be advective transport of substrate (Costerton et al., 1994) and also spatially heterogeneous in terms of composition and physical properties (Bishop and Rittmann, 1995). Therefore, two-dimensional and three-dimensional biofilm models allowing for spatial variability of biofilms considering the advection of substrate due to fluid flow and the detachment of biofilms due to interfacial shear stress are required to better understand biofilm structures.

The first comprehensive quantitative model structure for biofilm growth in two-dimensions was proposed by Picioreanu et al. (1999) and Picioreanu et al. (2000), which considered advection-diffusion of substrate in fluid media and diffusive growth of the biofilm. In this approach, the Lattice-Boltzmann method was used to compute the fluid flow velocity and substrate transport, while the cellular automaton method was used to simulate biofilm growth. Later, Picioreanu et al. (2001) incorporated erosive detachment based on the mechanical stress in the biofilm evaluated using the FEM. Although the discrete cellular automaton approach produced interesting results, the drawback is that it is somewhat *ad hoc* and does not give the exact location of the biofilm-fluid interface. On the other hand, a continuum model solved using the combined method does not rely on *ad hoc* rules to simulate growth and spreading; therefore, it is more amenable to mathematical analysis. Because biofilm growth is usually a slow process (on the order of a few millimeters per day) and the flow velocities are usually in the laminar regime (Reynolds number, $Re < 10$), a steady state Navier-Stokes solution obtained using the XFEM will be computationally more efficient than the Lattice-Boltzmann method. To simulate fluid flow and interface evolution consistently, we developed a two-dimensional continuum model for biofilm growth (Duddu et al., 2009) based on the initial work of Dockery and Klapper (2001), Chopp et al. (2002), Chopp et al. (2003), and Chopp (2007).

9.3.3 Two-Dimensional Model

We model the biofilm as a continuous medium made up of two components: active and inactive biomass. Active biomass consists of live bacterial cells, while the inactive biomass consists of extracellular polymeric substances (EPS) and dead cells. We employ a mixture theory approach, wherein we define the volume fraction of active biomass $f(\vec{x}, t)$ as a variable at a spatial location $\vec{x} = (x, y)$ and the volume fraction of inactive material is simply given by $1 - f(\vec{x}, t)$ for a binary mixture. The essential nutrient or substrate (oxygen) required for biofilm growth is dissolved in the fluid media, which is usually water mixed with minerals. The substrate diffuses from the fluid, through the biofilm-fluid interface, and into the biofilm itself where it is consumed only by the active biomass. Since substrate is consumed only inside the biofilm, the equations for substrate concentration are different inside and outside the biofilm. In this continuum model, there are five unknown (one vector and four scalar) fields: the steady state fluid flow velocity field $\vec{u}$, hydrostatic pressure p, substrate concentration S, biomass advection velocity potential Φ, and biomass volume fraction f. In this section, we briefly describe the governing equations and explain the BCs. For a more detailed account of the derivation of the one-dimensional model equations based on mass conservation, model assumptions, and values of biofilm parameters, we refer the reader to Chopp et al. (2002) and Chopp et al. (2003).

Consider a rectangular domain, Ω, of dimensions $L{\times}H$. The domain, Ω, contains the biofilm in Ω_b, and the fluid in Ω_f, as shown in Figure 9.4. The biofilm domain, $\Omega_b(t)$, which changes in time, is bounded by the biofilm-fluid interface, $\Gamma_{int}(t)$, at all times. The interface Γ_{int} is implicitly represented by a level set function $\phi(\vec{x}, t)$, such that $\phi < 0$ in the biofilm domain Ω_b and $\phi > 0$ in the fluid domain Ω_f. The north, south, east, and west boundaries of Ω are denoted by Γ_N, Γ_S, Γ_E, and Γ_W, respectively. The governing equations for the fluid flow velocity field $\vec{u}(\vec{x}, t) = (u_x, u_y)$ at steady state are given by the non-dimensionalized incompressible Navier-Stokes equations:

$$\left.\begin{array}{rl} \vec{u} \cdot \nabla\vec{u} + \nabla p = & Re^{-1}\nabla^2\vec{u}, \\ \nabla \cdot \vec{u} = & 0, \end{array}\right\} \text{ in } \Omega_f, \tag{9.18}$$

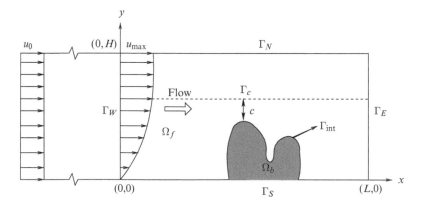

Figure 9.4 The domain and domain boundaries for the 2D biofilm growth model.

where p is the hydrostatic pressure and Re is the Reynolds number. The above equations are subject to BCs:

$$
\begin{aligned}
u_x &= u_{\max}\left(2y/H - y^2/H^2\right) && \text{on } \Gamma_W, \\
u_y &= 0 && \text{on } \Gamma_N \bigcup \Gamma_S \bigcup \Gamma_W, \\
u_x &= 0 && \text{on } \Gamma_S, \\
\frac{\partial u_x}{\partial x} &= \frac{\partial u_y}{\partial x} = 0 && \text{on } \Gamma_E,
\end{aligned}
\tag{9.19}
$$

and interface conditions:

$$
u_x = 0, \; u_y = 0 \; \text{ on } \Gamma_{\text{int}}.
\tag{9.20}
$$

The BCs state that u_x varies quadratically along the vertical y direction with a maximum flow speed $u_{\max}$ at the top, $u_y = 0$ at the inlet (west) boundary (i.e., flow is strictly in the x direction), there is no flow across the north (top) and south (bottom) boundaries, and that the flow is at a steady state sufficiently downstream, so it does not vary in the horizontal (x) direction. The interface conditions state that flow velocity is zero at the biofilm surface, thus, describing a no slip condition.

Let $S(\vec{x}, t)$ denote the concentration of the growth limiting substrate (oxygen) at a given point in the biofilm, and D_S^f and D_S^b denote the diffusion coefficients for substrate in the fluid and the biofilm, respectively. Here we assume that $D_S^b \approx 0.8 D_S^f$ (Williamson and McCarty, 1976). The governing equations for substrate transport at steady state are given by:

$$
\begin{cases}
D_S^b \nabla^2 S - f\mu_S &= 0 \quad \text{in } \Omega_b, \\
D_S^f \nabla^2 S - \vec{u} \cdot \nabla S &= 0 \quad \text{in } \Omega_f,
\end{cases}
\tag{9.21}
$$

where f is the active biomass volume fraction and μ_S is the specific substrate consumption rate given by:

$$
\mu_S(S) = \rho_x(\hat{q}_o + \gamma f_D b)\frac{S}{K_o + S},
\tag{9.22}
$$

and $\rho_x, \hat{q}_o, \gamma, f_D, b,$ and K_o are the biofilm model parameters defined in Chopp et al. (2003) and Duddu et al. (2008). Thus, in the biofilm domain, we solve a reaction-diffusion problem and in the fluid domain we solve an advection-diffusion problem. Equations (9.21) are subject to BCs:

$$
\begin{aligned}
S &= S_{\max} && \text{on } \Gamma_W \text{ or } \Gamma_C, \\
\frac{\partial S}{\partial y} &= 0 && \text{on } \Gamma_N \bigcup \Gamma_S, \\
\frac{\partial S}{\partial x} &= 0 && \text{on } \Gamma_E,
\end{aligned}
\tag{9.23}
$$

and interface jump conditions (IJCs):

$$
\begin{aligned}
[\![S]\!] &= (S^b - S^f) = 0 && \text{on } \Gamma_{\text{int}}, \\
[\![D_S\nabla S]\!] \cdot \vec{n} &= \left(D_S^b \nabla S^b - D_S^f \nabla S^f\right) \cdot \vec{n} = 0 && \text{on } \Gamma_{\text{int}},
\end{aligned}
\tag{9.24}
$$

where S^b and S^f are the limiting values of the substrate concentration at the interface when it is approached from the biofilm and fluid sides of the domain, respectively. The BCs state that substrate concentration S is set to $S_{\max}$ (solubility limit) at the inlet or at sufficiently far away distance from the biofilm that there is no loss of substrate through the north (top) and south (bottom) boundaries, and that S is at a steady state sufficiently downstream, so it does not vary in the horizontal (x) direction. The IJCs state that the substrate concentration S and the associated flux $D_S\nabla S \cdot \vec{n}$ are continuous across the interface, which arises from substrate mass balance. Because the diffusion coefficient D_S is discontinuous across the interface, the directional derivative of S in outward normal $\vec{n}$ direction will be discontinuous across the interface.

Next, we define a new quantity $\vec{U}(\vec{x}, t)$ denoting the advection velocity of biomass at a given point $\vec{x}$ in the biofilm at time t. Basically, this is the velocity with which the biomass moves due to cell reproduction and EPS production, occurring from the substratum up (Dockery and Keener, 2001). Thus, the velocity $\vec{U}$ is associated with the volumetric expansion or contraction of the biofilm and so it is reasonable to assume that the biofilm growth is irrotational, i.e., $\nabla \times \vec{U} = \vec{0}$. This implies that the velocity field can be derived from a potential function Φ defined in Ω such that $\vec{U} = \nabla \Phi$. The mass balance for total biomass can be written in terms of the growth velocity potential, Φ, as:

$$\begin{cases} \nabla^2 \Phi - f \mu_\Phi = 0 & \text{in } \Omega_b, \\ \Phi = 0 & \text{in } \Omega_f. \end{cases} \tag{9.25}$$

where $\mu_\Phi = (\mu_x + \mu_w)$, μ_x is the net rate of active biomass production, μ_w is the net rate of EPS production given by:

$$\begin{aligned} \mu_x &= (Y_{x/o} \hat{q}_o - b) \frac{S}{K_o + S}, \\ \mu_w &= \frac{\rho_x}{\rho_w}((1 - f_D)b + Y_{w/o}\hat{q}_o) \frac{S}{K_o + S}, \end{aligned} \tag{9.26}$$

and ρ_w, $Y_{x/o}$, and $Y_{w/o}$ are the biofilm model parameters defined in Chopp et al. (2003) and Duddu et al. (2008). Thus, in the domain, we solve an interior Poisson's equation, such that $\Phi < 0$ in the biofilm domain and $\Phi = 0$ in the fluid domain. Equations (9.25) are subject to the following interface and boundary conditions:

$$\begin{aligned} \Phi &= 0 & \text{on } \Gamma_{\text{int}}, \\ \frac{\partial \Phi}{\partial y} &= 0 & \Gamma_S \cap \Omega_b. \end{aligned} \tag{9.27}$$

The interface condition states that $\Phi = 0$ on the interface, to ensure its continuity across the interface; however, its directional derivative along the outward normal is discontinuous across the interface, which requires us to employ the discontinuous derivative enrichment function described in Section 9.3.6. The boundary condition states the vertical (y) component of the growth velocity $U_y = \dfrac{\partial \Phi}{\partial y} = 0$, as the biofilm is firmly attached to an impenetrable substratum at the bottom (south) boundary Γ_S.

In a continuum framework, the different components of multi-species biofilms are tracked by defining biomass volume fractions for each component (Alpkvist and Klapper, 2007; Chopp, 2007). In the case of a two-species biofilm with only one active species and one inactive components, it is sufficient to track only the active biomass volume fraction $f(\vec{x}, t)$. The mass balance equation for active biomass is given by:

$$\frac{\partial f}{\partial t} = f \mu_x - f^2(\mu_x + \mu_w) - \vec{U} \cdot \nabla f \quad \text{in } \Omega_b, \tag{9.28}$$

where $f \mu_x$ is the net production of active biomass and $f^2(\mu_x + \mu_w) + \vec{U} \cdot \nabla f$ is the net active biomass advected out of the control volume (Chopp et al., 2003). Thus, we solve an advection-reaction equation in the biofilm domain to calculate the active biomass volume fraction at every time step. This information is useful, for example, to evaluate the biofilm density at any point as $\rho(\vec{x}, t) = \rho_x f + \rho_w(1 - f)$. Equations (9.18)–(9.28) define the *two-dimensional biofilm growth problem*, which is a coupled system of nonlinear PDEs in six scalar variables u_x, u_y, p, S, Φ, and f. However, we note that the hydrostatic pressure p is essentially a Lagrange multiplier (LM) in this formulation and only the velocity $\vec{u}$ is required to establish the substrate concentration S in the domain.

9.3.4 Solution Strategy

We simulate the motion of the biofilm-fluid interface by coupling the XFEM and the level set method in a staggered fashion. The XFEM is used to solve the equations of fluid flow velocity $\vec{u}$, substrate diffusion S, and biofilm growth velocity potential Φ. The level set method is employed to update the geometry of the interface as given by the level set function, ϕ, and the active biomass fraction, f. The algorithm for solving the biofilm growth problem at time t is given as follows:

1. Compute the steady state fluid flow velocity field, $\vec{u}$, by solving the Navier-Stokes equations (9.18) for an incompressible fluid;
2. Calculate the steady state substrate concentration, S, and the velocity potential, Φ, from Equations (9.21) and Equation (9.25);
3. Determine the interface speed, F, on Γ_{int} using the domain-integral formulation described in Section 9.3.8;
4. Evaluate the extended interface speed function F^{ext} from Equation (9.12);
5. Update the active biomass volume fraction, f, using Equation (9.28);
6. Advance the biofilm interface by updating the level set function ϕ using the basic evolution Equation (9.7);
7. Extend the active biomass volume fraction, f, for the new points in the updated biofilm domain by considering a directional weighted average, similar to Equation (9.8).

The numerical implementation of the XFEM involves expressing the governing equations in the weak form, linearizing the nonlinear functions using Taylor's expansion and using the enriched approximation space to discretize the system of equations. The iterative procedure for solving the nonlinear biofilm model equations is briefly described in the following sections. The current two-dimensional implementation uses either: linear (three-node) triangles and a discontinuous derivative enrichment (Duddu et al., 2008) or bilinear (four-node) quadrilateral elements and a step enrichment function (Duddu et al., 2009). A brief description of these two enrichment functions is given in Section 9.3.6. In the elements that are not cut by the embedded interface, the domain integral terms obtained from the finite element approximation of the weak form are integrated using Gaussian quadrature. In the elements that are cut by the interface, integration is performed by subdividing the element into smaller triangles and/or quadrilaterals that conform with the interface approximated as a line segment (Vaughan et al., 2006). The Dirichlet boundary conditions are enforced on the moving boundary using either the penalty method or the LM method (see Belytschko et al. (2003) for details). The active biomass fraction is updated using a forward Euler scheme and extended using the fast marching algorithm using the finite difference method.

9.3.5 Variational Form

We employ the Bubnov-Galerkin procedure wherein the trial and test functions exist in the same finite element space. The trial functions for the unknown fields are to be chosen such that they satisfy all the Dirichlet boundary conditions and have the usual smoothness properties so that they are C^0 continuous in Ω. However, through the incorporation of different enrichment functions we can allow discontinuities in the value or gradient of the trial and test fields across the biofilm-fluid interface. The test functions are defined in the domain Ω as:

$$\begin{cases} \vec{u} \in \mathscr{U}^\mu, & \mathscr{U}^\mu = \{\vec{u} | \vec{u} \in C^0, \vec{u} = \vec{\bar{u}} \ \text{on} \ \Gamma_u^d\}, \\ p \in \mathscr{U}^p, & \mathscr{U}^p = \{p | p \in C^{-1}\}, \\ S \in \mathscr{U}^S, & \mathscr{U}^S = \{S | S \in C^0, S = \bar{S} \ \text{on} \ \Gamma_S^d\}, \\ \Phi \in \mathscr{U}^\Phi, & \mathscr{U}^\Phi = \{\Phi | \Phi \in C^0, \Phi = 0 \ \text{on} \ \Gamma_\Phi^d\}, \end{cases} \tag{9.29}$$

where Γ_u^d, Γ_S^d, Γ_Φ^d denote the Dirichlet boundaries for $\vec{u}$, S, and Φ, respectively; and $\vec{\bar{u}}$ and $\bar{S}$ are the imposed values of $\vec{u}$ and S. For the existence and uniqueness of a solution to the fluid flow equations, the function spaces $\mathscr{U}^\mu$ and $\mathscr{U}^p$ corresponding to the velocity $\vec{u}$ and the pressure p must satisfy the inf-sup or the LBB condition (Babuska, 1973; Brezzi, 1974; Ladyshenskaya, 1969). The respective test functions are defined as:

$$\begin{cases} \boldsymbol{w}^u \in \mathscr{U}_0^\mu, & \mathscr{U}_0^\mu = \{\boldsymbol{w}^u | \boldsymbol{w}^u \in C^0, \boldsymbol{w}^u = 0 \text{ on } \Gamma_u^d\}, \\ w^p \in \mathscr{U}_0^p, & \mathscr{U}_0^p = \{w^p | w^p \in C^{-1}\}, \\ w^S \in \mathscr{U}_0^S, & \mathscr{U}_0^S = \{w^S | w^S \in C^0, w^S = 0 \text{ on } \Gamma_S^d\}, \\ w^\Phi \in \mathscr{U}_0^\Phi, & \mathscr{U}_0^\Phi = \{w^\Phi | w^\Phi \in C^0, w^\Phi = 0 \text{ on } \Gamma_\Phi^d\}. \end{cases} \tag{9.30}$$

The weak or variational form of the two-dimensional biofilm growth problem in the domain Ω is: find $\vec{u} \in \mathscr{U}^\mu$, $S \in \mathscr{U}^S$, and $\Phi \in \mathscr{U}^\Phi$, such that for all test functions $\boldsymbol{w}^u \in \mathscr{U}_0^\mu$, $w^S \in \mathscr{U}_0^S$, and $w_\Phi \in \mathscr{U}_0^\Phi$, we have:

$$Re^{-1} \int_\Omega \nabla \boldsymbol{w}^u : \nabla \vec{u} \, \mathrm{d}\Omega + \int_\Omega \boldsymbol{w}^u \cdot (\vec{u} \cdot \nabla \vec{u}) \, \mathrm{d}\Omega + \int_\Omega \boldsymbol{w}^u \cdot \nabla p \, \mathrm{d}\Omega = 0, \tag{9.31}$$

$$\int_\Omega w^p (\nabla \cdot \vec{u}) \, \mathrm{d}\Omega = 0, \tag{9.32}$$

$$\int_\Omega D_S \nabla w^S \cdot \nabla S \, \mathrm{d}\Omega + \int_\Omega w^S (\vec{u} \cdot \nabla S) + \int_\Omega w^S f \mu_S \, \mathrm{d}\Omega = 0, \tag{9.33}$$

$$\int_\Omega \nabla w^\Phi \cdot \nabla \Phi \, \mathrm{d}\Omega + \int_\Omega w^\Phi f \mu_\Phi \, \mathrm{d}\Omega = 0. \tag{9.34}$$

In the above equations, the biomass volume fraction, f, is defined such that $f = 0$ in Ω_f and $0 \leq f \leq 1$ in Ω_b, and the fluid velocity is $\vec{u} = 0$ in Ω_b.

9.3.6 Enrichment Functions

In the XFEM, the physical domain, Ω, is divided into elements where an unknown field, generically named w in the following equation, is given by:

$$w(\vec{x}, t) = \sum_{I=1}^n N_I(\vec{x}) w_I(t) + \sum_{J=1}^{n_1(t)} \mathscr{S}_J(\vec{x}, t) a_J(t) + \sum_{K=1}^{n_2(t)} \mathscr{D}_K(\vec{x}, t) b_K(t), \tag{9.35}$$

where $N_I(\vec{x})$ are the standard finite element shape functions, w_I are nodal degrees of freedom (DOFs), n is the total number of nodes in the mesh, $\mathscr{S}$, $\mathscr{D}$ denotes the enrichment functions, a_J, b_K are the nodal enrichment DOFs (which are additional DOFs at the enriched nodes), and $n_1(t)$, $n_2(t)$ denote the total number of enriched nodes in the mesh at time step t. *One of the major advantages of the XFEM is the flexibility it offers in the choice of the enrichment functions used to model the problem at hand.* The choice of the enrichments depends on the behavior of the solution and multiple enrichment functions could be used. Below we describe two enrichment functions useful in simulating biofilm growth.

1. Discontinuous-derivative enrichment: A piecewise linear enrichment function $\mathscr{D}$, whose normal derivative is discontinuous through the interface, is used to allow a discontinuity in the gradient of the fields across the embedded interface. The function value is C^0 continuous. This enrichment function is defined by:

$$\mathscr{D}_J(\vec{x}, t) = N_J(\vec{x}) \Big(|\phi(\vec{x}, t)| - |\phi(\vec{x}_J, t)| \Big), \tag{9.36}$$

where $\phi(\vec{x}_J, t)$ are the values of the level set function at the enriched node J. The function, ϕ, is maintained as the signed distance from the interface Γ_{int}:

$$\phi(\vec{x}, 0) = \pm \min_{\vec{x}_0 \in \Gamma_{\text{int}}} \|\vec{x} - \vec{x}_0\|, \tag{9.37}$$

where the sign is positive/negative if the point $\vec{x}$ is outside/inside the region enclosed by the interface. The enrichment function, $\mathscr{D}$ in Equation (9.36), allows for a discontinuity in the gradients of the fields normal to Γ_{int}. The function, $\mathscr{D}$, is exactly reproduced in fully enriched elements and the enrichment vanishes at the edge of the support of the bisected nodes, providing a blending in partially enriched elements (Chessa and Belytschko, 2003). If only the absolute value of the level set function, $|\phi(\vec{x})|$, is used as enrichment, then in partially enriched elements there are spurious terms present as the enrichment does not vanish at the edge of the support of the bisected nodes and special treatment is required in blending elements (Chessa et al., 2003).

2. Step enrichment: A piecewise constant enrichment function, $\mathscr{S}$, can yield a continuous or discontinuous solution across the interface but requires LMs to apply the Dirichlet jump condition. This enrichment function is defined by:

$$\mathscr{S}_J(\vec{x}, t) = N_J(\vec{x})\Big(H(\phi(\vec{x}, t)) - H(-\phi(\vec{x}, t))\Big),$$ (9.38)

where $H(\phi)$ is the Heaviside step function given by:

$$H(\phi) = \begin{cases} 1 & \phi > 0, \\ 0 & \phi < 0. \end{cases}$$ (9.39)

This enrichment does not enforce the value of the jump in the solution or the normal derivative across the interface, so we need two separate interface conditions to maintain well-posedness of the boundary value problem. Here, we use LMs to enforce the jump condition across the interface in the solution, whereas the normal derivative condition across the interface is naturally satisfied (not enforced). A more detailed description of the implementation can be found in Vaughan et al. (2006).

9.3.7 Interface Conditions

In order for the biofilm growth problem to be well-posed, we need to enforce the Dirichlet conditions on the biofilm interface, which are the conditions of zero fluid flow velocity and zero growth velocity potential (i.e., $\vec{u} = 0$ and $\Phi = 0$ on Γ_{int}). In the case of biofilm growth under stagnant conditions, the Dirichlet condition for the substrate concentration $S = S_{\max}$ needs to be imposed on a moving interface Γ_c, as shown in Figure 9.4. In order to accurately and efficiently enforce such interface conditions on arbitrary embedded interfaces that are evolving with time, we use LMs. For example, consider the equation for the velocity potential, Φ, in the biofilm domain Ω_b given by:

$$\nabla^2 \Phi - f\mu_\Phi = 0,$$ (9.40)

subject to the constraint that $\Phi = 0$ on the biofilm interface Γ_{int}. The above equation can be rewritten to include the constraint on the interface as:

$$\nabla^2 \Phi - f\mu_\Phi + \lambda \Phi|_{\Gamma_{\text{int}}} = 0,$$ (9.41)

where λ is the LM function that enforces the constraint. To solve the above equation numerically and to avoid the stability issues associated with imposing stiff constraints, the following procedure is adopted (Vaughan et al., 2006):

1. Represent the 1D interface of a 2D biofilm using piecewise linear approximation by joining the interface line segments in each finite element of the 2D mesh containing the interface;
2. Place a 1D mesh along the piecewise linear interface with individual LM elements coinciding with the line segments;

3. Approximate the LM function, λ, using a standard finite element approximation as:

$$\lambda = \sum_{A=1}^{m(t)} \hat{N}^A \lambda^A,\tag{9.42}$$

where $\hat{N}^A$ are the 1D finite element shape functions, λ^A are LM nodal DOFs, and $m(t)$ is the total number of LM nodes at time step t;

4. Assemble the contributions of the LM matrix and the LM vector into the global tangent matrix and global residual vector, respectively.

The above strategy for generating the LM mesh has many similarities with that used in the mortared FEM (Kim et al., 2007). The LMs can be used to enforce jumps in the value or in the gradient of the unknown field on an arbitrary embedded interface. In the case of moving interface, one needs to generate the 1D LM mesh at each time step and additional DOFs need to be introduced. Because the LM mesh is one dimension less than the dimension of the problem domain, this procedure does not add to any significant computational burden. Alternatively, a penalty formulation can be used (Duddu et al., 2008), wherein the interface conditions are enforced by adding constraint terms multiplied by a large number called the penalty parameter. For accurate enforcement of the interfacial constraints, the penalty parameter is chosen to be orders of magnitude greater than the other parameters of the problem which can lead to a slow rate of convergence due to an ill-conditioned global tangent matrix or even to a lack of convergence of the numerical scheme. However, the penalty stiffness matrix and the penalty force term can be computed without having to introduce additional DOFs (unknowns) or to generate a 1D mesh at each time step, unlike when using the LM method.

A key challenge in using LMs for enforcing Dirichlet constraints on embedded interfaces in conjunction with the XFEM lies in the construction of a stable discrete LM space $\lambda \in \mathcal{L}$. A poor choice of $\mathcal{L}$ often leads to spurious oscillations in the multipliers referred to as boundary locking effect, which has been reported in several of the studies cited below. To resolve this issue, the vital-vertex algorithm was proposed in Moës et al. (2006), Béchet et al. (2009), and Nistor et al. (2009), which constructs a stable reduced LM space that passes the numerical inf-sup test; however, its implementation is relatively complex compared to our implementation. Although some oscillations in the LM values were noticed in the current implementation, the magnitude of these oscillations was relatively small; and by using a smoothing procedure based on the domain integral scheme (Ji and Dolbow, 2004) interface fluxes can be recovered accurately, as described in the next section. An alternative approach for imposing the interface Dirichlet condition involves the Nitsche method (Nitsche, 1971), which could be computationally more efficient and be readily extensible to three dimensions; moreover, it can lead to optimal convergence (Annavarapu et al., 2012a,b; Dolbow and Harari, 2009; Embar et al., 2010; Harari and Dolbow, 2010; Hautefeuille et al., 2012), and possibly, eliminate the need for the smoothing procedure.

9.3.8 Interface Speed Function

As noted earlier, the XFEM is used to compute the biofilm growth velocity potential Φ and the growth velocity $\vec{U}$ can be evaluated using the relation $\vec{U} = \nabla\Phi$. Note that $\vec{U}(\vec{x}, t)$ is only defined at points $\vec{x}$ lying in the biofilm domain $\Omega_b(t)$, and at the interface Γ_{int} we have the relation $\vec{v} = \vec{U}$ valid only on the biofilm side of the interface. The normal interface speed function F with which the biofilm interface evolves is equal to the normal component of $\vec{U}$ given by:

$$F = \vec{n} \cdot \vec{U}^b = \frac{\nabla\phi}{\|\nabla\phi\|} \cdot \vec{U}^b = \vec{n} \cdot \nabla\Phi^b = \frac{\nabla\phi}{\|\nabla\phi\|} \cdot \nabla\Phi^b,\tag{9.43}$$

where the superscript b denotes the limit of a field quantity as the interface is approached from the biofilm region, for example, $\nabla \Phi^b$ denotes the gradient when the interface is approached from biofilm side. However, for an accurate evaluation of F at the interface we employ a domain integral formulation to compute the interfacial jump quantity $[\![\nabla \Phi]\!] \cdot \vec{n}$ (Ji and Dolbow, 2004; Smith et al., 2007). Because the limiting value of the gradient $\nabla \Phi^f = 0$, if the interface is approached from the fluid side of the interface, we have:

$$[\![\nabla \Phi]\!] \cdot \vec{n} = \left(\nabla \Phi^f - \nabla \Phi^b\right) \cdot \vec{n} = -\nabla \Phi^b \cdot \vec{n}. \tag{9.44}$$

Thus, the magnitude of the jump quantity essentially gives the limit of the gradient as the interface is approached from the biofilm side. It is important to note that F, as defined in (9.43), is not a good choice of speed function away from the interface for reasons associated with numerical stability (Sethian, 1999b). For numerical stability, it is preferable to use a speed function that preserves the signed distance function while keeping the same speed value on the interface. This is effectively accomplished by using velocity extension as described in Section 9.2.3.

Let us now briefly review the procedure to approximate the jump quantity at a point $\vec{x}^d$ on the biofilm interface Γ_{int} using the domain integral scheme proposed in Ji and Dolbow (2004). We consider a section of the interface Γ_d containing the point $\vec{x}_d$, within compact support Ω_d of a scalar weight function w^d. Upon multiplying the velocity potential Equation (9.40) by w^d, integrating over Ω_d, and integrating by parts, we obtain:

$$\int_{\Omega_d} (\nabla w^\Phi \cdot \nabla \Phi + w^\Phi f \mu_\Phi)\, d\Omega = - \int_{\Gamma_d} w^d [\![\nabla \Phi]\!] \cdot \vec{n}\, d\Gamma_d. \tag{9.45}$$

where $w^\Phi \in \mathcal{U}_0^\Phi$ is the test function corresponding to the Φ introduced in Section 9.3.5. Next, we adopt the first-order approximation and assume that normal derivative jump is constant over Γ_d. Thus, we can bring the term $[\![\nabla \Phi]\!] \cdot \vec{n}$ out of the integral and obtain the following approximation:

$$F = \nabla \Phi^b \cdot \vec{n} = -[\![\nabla \Phi]\!] \cdot \vec{n} = \frac{1}{\int_{\Gamma_d} w^d\, d\Gamma_d} \int_{\Omega_d} (\nabla w^\Phi \cdot \nabla \Phi + w^\Phi f \mu_\Phi)\, d\Omega. \tag{9.46}$$

To determine the above jump quantity at the point $\vec{x}_d$ on the biofilm interface we first identify the nodes of the finite element containing it. We then evaluate the contribution of the finite element nodes to the jump in the normal derivative across the interface and interpolate the value at $\vec{x}_d$ using these nodal contributions using the standard finite element approximation (Ji and Dolbow, 2004). The above method is a nonlocal scheme for determining the velocity as opposed to the local scheme implemented in Duddu et al. (2008).

9.3.9 Accuracy and Convergence

Previously, in Duddu et al. (2008) we established the accuracy and efficiency of the combined method by comparing it numerically with the immersed interface method (LeVeque and Li, 1994). We demonstrated the first-order accuracy and convergence of the numerical scheme by comparing the numerical results for a simple two-dimensional biofilm growth problem under stagnant conditions obtained using different mesh sizes. Consider an isolated semi-circular biofilm with an initial radius of 0.1 mm in a domain of dimensions $L = 0.5$ mm $\times$ $H = 0.5$ mm. A constant substrate concentration of $S_{\text{max}} = 8.3 \times 10^{-6} \frac{\text{mgO}_2}{\text{mm}^3}$ is enforced on a horizontal moving boundary Γ_c with $c = 0.1$ mm measured from the topmost point of the biofilm, as indicated in Figure 9.4. For the purpose of the accuracy study, we identity 40 points equally spaced on the initial interface and compute the values of the unknowns S, Φ, and ϕ at these points using 50×50, 100×100, 200×200, and 400×400 finite element meshes consisting of linear triangles. The error between the values of the unknowns at these 40 points using 100×100 XFEM mesh and 50×50 XFEM mesh denoted by ϵ_{100-50} is calculated as given:

Table 9.1 Errors calculated on the interface of a semi-circular biofilm of radius 0.1 mm using different mesh resolutions.

Unknown	ϵ_{100-50}	$\epsilon_{200-100}$	$\epsilon_{400-200}$
ϕ	0.0039	0.0020	0.0010
S	2.5419×10^{-7}	1.2538×10^{-7}	6.6356×10^{-8}
Φ	1.3970×10^{-5}	8.5390×10^{-6}	4.2011×10^{-6}

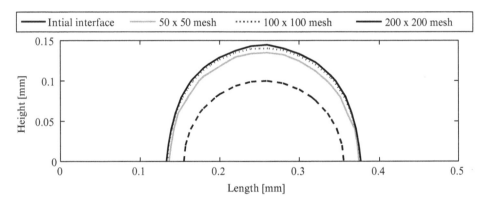

Figure 9.5 Convergence of the numerical scheme is established by comparing the final biofilm interface after five days of growth using different mesh resolutions.

$$\epsilon_{100-50} = \frac{1}{40}\left(\sum_{i=1}^{40}(X_i^{100} - X_i^{50})^2\right)^{1/2}. \tag{9.47}$$

The computed errors from different meshes are given in Table 9.1. We observe that value of $\epsilon_{200-100}$ is half that of ϵ_{100-50}, which indicates that the XFEM implementation is first-order accurate.

We next perform computations using the combined method to evolve the biofilm interface and compare the final interface, after five days of growth, obtained from three different meshes: Mesh(1) – 50 × 50 XFEM mesh coupled with 100 × 100 level set mesh ; Mesh(2) – 100 × 100 XFEM mesh coupled with 200 × 200 level set mesh ; Mesh(3) – 200 × 200 XFEM mesh coupled with 400 × 400 level set mesh. The final biofilm interface after five days of biofilm growth for the three different meshes is shown in Figure 9.5b. From this figure it is clear that final interfaces obtained using Mesh(3) and Mesh(2) agree very well and Mesh(1) result is only slightly off. For the sake of this convergence study, we used a constant time step Δt for the explicit time update of the biofilm interface such that it is always less than Δt_{max}, given by Equation (9.17) to ensure stability. For Mesh(1) we take the time step to be 0.1 day and a total of 50 time steps are required to simulate biofilm growth for five days. For Mesh(2) and Mesh(3) we take the time step to be 0.05 day and a total of 100 time steps are required to simulate the biofilm growth for five days. We observe that the numerical scheme is not sensitive to the time step used, provided it is less than maximum allowable time step Δt_{max}.

9.3.10 Numerical Results

In this section, we present two numerical examples to illustrate the ability of the combined method to simulate biofilm interface evolution. In the first example, we compare biofilm growth under stagnant (i.e., when fluid velocity $\vec{u} = 0$ in Ω) and flow conditions (i.e., when fluid velocity $\vec{u} \neq \vec{0}$ in Ω_f but $\vec{u} = \vec{0}$ in Ω_b). In the second example, we investigate biofilm growth at two different fluid flow rates but both in the laminar flow regime.

Example 1: We simulate the growth of multiple initially semi-circular biofilm colonies of radius 0.025 mm in a rectangular domain (flow cell) of dimensions $L = 2.0$ mm $\times H = 0.5$ mm, as described schematically in Figure 9.4. The biofilm colonies are assumed to be initially homogeneous containing 52% active biomass by volume, that is, $f(\vec{x}, 0) = 0.52$. In the stagnant case, S_{max} is prescribed on a horizontal moving boundary Γ_c such that the distance $c = 0.1$ mm measured from the topmost point of the biofilm. Note that choosing a smaller value of c will lead to a larger biofilm growth rate. In the case with fluid flow, the Reynolds number $Re = 0.83$ and the maximum substrate concentration $S_{max} = 8.3 \times 10^{-6} \frac{mgO_2}{mm^3}$ is imposed on the inlet boundary Γ_W. The results from the simulation are given in Figures 9.6 and 9.7.

It is evident from Figures 9.6a and 9.7a that the temporal evolution of biofilm morphology and growth rates are significantly different for the two cases. In the stagnant case, we see that all the biofilm colonies grow upward and the taller biofilms inhibit the growth of the neighboring biofilm colonies, as shown in Figure 9.6b. In the flow case, the shadowing effect of the upstream biofilm colonies on the downstream colonies is very apparent in Figure 9.7b. Also, the first biofilm colony (i.e., the colony closest to the inlet) grows upstream much faster and shadows the downstream biofilm colonies by depleting the nutrient concentration in the fluid. Although, at

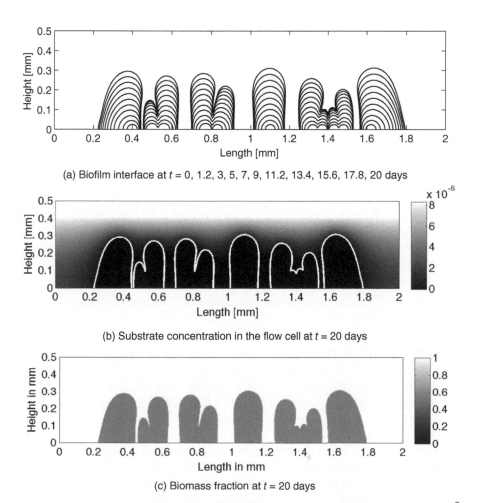

(a) Biofilm interface at $t = 0, 1.2, 3, 5, 7, 9, 11.2, 13.4, 15.6, 17.8, 20$ days

(b) Substrate concentration in the flow cell at $t = 20$ days

(c) Biomass fraction at $t = 20$ days

Figure 9.6 Simulated biofilm growth under stagnant condition with $S_{max} = 8.3 \times 10^{-6} \frac{mgO_2}{mm^3}$.

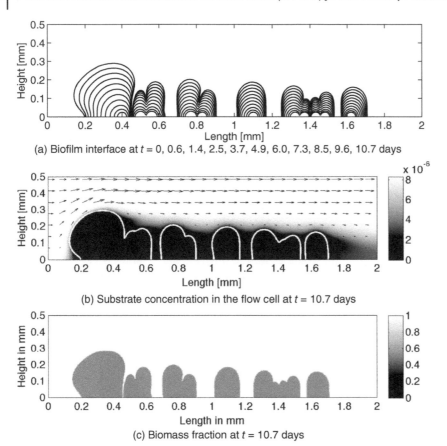

(a) Biofilm interface at t = 0, 0.6, 1.4, 2.5, 3.7, 4.9, 6.0, 7.3, 8.5, 9.6, 10.7 days

(b) Substrate concentration in the flow cell at t = 10.7 days

(c) Biomass fraction at t = 10.7 days

Figure 9.7 Simulated biofilm growth under flow condition with $Re = 0.83$, $S_{max} = 8.3 \times 10^{-6} \frac{\text{mgO}_2}{\text{mm}^3}$.

higher substrate concentrations this shadowing effect by the first colony is observed to be less significant. Note that in both cases, the active biomass fraction, f, at steady state is nearly spatially constant (≈ 0.52) throughout the domain, as evident from Figures 9.6c and 9.7c. The reason for this becomes clear if we examine the biomass update equation (9.28) at steady state. The terms $\frac{\partial f}{\partial t}$ and $\mathbf{U} \cdot \nabla f$ become insignificant as we approach steady state and $f \to \frac{\mu_x}{\mu_x + \mu_w} \simeq 0.52$. Thus, this kinetic model represents a situation wherein the live bacterial cells constituting the active biomass go into a stasis in substrate deficient zones, rather than dying; so the active biomass fraction at a point never decreases.

Example 2: Consider two laminar flow cases with different Reynold's numbers, $Re = 0.83$, 6.7 corresponding to average inlet flow velocities 0.83 ms^{-1} and 6.7 ms^{-1}, respectively. The values for the flow velocities are assumed from the numerical simulations conducted in Picioreanu et al. (2000). In both cases, we set $S_{max} = 8.3 \times 10^{-6} \frac{\text{mgO}_2}{\text{mm}^3}$ at the inlet Γ_W. The results from simulations are given in Figures 9.7 and 9.8. As expected, the substrate concentration near the interface is higher in the fast flow case, as evident from Figures 9.7b and 9.8b. In both cases, we observe merging of neighboring biofilm colonies at early stages of growth and preferential growth of the first biofilm colony; however, the shadowing effect of the first biofilm colony is slightly less at the higher flow rates because the substrate is replenished a bit faster from the bulk flow. This study shows that, while the fluid flow rate significantly affects biofilm growth rate, it does not effect the biofilm architecture much, as evident from Figures 9.7a and 9.8a.

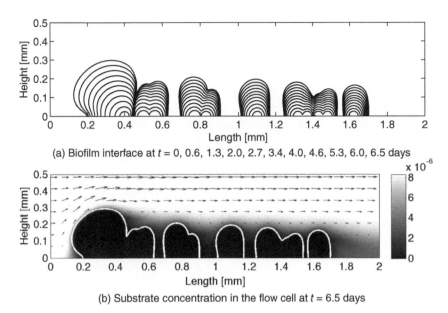

(a) Biofilm interface at t = 0, 0.6, 1.3, 2.0, 2.7, 3.4, 4.0, 4.6, 5.3, 6.0, 6.5 days

(b) Substrate concentration in the flow cell at t = 6.5 days

Figure 9.8 Simulated biofilm growth under flow conditions with $Re = 6.7$, $S_{max} = 8.3 \times 10^{-6} \frac{mgO_2}{mm^3}$.

9.4 Conclusion

The combination of the extended finite element (XFEM) and level set methods leads to a simple yet robust numerical method for simulating the evolution sharp material or phase interfaces. The XFEM allows for an accurate level set representation of the geometry of the free boundary (or interface) independent of the underlying finite element mesh. The enrichment of the standard finite element basis allows for the incorporation of jumps in both the unknown field and its normal derivative across the moving interface. The domain integral scheme yields a good approximation of the normal interface speed in conjunction with the Lagrange multiplier method. The advantage with using the level set method rather than an explicit tracking method (e.g., marker particle method; Du et al., 2006) is that it can handle arbitrary topological transitions such as interface merging and vanishing. This is important because free boundaries in general can evolve into complex shapes due to underlying physics of the problem, when phase volumes coalesce, transform, precipitate, or dissolve. Although the level set method is relatively expensive compared to the particle methods due to the solution of hyperbolic level set equation and the CFL condition for stability, it allows for a more robust handling of interface topology evolution. The combined XFE-LSM is stable, convergent, and first-order accurate, as demonstrated in Duddu et al. (2008, 2009).

Acknowledgment

Ravindra Duddu gratefully acknowledges the guidance provided by his Ph.D. advisors Prof. Brian Moran and Prof. David Chopp at Northwestern University, Evanston, Illinois. The contents of this chapter were presented in a PhD dissertation (Duddu, 2009).

References

Adalsteinsson, D. and Sethian, J. (1995). A fast level set method for propagating interfaces. *Journal of Computational Physics* 118 (2): 269–277.

Adalsteinsson, D. and Sethian, J. (1999). The fast construction of extension velocities in level set methods. *Journal of Computational Physics* 148: 2–22.

Alpkvist, E. and Klapper, I. (2007). A multidimensional multispecies continuum model for heterogeneous biofilm development. *Bulletin of Mathematical Biology* 69 (2): 765–789.

Annavarapu, C., Hautefeuille, M., and Dolbow, J. (2012a). A robust Nitsche's formulation for interface problems. *Computer Methods in Applied Mechanics and Engineering* 225–228: 44–54.

Annavarapu, C., Hautefeuille, M., and Dolbow, J. (2012b). Stable imposition of stiff constraints in explicit dynamics for embedded finite element methods. *International Journal for Numerical Methods in Engineering* 92 (2): 206–228.

Babuska, I. (1973). The finite element method with Lagrangian multipliers. *Numerische Mathematik* 20: 179–192.

Balay, S., Brown, J., Buschelman, K. et al. (2012). *PETSc Web page*. http://www.mcs.anl.gov/petsc.

Barth, T. J. and Sethian, J. A. (1998). Numerical schemes for the Hamilton-Jacobi and level set equations on triangulated domains. *Journal of Computational Physics* 145 (1): 1–40.

Béchet, E., Moës, N., and Wohlmuth, B. (2009). A stable Lagrange multiplier space for stiff interface conditions within the extended finite element method. *International Journal for Numerical Methods in Engineering* 78 (8): 931–954.

Beckett, G., Mackenzie, J., and Robertson, M. (2001). A moving mesh finite element method for the solution of two-dimensional Stefan problems. *Journal of Computational Physics* 168: 500–518.

Belytschko, T., Liu, W.K., and Moran, B. (2003). *Nonlinear Finite Elements for Continua and Structures*. John Wiley & Sons. LTD - ISBN: 0-471-98774-3.

Bishop, P.L. and Rittmann, B.E. (1995). Modeling heterogeneity in biofilms: Report of the discussion session. *Water Science and Technology* 32 (8): 263–265.

Bott, T.R. (1995). *Fouling of Heat Exchangers*. Amsterdam: Elsevier.

Brezzi, F. (1974). On the existence, uniqueness and approximation of saddle-point problems arising from Lagrangian multipliers. *ESAIM: Mathematical Modelling and Numerical Analysis – Modélisation Mathématique et Analyse Numérique* 8 (R2): 129–151.

Bryers, J. (2000). *Biofilms II Process Analysis and Applications, Chapter Biofilms: An Introduction*. New York: Wiley-Liss.

Chessa, J. and Belytschko, T. (2003). An extended finite element method for two-phase fluids. *Journal of Applied Mechanics - Transactions of the ASME* 70 (1): 10–17.

Chessa, J., Wang, H.W., and Belytschko, T. (2003). On the construction of blending elements for local partition of unity enriched finite elements. *International Journal for Numerical Methods in Engineering* 57 (7): 1015–1038.

Chopp, D. (2012). *Numerical methods for moving interfaces [lecture notes]*. Retrieved from http://people.esam.northwestern.edu/ chopp/course_notes/lecturenotes.html.

Chopp, D. and Sukumar, N. (2003). Fatigue crack propagation of multiple coplanar cracks with the coupled extended finite element/fast marching method. *International Journal of Engineering Science* 41 (8): 845–869.

Chopp, D.L. (1993). Computing minimal surfaces via level set curvature flow. *Journal of Computational Physics* 106: 77–91.

Chopp, D.L. (2001). Some improvements of the fast marching method. *SIAM Journal of Scientific Computing* 23 (1): 230–244.

Chopp, D.L. (2007). *Deformable Models: Biomedical and Clinical Applications, Chapter Simulating Bacterial Biofilms*. Springer.

Chopp, D.L., Kirisits, M.J., Moran, B., and Parsek, M.R. (2002). A mathematical model of quorum sensing in a growing biofilm. *Journal of Industrial Microbiology and Biotechnology* 29 (6): 339–346.

Chopp, D.L., Kirisits, M.J., Moran, B., and Parsek, M.R. (2003). The dependence of quorum sensing on the depth of a growing biofilm. *Bulletin of Mathematical Biology* 65 (6): 1053–1079.

Costerton, J.W., Lewandowski, Z., deBeer, D., et al. (1994). Biofilms, the customized microniche. *Journal of Bacteriology* 176 (8): 2137–2142.

Courant, R., Friedrichs, K., and Lewy, H. (1928). Über die partiellen Differenzengleichungen der mathematischen Physik. *Mathematische Annalen* 100: 32–74.

Dockery, J.D. and Keener, J.P. (2001). A mathematical model for quorum sensing in Pseudomonas aeruginosa. *Bulletin of Mathematical Biology* 63 (1): 95–116.

Dockery, J.D. and Klapper, J. (2001). Finger formation in biofilm layers. *SIAM Journal of Applied Mathematics* 62 (3): 853–869.

Dolbow, J. and Harari, I. (2009). An efficient finite element method for embedded interface problems. *International Journal for Numerical Methods in Engineering* 78 (2): 229–252.

Du, J., Fix, B., Glimm, J., et al. (2006). A simple package for front tracking. *Journal of Computational Physics* 213 (2): 613–628.

Duddu, R. (2014). Numerical modeling of corrosion pit propagation using the combined extended finite element and level set method. *Computational Mechanics* [online first] 1–15.

Duddu, R., Bordas, S., Chopp, D.L., and Moran, B. (2008). A combined extended finite element and level set method for biofilm growth. *International Journal for Numerical Methods in Engineering* 74 (5): 848–870.

Duddu, R., Chopp, D.L., and Moran, B. (2009). A two-dimensional continuum model of biofilm growth incorporating fluid flow and shear stress based detachment. *Biotechnology and Bioengineering* 103 (1): 92–104.

Duddu, R., Chopp, D.L., Voorhees, P.W., and Moran, B. (2011). Diffusional evolution of precipitates in elastic media using the extended finite element and the level set methods. *Journal of Computational Physics* 230 (4): 1249–1264.

Embar, A., Dolbow, J., and Harari, I. (2010). Imposing Dirichlet boundary conditions with Nitsche's method and spline-based finite elements. *International Journal for Numerical Methods in Engineering* 83 (7): 877–898.

Evans, L.C. (1998). *Partial Differential Equations*. Providence, RI: American Mathematical Society.

Fedkiw, R., Aslam, T., Merriman, B., and Osher, S. (1999). A non-oscillatory Eulerian approach to interfaces in multimaterial flows (the ghost fluid method). *Journal of Computational Physics* 152: 457–492.

Friedman, A. (1982). *Variational Principles and Free Boundary Problems*. New York, US: Wiley-Interscience Publication.

Harari, I. and Dolbow, J. (2010). Analysis of an efficient finite element method for embedded interface problems. *Computational Mechanics* 46 (1): 205–211.

Hautefeuille, M., Annavarapu, C., and Dolbow, J. (2012). Robust imposition of Dirichlet boundary conditions on embedded surfaces. *International Journal for Numerical Methods in Engineering* 90 (1): 40–64.

Huang, W. and Russell, R. (1999). Moving mesh strategy based upon a gradient flow equation for two dimensional problems. *SIAM Journal of Scientific Computing* 20: 998–1015.

Ji, H. and Dolbow, J.E. (2004). On strategies for enforcing interfacial constraints and evaluating jump conditions with the extended finite element method. *International Journal for Numerical Methods in Engineering* 61 (14): 2508–2535.

Kim, T.-Y., Dolbow, J., and Laursen, T. (2007). A mortared finite element method for frictional contact on arbitrary interfaces. *Computational Mechanics* 39 (3): 223–235.

Ladyshenskaya, O.A. (1969). *The Mathematical Theory of Viscous Incompressible Flow*. Gordon and Breach Science Publishers, Inc.

LeVeque, R. and Li, Z. (1994). The immersed interface method for elliptic equations with discontinuous coefficients and singular sources. *SIAM Journal of Numerical Analysis* 31: 1019–1044.

Li, S. and Petzold, L. (1997). Moving mesh method with upwinding schemes for time-dependent PDEs. *Journal of Computational Physics* 131: 368–377.

Min, C. and Gibou, F. (2007). A second order accurate level set method on non-graded adaptive cartesian grids. *Journal of Computational Physics* 225 (1): 300–321.

Moës, N., Béchet, E., and Tourbier, M. (2006). Imposing Dirichlet boundary conditions in the extended finite element method. *International Journal for Numerical Methods in Engineering* 67 (12): 1641–1669.

Nistor, I., Guiton, M.L.E., and Massin, P. (2009). An X-FEM approach for large sliding contact along discontinuities. *International Journal for Numerical Methods in Engineering* 78 (12): 1407–1435.

Nitsche, J.A. (1970–1971). Über ein Variationsprinzip zur lösung von Dirichlet-Problemen bei Verwendung von Teilräumen, die keinen Randbedingungen unterworfen sind. *Abhandlungen aus dem Mathematischen Seminar der Universität Hamburg* 36: 9–15.

Osher, S. and Sethian, J.A. (1988). Fronts propagating with curvature-dependent speed: Algorithms based on Hamilton-Jacobi formulations. *Journal of Computational Physics* 79 (1): 12–49.

Peskin, C.S. (1977). Numerical analysis of blood flow in the heart. *Journal of Computational Physics* 25: 220–252.

Picioreanu, C., Loosdrecht, M.v., and Heijnen, J. (1999). Discrete-differential modelling of biofilm structure. *Water Science and Technology* 39 (7): 115–122.

Picioreanu, C., Loosdrecht, M.v., and Heijnen, J. (2000). Effect of diffusive and convective substrate transport on biofilm structure formation: A two-dimensional modeling study. *Biotechnology and Bioengineering* 69 (5): 504–515.

Picioreanu, C., Loosdrecht, M.v., and Heijnen, J. (2001). Two-dimensional model of biofilm detachment caused by internal stress from liquid flow. *Biotechnology and Bioengineering* 72 (2): 205–218.

Press, W., Teukolsky, S., Vetterling, W., and Flannery, B. (1992). *Numerical Recipes in C (2nd ed.): The Art of Scientific Computing*. New York, NY, USA: Cambridge University Press.

Rittmann, B. and McCarty, P. (2001). *Environmental Biotechnology: Principles and Applications*. New York, NY, USA: Mc-Graw Hill.

Rittmann, B.E. and Manem, J.A. (1992). Development and experimental evaluation of a steady-state, multispecies biofilm model. *Biotechnology and Bioengineering* 39 (9): 914–922.

Rittmann, R., Harley Davidson, W., Doz, Y.L., e al. (2000 July). *Environmental Biotechnology*. McGraw-Hill Higher Education. ISBN: 0072345535.

Schenk, O., Gärtner, K., Fichtner, W., and Stricker, A. (2001). PARDISO: A high-performance serial and parallel sparse linear solver in semiconductor device simulation. *Future Generation Computer Systems* 18 (1): 69–78.

Sethian, J.A. (1996). A marching level set method for monotonically advancing fronts. *Proceedings of the National Academy of Sciences* 93 (4): 1591–1595.

Sethian, J.A. (1999a). Fast marching methods. *Siam Review* 41 (2): 199–235.

Sethian, J.A. (1999b). *Level Set Methods & Fast Marching Methods: Evolving Interfaces in Computational Geometry, Fluid Mechanics, Computer Vision, and Materials Science*. Cambridge, UK: Cambridge University Press.

Shepel, S.V., Paolucci, S., and Smith, B.L. (2005). Implementation of a level set interface tracking method in the FIDAP and CFX-4 codes. *Journal of Fluids Engineering* 127 (4): 674–686.

Smith, B.G., Vaughan, B.L., and Chopp, D.L. (2007). The extended finite element method for boundary layer problems in biofilm growth. *Communications in Applied Mathematics and Computational Science* 2 (1): 35–56.

Stewart, P.S. (1997). Biofilm accumulation model that predicts antibiotic resistance of Pseudomonas aeruginosa biofilms. *Antimicrobial Agents & Chemotherapy* 38 (5): 1052–1058.

Sussman, M., Almgren, A., Bell, J.B., et al. (1999). An adaptive level set approach for incompressible fluid flow. *Computational Physics* 148: 81–124.

Sussman, M. and Fatemi, E. (1999). An efficient interface preserving level set redistancing algorithm and its application to interfacial incompressible fluid flow. *Journal of Scientific Computing* 20 (4): 1165–1191.

Sussman, M., Fatemi, E., Smereka, P., and Osher, S. (1997). An improved level set method for incompressible two-phase flows. *Computers and Fluids* 27 (5): 663–680.

Sussman, M. and Puckett, E.G. (2000). A coupled level set and volume-of-fluid method for computing 3D and axisymmetric incompressible two-phase flows. *Journal of Computational Physics* 162 (2): 301–337.

Sussman, M., Smereka, P., and Osher, S. (1994). A level set approach for computing solutions to incompressible two-phase flows. *Journal of Computational Physics* 114: 146–159.

Tornberg, A.-K. and Engquist, B. (2000). A finite element based level-set method for multiphase flow applications. *Computing and Visualization in Science* 3: 93–101.

Vaughan, B.L., Smith, B.G., and Chopp, D.L. (2006). A comparison of the extended finite element method with the immersed interface method for elliptic equations with discontinuous coefficients and singular sources. *Communications in Applied Mathematics and Computational Science* 1 (1): 207–228.

Wanner, O. and Gujer, W. (1986). A multispecies biofilm model. *Biotechnology & Bioengineering* 28 (3): 314–328.

Williamson, K. and McCarty, P. (1976). Verification studies of biofilm model for bacterial substrate utilization. *Journal of Water Pollution Control Federation* 48 (2): 281–296.

Yan, J. and Osher, S. (2011). A local discontinuous Galerkin method for directly solving Hamilton-Jacobi equations. *Journal of Computational Physics* 230 (1): 232–244.

Zhao, X., Duddu, R., Bordas, S., and Qu, J. (2013). Effects of elastic strain energy and interfacial stress on the equilibrium morphology of misfit particles in heterogeneous solids. *Journal of the Mechanics and Physics of Solids* 61 (6): 1433–1445.

10

XFEM for 3D Fracture Simulation

Indra Vir Singh

Department of Mechanical and Industrial Engineering, Indian Institute of Technology Roorkee, Roorkee, Uttarakhand, India

10.1 Introduction

Two-dimensional (2D) analysis can be useful only for initial design phase and preliminary prototypes. However, 2D realization of every component is not achievable and it can lead to inaccurate designs. The failure prediction of various components features like pocket slots, spherical cavities, and many more, which are not through-thickness, is not possible with the 2D analysis. Even the 2D realization of loading is not feasible in various cases. The 2D realization process becomes impractical in the presence of the cracks for which an assumption of through-thickness has to be considered. Therefore, the 3D numerical analysis of the machine components is an unavoidable process before finalizing the design and start of mass production. The 3D numerical analysis provides more accurate results and realistic conditions. It is beneficial for the cracked components as it can account for the effect of crack front shape.

The extended finite element method (XFEM) has been proved effective for the analysis of the 2D cracked domain as explained in previous chapters. This method can be extended for the 3D analysis also by converting crack from a line to a surface and crack front from a point to a line. The constant efforts from various researchers make it possible to perform the numerical analysis of the 3D cracked domain. The non-planar crack growth simulations were performed using the coupled 3D XFEM and level set method by Moës et al. (2002) and Gravouil et al. (2002). The level set method permits the seamless merging of cracks with boundaries. XFEM was coupled with the level set in the commercial finite element method (FEM) package I-DEAS by Bordas and Moran (2006) to solve complex 3D industrial fracture mechanics problems. The 3D crack in the domain is defined by two orthogonal level sets that are derived from the signed distance functions. Among these two level sets, one function describes the crack surface in 3D space and the other is used to describe the crack front. This is the implicit description of the crack front that enables the user to capture an arbitrary evolving crack without explicit geometric representation of the crack. Alternatively, the 3D XFEM was then coupled with the fast marching method (Sethian, 1999), a more efficient level set algorithm, to solve single crack (Sukumar and Prévost, 2003), multiple planar cracks (Chopp and Sukumar, 2003), non-planar crack growth (Sukumar et al., 2008), and industrial component (Shi et al., 2010). The corrected XFEM is extended for the 3D cracked domain by Loehnert et al. (2011) and addressed the various computational issues like quadrature rules for elements with discontinuities. The three level set functions were proposed by Fries and Baydoun (2012), which imply a coordinate system at the crack front and provide a basis for the enrichment. The 3D

Partition of Unity Methods, First Edition. Stéphane P. A. Bordas, Alexander Menk, and Sundararajan Natarajan.
© 2024 John Wiley & Sons Ltd. Published 2024 by John Wiley & Sons Ltd.

fatigue crack growth simulations were performed by Pathak et al. (2015a) using XFEM under the cyclic thermal loading. Different crack geometries such as planer, non-planar, and arbitrary spline shape cracks were considered under thermal shock, adiabatic, and isothermal loads for the simulations. A vector level set method was coupled with stable XFEM to perform the numerical analysis (Agathos et al., 2018a). This coupled method provided the optimal convergence rate similar to the standard FEM. A set of heuristic optimization algorithms are recombined into a multiscale optimization scheme by Agathos et al. (2018b) to tackle the multiple flaws in the domain.

10.2 Governing Equations

A residual stress-free domain of isotropic homogeneous material is considered for the formulation. The domain is assumed of volume Ω and bounded by the surface Γ as shown in Figure 10.1. Prescribed traction and displacement are applied to the surface Γ_t and Γ_u of the domain, respectively. A sharp traction-free crack in the domain is also considered and denoted by the Γ_c in Figure 10.1. The equilibrium equation and the associated boundary conditions for the domain are defined as:

$$\nabla : \sigma + \mathbf{b} = 0 \quad \text{in} \quad \Omega \tag{10.1}$$

$$\sigma.\mathbf{n} = \bar{\mathbf{t}} \quad \text{on} \quad \Gamma_t \tag{10.2}$$

$$\sigma.\mathbf{n} = 0 \quad \text{on} \quad \Gamma_c \tag{10.3}$$

$$\boldsymbol{u} = \bar{\mathbf{u}} \quad \text{on} \quad \Gamma_u \tag{10.4}$$

where σ is Cauchy stress tensor, $\mathbf{b}$ is the body forces per unit volume, $\mathbf{n}$ is the unit outward normal on the boundaries, $\boldsymbol{u}$ is the displacement field, and $\bar{\mathbf{t}}$ and $\bar{\mathbf{u}}$ are the prescribed traction and displacement on the surface Γ_t and Γ_u, respectively. For the linear elastic material, Cauchy stress tensor can be written in terms of strain as:

$$\sigma = \mathbf{C}\varepsilon \tag{10.5}$$

where $\mathbf{C}$ is elastic fourth-order tensor for constitutive relation and ε is the strain vector.

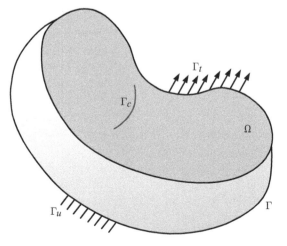

Figure 10.1 An illustration of the 3D cracked domain along with boundary conditions.

The weak form of governing equation (Equation 10.1), after introducing constitutive relation, can be written as:

$$\int_{\Omega} \varepsilon(\boldsymbol{u}) : \mathbf{C} : \varepsilon(\mathbf{v}) \, d\Omega - \int_{\Omega} \mathbf{b}.\mathbf{v} d\Omega - \int_{\Gamma_t} \bar{\mathbf{t}}.\mathbf{v} d\Gamma = 0 \quad \forall \mathbf{v} \in \mathbf{u_o} \tag{10.6}$$

where $\mathbf{v}$ is the test function space that is defined as:

$$\boldsymbol{u_o} = \{\mathbf{v} \in \Re : \mathbf{v} = 0 \text{ on } \Gamma_c, \mathbf{v} \text{ possibly discontinuous on } \Gamma_c\} \tag{10.7}$$

The discrete equations are obtained by introducing displacement approximations, weight, and shape functions in Equation 10.6. After introducing the displacement-based extrinsically enriched approximation, trial and test functions, and using the arbitrariness of nodal variation, a set of discrete equations can be written as:

$$\left[\boldsymbol{K}_{ij}^e\right]\{\boldsymbol{u}^e\} = \{\boldsymbol{F}^e\} \tag{10.8}$$

where K^e, u^e and F^e represent elemental stiffness matrix, elemental displacement matrix and elemental force matrix respectively.

10.3 XFEM Enrichment Approximation

In XFEM, geometrical discontinuity is modeled through enriched approximation technique. Geometric discontinuities such as cracks, inclusions, and holes are not considered a part of the finite element mesh/domain as shown in Figure 10.2. The local region around the geometrical discontinuity is modeled by enriching the approximation with additional functions and the rest of the domain is solved by the classical FEM. These enrichment functions are obtained from the theoretical background of the problem. An enriched displacement approximation for the crack in the homogeneous material domain can be written as (Pathak et al., 2017):

$$\boldsymbol{u}^h(\boldsymbol{x}) = \sum_{j=1}^{n} N_j(\mathbf{x}) \left[\boldsymbol{u}_j + \underbrace{(H(\boldsymbol{x}) - H(\boldsymbol{x}_j))\, a_j}_{j \in n_s} + \underbrace{\sum_{\ell=1}^{4} (\beta_\ell(\boldsymbol{x}) - \beta_\ell(\boldsymbol{x}_j))\, b_j^\ell}_{j \in n_f} \right] \tag{10.9}$$

where $\mathbf{u}_j$ is the nodal displacement vector associated with the continuous part of the finite element solution; N_j Langrage basis standard FEM shape function of the element; $H(\boldsymbol{x})$ is the additional mathematical function to capture discontinuity across the crack surface; $\beta_\ell(\boldsymbol{x})$ is the branch enrichment function to model stress field singularity near to the crack front; a_j is the additional degrees of freedom vector associated with the Heaviside function $H(\boldsymbol{x})$; b_j^ℓ is the additional degrees of freedom vector associated with the asymptotic functions; n is the set of all nodes in the mesh; n_s is the set of nodes whose shape function is completely cut by the crack surface; and n_f is the set of nodes whose shape function support is partially cut by the crack front.

Further, the enriched approximation for an interfacial crack domain can be written as:

$$\boldsymbol{u}^h(\boldsymbol{x}) = \sum_{j=1}^{n} N_j(\mathbf{x}) \left[\boldsymbol{u}_j + \underbrace{(H(\boldsymbol{x}) - H(\boldsymbol{x}_j))\, a_j}_{j \in n_s} + \underbrace{\sum_{\ell=1}^{12} (\beta_\ell(\boldsymbol{x}) - \beta_\ell(\boldsymbol{x}_j))\, b_j^\ell}_{j \in n_f} + \underbrace{(\chi(\boldsymbol{x}) - \chi(\boldsymbol{x}_j))\, c_j}_{j \varepsilon n_b} \right] \tag{10.10}$$

where c_j is the additional degree of freedom vector associated with the level set enrichment function, $\chi(\boldsymbol{x})$ is the level set enrichment function, and n_b is the set of nodes whose shape function are completely cut by the material interface. Heaviside function, $H(\mathbf{x})$, is defined for those elements which are completely cut by the crack

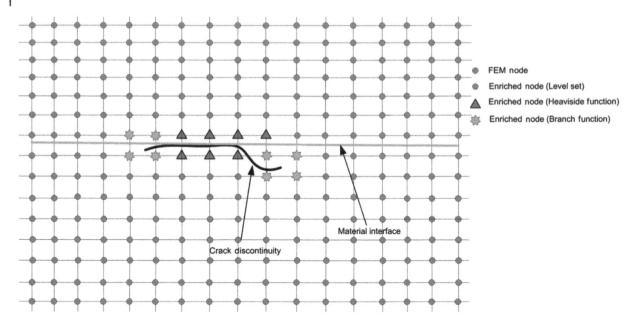

Figure 10.2 XFEM discretized domain with enriched nodes.

surface, level set enrichment function $\chi(x)$ employed over the material interface elements, whereas asymptotic functions, $\beta_\ell(x)$, are defined for those elements which are partially cut by the crack front. Four asymptotic enrichment functions are used to model the radial as well as the angular behavior of asymptotic crack-tip stress fields in the homogeneous material domain. For interfacial crack problems, 12 asymptotic enrichment functions with oscillatory nature are employed. Heaviside function used for the modeling of discontinuities in displacement field is given as:

$$H(x) = \begin{cases} 1 & if(x - x^*).n \geq 0 \\ -1 & \text{otherwise} \end{cases} \tag{10.11}$$

where x^* be the closest point of projection to x (evaluation point) on the crack surface and n is normal to the crack plane. The level set approach is used to capture the weak discontinuity, which arises due to the material interface within the element. An enrichment function for the material interface is defined as:

$$\chi(x) = \sum_I N_I |\zeta_I| - \left| \sum_I N_I \zeta_I \right| \tag{10.12}$$

where $\zeta\left(\Phi(x)\right) = |\Phi(x)|$ is the signed distance function and $\Phi(x)$ is the level set function, which is defined as:

$$\Phi(x) = \pm\min \|x - x_\Gamma\| \tag{10.13}$$

The branch enrichment functions $\beta_\ell(x)$ are used to model the singularity at the crack tip. This allows an effective representation of the crack front fields. To capture the singularity at the crack tip, the enrichment functions used in the solution for displacement field are defined as:

$$\beta_\ell(\boldsymbol{x}) = \left[\sqrt{R} \sin\frac{\theta}{2}, \sqrt{R} \cos\frac{\theta}{2}, \sqrt{R} \cos\frac{\theta}{2} \sin\theta, \sqrt{R} \sin\frac{\theta}{2} \sin\theta \right] \tag{10.14}$$

where R and θ are the polar coordinates with respect to the crack tip. For orthotropic media, crack front enrichment function can be employed in such a way that includes all possible displacement fields at the vicinity of the crack tip and expressed as:

$$\beta_\ell(\boldsymbol{x}) = \left[\sqrt{R} \cos\frac{\theta_1}{2}\sqrt{g_1(\theta)}, \sqrt{R} \cos\frac{\theta_2}{2}\sqrt{g_2(\theta)}, \sqrt{R} \sin\frac{\theta_1}{2}\sqrt{g_1(\theta)}, \sqrt{R} \sin\frac{\theta_2}{2}\sqrt{g_2(\theta)} \right] \tag{10.15}$$

$$\text{where, } \theta_j = \arctan\left(\frac{\mu_{jy} \sin\theta}{\cos\theta + \mu_{jx} \sin\theta} \right) \tag{10.16}$$

$$g_j(\theta) = \sqrt{\left(\cos\theta + \mu_{jx} \sin\theta\right)^2 + \left(\mu_{jy} \sin\theta\right)^2} \tag{10.17}$$

$j = 1, 2$ and in the above equations μ_{jx}, μ_{jy} are the real and imaginary parts of the roots of the characteristic equation. After implementing equilibrium and compatibility conditions, the fourth-order partial differential equation with the following characteristic equation has been obtained as follows:

$$C_{11} - 2C_{16}\mu^3 + (2C_{12} + C_{66})\mu^2 - 2C_{26}\mu + C_{22} = 0 \tag{10.18}$$

$$C_{44}\mu^2 + 2C_{45}\mu + C_{55} = 0 \tag{10.19}$$

For an interfacial crack problem, crack front enrichment functions employed to include all possible oscillatory nature of displacement fields at the vicinity of the interfacial crack front are given as:

$$\beta_\ell(\boldsymbol{x}) = \begin{bmatrix} \sqrt{R} \cos(\varsigma \log R) e^{-\varsigma\theta} \sin\left(\frac{\theta}{2}\right), & \sqrt{R} \cos(\varsigma \log R) e^{-\varsigma\theta} \cos\left(\frac{\theta}{2}\right), \\ \sqrt{R} \cos(\varsigma \log R) e^{\varsigma\theta} \sin\left(\frac{\theta}{2}\right), & \sqrt{R} \cos(\varsigma \log R) e^{\varsigma\theta} \cos\left(\frac{\theta}{2}\right), \\ \sqrt{R} \cos(\varsigma \log R) e^{\varsigma\theta} \sin\left(\frac{\theta}{2}\right) \sin(\theta), & \sqrt{R} \cos(\varsigma \log R) e^{\varsigma\theta} \cos\left(\frac{\theta}{2}\right) \sin(\theta), \\ \sqrt{R} \sin(\varsigma \log R) e^{-\varsigma\theta} \sin\left(\frac{\theta}{2}\right), & \sqrt{R} \sin(\varsigma \log R) e^{-\varsigma\theta} \cos\left(\frac{\theta}{2}\right), \\ \sqrt{R} \sin(\varsigma \log R) e^{\varsigma\theta} \sin\left(\frac{\theta}{2}\right), & \sqrt{R} \sin(\varsigma \log R) e^{\varsigma\theta} \cos\left(\frac{\theta}{2}\right), \\ \sqrt{R} \sin(\varsigma \log R) e^{\varsigma\theta} \sin\left(\frac{\theta}{2}\right) \sin(\theta), & \sqrt{R} \sin(\varsigma \log R) e^{\varsigma\theta} \cos\left(\frac{\theta}{2}\right) \sin(\theta) \end{bmatrix} \tag{10.20}$$

The bi-material constant ς is a function of material interface property and can be defined as:

$$\varsigma = \frac{1}{2\pi} \log\left(\frac{1-\bar{\beta}}{1+\bar{\beta}}\right) \text{ and } \bar{\beta} = \frac{\mu_1(\kappa_2 - 1) - \mu_2(\kappa_1 - 1)}{\mu_1(\kappa_2 + 1) + \mu_2(\kappa_1 + 1)} \tag{10.21}$$

The additional degrees of freedom due to the enrichment are handled by considering the fictitious nodes. The elemental contributions of stiffness matrix for extrinsically enriched approximation are given as (Pathak et al., 2013a):

$$K_{ij}^e = \begin{bmatrix} K_{ij}^{uu} & K_{ij}^{ua} & K_{ij}^{ub} \\ K_{ij}^{au} & K_{ij}^{aa} & K_{ij}^{ab} \\ K_{ij}^{bu} & K_{ij}^{ba} & K_{ij}^{bb} \end{bmatrix} \tag{10.22}$$

$$K_{ij}^e = K_{ij}^{rs} = \int_{\Omega^e} \left(B_i^r\right)^T \mathbf{C} B_j^s d\Omega \text{ and } r, s = u, a, b \tag{10.23}$$

$$F^e = \left\{ \begin{matrix} F_i^u & F_i^a & F_i^b \end{matrix} \right\} \tag{10.24}$$

$$F_i^u = \int_{\Gamma^t} N_i \bar{t} d\Gamma + \int_{\Omega^e} N_i \mathbf{b} d\Omega \tag{10.25}$$

$$F_i^a = \int_{\Gamma^t} N_i \{H(\mathbf{x}) - H(\mathbf{x}_I)\} \bar{t} d\Gamma + \int_{\Omega^e} N_i \{H(\mathbf{x}) - H(\mathbf{x}_I)\} \mathbf{b} d\Omega \tag{10.26}$$

$$F_i^b = \int_{\Gamma^t} N_i \{\beta_k(\mathbf{x}) - \beta_k(\mathbf{x}_I)\} \bar{t} d\Gamma + \int_{\Omega^e} N_i \{\beta_k(\mathbf{x}) - \beta_k(\mathbf{x}_I)\} \mathbf{b} d\Omega; \quad k = 1, 2, 3, 4 \tag{10.27}$$

$$B_i^u = \begin{bmatrix} N_{i,x} & 0 & 0 \\ 0 & N_{i,y} & 0 \\ 0 & 0 & N_{i,z} \\ N_{i,y} & N_{i,x} & 0 \\ 0 & N_{i,z} & N_{i,y} \\ N_{i,z} & 0 & N_{i,x} \end{bmatrix} \tag{10.28}$$

$$B_i^a = \begin{bmatrix} (N\{H(\mathbf{x}) - H(\mathbf{x}_I)\})_{i,x} & 0 & 0 \\ 0 & (N\{H(\mathbf{x}) - H(\mathbf{x}_I)\})_{i,y} & 0 \\ 0 & 0 & (N\{H(\mathbf{x}) - H(\mathbf{x}_I)\})_{i,z} \\ (N\{H(\mathbf{x}) - H(\mathbf{x}_I)\})_{i,y} & (N\{H(\mathbf{x}) - H(\mathbf{x}_I)\})_{i,x} & 0 \\ 0 & (N\{H(\mathbf{x}) - H(\mathbf{x}_I)\})_{i,z} & (N\{H(\mathbf{x}) - H(\mathbf{x}_I)\})_{i,y} \\ (N\{H(\mathbf{x}) - H(\mathbf{x}_I)\})_{i,z} & 0 & (N\{H(\mathbf{x}) - H(\mathbf{x}_I)\})_{i,x} \end{bmatrix} \tag{10.29}$$

$$B_i^b = \begin{bmatrix} (N\{\beta_\ell(\mathbf{x}) - \beta_\ell(\mathbf{x}_I)\})_{i,x} & 0 & 0 \\ 0 & (N\{\beta_\ell(\mathbf{x}) - \beta_\ell(\mathbf{x}_I)\})_{i,y} & 0 \\ 0 & 0 & (N\{\beta_\ell(\mathbf{x}) - \beta_\ell(\mathbf{x}_I)\})_{i,z} \\ (N\{\beta_\ell(\mathbf{x}) - \beta_\ell(\mathbf{x}_I)\})_{i,y} & (N\{\beta_\ell(\mathbf{x}) - \beta_\ell(\mathbf{x}_I)\})_{i,x} & 0 \\ 0 & (N\{\beta_\ell(\mathbf{x}) - \beta_\ell(\mathbf{x}_I)\})_{i,z} & (N\{\beta_\ell(\mathbf{x}) - \beta_\ell(\mathbf{x}_I)\})_{i,y} \\ (N\{\beta_\ell(\mathbf{x}) - \beta_\ell(\mathbf{x}_I)\})_{i,z} & 0 & (N\{\beta_\ell(\mathbf{x}) - \beta_\ell(\mathbf{x}_I)\})_{i,x} \end{bmatrix} \tag{10.30}$$

where N is Langrage finite element shape function and B is the derivative matrix of the shape functions. The constitutive matrix $\mathbf{C}$ for a 3D linear isotropic elastic material can be written as:

$$C = \frac{E}{(1 + \nu)(1 - 2\nu)} \begin{bmatrix} (1 - \nu) & \nu & \nu & 0 & 0 & 0 \\ \nu & (1 - \nu) & \nu & 0 & 0 & 0 \\ \nu & \nu & (1 - \nu) & 0 & 0 & 0 \\ 0 & 0 & 0 & (0.5 - \nu) & 0 & 0 \\ 0 & 0 & 0 & 0 & (0.5 - \nu) & 0 \\ 0 & 0 & 0 & 0 & 0 & (0.5 - \nu) \end{bmatrix} \tag{10.31}$$

where E is Young's modulus and ν is the Poisson's ratio of isotropic material. In a similar fashion, the constitutive matrix for 3D linear orthotropic elastic materials can be written as:

$$C = \begin{bmatrix} \bar{C}_{11} & \bar{C}_{12} & \bar{C}_{13} & 0 & 0 & \bar{C}_{16} \\ \bar{C}_{21} & \bar{C}_{22} & \bar{C}_{23} & 0 & 0 & \bar{C}_{26} \\ \bar{C}_{31} & \bar{C}_{32} & \bar{C}_{33} & 0 & 0 & \bar{C}_{36} \\ 0 & 0 & 0 & \bar{C}_{44} & \bar{C}_{45} & 0 \\ 0 & 0 & 0 & \bar{C}_{54} & \bar{C}_{55} & 0 \\ \bar{C}_{61} & \bar{C}_{62} & \bar{C}_{63} & 0 & 0 & \bar{C}_{66} \end{bmatrix} \tag{10.32}$$

The components of the global constitutive matrix are obtained by tensor transformation and the explicit expressions that can be written as:

$$\bar{C}_{11} = C_{11} \cos^4 \theta + 2(C_{12} + 2C_{66}) \sin^2 \theta \cos^2 \theta + C_{22} \sin^4 \theta \tag{10.33}$$

$$\bar{C}_{12} = (C_{11} + C_{22} - 4C_{66}) \sin^2 \theta \cos^2 \theta + C_{12} \left(\sin^4 \theta + \cos^4 \theta \right) \tag{10.34}$$

$$\bar{C}_{13} = C_{13} \cos^2 \theta + C_{23} \sin^2 \theta \tag{10.35}$$

$$\bar{C}_{16} = (C_{11} - C_{12} - 2C_{66}) \sin \theta \cos^3 \theta + (C_{12} - C_{22} + 2C_{66}) \cos \theta \sin^3 \theta \tag{10.36}$$

$$\bar{C}_{22} = C_{11} \sin^4 \theta + 2(C_{12} + 2C_{66}) \sin^2 \theta \cos^2 \theta + C_{22} \cos^4 \theta \tag{10.37}$$

$$\bar{C}_{23} = C_{13} \sin^2 \theta + C_{23} \cos^2 \theta \tag{10.38}$$

$$\bar{C}_{26} = (C_{11} - C_{12} - 2C_{66}) \cos \theta \sin^3 \theta + (C_{12} - C_{22} + 2C_{66}) \sin \theta \cos^3 \theta \tag{10.39}$$

$$\bar{C}_{33} = C_{33} \tag{10.40}$$

$$\bar{C}_{36} = (C_{13} - C_{23}) \sin \theta \cos \theta \tag{10.41}$$

$$\bar{C}_{44} = C_{44} \cos^2 \theta + C_{55} \sin^2 \theta \tag{10.42}$$

$$\bar{C}_{45} = (C_{55} - C_{44}) \sin \theta \cos \theta \tag{10.43}$$

$$\bar{C}_{55} = C_{55} \cos^2 \theta + C_{44} \sin^2 \theta \tag{10.44}$$

$$\bar{C}_{66} = (C_{11} + C_{22} - 2C_{12} - 2C_{66}) \sin^2 \theta \cos^2 \theta + C_{66} \left(\sin^4 \theta + \cos^4 \theta \right) \tag{10.45}$$

where C_{ij} is the component of the constitutive matrix in material coordinate direction and θ is the lamina orientation with global coordinate. The relation between the C_{ij} and engineering constants for orthotropic material can be written as Equations 10.46–10.53:

$$C_{11} = E_{11} (1 - \nu_{23} \nu_{32}) / \Delta \tag{10.46}$$

$$C_{22} = E_{22} (1 - \nu_{13} \nu_{31}) / \Delta \tag{10.47}$$

$$C_{33} = E_{33} (1 - \nu_{12} \nu_{21}) / \Delta \tag{10.48}$$

$$C_{44} = G_{23}, C_{55} = G_{13}, C_{66} = G_{12} \tag{10.49}$$

$$C_{12} = E_{11} (\nu_{21} + \nu_{23} \nu_{31}) / \Delta \tag{10.50}$$

$$C_{13} = E_{11} (\nu_{31} + \nu_{21} \nu_{32}) / \Delta \tag{10.51}$$

$$C_{23} = E_{11} (\nu_{32} + \nu_{12} \nu_{31}) / \Delta \tag{10.52}$$

$$\Delta = 1 - \nu_{12} \nu_{21} - \nu_{23} \nu_{32} - \nu_{13} \nu_{31} - 2\nu_{13} \nu_{21} \nu_{32} \tag{10.53}$$

10.4 Vector Level Set

A vector level set is a good tool to locate and model the geometry of the crack surface as well as the crack front. In the vector level set, a crack is defined by two vector functions. These functions are calculated at nodes so that they can be easily approximated by finite element interpolation functions. Geometry and solution can be discretized by linear or higher order solid elements. However, for the simplicity the discretization of problem domain by eight-node linear solid elements, and the approximation of crack surface using quadratic 20-node hexahedral elements, is a preferable option. Quadratic approximation of crack surface provides a better representation of the curved crack. The vector level set can be defined in the following way. Consider a 2D domain Ω divided by an interface Γ. This interface Γ subdivided the domain into two non-overlapping domain Ω_1 and Ω_2 as shown in Figure 10.3 and expressed in the form of an equation as:

$$\Phi(\boldsymbol{x}) = \begin{cases} > 0 & \boldsymbol{x} \in \Omega_1 \\ = 0 & \boldsymbol{x} \in \Gamma \\ < 0 & \boldsymbol{x} \in \Omega_2 \end{cases} \tag{10.54}$$

In a general way, the signed distance function is widely used as a level set function and defined as:

$$\Phi(\mathbf{x}) = \begin{cases} \rho & \boldsymbol{x} \in \Omega_1 \\ -\rho & \boldsymbol{x} \in \Omega_2 \end{cases} \tag{10.55}$$

where ρ is the normal distance from a point to the interface Γ. This signed distance function must satisfy the unit property as:

$$\|\nabla\Phi(\boldsymbol{x})\| = 1 \tag{10.56}$$

A signed distance function can be computed from $\mathbf{f}(\boldsymbol{x})$ by the following formula $\varphi(\boldsymbol{x}) = \|\mathbf{f}(\boldsymbol{x})\| \, H(\boldsymbol{\xi}_2.\mathbf{f}(\boldsymbol{x}))$, where $H(\mathbf{x})$ is the Heaviside function which takes $+1$ above the crack surface and -1 below the crack surface. Here, $\mathbf{f}(\boldsymbol{x})$ is the vector between a point $\boldsymbol{x}$ and its projection on the crack surface (Γ_c) and is directed from $\boldsymbol{x}$ to the crack surface, i.e., $\mathbf{f}(\mathbf{x}) = \boldsymbol{x}_f - \boldsymbol{x}$, where $\boldsymbol{x}_f$ is the closest point projection of $\boldsymbol{x}$ on the crack surface (Γ_c). The second level set function is a signed distance function with respect to the crack front. It is computed using the following

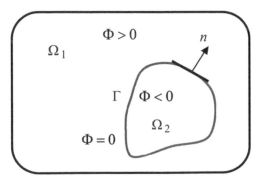

Figure 10.3 Definition of level set method.

formula $\psi(\boldsymbol{x}) = \boldsymbol{\xi}_1.\mathbf{f}(\boldsymbol{x})$. A crack surface within the finite elements is represented by:

$$\varphi_{approx}(\boldsymbol{x}) = \sum_{i=1}^{n} N_i(\boldsymbol{x})\varphi(\boldsymbol{x}_i) \tag{10.57}$$

$$\psi_{approx}(\boldsymbol{x}) = \sum_{i=1}^{n} N_i(\boldsymbol{x})\psi(\boldsymbol{x}_i) \tag{10.58}$$

The criterion for the finding whether an element is crack front or crack surface enriched is given as:

For crack front enriched elements: $\psi_{approx}(\boldsymbol{x}) = 0$ and $\varphi_{approx}(\boldsymbol{x}) = 0$ (10.59)

For crack surface enriched elements: $\psi_{approx}(\boldsymbol{x}) < 0$ and $\varphi_{approx}(\boldsymbol{x}) = 0$ (10.60)

The crack enrichment procedure in 3D XFEM is relatively a tedious task as compared to 2D XFEM; however, the implementing approach remains the same. In general, it is quite tedious to find the nearest point on a curve from an arbitrary point. A crack front is divided into finite numbers of piecewise spline curves (Pathak et al., 2013b) to evaluate level set values as shown in Figure 10.4. The approach is known as piecewise crack segment approach. The points on the piecewise spline curve are given by $\mathbf{f}_1, \mathbf{f}_2, \mathbf{f}_3, \mathbf{f}_4 \ldots\ldots\ldots \mathbf{f}_{i-2}, \mathbf{f}_{i-1}, \mathbf{f}_i$.

In general, a point on the spline segment of the crack front is defined by $\mathbf{f}_{i,t} = t \times \mathbf{f}_i + (1-t) \times \mathbf{f}_{i+1}$, where t is the fractional length on the interpolating lines. In this piecewise crack segment approach, the crack front is divided into finite numbers of piecewise linear or curved spline segments, then for each crack segment, a nearest point on the crack front from Gauss point has been selected using a local coordinate system. An orthogonal curvilinear coordinate system has been used to define the nearest point on the crack front, as it is convenient to define enrichment functions in the coordinate system. The derivatives of the level set function are required with respect to the global coordinate system. To explain this, consider an arbitrary 3D crack as shown in Figure 10.4. The geometry of the crack front can be represented by the position vector $r(\boldsymbol{x_f})$ pointing from the origin of the global Cartesian coordinate system (X, Y, Z) to some point t on the crack front. The mathematical description of the crack front and the crack surface can be written as:

$$r(\boldsymbol{x_f}) = 0 \tag{10.61}$$

$$f(\boldsymbol{x_f}) = 0 \tag{10.62}$$

A curvilinear coordinate system on the curve is denoted by $\boldsymbol{\xi}_1, \boldsymbol{\xi}_2$, and $\boldsymbol{\xi}_3$. The direction of $\boldsymbol{\xi}_3$ is taken parallel to the tangent of the curve, $\boldsymbol{\xi}_1$ is the outward normal to the crack front and it lies in the plane of the crack, $\boldsymbol{\xi}_2$ is perpendicular to the crack surface and is named as the gradient of the crack surface. $\boldsymbol{\xi}_3$ is tangent to the crack

Figure 10.4 Arbitrary 3D crack surface.

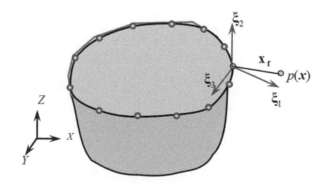

front and it is calculated by vector multiplication of $\boldsymbol{\xi}_1$ and $\boldsymbol{\xi}_2$ as written:

$$\boldsymbol{\xi}_3 = \boldsymbol{\xi}_1 \times \boldsymbol{\xi}_2 \tag{10.63}$$

Let $\boldsymbol{\xi}_1 = a_1\mathbf{i} + b_1\mathbf{j} + c_1\mathbf{k}$, $\boldsymbol{\xi}_2 = a_2\mathbf{i} + b_2\mathbf{j} + c_2\mathbf{k}$, and $\boldsymbol{\xi}_3 = a_3\mathbf{i} + b_3\mathbf{j} + c_3\mathbf{k}$, where $\mathbf{i}, \mathbf{j}$, and $\mathbf{k}$ are unit vectors in X, Y, and Z directions, respectively. The foot of the perpendicular from a point/node on the surface is denoted by $\boldsymbol{x}_f$. Now, a local vector from the foot to a point $p(\boldsymbol{x})$ can be written as:

$$\mathbf{f}(\boldsymbol{x}) = (\boldsymbol{x}_f - \boldsymbol{x}) \tag{10.64}$$

The local coordinate of a point with respect to the crack front can be represented by:

$$\tilde{x} = \mathbf{f}(\boldsymbol{x}).\boldsymbol{\xi}_1, \tilde{y} = \mathbf{f}(\boldsymbol{x}).\boldsymbol{\xi}_2, \tilde{z} = \mathbf{f}(\boldsymbol{x}).\boldsymbol{\xi}_3 \tag{10.65}$$

Due to the orthogonal coordinate system, $\tilde{z}$ automatically becomes zero, so R and θ can be calculated in terms of $\tilde{x}$ and $\tilde{y}$ as:

$$R = \sqrt{(\tilde{x}^2 + \tilde{y}^2)}, \theta = \tan^{-1}\left(\frac{\tilde{y}}{\tilde{x}}\right) \tag{10.66}$$

where $\tilde{x}$ and $\tilde{y}$ are the function of (X, Y, Z). After finding the values of $\tilde{x}$ and $\tilde{y}$ at each node, it can be easily approximated using a partition of unity. This approximation is performed in a small band of enriched elements as described below:

$$\tilde{x}_{approx}(\xi, \eta, \zeta) = \sum_{i=1}^{i=n} \tilde{x}_i N_i(\xi, \eta, \zeta) \tag{10.67}$$

$$\tilde{y}_{approx}(\xi, \eta, \zeta) = \sum_{i=1}^{i=n} \tilde{y}_i N_i(\xi, \eta, \zeta) \tag{10.68}$$

where (ξ, η, ζ) are the coordinates of a point in the natural coordinate system. After creating a local polar coordinate system R and θ as:

$$R = \sqrt{\left(\tilde{x}^2_{approx}(\xi, \eta, \zeta) + \tilde{y}^2_{approx}(\xi, \eta, \zeta)\right)}, \quad \theta = \tan^{-1}\left(\frac{\tilde{y}_{approx}(\xi, \eta, \zeta)}{\tilde{x}_{approx}(\xi, \eta, \zeta)}\right) \tag{10.69}$$

It can be easily differentiated using the derivatives of shape functions. However, the quadratic shape functions are used to approximate level set functions while the linear shape functions are used to approximate the solution.

10.5 Computation of Stress Intensity Factor

10.5.1 Brittle Material

The domain-based interaction integral approach provides the mixed-mode stress intensity factors (SIFs) in the post-processing phase. In the 3D cracked domain, the surface contour integral is virtually extended along the crack front as shown in Figure 10.5. As the crack surface is traction free hence J-integral will not be evaluated on the crack surface. The J-integral at location $\mathbf{x}$ along a 3D crack front can be defined as:

$$J(\mathbf{x}) = \lim_{\Gamma \to 0} \int_\Gamma \left(W\delta_{1i} - \sigma_{ij}\frac{\partial u_j}{\partial x_i}\right) n_i d\Gamma \tag{10.70}$$

where W is strain energy density, σ_{ij} is the element of Cauchy stress tensor, δ_{1i} is Kronecker delta, and u_j is a displacement component. The energy released $J'(\mathbf{x})$ per unit advance of crack front segment or length of the

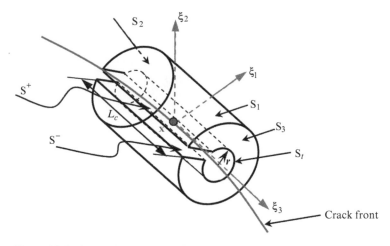

Figure 10.5 Virtually extended 3D contour integral.

virtual integral domain L_c is defined in the following way:

$$J'(\mathbf{x}) = \int_V \left(\sigma_{ij} \frac{\partial u_j}{\partial x_i} - W\delta_{1i} \right) \frac{\partial q}{\partial x_i} dV + \int_V \frac{\partial}{\partial x_i} \left(\sigma_{ij} \frac{\partial u_j}{\partial x_i} - W\delta_{1i} \right) q dV - \int_{S^+ + S^-} t_j \frac{\partial u_j}{\partial x_i} q ds \tag{10.71}$$

where t_j is component of traction acting on the crack-face, surfaces S^+, S^-, S_1, S_2, S_t, and S_3 are the enclosed surfaces for integral contour volume V, and surface S_t shrinks to the crack front, i.e., $(r \rightarrow 0)$.

In the above equation, the term q is defined as a scalar weight function and varies smoothly within volume contour V as follows:

$$q = \begin{cases} 0 & \text{at outer surface } S_1, S_2, \text{ and } S_3 \\ 1 & \text{at } S_t \\ \text{arbitrary} & \text{otherwise} \end{cases} \tag{10.72}$$

q is defined in such a way so that it can be easily interpolated using shape functions:

$$q = \sum_{i=1}^N q_i N_i \tag{10.73}$$

The variation of weight function q can be taken linear or higher order along the length of J-contour. In 3D crack problems, an independent mesh topology is needed to create around the crack front for the purpose to compute SIFs. This mesh topology in 3D hollow shape would act as a domain for interaction integral. For homogeneous linear elastic material under quasi-static loading condition, the second and third terms of Equation 10.71 vanish. Hence, it is now expressed as:

$$J'(\mathbf{x}) = \int_V \left(\sigma_{ij} \frac{\partial u_j}{\partial x_i} - W\delta_{1i} \right) \frac{\partial q}{\partial x_i} dV \tag{10.74}$$

However, for the curved crack front, the second term will not vanish. The energy release rate along the crack segment or length of the virtual integral domain L_c can be defined as:

$$J(\mathbf{x}) = \frac{J'(\mathbf{x})}{\int_{L_c} q(\mathbf{x}) ds} \tag{10.75}$$

After superimposing actual and auxiliary fields in energy released rate equation:

$$J'(\mathbf{x}) = \int_V \left[\left(\sigma_{ij}^{(1)} + \sigma_{ij}^{(2)} \right) \left(\frac{\partial u_j^{(1)}}{\partial x_i} + \frac{\partial u_j^{(2)}}{\partial x_i} \right) - W^{(T)} \delta_{1i} \right] \frac{\partial q}{\partial x_i} dV +$$

$$\int_V \frac{\partial}{\partial x_i} \left[\left(\sigma_{ij}^{(1)} + \sigma_{ij}^{(2)} \right) \left(\frac{\partial u_j^{(1)}}{\partial x_i} + \frac{\partial u_j^{(2)}}{\partial x_i} \right) - W^{(T)} \delta_{1i} \right] q dV - \qquad (10.76)$$

$$\int_{S^+ + S^-} \left(t_j^{(1)} + t_j^{(2)} \right) \left(\frac{\partial u_j^{(1)}}{\partial x_i} + \frac{\partial u_j^{(2)}}{\partial x_i} \right) q ds$$

In superimposed state, strain energy density is defined as:

$$W^{(T)} = \frac{1}{2} \left(\sigma_{ij}^{(1)} + \sigma_{ij}^{(2)} \right) \left(\varepsilon_{ij}^{(1)} + \varepsilon_{ij}^{(2)} \right) = W^{(1)} + W^{(2)} + W^{(12)} \qquad (10.77)$$

$$W^{(12)} = \frac{1}{2} \left(\sigma_{ij}^{(1)} \varepsilon_{ij}^{(2)} + \sigma_{ij}^{(2)} \varepsilon_{ij}^{(1)} \right) \qquad (10.78)$$

Equation 10.78 can be written as:

$$J'^{(T)}(\mathbf{x}) = J'^{(1)}(\mathbf{x}) + J'^{(2)}(\mathbf{x}) + M'^{(12)}(\mathbf{x}) \qquad (10.79)$$

Hence, the interaction integral term is broadly defined as:

$$M'^{(12)}(\mathbf{x}) = \int_V \left[\sigma_{ij}^{(1)} \frac{\partial u_j^{(2)}}{\partial x_i} + \sigma_{ij}^{(2)} \frac{\partial u_j^{(1)}}{\partial x_i} - \frac{1}{2} \left(\sigma_{jk}^{(1)} \varepsilon_{jk}^{(2)} + \sigma_{jk}^{(2)} \varepsilon_{jk}^{(1)} \right) \delta_{1i} \right] \frac{\partial q}{\partial x_i} dV +$$

$$\int_V \frac{\partial}{\partial x_i} \left[\sigma_{ij}^{(1)} \frac{\partial u_j^{(2)}}{\partial x_i} + \sigma_{ij}^{(2)} \frac{\partial u_j^{(1)}}{\partial x_i} - \frac{1}{2} \left(\sigma_{jk}^{(1)} \varepsilon_{jk}^{(2)} + \sigma_{jk}^{(2)} \varepsilon_{jk}^{(1)} \right) \delta_{1i} \right] q dV - \qquad (10.80)$$

$$\int_{S^+ + S^-} \left(t_j^{(1)} \frac{\partial u_j^{(2)}}{\partial x_i} + t_j^{(2)} \frac{\partial u_j^{(1)}}{\partial x_i} \right) q ds$$

For a straight crack front with traction-free crack surface, the second and third term of the above equation will vanish:

$$M'^{(12)}(\mathbf{x}) = \int_V \left[\sigma_{ij}^{(1)} \frac{\partial u_j^{(2)}}{\partial x_i} + \sigma_{ij}^{(2)} \frac{\partial u_j^{(1)}}{\partial x_i} - \frac{1}{2} \left(\sigma_{jk}^{(1)} \varepsilon_{jk}^{(2)} + \sigma_{jk}^{(2)} \varepsilon_{jk}^{(1)} \right) \delta_{1i} \right] \frac{\partial q}{\partial x_i} dV \qquad (10.81)$$

For thermo-elastic loading, the interaction integral is modified into:

$$M'^{(12)}(\mathbf{x}) = \int_V \left[\sigma_{ij}^{(1)} \frac{\partial u_j^{(2)}}{\partial x_i} + \sigma_{ij}^{(2)} \frac{\partial u_j^{(1)}}{\partial x_i} - \frac{1}{2} \left(\sigma_{jk}^{(1)} \varepsilon_{jk}^{(2)} + \sigma_{jk}^{(2)} \varepsilon_{jk}^{(1)} \right) \delta_{1i} \right] \frac{\partial q}{\partial x_i} dV + \alpha \int_V \frac{\partial T}{\partial x_i} \sigma_{kk}^{(2)} q dV \qquad (10.82)$$

For a bi-material interface crack, if crack and interface along ξ direction, the above expression of interaction integral gets simplified to:

$$M'^{(12)}(\mathbf{x}) = \sum_{k=1}^{2} \int_{V_k} \left[\sigma_{ij}^{(1)} \frac{\partial u_j^{(2)}}{\partial x_i} + \sigma_{ij}^{(2)} \frac{\partial u_j^{(1)}}{\partial x_i} - \frac{1}{2} \left(\sigma_{jk}^{(1)} \varepsilon_{jk}^{(2)} + \sigma_{jk}^{(2)} \varepsilon_{jk}^{(1)} \right) \delta_{1i} \right] \frac{\partial q}{\partial x_i} dV + \sum_{k=1}^{2} \alpha \int_{V_k} \frac{\partial T}{\partial x_i} \sigma_{kk}^{(2)} q dV \qquad (10.83)$$

where $k = 1, 2$ represents the two materials of the bi-material domain. For linear-elastic solids under mixed-mode loading conditions, the energy release rate can be defined in terms of mixed-mode SIFs:

$$J'(\mathbf{x}) = \frac{K_I^2 + K_{II}^2}{\bar{E}} + \frac{1 + \upsilon}{\bar{E}} K_{III}^2 \tag{10.84}$$

where $\bar{E} = \frac{E}{1 - \upsilon^2}$. For superimposed state, energy release rate may be defined in terms of mixed-mode SIFs as:

$$J'^{(T)}(\mathbf{x}) = \frac{1}{\bar{E}} \left[\left(K_I^{(1)} + K_I^{(2)} \right)^2 + \left(K_{II}^{(1)} + K_{II}^{(2)} \right)^2 \right] + \frac{1 + \upsilon}{\bar{E}} \left(K_{III}^{(1)} + K_{III}^{(2)} \right)^2 \tag{10.85}$$

$$J'^{(T)}(\mathbf{x}) = J'^{(1)}(\mathbf{x}) + J'^{(2)}(\mathbf{x}) + M'^{(12)}(\mathbf{x}) \tag{10.86}$$

Comparing Equation 10.85 with Equation 10.86, we get:

$$M'^{(12)}(\mathbf{x}) = \frac{(1 - \nu^2)}{E} \left(2K_I^{(1)} K_I^{(2)} + 2K_{II}^{(1)} K_{II}^{(2)} \right) + \frac{1}{\mu} \left(K_{III}^{(1)} K_{III}^{(2)} \right) \tag{10.87}$$

For bi-materials:

$$M'^{(12)}(\mathbf{x}) = \frac{2}{E^* \left(\cosh^2 \pi \tilde{\varepsilon} \right)} \left(2K_I^{(1)} K_I^{(2)} + 2K_{II}^{(1)} K_{II}^{(2)} \right) + \frac{1}{\mu^*} \left(K_{III}^{(1)} K_{III}^{(2)} \right) \tag{10.88}$$

where $\tilde{\varepsilon}$ is bi-material constant and E^* and μ^* are equivalent Young's modulus and shear modulus for bi-materials, respectively. These material parameters are obtained by the following relations:

Oscillatory index, $\tilde{\varepsilon} = \dfrac{1}{2\pi} \ln \left[\dfrac{\kappa_1 \mu_2 + \mu_1}{\kappa_2 \mu_1 + \mu_2} \right]$ \hfill (10.89)

Kolosov constant, $\kappa_i = 3 - 4\upsilon_i$ \hfill (10.90)

Equivalent Young's modulus, $\dfrac{1}{E^*} = \dfrac{1}{2} \left(\dfrac{1 - \upsilon_1^2}{E_1} + \dfrac{1 - \upsilon_2^2}{E_2} \right)$ \hfill (10.91)

Equivalent shear modulus, $\dfrac{1}{\mu^*} = \dfrac{1}{2} \left(\dfrac{1}{\mu_1} + \dfrac{1}{\mu_2} \right)$ \hfill (10.92)

Dundurs' parameter (Tensile modulus) $\tilde{\alpha} = \left[\dfrac{\mu_1 (\kappa_2 + 1) - \mu_2 (\kappa_1 + 1)}{\mu_2 (\kappa_1 + 1) + \mu_1 (\kappa_2 + 1)} \right]$ \hfill (10.93)

(Bulk modulus) $\tilde{\beta} = \left[\dfrac{\mu_1 (\kappa_2 - 1) - \mu_2 (\kappa_1 - 1)}{\mu_2 (\kappa_1 + 1) + \mu_1 (\kappa_2 + 1)} \right]$ \hfill (10.94)

For the orthotropic material, interaction integral relation can be written as:

$$J_1 = \frac{-\pi K_1^2}{2} C_{22} \text{Im} \left[\frac{\mu_1 + \mu_2}{\mu_1 \mu_2} \right] \tag{10.95}$$

$$J_2 = \frac{\pi K_2^2}{2} C_{11} \text{Im} \left[\mu_1 + \mu_2 \right] \tag{10.96}$$

$$J_3 = \frac{\pi K_3^2}{2} \frac{\text{Im} \left[a_{45} + \mu_3 a_{44} \right]}{a_{44} a_{55}} \tag{10.97}$$

Thus, the individual SIF for the actual state can be obtained by judiciously choosing the auxiliary state (state 2). The main difficulty in calculating the interaction integral lies in the evaluation of the gradients and higher order derivatives of the auxiliary fields for the curved crack front. The auxiliary field can be calculated by two methods, first one is an analytical approach which is possible only for some simple shape cracks, and the second one is an approximate approach which is based on the partition of unity. The approximated auxiliary field using the partition of unity approach can be written as:

$$\sigma^{(2)}(j,k) \approx \sum_{i=1}^{n} \sigma_i^{(2)}(j,k) N_i(X,Y,Z) \tag{10.98}$$

where i is the node number:

$$u^{(2)} \approx \sum_{i=1}^{n} u_i^{(2)} N_i(X,Y,Z) \tag{10.99}$$

After approximation of the auxiliary field, the gradient of the auxiliary field can be easily evaluated.

10.5.2 Ductile Material

The individual modes of SIFs can be evaluated from J-integral, which is calculated at the ends of the crack front line segments using the J-integral decomposition approach for ductile material (Kumar et al., 2017). A virtual cylindrical domain as shown in Figure 10.6 is created at these ends of crack front to perform the J-integral computation.

The decomposed form of J-integral at these ends is defined as:

$$
\begin{aligned}
J^N = \ &\frac{1}{2L_e}\Big(\int_\Lambda \big(\sigma_{ij}^N (u_{i,1})^N - W^N \delta_{1j} \big) q_{,j} d\Lambda + \int_\Lambda \big(\sigma_{ij}^N (u_{i,1j})^N - W_{,1}^N \big) q d\Lambda \\
&+ \int_\Lambda \big(\sigma_{i1,1}^N + \sigma_{i2,2}^N \big) (u_{i,1})^N d\Lambda - \int_\Lambda \sigma_{i3}^N (u_{i,13})^N d\Lambda \Big) \forall N \in \{I, II, III\}
\end{aligned}
\tag{10.100}
$$

where, I, II, III represent the mode-I, mode-II, and mode-III, respectively, W is the strain energy density, L_e is the length of the virtual domain along the crack front, and q is a function having value one at the crack front and zeros at the boundary of the virtual domain. The symmetric portion of fields provides the mode-I SIF, whereas the anti-symmetric portion is further divided to obtain the mode-II and mode-III SIFs. In order to compute the J-integral by this approach, the required fields are interpolated from the original mesh to the virtual cylindrical domain via shape function interpolation and decomposed into symmetric and anti-symmetric portions across the crack surface. For the decomposition of fields, all the fields are required at the mirror point of the integration point

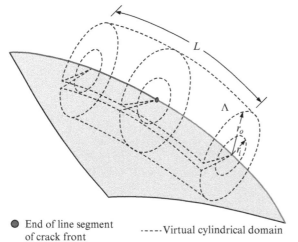

Figure 10.6 A virtual domain at the end of the crack front line segment for the calculation of J-integral.

● End of line segment of crack front   ----- Virtual cylindrical domain

about crack surface that can be calculated from the nodal data by interpolation functions. Due to the presence of plasticity, the stress field cannot be directly obtained at the virtual domain and mirrored point from the displacement field. Therefore, to calculate the stress field at the required point, a data transfer scheme is utilized. In this scheme, the stress field is transferred from the integration points to the nodes (Durand and Farias, 2014) by:

$$\chi_n = \left(\tilde{\mathbf{N}}^T \tilde{\mathbf{N}} \right)^{-1} \tilde{\mathbf{N}}^T \chi_{ip} \tag{10.101}$$

where $\tilde{\mathbf{N}}^T$ is the matrix that contains the value of shape functions at the integration points and χ_n and χ_{ip} are the fields at node and integration point, respectively. After that, the nodal stress is interpolated at the required point using the shape functions of the element. Due to the use of sub-tetrahedralization for enriched elements, the stress field is extrapolated at the nodes of sub-tetrahedron using Equation 10.101 and stored for each tetrahedron separately. The interpolation of the stress field in the enriched element is performed in two steps. In the first step, sub-tetrahedron is identified that contains the mirrored point, while in the second step, the stress field is interpolated using the extrapolated nodal stress and shape functions of the identified sub-tetrahedron. The decomposed stress field for all the modes at the spatial point P across the crack surface 1–3 (as shown in Figure 10.7) is expressed as:

$$\sigma_{ij\,P}^I = \left\{ \begin{array}{c} \sigma_{11\,P}^S \\ \sigma_{22\,P}^S \\ \sigma_{33\,P}^S \\ \sigma_{12\,P}^S \\ \sigma_{23\,P}^S \\ \sigma_{13\,P}^S \end{array} \right\} = \frac{1}{2} \left\{ \begin{array}{c} \sigma_{11\,P} + \sigma_{11\,P'} \\ \sigma_{22\,P} + \sigma_{22\,P'} \\ \sigma_{33\,P} + \sigma_{33\,P'} \\ \sigma_{12\,P} - \sigma_{12\,P'} \\ \sigma_{23\,P} - \sigma_{23\,P'} \\ \sigma_{13\,P} + \sigma_{13\,P'} \end{array} \right\} \tag{10.102}$$

$$\sigma_{ij\,P}^{II} + \sigma_{ij\,P}^{III} = \left\{ \begin{array}{c} \sigma_{11\,P}^{AS} \\ \sigma_{22\,P}^{AS} \\ \sigma_{33\,P}^{AS} \\ \sigma_{12\,P}^{AS} \\ 0 \\ 0 \end{array} \right\} + \left\{ \begin{array}{c} 0 \\ 0 \\ 0 \\ 0 \\ \sigma_{23\,P}^{AS} \\ \sigma_{13\,P}^{AS} \end{array} \right\} = \frac{1}{2} \left\{ \begin{array}{c} \sigma_{11\,P} - \sigma_{11\,P'} \\ \sigma_{22\,P} - \sigma_{22\,P'} \\ \sigma_{33\,P} - \sigma_{33\,P'} \\ \sigma_{12\,P} + \sigma_{12\,P'} \\ 0 \\ 0 \end{array} \right\} + \frac{1}{2} \left\{ \begin{array}{c} 0 \\ 0 \\ 0 \\ 0 \\ \sigma_{23\,P} + \sigma_{23\,P'} \\ \sigma_{13\,P} - \sigma_{13\,P'} \end{array} \right\} \tag{10.103}$$

In a similar way, other fields can also be decomposed (Kumar et al., 2019). The analytical derivatives of strain energy density are not possible due to the plasticity; hence, it is evaluated by the function approximation. The required field of all the integration points of an element is fitted into a quadratic function by nonlinear least squares method as:

$$\varpi^N = f_1 + f_2\xi + f_3\eta + f_4\zeta + f_5\xi^2 + f_6\eta^2 + f_7\zeta^2 + f_8\xi\eta + f_9\eta\zeta + f_{10}\xi\zeta \tag{10.104}$$

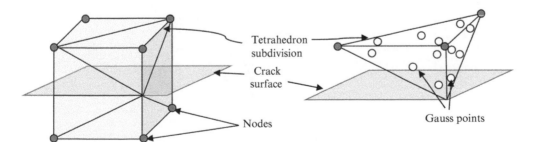

Figure 10.7 Tetrahedral subdivision of split element.

where ξ, η, ζ are the local coordinates of the integration points and f_1, f_2, f_3, f_4, f_5, f_6, f_7, f_8, f_9, and f_{10} are the fitting constants. The derivative of Equation 10.104 is used to compute the derivative of strain energy density.

10.6   Numerical Simulations

10.6.1   Computation of Fracture Parameters

For the computations of the fracture parameters, the 3D numerical simulations are performed for different types of crack fronts subjected to mechanical as well as thermo-elastic loading conditions. All crack domains are uniformly discretized by eight-noded hexahedral elements without considering the crack discontinuity in the geometry. The crack surface and crack front are numerically traced by the vector level set method. For the calculation of level set functions, the information of the closest point from the nodes to the crack front and crack surface must be known. In the simulations, initially a crack front is divided into hundreds of curved crack segments, then for each crack segment, the nearest point from an evaluation/Gauss point is obtained using an orthogonal local coordinate system as shown in Figure 10.4. For numerical integration, discontinuous elements are subdivided into tetrahedron in such a way one of the tetrahedron division should align with the crack surface/front as shown in Figures 10.7 and Figure 10.8 (Pathak et al., 2015b). Further, higher order Gauss quadrature has been employed in enriched elements.

During computations, three and seven points Gauss quadrature have been employed in each tetrahedral subdivision for spilt and crack front enriched elements, respectively. However, classical finite elements are numerically integrated by two-point Gauss quadrature scheme. Both planar and non-planar crack geometries have been analyzed under mechanical and thermal loading environment. For mechanical loading cases, mechanical traction is imposed over top surface while bottom surface of problem geometry is fixed. In thermal shock load, uniform temperature change is imposed throughout the problem domain. Top and bottom surfaces are constrained which results in the thermal stress. The problems are divided into three material-dependent problems, namely, "Isotropic homogeneous crack geometry," "Orthotropic homogeneous crack geometry," and "Isotropic bi-material crack geometry."

10.6.1.1   Isotropic Homogeneous Crack Geometry

Three cracked geometries are taken to evaluate fracture parameter (SIFs) under mechanical load and thermal shock load separately. At first, a cuboid of 2 m × 2 m × 2 m with an embedded elliptical shape crack of major axis radius of 0.4 m with eccentricity 0.7 m has been simulated. The geometry along with other boundary conditions is shown in Figure 10.9. The crack geometry is uniformly discretized by 24 × 24 × 24 nodes in x, y, and z -directions.

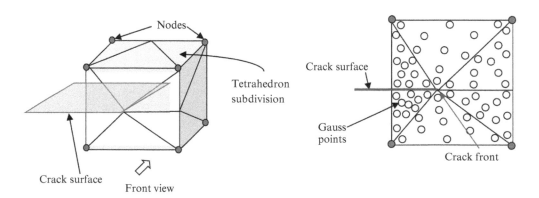

Figure 10.8   Tetrahedral subdivision for crack front element and Gauss point distribution.

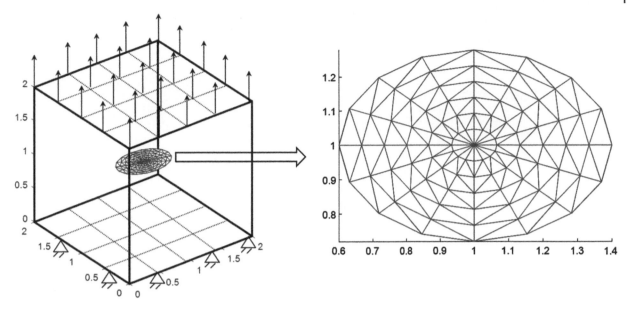

Figure 10.9 A cuboid with an embedded elliptical shape crack under mechanical load.

Figure 10.10 A cuboid with an embedded elliptical shape crack under thermal shock.

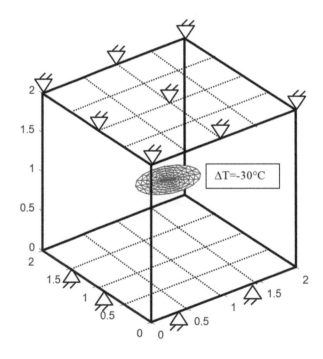

The material properties used for this simulation are $E = 200$ GPa, $v = 0.3$, and $\alpha = 11.7 \times 10^{-6}/°$C. To apply mechanical boundary conditions, a constant amplitude mechanical traction of $\sigma = 75$ MPa is applied at top surface, whereas all degrees of freedom associated with the bottom surface are constrained. In thermal shock load, uniform temperature change is imposed throughout the domain. Top and bottom surfaces are constrained which results in the thermal stress as shown in Figure 10.10. The mixed-mode SIFs are computed along the crack front, the location of J-domain along crack front is shown in Figure 10.11.

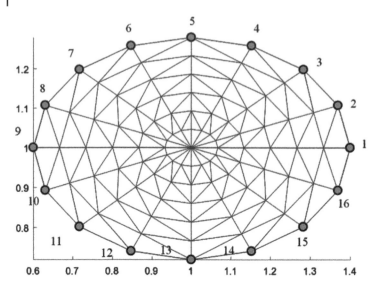

Figure 10.11 J-domain position along crack front for SIFs computation.

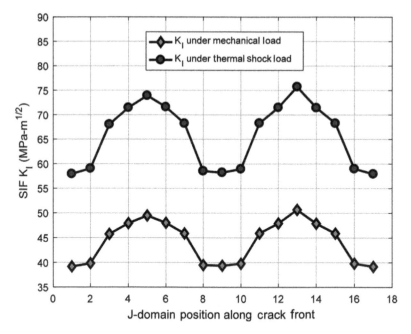

Figure 10.12 K_I variation along crack front for embedded elliptical shape crack under mechanical and thermal shock load.

The SIFs computed for the embedded elliptical shape crack under both mechanical and thermal shock loading conditions are presented in Figure 10.12.

From the computed SIFs, it has been found that K_I value is maximum at minor axis quadrant J-domain position and minimum at major axis quadrant J-domain position. It has been also observed that K_I values dominate as

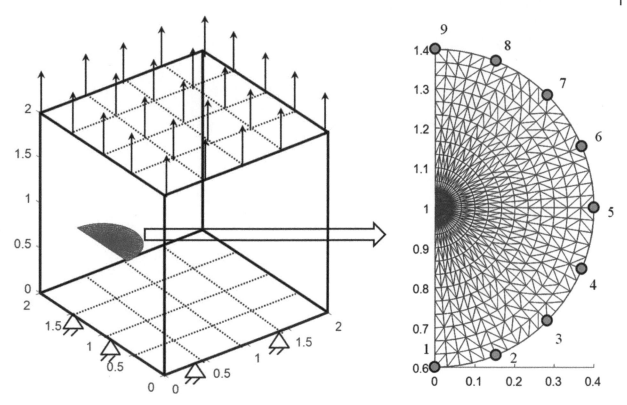

Figure 10.13 A cuboid with a surface semi-penny shape crack under mechanical load along with magnified view of surface semi-penny shape crack with *J*-domain position.

compared to K_{II} and K_{III} for both considered mechanical and thermal shock loads. Further the technique has been employed to solve semi-penny shape crack and embedded lens shape crack (non-planar). A cuboid of 2 m × 2 m × 2 m with a semi-penny shape crack with radius of 0.4 m has been simulated for SIF computation. The problem is solved for both mechanical and thermal shock loading conditions. The geometry along with other boundary conditions is shown in Figure 10.13. The SIFs are computed along the crack front at specified *J*-domain position. These *J*-domain positions are shown by the enlarge view in Figure 10.13 for better understanding. To impose mechanical boundary conditions, a constant amplitude mechanical traction of $\sigma = 75$ MPa is applied at top surface, whereas all degrees of freedom associated with the bottom surface are constrained. In thermal shock load, uniform temperature change is imposed throughout the problem domain; in addition top and bottom surfaces are making fixed in all x, y, and z translation degree of freedom.

For the mechanical loading conditions, it can be seen that maximum K_I has been observed at surface side and gradually decreases toward the embedded crack front quadrant, whereas in the case of thermal shock condition, minimum SIF has been observed at surface side and gradually increases toward the embedded crack front quadrant as shown in Figure 10.14.

Further, a non-planar crack has been simulated where a cuboid with embedded lens shape crack has been simulated for both mechanical and thermal shock loading conditions. Complete problem geometry along with *J*-domain position is shown in Figure 10.15 and Figure 10.16. From the SIFs plot (Figure 10.17), it can be concluded that K_I values are almost constant throughout the crack front. However, thermal shock load induce higher K_I as compared to mechanical traction load.

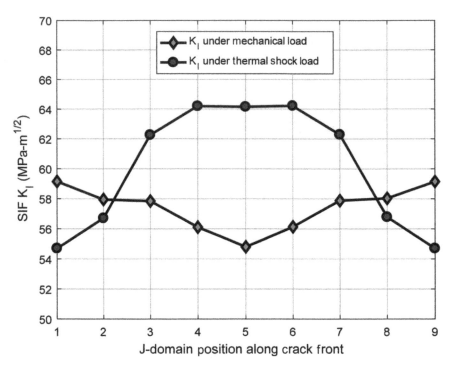

Figure 10.14 K_I variation along crack front for surface semi-penny shape crack under mechanical and thermal shock load.

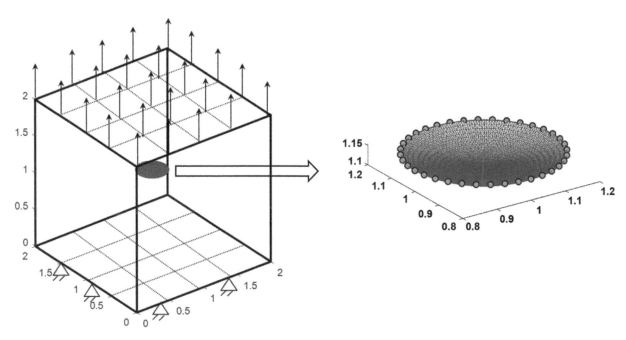

Figure 10.15   A cuboid with an embedded lens shape crack under mechanical load.

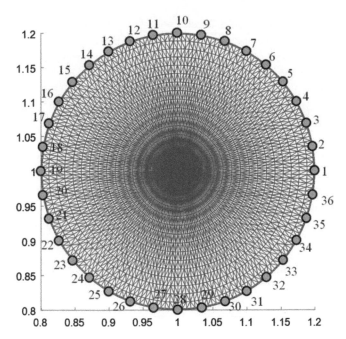

Figure 10.16   Top view of embedded lens shape crack with J-domain position.

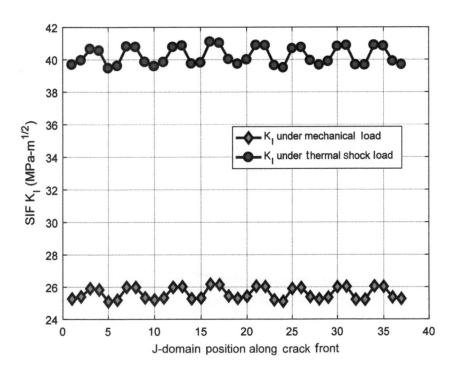

Figure 10.17 K_I variation along crack front for embedded lens shape crack under mechanical and thermal shock load.

10.6.1.2 Orthotropic Homogeneous Crack Geometry

The fracture parameter (SIF) is computed for orthotropic homogeneous crack geometry. Effect of orthotropic orientation has been analyzed with reference to SIFs. Crack surface discontinuity is modeled by Heaviside enrichment function, whereas singular stress field due to crack front is approximated by angular orthotropic branch enrichment functions. All considered numerical cases are solved for mesh data of 24 × 24 × 24 nodal points, which is uniformly created within the cracked domain by using eight-noded hexahedral elements. Two different cases of 3D elastic orthotropic crack problems, namely, "embedded penny shape crack" and "surface semi-penny shape crack" have been considered and solved by the XFEM. Numerical results are presented in the form of SIFs with orthotropic orientation angle. Material properties used in the simulation are given in Table 10.1.

An embedded penny shape crack in cuboid domain has been numerically simulated for different orthotropic angles. From the SIFs plot (Figure 10.18), it is observed that K_I decreases with lamina orientation, which is due to the alignment of material principal axis with the loading direction (z-axis, vertically aligned). Stress contours along the loading direction (σ_{zz}) are also computed at crack surface and presented for different orthotropic angle as shown in Figure 10.19. Further a surface semi-penny shape crack (as shown in Figure 10.13) has been numerically

Table 10.1 Orthotropic material properties for numerical simulation.

Material	E_1 (GPa)	E_2, E_3 (GPa)	v_{23}, v_{13}	v_{12}	G_{12}, G_{23} (GPa)	G_{31} (GPa)	α (/°C)
Elastic orthotropic	38.6	8.27	0.168	0.035	3.14	4.14	11.7×10^{-6}

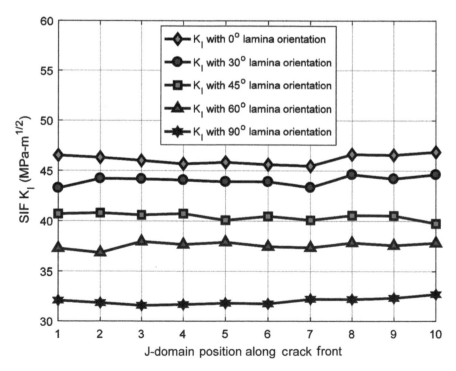

Figure 10.18 K_I variation with orthotropic angle for embedded penny shape crack under mechanical load.

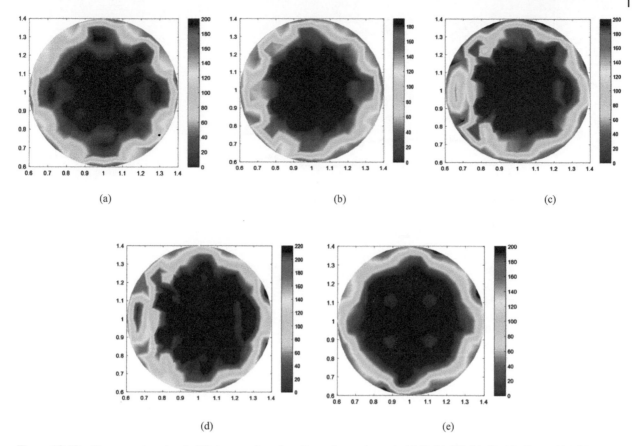

Figure 10.19 Stress contour (σ_{zz} in MPa) at crack surface for orthotropic angle (a) 0° (b) 30° (c) 45° (d) 60° and (e) 90°.

simulated under mechanical traction load. Numerical SIFs are obtained and presented for different orthotropic orientation as shown in Figure 10.20. From Figure 10.20, it can be seen that SIFs decrease with lamina orientation. The obtained SIFs (K_I) is higher in the case of surface penny shape crack as compared to embedded penny shape crack.

10.6.1.3 Bi-Material Crack Geometry

An interfacial crack in bi-material domain has been numerically simulated for mixed-mode SIFs. A cuboid with interfacial penny shape crack and surface semi-penny shape crack is numerically simulated under mechanical loading. A complete problem geometry is shown in Figure 10.21. This 3D bi-material crack simulation is performed by considering all interface crack modeling issues like logarithmic singularity and material discontinuity. Logarithmic singularity based 12 branch enrichment functions are employed to capture oscillatory nature of stress field at interfacial crack front. The material properties used in the present simulations are tabulated in Table 10.2. The results are presented in the form of mixed-mode SIFs in Figure 10.22.

From the SIFs plot (Figure 10.22), it has been observed that K_I and K_{II} vary symmetrically about the crack front quadrant, whereas K$_{II}$ values are negative and K$_{III}$ values are oscillating. Further, a surface semi-penny shape crack has been simulated in bi-material cuboid domain. All three modes of SIFs are computed and presented in

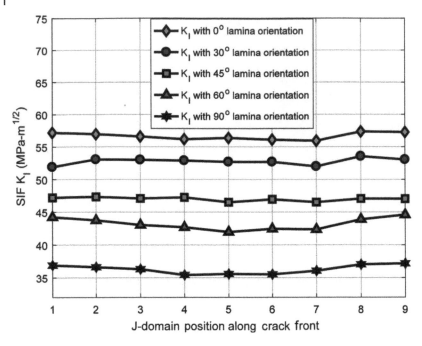

Figure 10.20 K_I variation with orthotropic angle for surface penny shape crack under mechanical load.

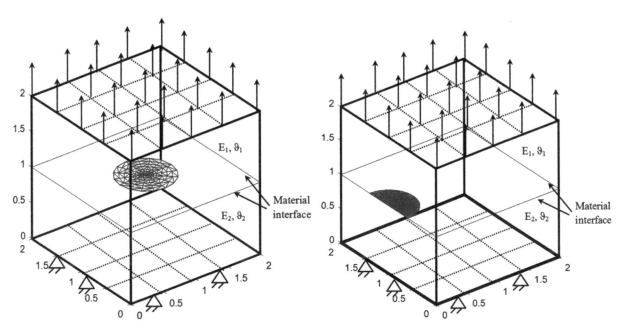

Figure 10.21   A bi-material cuboid under mechanical load with (a) an embedded interfacial penny shape crack and (b) a surface interfacial semi-penny shape crack.

Table 10.2 Bi-material properties for numerical simulation.

S. No.	Parameter	Numerical value
1	Elastic modulus of stiff material (E_1)	200 GPa
2	Elastic modulus of soft material (E_2)	70 GPa
3	Poisson's ratio for stiff material (ϑ_1)	0.25
4	Poisson's ratio for soft material (ϑ_2)	0.3

Table 10.3 Material properties of Ni-based superalloy at 650 °C.

Mechanical Properties	650 °C Value
Young's modulus, E (GPa)	180
Poisson ratio, v	0.33
Yield strength, σ_{yts} (MPa)	653
Ultimate tensile strength, σ_{uts} (MPa)	987
Paris Law constant, C	1.78×10^{-8}
Paris Law constant, m	2.89

Figure 10.23. From the SIFs plot, maximum K_I value is observed at surface and further decreases along the crack front toward the inner side. K_{II} values increase with the crack front toward the inner side, whereas K_{III} values remain nearly constant.

10.6.2 Fatigue Crack Growth in Compact Tension Specimen

A compact tensile specimen consists of Ni-based superalloy having 32 mm width and 6 mm thickness is considered for the fatigue crack growth (FCG) simulation. A through crack of 7.2 mm is considered in the specimen as shown in Figure 10.24 and the fatigue load of $F_{max} = 3500\,\text{N}$ $(R = 0.1)$ is applied at the elevated temperature of 650°C. The mechanical and fatigue properties of Ni-based superalloy at the elevated temperature are provided in Table 10.3. The specimen is discretized into $15 \times 15 \times 3$ elements while the crack front is divided into eight line segments. A virtual cylindrical domain is created at the ends of the line segments of crack front except for the corner points of the crack front. The radius and length of the virtual cylindrical domain are taken as 1 mm and 0.5 mm, respectively. The decomposed fields are used to calculate the individual modes of SIFs as described in the previous sections. The crack growth is evaluated from the crack growth rate at each endpoint of line segments of crack front for a particular number of cycles. Initially, the number of cycles is kept high but when the crack growth is in the range of element size then the number of cycles is reduced to capture the very high rate of crack growth. The numerically computed FCG is shown in Figure 10.25 and compared with the experimental results (Kumar et al., 2018). The numerical values are in a good agreement with the experimental results. The numerically obtained crack front at different stages of the simulation is also presented in Figure 10.26. The predicted growth of the crack front in the middle of the specimen is high as compared to the surface of the specimen, which is consistent with the theoretical expectations.

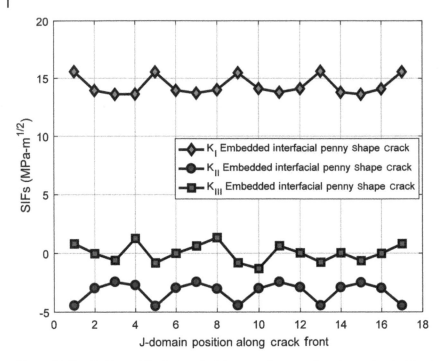

Figure 10.22 K_I, K_{II}, and K_{III} for interfacial penny shape crack under mechanical load.

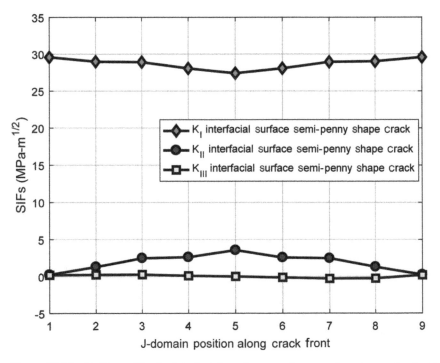

Figure 10.23 K_I, K_{II}, and K_{III} for interfacial semi-penny shape crack under mechanical load.

Figure 10.24 A schematic of compact tensile specimen considered for simulation.

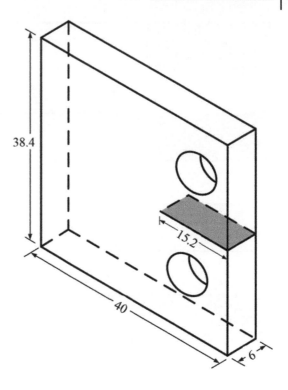

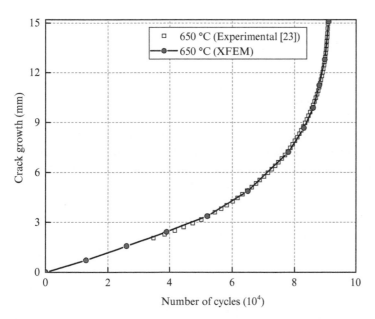

Figure 10.25 A comparison of numerically predicted fatigue crack growth and experimental results for Ni-based superalloy at elevated temperature.

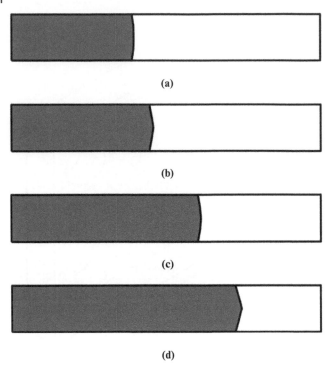

Figure 10.26 Numerically predicted crack path at different stages of the simulation after (a) 13 000 cycles (b) 52 000 cycles (c) 86 000 cycles (d) 91 000 cycles.

10.7 Summary

This chapter presents 3D fracture modeling using extrinsically enriched XFEM. Complete steps for computational modeling like geometrical modeling of crack by using vector level set, enrichment procedure for crack front and crack surface, and enriched approximation technique are discussed in detail. Enrichment functions for crack surface discontinuity and crack front singularity are explained for homogeneous isotropic crack domain, orthotropic crack domain, and bi-material isotropic crack domain, respectively. Numerical integration issues for discontinuous elements are discussed in detail with possible solution strategy. Vector level set and domain-based interaction integral technique is employed to trace crack geometry as well as to predict fracture parameters in terms of mixed-mode SIFs. Various 3D numerical simulations are presented to compute fracture parameters under mechanical as well as thermo-elastic loading conditions for the better understanding of technique. The 3D fatigue crack growth simulation in compact tension specimen made of ductile material is also presented.

References

Agathos, K., Ventura, G., Chatzi, E., and Bordas, S.P. (2018a). Stable 3D XFEM/vector level sets for non-planar 3D crack propagation and comparison of enrichment schemes. *International Journal for Numerical Methods in Engineering* 113 (2): 252–276.

Agathos, K., Chatzi, E., and Bordas, S.P. (2018b). Multiple crack detection in 3D using a stable XFEM and global optimization. *Computational Mechanics* 62 (4): 835–852.

Bordas, S. and Moran, B. (2006). Enriched finite elements and level sets for damage tolerance assessment of complex structures. *Engineering Fracture Mechanics* 73 (9): 1176–1201.

Chopp, D.L. and Sukumar, N. (2003). Fatigue crack propagation of multiple coplanar cracks with the coupled extended finite element/fast marching method. *International Journal of Engineering Science* 41 (8): 845–869.

Durand, R. and Farias, M.M. (2014). A local extrapolation method for finite elements. *Advances in Engineering Software* 67: 1–9.

Fries, T.P. and Baydoun, M. (2012). Crack propagation with the extended finite element method and a hybrid explicit–implicit crack description. *International Journal for Numerical Methods in Engineering* 89 (12): 1527–1558.

Gravouil, A., Moës, N., and Belytschko, T. (2002). Non-planar 3D crack growth by the extended finite element and level sets-Part II: Level set update. *International Journal for Numerical Methods in Engineering* 53 (11): 2569–2586.*

Kumar, M., Bhuwal, A.S, Singh, I.V., et al. (2017). Nonlinear fatigue crack growth simulations using J-integral decomposition and XFEM. *Procedia Engineering* 173: 1209–1214.

Kumar, M., Ahmad, S., Singh, I.V., et al. (2018). Experimental and numerical studies to estimate fatigue crack growth behavior of Ni-based super alloy. *Theoretical and Applied Fracture Mechanics* 96: 604–616.

Kumar, M., Singh, I.V., and Mishra, B.K. (2019). Fatigue crack growth simulations of plastically graded materials using XFEM and J-integral decomposition approach. *Engineering Fracture Mechanics* 216: 106470.

Loehnert, S., Mueller-Hoeppe, D.S., and Wriggers, P. (2011). 3D corrected XFEM approach and extension to finite deformation theory. *International Journal for Numerical Methods in Engineering* 86 (4–5): 431–452.

Moës, N., Gravouil, A., and Belytschko, T. (2002). Non-planar 3D crack growth by the extended finite element and level sets-Part I: Mechanical model. *International Journal for Numerical Methods in Engineering* 53 (11): 2549–2568.

Pathak, H., Singh, A., and Singh, I.V. (2013a). Fatigue crack growth simulations of 3-D problems using XFEM. *International Journal of Mechanical Sciences* 76: 112–131.

Pathak, H., Singh, A., Singh, I.V., and Yadav, S.K. (2013b). A simple and efficient XFEM approach for 3-D cracks simulations. *International Journal of Fracture* 181 (2): 189–208.

Pathak, H., Singh, A., Singh, I.V., and Yadav, S.K. (2015a). Fatigue crack growth simulations of 3-D linear elastic cracks under thermal load by XFEM. *Frontiers of Structural and Civil Engineering* 9 (4): 359–382.

Pathak, H., Singh, A., Singh, I.V., and Brahmankar, M. (2015b). Three-dimensional stochastic quasi-static fatigue crack growth simulations using coupled FE-EFG approach. *Computers & Structures* 160: 1–19.

Pathak, H., Singh, A., and Singh, I.V. (2017). Numerical simulation of 3D thermo-elastic fatigue crack growth problems using coupled FE-EFG approach. *Journal of the Institution of Engineers (India): Series C* 98 (3): 295–312.

Sethian, J.A. (1999). Fast marching methods. *SIAM Review* 41 (2): 199–235.

Shi, J., Chopp, D., Lua, J., et al. (2010). Abaqus implementation of extended finite element method using a level set representation for three-dimensional fatigue crack growth and life predictions. *Engineering Fracture Mechanics* 77 (14): 2840–2863.

Sukumar, N. and Prévost, J.H. (2003). Modeling quasi-static crack growth with the extended finite element method Part I: Computer implementation. *International Journal of Solids and Structures* 40 (26): 7513–7537.

Sukumar, N., Chopp, D.L., Béchet, E., and Moës, N. (2008). Three-dimensional non-planar crack growth by a coupled extended finite element and fast marching method. *International Journal for Numerical Methods in Engineering* 76 (5): 727–748.

11

XFEM Modeling of Cracked Elastic-Plastic Solids

Emilio Martínez-Pañeda

Department of Civil and Environmental Engineering, Imperial College London, London, UK

11.1 Introduction

The eXtended finite element method (XFEM) was developed initially for elastic solids and its applications have been mostly limited to the context of linear elastic fracture mechanics (Li et al., 2018; Moës et al., 1999). Previous chapters exploited this approximation. However, most materials exhibit inelastic behavior, particularly in the vicinity of cracks and other stress concentrators. For example, characterizing the local crack tip behavior in metals inevitably requires taking into consideration the role of dislocation densities and plastic deformations (Martínez-Pañeda et al., 2019). A variety of material models can be used to characterize metallic fracture and several of them have been successfully coupled with the XFEM, including conventional von Mises plasticity (Elguedj et al., 2006; Martin et al., 2015), Lemaitre's damage-plasticity model (Broumand and Khoei, 2013), and strain gradient plasticity (Martínez-Pañeda et al., 2017). This chapter describes some of these endeavors and provides an introduction to the modeling of fracture problems in elastic-plastic solids using XFEM.

11.2 Conventional von Mises Plasticity

We shall start by the constitutive model that is arguably the most widely used to describe material deformation in elastic-plastic solids: von Mises J2 plasticity theory (Bower, 2009; Dunne and Petrinic, 2005). First, the main constitutive features of J2 flow theory are presented in Section 11.2.1. Then, the resulting asymptotic crack tip characterization is given in Section 11.2.2. The XFEM enrichment follows in Section 11.2.3 and relevant details of the numerical implementation are given in Section 11.2.4. Finally, representative results are provided in Section 11.2.5.

11.2.1 Constitutive Model

A summary of the main features of conventional von Mises plasticity is summarized below, the reader is referred to standard textbooks such as Bower (2009) and Dunne and Petrinic (2005) for more details. Small strains will be assumed for simplicity.

Partition of Unity Methods, First Edition. Stéphane P. A. Bordas, Alexander Menk, and Sundararajan Natarajan.
© 2024 John Wiley & Sons Ltd. Published 2024 by John Wiley & Sons Ltd.

Strain decomposition. The total strains ε_{ij} are additively decomposed into the elastic ε_{ij}^e and plastic ε_{ij}^p strain tensors:

$$\varepsilon_{ij} = \varepsilon_{ij}^e + \varepsilon_{ij}^p \tag{11.1}$$

Incompressibility condition. Plastic deformation takes place without volume change, implying that the sum of the axial plastic strain rate components is zero:

$$\varepsilon_{kk} = \dot{\varepsilon}_{11}^p + \dot{\varepsilon}_{22}^p + \dot{\varepsilon}_{33}^p = 0 \tag{11.2}$$

Yield condition. A yield condition is needed to determine if we are in the elastic or plastic regime. For a given material yield stress σ_y and effective stress σ_e, plasticity will take place when the following yield function is equal to zero.

$$f = \sigma_e - \sigma_y \tag{11.3}$$

It remains to define the effective stress σ_e. In the context of von Mises plasticity theory, this is done taking into consideration that: (i) yield is independent of the hydrostatic stress, (ii) yield in polycrystalline metals can be taken to be isotropic, and (iii) the yield condition is the same for compression and tension. In von Mises plasticity, the effective stress is defined assuming that yielding occurs when the distortion energy w_d reaches a critical value. w_d is the deviatoric part of the elastic strain energy density:

$$w_d = \frac{1+\nu}{2E} \mathrm{tr}\left(\left(\sigma_{ij}'\right)^2\right) \tag{11.4}$$

where σ_{ij}' is the deviatoric stress tensor, E is Young's modulus, and ν is Poisson's ratio. Using the principal stress components:

$$w_d = \frac{1+\nu}{6E}\left((\sigma_I - \sigma_{II})^2 + (\sigma_{II} - \sigma_{III})^2 + (\sigma_{III} - \sigma_I)^2\right) \tag{11.5}$$

In uniaxial tension, it reads:

$$w_d = \frac{1+\nu}{6E}\sigma_e^2 \tag{11.6}$$

Substituting in (11.3) and rearranging we reach the yielding function:

$$f\left(\sigma_{ij}\right) = \frac{1}{\sqrt{2}}\left[(\sigma_I - \sigma_{II})^2 + (\sigma_{II} - \sigma_{II})^2 + (\sigma_{III} - \sigma_I)^2\right]^{1/2} = \sigma_y \tag{11.7}$$

Thus, one can define the effective stress as:

$$\sigma_e = \sqrt{\frac{1}{2}\left[(\sigma_I - \sigma_{II})^2 + (\sigma_I - \sigma_{III})^2 + (\sigma_{II} - \sigma_{III})^2\right]} = \sqrt{\frac{3}{2}\sigma_{ij}'\sigma_{ij}'} \tag{11.8}$$

$$= \sqrt{\frac{1}{2}\left[(\sigma_{11} - \sigma_{22})^2 + (\sigma_{11} - \sigma_{33})^2 + (\sigma_{22} - \sigma_{33})^2 + 6\sigma_{12}^2 + 6\sigma_{13}^2 + 6\sigma_{23}^2\right]}$$

Plastic flow rule. The normality condition tells us the *direction* of plastic flow after yield. As sketched in Figure 11.1, assuming von Mises plasticity (associated flow), the increment in the plastic strain tensor is in a direction which is normal to the yield surface.

The flow rule is then given as a function of the yield function f and the plastic multiplier λ:

$$\dot{\varepsilon}_{ij}^p = \lambda \frac{\partial f}{\partial \sigma_{ij}} \tag{11.9}$$

where λ gives the magnitude of the plastic strain rate and $\partial f / \partial \sigma_{ij}$ gives the direction of the plastic strain increment. The plastic multiplier is obtained from the consistency condition: the requirement for the load point to remain in

Figure 11.1 The von Mises yield surface for conditions of plane stress, showing the increment in plastic strain $d\varepsilon_{ij}^p$, in a direction normal to the tangent to the surface.

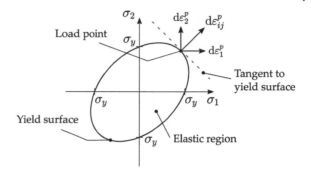

the yield surface during plastic deformation. In von Mises plasticity, the plastic multiplier is equivalent to the effective plastic strain, which is defined as:

$$\varepsilon_p = \sqrt{\frac{2}{3}\varepsilon_{ij}^p\varepsilon_{ij}^p} \tag{11.10}$$

and accordingly the plastic flow law reads:

$$\dot{\varepsilon}_{ij}^p = \frac{3}{2}\dot{\varepsilon}_p \frac{\sigma_{ij}'}{\sigma_e} \tag{11.11}$$

Hardening behavior. Most metals harden when deformed plastically; that is, the stress required to cause further plastic deformation increases, often as a function of accumulated plastic strain. If the yield surface expands uniformly in all directions in the stress space, the work hardening behavior is referred to as *isotropic*. On the other hand, the yield surface might translate in the stress space, in what is referred to as *kinematic* hardening. Notable differences are observed between isotropic and kinematic work hardening behaviors under reverse loading conditions. Accordingly, kinematic hardening or combined isotropic-kinematic hardening models are commonly used in the analysis of fatigue cracking. This is not the case of monotonic, static fracture as both are equivalent in a deformation theory solid (proportional loading). However, one must note that even if the case of monotonic fracture, kinematic hardening effects should not be neglected as they become significant with crack advance and the associated non-proportional straining (Juul et al., 2019; Martínez-Pañeda and Fleck, 2018). Nevertheless, isotropic hardening behavior will be assumed here for simplicity.

Within the realm of isotropic hardening behavior, different hardening laws can be used to relate the flow stress to the equivalent strain. A popular choice is the so-called Ramberg-Osgood power law, which in a uniaxial stress-strain ($\sigma - \varepsilon$) setting reads:

$$\frac{\varepsilon}{\varepsilon_y} = \frac{\sigma}{\sigma_y} + \alpha \left(\frac{\sigma}{\sigma_y}\right)^n \tag{11.12}$$

where $\varepsilon_y = \sigma_y/E$ is the yield strain, α is a dimensionless constant, and n is the strain hardening exponent.

11.2.2 Asymptotic Crack Tip Fields

The first step in providing an enriched description of crack tip fields is to characterize the asymptotic nature of crack tip fields. In solids characterized by conventional von Mises plasticity, this can be achieved following the pioneering work by Hutchinson (1968) and Rice and Rosengren (1968), in what is commonly referred to as the HRR singularity analysis.

Consider a polar coordinate system (r, θ) centered at the crack tip. For a solid with strain energy density w and a hardening behavior characterized by Ramberg-Osgood's power law (11.12), the HRR analysis leads to the

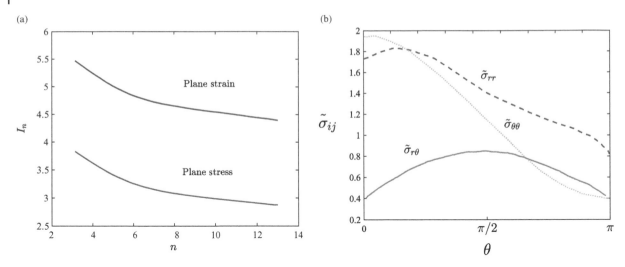

Figure 11.2 HHR solution (Hutchinson, 1968; Rice and Rosengren, 1968): (a) effect of the strain hardening exponent on the integration constant, and (b) angular variation of the dimensionless stress $\tilde{\sigma}_{ij}$ for $n = 3$ and plane strain.

following scaling of the stresses, strains, and displacements with distance r ahead of the crack:

$$\sigma_{ij} = \sigma_y \left(\frac{J}{\alpha \sigma_y \varepsilon_y I_n r} \right)^{1/(n+1)} \tilde{\sigma}_{ij}(\theta, n) \tag{11.13}$$

$$\varepsilon_{ij} = \alpha \varepsilon_y \left(\frac{J}{\alpha \sigma_y \varepsilon_y I_n r} \right)^{n/(n+1)} \tilde{\varepsilon}_{ij}(\theta, n) \tag{11.14}$$

$$u_i = \alpha \varepsilon_y r \left(\frac{J}{\alpha \sigma_y \varepsilon_y I_n r} \right)^{n/(n+1)} \tilde{u}_i(\theta, n) \tag{11.15}$$

where I_n is an integration constant that depends on n, and $\tilde{\sigma}_{ij}$, $\tilde{\varepsilon}_{ij}$, and $\tilde{u}_i$ are dimensionless angular functions of n and θ. The sensitivity of I_n to the strain hardening exponent is shown in Figure 11.2a for both plane stress and plane strain, while the angular functions are shown in Figure 11.2b for the specific choice of $n = 3$ and plane strain.

The HRR singularity analysis builds upon the assumption that elastic strains are negligible at the crack tip and upon Rice's J-integral (Eshelby, 1956; Rice and Rosengren, 1968). For a solid with strain energy density w, the J-integral reads:

$$J = \int_\Gamma \left(w n_1 - \sigma_{ij} n_j \frac{\partial u_i}{\partial x_1} \right) ds \tag{11.16}$$

where, as sketched in Figure 11.3, Γ is an arbitrary counter-clockwise path around the tip of a crack and n_i is the unit outward normal to Γ.

A detailed derivation of the HRR singularity is out of the scope (the interested reader is referred to Hutchinson (1968) and Rice and Rosengren (1968)) but the J-integral concept can be easily used to show that the power of the stress singularity is given by $1/(n+1)$ in the vicinity of the crack tip. Consider a solid characterized by a power law hardening rule such as $\varepsilon/\varepsilon_y = (\sigma/\sigma_y)^n$; the strain energy density reads:

$$w = \int_0^\varepsilon \sigma \, d\varepsilon = \frac{n}{n+1} \sigma_y \varepsilon_y \left(\frac{\sigma}{\sigma_y} \right)^{n+1} \tag{11.17}$$

Figure 11.3 Arbitrary contour around the tip of a crack.

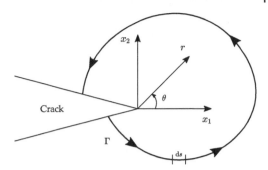

The existence of a J-integral, (11.16), implies that the strain energy density of the solid will asymptotically behave as $w \sim J/r$ so as to give a finite energy release rate J at the crack tip. Accordingly,

$$w = c\frac{J}{r} \tag{11.18}$$

where c is a constant. Consequently,

$$\left(\frac{\sigma}{\sigma_y}\right)^{n+1} = \frac{n+1}{n}\frac{c}{\sigma_y \varepsilon_y}\frac{J}{r} \tag{11.19}$$

Leading to a stress and strain singularity of the type:

$$\sigma \propto \left(\frac{J}{r}\right)^{1/(n+1)} \quad \text{and} \quad \varepsilon \propto \left(\frac{J}{r}\right)^{n/(n+1)} \tag{11.20}$$

11.2.3   XFEM Enrichment

Once adequate asymptotic solutions have been identified, one can proceed to enrich the shape function basis. The following is based on the enrichment basis proposed by Elguedj et al. (2006).

First, recall that the XFEM builds upon the partition of unity method to enrich the finite element (FE) approximation by incorporating the nature of the solution. Thus, the displacement approximation can be decomposed into a standard part and an enriched part:

$$u_i^h = \underbrace{\sum_{I\in\mathcal{N}^{\text{fem}}} N_i^I u_i^I}_{\text{Standard}} + \underbrace{\sum_{J\in\mathcal{N}^c} N_i^J H(\phi)a_i^J + \sum_{K\in\mathcal{N}^{\text{f}}} N_i^K \sum_{\alpha=1}^{n} F_\alpha(r,\theta)b_i^{K\alpha}}_{\text{Enriched}} \tag{11.21}$$

where $\mathcal{N}^{\text{fem}}$ is the set of all nodes in the FE mesh, $\mathcal{N}^c$ is the set of nodes whose shape function support is cut by the crack interior, and $\mathcal{N}^{\text{f}}$ is the set of nodes whose shape function support is cut by the crack tip. $H(\phi)$ and $F_\alpha(r,\theta)$ are the enrichment functions chosen to respectively capture the displacement jump across the crack surface and the singularity at the crack tip, with a_i^J and $b_i^{K\alpha}$ being their associated degrees of freedom (DOFs). The Heaviside jump function $H(\phi)$ has been described in detail elsewhere in this book (see Chapters 2 and 3) and will not be addressed here. The focus is on the functions with singular derivative near the crack that spans the near-tip stress field, $F_\alpha(r,\theta)$. Recall that for the linear elastic case and plane strain conditions, the crack tip asymptotic displacement solutions are given by:

$$u_1(r,\theta) = \frac{1+\nu}{E}\sqrt{\frac{r}{2\pi}}\left(K_I \cos\left(\frac{\theta}{2}\right)(3 - 4\nu - \cos\theta) + K_{II}\sin\left(\frac{\theta}{2}\right)(5 - 4\nu + \cos\theta)\right) \tag{11.22}$$

$$u_2(r,\theta) = \frac{1+\nu}{E}\sqrt{\frac{r}{2\pi}}\left(K_I\sin\left(\frac{\theta}{2}\right)(3-4\nu-\cos\theta)-K_{II}\cos\frac{\theta}{2}(1-4\nu\cos\theta)\right) \qquad (11.23)$$

where K_I and K_{II} respectively denote the mode I and mode II stress intensity factors. Accordingly, the following basis is used:

$$F_\alpha(r,\theta) = r^{1/2}\left\{\sin\frac{\theta}{2}, \cos\frac{\theta}{2}, \sin\frac{\theta}{2}\sin\theta, \cos\frac{\theta}{2}\sin\theta\right\} \qquad (11.24)$$

In the case of elastic-plastic solids characterized by conventional von Mises plasticity, evaluating the asymptotic crack tip fields requires the use of numerical methods (Symington et al., 1990). Alternatively, the HRR singular fields can be approximated by the use of Fourier series (Elguedj et al., 2006). Thus, consider the HRR solution (11.15), a Fourier decomposition of the functions $\tilde{u}_i(\theta, n)$ is carried out for both mode I and mode II. As described in Elguedj et al. (2006), these functions are periodized on the interval $[0; 4\pi]$ by conserving the symmetry and anti-symmetry properties of the linear elastic fields, and considering the variable to be $\theta/2$ instead of θ. A standard Fourier analysis follows, by which a function f of a real parameter t with period T is approximated as:

$$f(t) = \frac{1}{T}\int_\alpha^{\alpha+T} f(t)\,\mathrm{d}t + \sum_{n=1}^\infty\left(\cos(n\omega t)\frac{2}{T}\int_\alpha^{\alpha+T} f(t)\cos(n\omega t)\,\mathrm{d}t + \sin(n\omega t)\frac{2}{T}\int_\alpha^{\alpha+T} f(t)\cos(n\omega t)\,\mathrm{d}t\right)$$

$$(11.25)$$

where α is a given real and n a positive integer. It is found that the only non-zero harmonics are given by $\cos(k\theta/2)$ and $\sin(k\theta/2)$, where k is a natural number. Also, the Fourier decomposition of solutions with n values between 1 and 100 shows that the HRR fields can be accurately approximated by using a truncated Fourier expansion with only the first four non-zero harmonics for each function (Elguedj et al., 2006). Accordingly, the following basis can be used to approximate the HRR displacement field:

$$F_\alpha(r,\theta) = r^{1/(n+1)}\left\{\left(\cos\frac{k\theta}{2}, \sin\frac{k\theta}{2}\right); k \in [1, 3, 5, 7]\right\} \qquad (11.26)$$

Three bases are examined in Elguedj et al. (2006), resulting in the choice of the following:

$$F_\alpha(r,\theta) = r^{1/(n+1)}\left\{\sin\frac{\theta}{2}, \cos\frac{\theta}{2}, \sin\frac{\theta}{2}\sin\theta, \cos\frac{\theta}{2}\sin\theta, \sin\frac{\theta}{2}\sin 3\theta, \cos\frac{\theta}{2}\sin 3\theta\right\} \qquad (11.27)$$

as done by Rao and Rahman (2004) in the context of the element-free Galerkin method.

11.2.4 Numerical Implementation

Several aspects have to be taken into consideration in regard to the numerical implementation, many of which differ from a standard XFEM implementation for linear elastic solids. First, the nonlinear nature of the elastic-plastic problem has to be resolved. This is typically addressed using the Newton-Raphson method in combination with a radial return mapping algorithm; the interested reader is referred to classic textbooks in computational inelasticity such as Bonet and Wood (1997) and Simo and Hughes (2006) for further details.

Second, there are considerations inherent to the nature of the enrichment that need to be taken care of. As elaborated in Elguedj et al. (2006), trigonometric identities are used to employ only one basis function ($\sin\theta/2$) with discontinuity between $\theta = \pi$ and $\theta = -\pi$. The dimension of the enrichment basis is limited to the choice (11.27) and the integration scheme should be improved to deal with the resulting ill-conditioning of the stiffness matrix. Only one element, the one containing the crack tip, is enriched. This limits the applicability to conditions of confined plasticity, where the plastic zone is notably small. However, these conditions are relevant to a number of technologically-relevant phenomena such as fatigue damage or stress corrosion cracking in high-strength alloys (Martínez-Pañeda et al., 2016; Turnbull, 1993).

Unlike in the standard elasticity case, there is a need for an enhanced integration scheme due to the high-order trigonometric Fourier terms present in the elastic-plastic basis functions. Also, one must take care of projecting history-dependent variables in the case of crack growth and the associated re-meshing. One possibility is to partition cut elements into sub-quadrilaterals, with a large number of integration points in each sub-quadrilateral (Ji et al., 2002). These sub-elements can be classified into two sets to address the compatibility issues arising due to the discontinuity of the enrichment functions on the crack faces (Elguedj et al., 2006).

11.2.5 Representative Results

Representative results from the pioneering work by Elguedj et al. (2006) are shown below to showcase the benefit of the enrichment strategy. Both mode I (Section 11.2.5.1) and mixed-mode (Section 11.2.5.2) fracture conditions are assessed and compared with a reference FE solution. In both cases, the role of material hardening is investigated by considering two materials, one with a small degree of hardening ($n = 30$) and another one with a relative high sensitivity to work hardening ($n = 3.7$). The predictions from the standard and enriched models are compared by plotting the evolution of two different quantities as a function of the loading factor (computational step). One variable is the displacement jump between the crack faces $[u]$, which is evaluated at the intersection between the crack faces and the boundary of the crack tip element, in the normal direction to the crack faces. The second variable is the J-integral, see Equation (11.16).

11.2.5.1 Mode I Fracture

The mode I fracture example deals with a single-edge notched specimen subjected to a remote traction σ^∞. The geometry and dimensions are shown in Figure 11.4, where the width equals $W = 200$ mm, the height is $2L = 500$ mm, and the length of the initial crack is $a = 100$ mm. The sample is first subjected to a monotonically increasing load that is then reversed in the same number of steps. As shown in Figure 11.4, the mesh for the XFEM analysis consists of 380 linear quadrilateral elements, with a total of 548 DOFs. The reference FE result is computed with a much finer mesh, also shown in Figure 11.4, consisting of 892 quadratic triangular elements and 4106 DOFs.

The results obtained for the evolution of the displacement jump $[u]$ are shown in Figure 11.5, for both the cases of $n = 3.7$ and $n = 30$. The reference FE result is shown, together with the XFEM prediction and the percentage difference between them. A very good agreement between the enriched formulation and the reference FE result is observed, despite the notable differences in mesh discretization. The agreement is particularly good for the case of a material with substantial strain hardening ($n = 3.7$), where the percentage difference is below 1% in all cases. For the case of $n = 30$, these differences increase up to a maximum of roughly 3%.

Figure 11.4 Mode I fracture case study: geometry and finite element mesh, for both the XFEM and standard finite element cases. *Source:* Adapted from Elguedj et al. (2006).

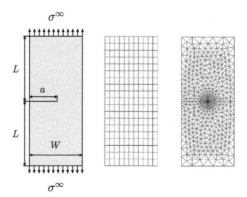

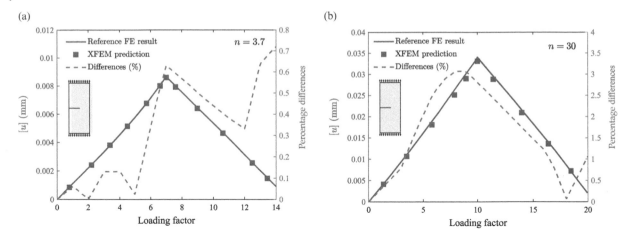

Figure 11.5 Mode I fracture. Displacement jump [u] predictions for (a) $n = 3.7$ and (b) $n = 30$. Data digitized from Elguedj et al. (2006).

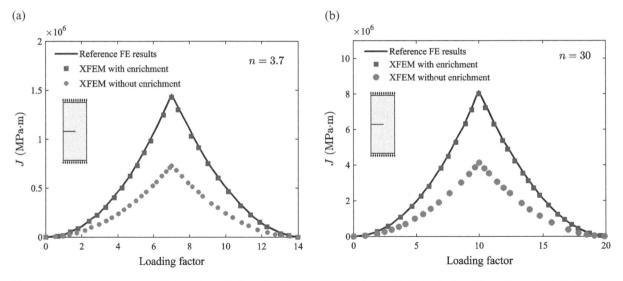

Figure 11.6 Mode I fracture. J-integral predictions for (a) $n = 3.7$ and (b) $n = 30$. Data digitized from Elguedj et al. (2006).

The predicted evolution of the J-integral is shown in Figure 11.6. In this case, the results corresponding to an XFEM analysis *without* the HRR enrichment are also included, in addition, those relevant to the reference FE model and the enriched solution. For both $n = 3.7$ and $n = 30$, a very good agreement is observed between the HRR-enriched XFEM solution and the reference result throughout the loading history. However, the XFEM analysis without enrichment fails to capture the correct evolution of J, revealing significant differences and evidencing the benefits of the enrichment strategy presented in Section 11.2.3.

11.2.5.2 Mixed Mode Fracture

The capabilities of the enrichment in capturing the local crack tip behavior have also been demonstrated under mixed mode fracture. As shown in Figure 11.7, a plate of dimensions $W = 200$ mm and $L = 250$ mm is subjected to a monotonically increasing remote tensile stress σ^∞. To induce crack tip mixed-mode conditions, the initial crack (of length $a = 111.8$ mm) is inclined $\beta = 26.5°$ relative to the direction perpendicular to the applied load. The FE meshes employed for the reference FE case and the XFEM analysis are also shown in Figure 11.7.

Figure 11.7 Mixed-mode fracture case study: geometry and finite element mesh, for both the XFEM and standard finite element cases. *Source:* Adapted from Elguedj et al. (2006).

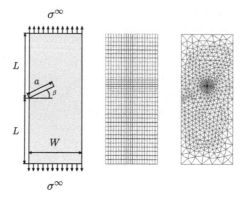

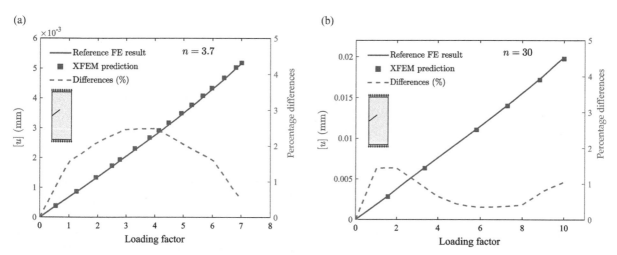

Figure 11.8 Mixed-mode fracture. Displacement jump [*u*] predictions for (a) *n* = 3.7 and (b) *n* = 30. Data digitized from Elguedj et al. (2006).

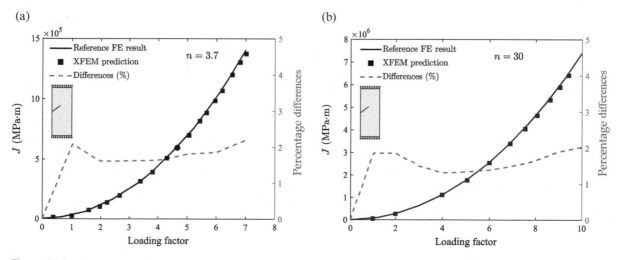

Figure 11.9 Mixed-mode fracture. *J*-integral predictions for (a) *n* = 3.7 and (b) *n* = 30. Data digitized from Elguedj et al. (2006).

The results obtained by Elguedj et al. (2006) for the mixed-mode analysis are reported in Figures 11.8 and 11.9. First, consider the case of the displacement jump $[u]$ predictions, Figure 11.8. Results show a very good agreement with the reference solution for both the cases of a material with low hardening capacity ($n = 30$) and a material with a higher one ($n = 3.7$). The % difference with the reference FE solution is higher than the one observed under mode I conditions for $n = 3.7$ but smaller for $n = 30$. In all cases, it remains below 3%.

Finally, the results obtained in terms of the J-integral are shown in Figure 11.9. As for the displacement jump, a very good agreement with the reference FE solution is observed for both $n = 3.7$ and $n = 30$, with differences remaining below 3% throughout the loading history.

11.3 Strain Gradient Plasticity

Conventional plasticity models fail to capture the dislocation hardening mechanisms taking place in the vicinity of the crack tip (Komaragiri et al., 2008; Martínez-Pañeda and Niordson, 2016; Wei and Hutchinson, 1997). It is now widely accepted that the presence of plastic strain gradients in small volumes leads to additional material hardening, as shown in experiments such as the torsion of thin wires (Bardella and Panteghini, 2015; Fleck et al., 1994; Guo et al., 2017), the bending of foils (Martínez-Pañeda et al., 2016; Stölken and Evans, 1998), the shear of constrained films (Dahlberg and Ortiz, 2019; Mu et al., 2014) or micro/nano-indentation (Nix and Gao, 1998; Poole et al., 1996), where conditions resemble those ahead of a crack tip. A threefold increase in the effective flow stress is observed in these experiments relative to the case of negligible strain gradients. Plastic strain gradients are commonly very large in the vicinity of cracks due to the small physical size of the plastic zone adjacent to the crack tip. Thus, it is expected that the aforementioned hardening mechanisms will be present to a great extent, elevating crack tip stresses (Hutchinson, 1997; Martínez-Pañeda and Betegón, 2015). In fact, this stress elevation, not accounted for in conventional plasticity models, has been used to rationalize quasi-brittle failure phenomena in metals, which requires attaining very high stresses, exceeding the material cleavage strength, over a physically meaningful distance. Examples include atomic decohesion at bi-material interfaces (Fuentes-Alonso and Martínez-Pañeda, 2020; Jiang et al., 2010), hydrogen embrittlement (Martínez-Pañeda et al., 2016, 2020) and low-temperature cleavage of ferritic steels (Martínez-Pañeda et al., 2019; Qian et al., 2011).

To capture the stress elevation associated with plastic strain gradients, significant efforts have been allocated to the development of so-called strain gradient plasticity theories (see, e.g., Fleck and Hutchinson (1997), Fleck and Hutchinson (2001), Gao et al. (1999), Gudmundson (2004), and Gurtin and Anand (2005) and references therein), which incorporate relevant gradient quantities in their constitutive equations, along with one or more associated length scale parameters (as required due to dimensional consistency). Strain gradient plasticity models have been extensively used to accurately characterize crack tip mechanics and develop enriched damage models for phenomena that require resolving the microscale (see, e.g., Brinckmann and Siegmund (2008), Kristensen et al. (2020), Martínez-Pañeda et al. (2016), Pribe et al. (2019), Srinivasan et al. (2008), and Tvergaard and Niordson (2004) and references therein). However, the additional physical insight provided by strain gradient plasticity comes at a cost; an appropriate characterization of gradient effects ahead of a crack requires the use of extremely refined meshes, with a characteristic element length of a few nanometers in the vicinity of the crack. This in turn leads to convergence issues as these small elements are more prone to distortion in finite strain analyses involving loads of sufficiently high magnitude (Pan and Yuan, 2011a,b). But if the singular nature of the asymptotic crack tip fields is known, the XFEM can be used to enrich the solution and significantly alleviate mesh refinement. This section is based on the XFEM implementation for strain gradient plasticity developed by Martínez-Pañeda et al. (2017) and aims at describing both relevant results and the numerical implementation, including the constitutive model, the asymptotic crack tip characterization, and the XFEM enrichment.

11.3.1 Constitutive Model

The potential of XFEM in optimizing the numerical characterization of crack tip fields using strain gradient plasticity will be demonstrated using the mechanism-based strain gradient (MSG) plasticity theory developed by Gao et al. (1999). The MSG model is based on a multiscale framework linking the microscale concept of statistically stored dislocations (SSDs) and geometrically necessary dislocations (GNDs) to the mesoscale notion of plastic strains and strain gradients. Unlike other gradient plasticity formulations, MSG plasticity introduces a linear dependence of the square of plastic flow stress on strain gradients. This linear dependence is motivated by the nano-indentation experiments by Nix and Gao (1998) and comes out naturally from Taylor's dislocation model (Taylor, 1938), on which MSG plasticity is built. Therefore, while all continuum formulations have a strong phenomenological component, MSG plasticity differs from most gradient plasticity models in its mechanism-based guiding principles. The constitutive equations are summarized below, more details can be found in Gao et al. (1999).

In MSG plasticity, since the Taylor model is adopted as a founding principle, the shear flow stress τ is formulated in terms of the dislocation density ρ as:

$$\tau = \alpha \mu b \sqrt{\rho} \tag{11.28}$$

where μ is the shear modulus, b is the magnitude of the Burgers vector, and α is an empirical coefficient which takes values between 0.3 and 0.5. The dislocation density is composed of the sum of the density ρ_S for SSDs and the density ρ_G for GNDs, such that:

$$\rho = \rho_S + \rho_G \tag{11.29}$$

The GND density ρ_G is related to the effective plastic strain gradient η^p by:

$$\rho_G = \bar{r} \frac{\eta^p}{b} \tag{11.30}$$

where $\bar{r}$ is the Nye-factor which is assumed to be 1.90 for face-centered-cubic (fcc) polycrystals. Following Fleck and Hutchinson (1997), three quadratic invariants of the plastic strain gradient tensor are used to represent the effective plastic strain gradient η^p as:

$$\eta^p = \sqrt{c_1 \eta^p_{iik} \eta^p_{jjk} + c_2 \eta^p_{ijk} \eta^p_{ijk} + c_3 \eta^p_{ijk} \eta^p_{kji}} \tag{11.31}$$

The coefficients have been determined to be equal to $c_1 = 0$, $c_2 = 1/4$, and $c_3 = 0$ from three dislocation models for bending, torsion, and void growth, leading to:

$$\eta^p = \sqrt{\frac{1}{4} \eta^p_{ijk} \eta^p_{ijk}} \tag{11.32}$$

where the components of the strain gradient tensor are obtained by $\eta^p_{ijk} = \varepsilon^p_{ik,j} + \varepsilon^p_{jk,i} - \varepsilon^p_{ij,k}$. The tensile flow stress σ_{flow} is related to the shear flow stress τ by:

$$\sigma_{flow} = M\tau \tag{11.33}$$

where M is the Taylor factor, taken to be 3.06 for fcc metals. Rearranging Equations (11.28–11.30) and Equation (11.33) yields:

$$\sigma_{flow} = M\alpha \mu b \sqrt{\rho_S + \bar{r} \frac{\eta^p}{b}} \tag{11.34}$$

The SSD density ρ_S can be determined from (11.34) knowing the relation in uniaxial tension between the flow stress and the material stress-strain curve as follows:

$$\rho_S = \left(\frac{\sigma_{ref} f(\varepsilon^p)}{M \alpha \mu b} \right)^2 \tag{11.35}$$

Here, σ_{ref} is a reference stress and f is a non-dimensional function of the plastic strain ε^p determined from the uniaxial stress-strain curve. Substituting back into (11.34), σ_{flow} yields:

$$\sigma_{flow} = \sigma_{ref} \sqrt{f^2(\varepsilon^p) + \ell \eta^p} \tag{11.36}$$

where ℓ is the intrinsic material length. It can be readily seen that if the characteristic length of plastic deformation is much larger than ℓ, the gradient-related term $\ell \eta^p$ becomes negligible and the flow stress degenerates to $\sigma_{ref} f(\varepsilon^p)$, as in conventional plasticity. That is, making $\ell = 0$ recovers the conventional von Mises plasticity result. Also, one should note that the model is built on the assumption of a collective behavior of dislocations, implying that it is only applicable at a scale much larger than the average dislocation spacing, i.e., distances of 100 nm or larger.

11.3.2 Asymptotic Crack Tip Fields

As shown semi-analytically by Shi et al. (2001) and numerically by Jiang et al. (2001), solids characterized by MSG plasticity theory exhibit a distinct stress singularity ahead of crack tips, different from those intrinsic to linear elastic or conventional elastic-plastic solids. The asymptotic crack tip solution for MSG plasticity is here derived, following Shi et al. (2001).

The following asymptotic crack tip analysis builds upon the assumption that plastic strains dominate close to the crack tip ($\varepsilon_{ij} \equiv \varepsilon_{ij}^p$). One should note that the $\varepsilon_{ij}^p \gg \varepsilon_{ij}^e$ assumption at the crack tip has proven to be inappropriate for the Gudmundson (2004) and Gurtin (2004) classes of strain gradient plasticity theories (Fuentes-Alonso and Martínez-Pañeda, 2020; Martínez-Pañeda and Fleck, 2019). Dropping the p superscript, the strains and plastic strain gradients are related to the displacement field u_i as follows:

$$\varepsilon_{ij} = \frac{1}{2} \left(u_{i,j} + u_{j,i} \right), \qquad \eta_{ijk} = u_{k,ij} \tag{11.37}$$

If, as here, elastic deformation is assumed to be negligible at the crack tip, the incompressibility condition requires $\varepsilon_{ii} = 0$ and $\eta_{kii} = 0$. Also, symmetry considerations dictate $\varepsilon_{ij} = \varepsilon_{ji}$ and $\eta_{ijk} = \eta_{jik}$. In its higher order version, the model includes higher order stresses τ_{ijk}, work conjugate to plastic strain gradients, and the equilibrium equations for an incompressible solid are given by:

$$\sigma'_{ik,i} - \tau'_{ijk,ij} + H_{,k} = 0 \tag{11.38}$$

where body forces have been neglected and H is the combined measure of the hydrostatic stress and the hydrostatic higher order stress. Higher order stresses dominate in a region adjacent to the crack tip, where crack tip fields are non-separable (Shi et al., 2001). However, this region is smaller than the domain of physical validity of strain gradient plasticity theories, and the stress field becomes separable outside of it ($r > 0.1$ μm). Thus, we disregard the inner annular region where τ_{ijk} dominates the response and focus on the domain $r > 0.1$ μm, common to both lower order and higher order versions of MSG plasticity. Within this regime of gradient dominance and separable stress field, the deviatoric part of the Cauchy stress is related to the flow stress by:

$$\sigma'_{ij} = \frac{2\varepsilon_{ij}}{3\varepsilon} \sigma_{flow} \tag{11.39}$$

Now consider a mode I crack, with a polar coordinate system centered at the crack tip. As deformation is incompressible, we can define a displacement potential ϕ such that the displacement components are given by:

$$u_r = -\frac{1}{r} \frac{\partial \phi}{\partial \theta}, \qquad u_\theta = \frac{\partial \phi}{\partial r} \tag{11.40}$$

Following Shi et al. (2001), we assume a separable form for ϕ such that:

$$\phi = r^{3-2\lambda}\tilde{\phi}(\theta) \tag{11.41}$$

where λ is the power of the stress singularity (to be determined) and $\tilde{\phi}(\theta)$ is the angular distribution. Accordingly, the effective quantities and tensors related to strains and strain gradients read:

$$\varepsilon_{\alpha\beta} = r^{1-2\lambda}\tilde{\varepsilon}_{\alpha\beta}(\theta), \qquad \varepsilon = r^{1-2\lambda}\tilde{\varepsilon}(\theta) \tag{11.42}$$

$$\eta_{\alpha\beta\gamma} = r^{-2\lambda}\tilde{\eta}_{\alpha\beta\gamma}(\theta), \qquad \eta = r^{-2\lambda}\tilde{\eta}(\theta) \tag{11.43}$$

Here, the angular functions are determined in terms of $\tilde{\phi}(\theta)$ in agreement with (11.37). Considering the constitutive relations (11.36) and (11.39), the flow stress and the deviatoric stress tensor read:

$$\sigma_{flow} = \frac{3\alpha\mu\sqrt{2b}}{r^\lambda}\tilde{\eta}^{1/2}(\theta), \qquad \sigma'_{\alpha\beta} = \frac{2\alpha\mu\sqrt{2b}}{r^\lambda}\frac{\tilde{\eta}^{1/2}(\theta)}{\tilde{\varepsilon}(\theta)}\tilde{\varepsilon}_{\alpha\beta}(\theta) \tag{11.44}$$

and the hydrostatic stress is given by:

$$H = \frac{2\alpha\mu\sqrt{2b}}{r^\lambda}\frac{\tilde{\eta}^{1/2}(\theta)}{\tilde{\varepsilon}(\theta)}\tilde{H}(\theta) \tag{11.45}$$

where $\tilde{H}(\theta)$ is yet to be determined. Neglecting the role of the higher order stress ($r > 0.1\ \mu\text{m}$), the equilibrium Equation (11.38) translates into the following two ordinary different equations (ODEs) for the angular functions $\tilde{\phi}(\theta)$ and $\tilde{H}(\theta)$:

$$-\frac{\tilde{\eta}^{1/2}}{\tilde{\varepsilon}}\left[(2-\lambda)\tilde{\varepsilon}_{\theta\theta} + \lambda\tilde{H}\right] + \frac{d}{d\theta}\left[\frac{\tilde{\eta}^{1/2}}{\tilde{\varepsilon}}\tilde{\varepsilon}_{r\theta}\right] = 0$$

$$(2-\lambda)\frac{\tilde{\eta}^{1/2}}{\tilde{\varepsilon}}\tilde{\varepsilon}_{r\theta} + \frac{d}{d\theta}\left[\frac{\tilde{\eta}^{1/2}}{\tilde{\varepsilon}}(\tilde{\varepsilon}_{\theta\theta} + \tilde{H})\right] = 0 \tag{11.46}$$

From the symmetry condition at $\theta = 0$,

$$\tilde{\phi}|_{\theta=0} = 0, \qquad \tilde{\phi}'|_{\theta=0} = 0 \tag{11.47}$$

and all angular functions and their derivatives are bounded at $\theta = 0$, such that (11.46a) gives an additional boundary condition at $\theta = 0$:

$$-(2-\lambda)\tilde{\varepsilon}_{\theta\theta}|_{\theta=0} - \lambda\tilde{H}|_{\theta=0} + \frac{d\tilde{\varepsilon}_{r\theta}}{d\theta}|_{\theta=0} = 0 \tag{11.48}$$

On the crack faces, the stress tractions vanish, giving:

$$(\tilde{\varepsilon}_{\theta\theta} + \tilde{H})|_{\theta=\pi} = 0, \qquad \tilde{\varepsilon}_{r\theta}|_{\theta=\pi} = 0 \tag{11.49}$$

A normalization condition can be imposed such that:

$$\tilde{\phi}'|_{\theta=0} = \sin\beta_0, \qquad \tilde{\phi}''|_{\theta=0} = \cos\beta_0, \quad \text{with } (0 \leq \beta_0 \leq \pi) \tag{11.50}$$

such that:

$$\sqrt{\left(\tilde{\phi}'|_{\theta=0}\right)^2 + \left(\tilde{\phi}''|_{\theta=0}\right)^2} = 1 \tag{11.51}$$

Thus, for a given λ and β_0, the fifth-order ODE in (11.46) can be used using numerical methods since (11.47), (11.48), and (11.50) give five boundary conditions at $\theta = 0$; Runge-Kutta is used in Shi et al. (2001). The parameters

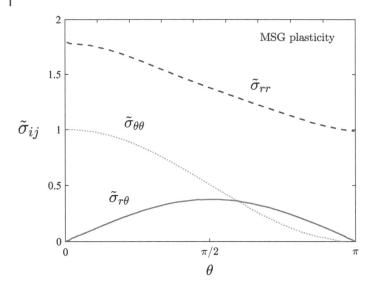

Figure 11.10 MSG plasticity. Angular variation of the dimensionless stress $\tilde{\sigma}_{ij}$ for mode I fracture and plane strain conditions. Values have been normalized such that $\tilde{\sigma}_{\theta\theta}\,|_{\theta=0}= 1$.

λ and β_0 can be determined by ensuring that the two crack-face traction-free boundary conditions (11.49) are met; Shi et al. (2001) used an iterative shooting method, obtaining:

$$\sigma_{\alpha\beta} = \frac{A}{r^\lambda}\tilde{\sigma}_{\alpha\beta}\left(\theta\right), \quad \text{with } \lambda = 0.63837 \tag{11.52}$$

where A is an amplitude factor that depends on the loading, specimen geometry and material properties. Thus, $\lambda \approx 2/3$ and the stress singularity is stronger not only than that of conventional plasticity, see (11.26), but also than linear elasticity (11.24). Also, it is worth emphasizing that, unlike the HRR field, the power of the stress singularity is independent of the strain hardening exponent n, albeit recent FE calculations show some sensitivity (Shlyannikov et al., 2021). From a physical viewpoint, an n-independent singularity indicates that the density of GNDs ρ_G in the vicinity of the crack tip is significantly larger than the density of SSDs ρ_S, and consequently the strain gradient term dominates the contribution to the flow stress (11.36). The angular functions are shown in Figure 11.10, where values have been normalized such that $\tilde{\sigma}_{\theta\theta}\,|_{\theta=0}= 1$.

11.3.3 XFEM Enrichment

Crack tip fields in elastic-plastic solids can be divided into three domains, as sketched in Figure 11.11. Far away from the crack tip, the deformation is purely elastic and the asymptotic stress field is governed by the linear elastic singularity: $\sigma_{ij} \sim r^{-1/2}$. When the effective stress overcomes the initial yield stress σ_Y, plastic deformations occur and the stress field is characterized by the HRR solution: $\sigma_{ij} \sim r^{-1/(n+1)}$. As the distance to the crack tip decreases to the order of the plastic length scale ℓ (1–10 μm for most metals), the presence of large plastic strain gradients promotes dislocation hardening and the stress field is described by the asymptotic stress singularity of strain gradient plasticity. In the case of MSG plasticity, as derived in Section 11.3.2, $\sigma_{ij} \sim r^{-2/3}$.

The XFEM enrichment basis presented here for MSG plasticity corresponds to that presented by Martínez-Pañeda et al. (2017). First, note that the angular functions of MSG plasticity (Figure 11.10) are similar to those of linear elasticity, particularly for the terms $\tilde{\sigma}_{\theta\theta}$ and $\tilde{\sigma}_{r\theta}$. Also, one should note that the angular functions play a negligible role in the overall representation of the asymptotic field (Duflot and Bordas, 2008). Thus, the functions used in the linear elastic fracture mechanics approximation are used in the enrichment basis. However, the power

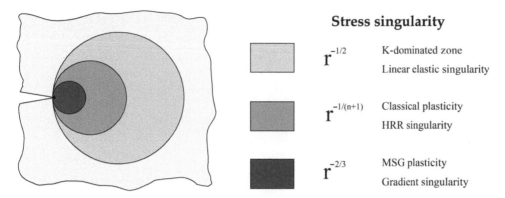

Stress singularity

$r^{-1/2}$ K-dominated zone
                Linear elastic singularity

$r^{-1/(n+1)}$ Classical plasticity
                HRR singularity

$r^{-2/3}$ MSG plasticity
                Gradient singularity

Figure 11.11 Schematic diagram of the different domains surrounding the crack tip. Three regions are identified as a function of asymptotic stress fields: the linear elastic solution, the HRR solution, and the MSG plasticity solution. *Source: Reproduced from Martínez-Pañeda et al. (2017)/Springer Nature.*

of the singularity does play a significant role and is chosen accordingly to the singular behavior exhibited by MSG plasticity solids – see Figure 11.11.

11.3.4   Numerical Implementation

As an example of numerical implementation of an XFEM code for strain gradient plasticity, the work by Martínez-Pañeda et al. (2017) is described. Two aspects of the implementation should be emphasized: the implementation of the MSG plasticity constitutive model and the numerical treatment of the enrichment region.

11.3.4.1   CMSG Plasticity: A Low-Order Version of MSG Plasticity

Since the influence of higher order stresses in MSG plasticity is restricted to a very small region falling outside of the domain of physical validity of strain gradient plasticity theories, a lower order version of MSG plasticity is considered for numerical implementation purposes. This is the so-called conventional mechanism-based strain gradient (CMSG) plasticity model (Huang et al., 2004), which employs the same equilibrium equations and boundary conditions as conventional continuum models.

Recall that Taylor's dislocation model gives the flow stress dependent on both the equivalent plastic strain ε^p and effective plastic strain gradient η^p

$$\dot{\sigma} = \frac{\partial \sigma}{\partial \varepsilon^p} \dot{\varepsilon}^p + \frac{\partial \sigma}{\partial \eta^p} \dot{\eta}^p \tag{11.53}$$

such that, for a plastic strain rate ε^p_{ij} proportional to the deviatoric stress σ'_{ij}, a self contained constitutive model cannot be obtained due to the term $\dot{\eta}^p$. In order to overcome this situation without employing higher order stresses, Huang et al. (2004) adopted a viscoplastic formulation to obtain $\dot{\varepsilon}^p$ in terms of the effective stress σ_e rather than its rate $\dot{\sigma}_e$:

$$\dot{\varepsilon}^p = \dot{\varepsilon} \left[\frac{\sigma_e}{\sigma_{ref} \sqrt{f^2(\varepsilon^p) + \ell \eta^p}} \right]^m \tag{11.54}$$

The viscoplastic-limit approach developed by Kok et al. (2002) is used to suppress strain rate and time dependence by replacing the reference strain rate $\dot{\varepsilon}_0$ with the effective strain rate $\dot{\varepsilon}$. The exponent is taken to fairly large values ($m \geq 20$), which in Kok et al. (2002) scheme is sufficient to reproduce the rate-independent behavior given by the viscoplastic limit in a conventional power law (see Huang et al. (2004). Taking into account that the volumetric ($\dot{\varepsilon}_{kk}$) and deviatoric ($\dot{\varepsilon}'_{ij}$) strain rates are related to the stress rate in the same way as in classic plasticity, the constitutive equation yields:

$$\dot{\sigma}_{ij} = K\dot{\varepsilon}_{kk}\delta_{ij} + 2\mu\left\{\dot{\varepsilon}'_{ij} - \frac{3\dot{\varepsilon}}{2\sigma_e}\left[\frac{\sigma_e}{\sigma_{ref}\sqrt{f^2(\varepsilon^p) + \ell\eta^p}}\right]^m \dot{\sigma}'_{ij}\right\} \tag{11.55}$$

where K is the bulk modulus. The reference stress σ_{ref} and the non-dimensional function $f(\varepsilon^p)$ are given in agreement with the following isotropic power-law hardening function:

$$\sigma = \sigma_Y\left(1 + \frac{E\varepsilon^p}{\sigma_Y}\right)^{(1/n)} \tag{11.56}$$

such that:

$$\sigma_{ref} = \sigma_Y\left(\frac{E}{\sigma_Y}\right)^{(1/n)}, \quad \text{and} \quad f(\varepsilon^p) = \left(\varepsilon^p + \frac{\sigma_Y}{E}\right)^{(1/n)} \tag{11.57}$$

As described in Martínez-Pañeda et al. (2017), the effective plastic strain gradient η^p is computed by numerical differentiation within the element using the shape functions. In order to do so, a surface is first created by linearly interpolating the incremental values of the plastic strains $\Delta\varepsilon^p_{ij}$ at the Gauss integration points in the entire model. Subsequently, the values of $\Delta\varepsilon^p_{ij}$ are sampled at the nodal locations. The reader is referred to Martínez-Pañeda et al. (2017) and Mathew et al. (2018) for alternative procedures to estimate η^p, which might be more suitable for commercial FE codes.

11.3.4.2 XFEM Details

A direct consequence of the enrichment strategy presented for MSG plasticity is the possibility of employing simpler meshes that do not need to conform to the crack geometry. The crack can be represented through level sets (Duflot, 2007) or a hybrid explicit implicit representation (Fries and Baydoun, 2012; Moumnassi et al., 2011). In Martínez-Pañeda et al. (2017), a level set representation is used and the enrichment functions at any point of interest are computed using the FE approximation of the level set functions.

A typical XFEM mesh with an arbitrary crack is shown in Figure 11.12. The enrichment domain is restricted to the vicinity of the crack tip. As sketched in the figure, elements can be classified into four categories: (i) standard, (ii) tip enriched, (iii) split enriched, and (iv) blending elements. Blending elements are those at the interface of the standard and enriched elements, where the partition of unity is not satisfied and oscillations in

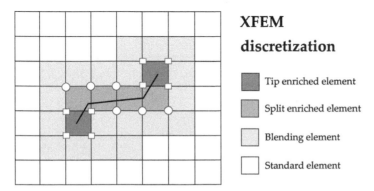

Figure 11.12 Typical XFEM mesh with an arbitrary crack. Circled nodes are enriched with the discontinuous function while squared nodes are enriched with near-tip asymptotic fields. *Source:* Reproduced from Martínez-Pañeda et al. (2017)/Springer Nature.

the results can be observed. This pathological behavior has attracted a considerable research effort; some of the proposed solutions include: assumed strain blending elements (Gracie et al., 2008), corrected or weighted XFEM (Fries, 2008; Ventura et al., 2009), hybrid-crack elements (Xiao and Karihaloo, 2007), semi-analytical approaches (Natarajan and Song, 2013; Réthoré et al., 2010), and spectral functions (Legay et al., 2005). In Laborde et al. (2005), it was numerically observed that to achieve an optimal convergence rate, a fixed area around the crack tip should be enriched with singular functions. This was referred to as *geometrical* enrichment, as opposed to *topological* enrichment, where only one layer of elements around the crack tip is enriched. As detailed below, both topological and geometrical enrichment strategies have been considered here. As proposed by Fries (2008), a linear weighting function is employed to suppress the oscillatory behavior in the partially enriched elements.

Another numerical issue inherent to XFEM is the integration of singular and discontinuous integrands. One compelling and yet simple solution is to partition the elements into triangles. The numerical integration of singular and discontinuous integrands can also be undertaken through other techniques such as polar integration (Laborde et al., 2005), complex mapping (Natarajan et al., 2009), equivalent polynomials (Ventura, 2006), generalized quadrature (Mousavi and Sukumar, 2010), smoothed XFEM (Bordas et al., 2011), or adaptive integration schemes (Xiao and Karihaloo, 2006), among others. Recently, Chin (2017) employed the method of numerical integration of homogeneous functions to integrate discontinuous and weakly singular functions. In the present study, elements are partitioned into triangles and the triangular quadrature rule is employed to integrate the terms in the stiffness matrix. Finally, to compute the stress contours in the vicinity of the crack, Delaunay triangulation is used to connect the integration points, and the stress values within each triangle are inferred using linear interpolation.

11.3.5 Representative Results

Representative results are shown, as obtained from the MATLAB nonlinear finite element method (FEM) and XFEM code developed in Martínez-Pañeda et al. (2017). All results refer to the same boundary value problem: a cracked plate of width $W = 35$ mm and height $H = 100$ mm subjected to a remote uniaxial displacement, see Figure 11.13. Plane strain conditions are assumed and the horizontal displacement is restricted in the node located at $x_1 = W$ and $x_2 = H/2$ to prevent rigid body motion. The following material properties are adopted throughout the analysis: $E = 260$ GPa, $\nu = 0.3$, $\sigma_Y = 200$ MPa, and $n = 5$, with material work hardening behavior being characterized by (11.56). The strain gradient length scale is taken to be equal to $\ell = 5$ μm, a typical estimate for Ni (Stölken and Evans, 1998) and an intermediate value within the range of experimentally observed material length scales reported in the literature (Fuentes-Alonso and Martínez-Pañeda, 2020).

Figure 11.13 Boundary value problem: dimensions and boundary conditions for the single edge cracked plate considered in the XFEM and strain gradient plasticity analysis. *Source:* Adapted from Martínez-Pañeda et al. (2017).

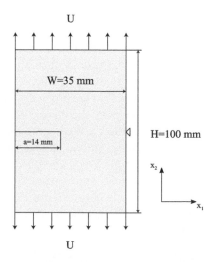

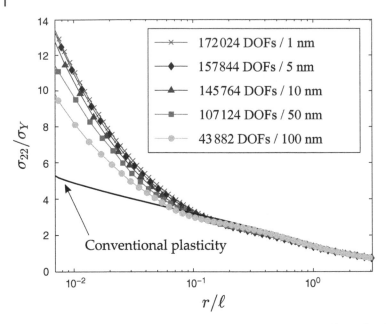

Figure 11.14 Mesh sensitivity analysis for the standard FE case. Normalized tensile stress distribution ahead of the crack tip for different mesh densities, identified as a function of the total number of degrees of freedom (DOFs) and the characteristic element length near the crack tip. The figure shows results along the extended crack plane ($\theta = 0°$) with the normalized distance to the crack tip r/ℓ in log scale. *Source:* Reproduced from Martínez-Pañeda et al. (2017)/Springer Nature.

First, we shall show how an accurate characterization of crack tip fields in strain gradient plasticity solids comes at a significant computational cost in the context of a standard FE analysis. A remote displacement of $U = 0.0011$ mm is applied and the tensile stress σ_{22} distribution ahead of the crack ($\theta = 0°$) is computed. The results obtained using the standard FEM are shown in Figure 11.14 for increasing levels of mesh refinement. Both the total number of DOFs and the characteristic element size in the crack tip region h are reported. Quadratic elements with reduced integration have been employed in all cases. This mesh sensitivity analysis reveals that convergence is achieved for a mesh with 157 844 DOFs and a characteristic element size of 5 nm, as further refinement in the crack tip region leads to almost identical results. This will be considered as the reference FE solution. The result obtained for conventional plasticity is also shown; as expected, it can be seen that far away from the crack tip both conventional plasticity and MSG plasticity agree, but plastic strain gradients become increasingly important as we approach the crack tip and lead to a stress elevation relative to the $\ell = 0$ result.

A representative illustration of the mesh employed is shown in Figure 11.15, where only half of the model is shown, taking advantage of symmetry. As it can be seen in the figure, special care is taken so as to keep an element ratio of 1 close to the crack tip while the mesh gets gradually coarser as we move away from the crack. Accurately characterizing the role of plastic strain gradients at the crack tip requires a very fine mesh and small elements in the crack region that are prone to distortion errors.

Once a reference FE solution has been obtained, we shall proceed to showcase the results obtained with the XFEM formulation and compare the two of them. As shown in Figure 11.16, a much coarser mesh is used for the XFEM analysis. Both topological and geometrical enrichment are considered. As shown in Figure 11.14, the distance over which gradient effects are relevant equals $r = 0.1\ell = 0.5$ µm for the boundary value problem, material properties and applied load considered. Thus, in the topological enrichment case, a tip element with a characteristic length of 1 µm is adopted to ensure that the enriched region engulfs the gradient dominated zone. While in the geometrical enrichment case, the characteristic length of the enriched region is chosen to be of 0.5 µm.

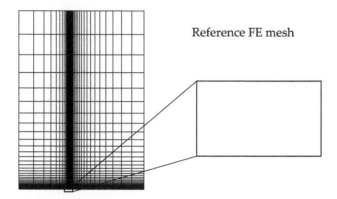

Reference FE mesh

Figure 11.15 Representative finite element mesh, only the upper half of the model is shown due to symmetry. *Source:* Reproduced from Martínez-Pañeda et al. (2017)/Springer Nature.

Figure 11.16 Mesh employed in the XFEM calculations, schematic view and detail of the topological (top) and geometrical (bottom) enrichment regions. *Source:* Reproduced from Martínez-Pañeda et al. (2017)/Springer Nature.

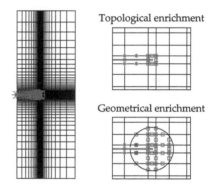

Topological enrichment

Geometrical enrichment

The results obtained with the XFEM formulation are shown below. First, consider the case of topological enrichment using both quadratic and linear elements – Figure 11.17. As in the conventional FE case, the normalized tensile stress σ_{22}/σ_Y is plotted as a function of the normalized distance r/ℓ, the latter being in logarithmic scale. For each solution scheme, the number of DOFs employed and the size of the characteristic element are also reported. XFEM predictions reveal a good agreement with the reference FE solution, despite the very significant differences in the number of DOFs – one order of magnitude relative to the linear element case. At the local level, small differences are observed in the blending elements, despite the corrected XFEM approximation adopted. Overall, the enriched formulation appears to provide a very accurate characterization of crack tip fields, despite using a characteristic element size that is two orders of magnitude larger than in the reference FE mesh. Moreover, in the XFEM formulation, gradient effects can also be captured by means of linear quadrilateral elements. This enrichment-enabled capability permits using lower order displacement elements, further reducing computational cost and maximizing user versatility. Accordingly, linear elements have been used in the remaining computations below.

The results obtained for a fixed geometrical enrichment radius and varying mesh densities are given in Figure 11.18. As in the topological case, a very promising agreement can be observed, with mesh densities being significantly smaller than the reference FE solution, and computation times varying accordingly. The agreement improves by refining the mesh but all cases reveal a satisfactory degree of accuracy at a fraction of the computational cost of the FE solution. By changing the enrichment radius r_e (results not shown), it is observed that the highest accuracy is achieved for $r_e = 0.5$ μm, when the enriched area and the gradient-dominated region coincide. As discussed above, see Figure 11.11, crack tip fields are characterized by three annular domains: elastic,

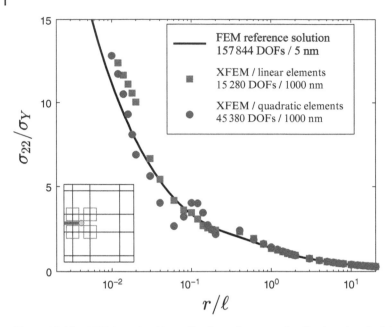

Figure 11.17 XFEM results. Normalized tensile stress distribution ahead of the crack tip for topological enrichment, with both linear and quadratic elements, and the reference FE solution, with mesh densities identified as a function of the total number of DOFs and the characteristic length of the element at the crack tip. The figure shows results along the extended crack plane ($\theta = 0°$) with the normalized distance to the crack tip r/ℓ in log scale. *Source:* Reproduced from Martínez-Pañeda et al. (2017)/Springer Nature.

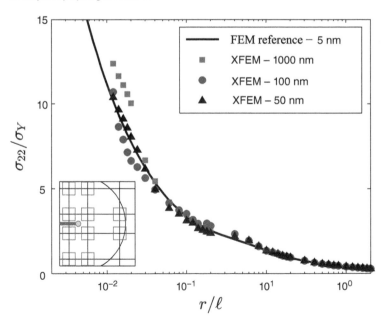

Figure 11.18 XFEM results. Normalized tensile stress distribution ahead of the crack tip for geometrical enrichment with enrichment radius $r_e = 0.5$ μm. Linear elements are used with different mesh densities, identified as a function of the characteristic length of the element at the crack tip. The figure shows results along the extended crack plane with the normalized distance to the crack tip r/ℓ in log scale. *Source:* Reproduced from Martínez-Pañeda et al. (2017)/Springer Nature.

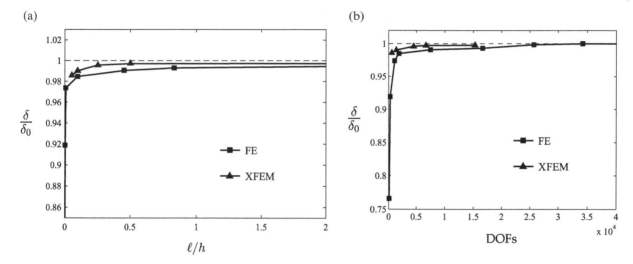

Figure 11.19 Crack opening displacement δ for both XFEM and standard FE calculations using various mesh densities. The results are normalized by the crack opening displacement of the reference FEM solution δ_0, and are reported as a function of (a) the characteristic element size h, and (b) the number of DOFs. The XFEM results have been obtained with topological enrichment and linear elements.

conventional elastic-plastic, and gradient elastic-plastic, each with its own singular solution. Thus, a more elaborated scheme could be developed, in which three classes of asymptotically enriched nodes are defined. In the present case, achieving mesh-independent results with coarser meshes requires aligning the enriched domain with the size of the gradient-dominated region. This limitation is also common to the conventional plasticity analysis, see Section 11.2 and Elguedj et al. (2006), where plasticity was assumed to be confined to the crack tip element. However, note that the size of the gradient-dominated region is much less sensitive to material properties or the magnitude of the applied load than the size of the plastic zone, and can be properly chosen based on previous parametric studies (Martínez-Pañeda and Betegón, 2015).

Finally, we shall showcase the potential of strain gradient XFEM formulations for efficiently computing relevant fracture parameters. Specifically, the focus is on the crack opening displacement δ. The magnitude of δ at the crack mouth is measured for various mesh densities and reported normalized by the reference FE solution: δ/δ_0. Figure 11.19 shows the results as a function of the characteristic element length h, while Figure 11.19b gives the results as a function of the number of DOFs. Results are shown for both the standard FE solution and the XFEM scheme with topological enrichment and linear elements.

The results shown in Figure 11.19 reveal a very good performance of the gradient-enhanced enrichment strategy presented in Section 11.3.3; an excellent agreement with the reference FE solution can be attained with very coarse meshes. The XFEM formulation is able to efficiently predict crack tip blunting even when the enriched domain goes far beyond the gradient dominated zone. Thus, enriching the solution strategy enables fast predictions of relevant fracture parameters, which could be of particular relevance for engineering assessment of structural integrity.

11.4 Conclusions

The aim of this chapter is to showcase the capabilities of the XFEM in predicting fracture in scenarios where material nonlinearities and inelastic deformation govern crack tip behavior, as in the case of metals and many other engineering materials.

First, following the work by Elguedj et al. (2006), an enrichment basis is presented for conventional von Mises plasticity, which builds on the HRR asymptotic crack tip analysis (Hutchinson, 1968; Rice and Rosengren, 1968). Both mode I and mixed-mode fracture conditions are considered. The analysis of the evolution of the displacement jump and the *J*-integral with remote load reveals a very good agreement between the XFEM predictions and the reference FE result. This agreement becomes notably worse if the *ad hoc* HRR enrichment is not accounted for, showcasing the benefit of the formulation presented.

Second, an MSG plasticity formulation is used to provide a more accurate characterization of crack tip fields in elastic-plastic solids. Thus, the role of GNDs and the associated plastic strain gradients in elevating crack tip stresses is accounted for in an implicitly multiscale approach. Following a crack tip asymptotic analysis (Shi et al., 2001), crack tip stresses are shown to exhibit a singular behavior of the type $\sigma \sim r^{-2/3}$. The XFEM solution is subsequently enriched following this asymptotic crack tip analysis and the work by Martínez-Pañeda et al. (2017). The computation of crack tip stresses and fracture parameters reveals that the strain-gradient enrichment provides accurate solutions with an order of magnitude reduction in the computational cost, relative to the reference FE analysis. The implications are important, as a wider embrace of strain gradient models for crack tip assessment is hindered by the large computational cost required for resolving the microscale physical mechanisms that govern fracture behavior. Thus, another area where XFEM could be of notable benefit has been identified. Similar approaches could be pursued to enrich the numerical solution with the underlying physical mechanisms as dictated by mathematical analysis of asymptotic crack tip behavior using other mechanistic, multiscale models.

References

Bardella, L. and Panteghini, A. (2015). Modelling the torsion of thin metal wires by distortion gradient plasticity. *Journal of the Mechanics and Physics of Solids* 78: 467–492.

Bonet, J. and Wood, R.D. (1997). *Nonlinear Continuum Mechanics for Finite Element Analysis*. Cambridge, UK: Cambridge University Press.

Bordas, S.P.A., Natarajan, S., Kerfriden, P., et al. (2011). On the performance of strain smoothing for quadratic and enriched finite element approximations (XFEM/GFEM/PUFEM). *International Journal for Numerical Methods in Engineering* 86: 637–666.

Bower, A.F. (2009). *Applied Mechanics of Solids*. Taylor & Francis, Boca Raton: CRC Press.

Brinckmann, S. and Siegmund, T. (2008). A cohesive zone model based on the micromechanics of dislocations. *Modelling and Simulation in Materials Science & Engineering* 16: 065003 (19pp).

Broumand, P. and Khoei, A.R. (2013). The extended finite element method for large deformation ductile fracture problems with a non-local damageplasticity model. *Engineering Fracture Mechanics* 112–113: 97–125.

Chin, E.B., Lasserre, J.B., and Sukumar, N. (2017). Modeling crack discontinuities without element-partitioning in the extended finite element method. *International Journal for Numerical Methods in Engineering* 110 (11): 1021–1048.

Dahlberg, C.F. and Ortiz, M. (2019). Fractional strain-gradient plasticity. *European Journal of Mechanics A/Solids* 75: 348–354.

Duflot, M. (2007). A study of the representation of cracks with level sets. *International Journal for Numerical Methods in Engineering* 70: 1261–1302.

Duflot, M. and Bordas, S. (2008). A posteriori error estimation for extended finite elements by an extended global recovery. *International Journal for Numerical Methods in Engineering* 76: 1123–1138.

Dunne, F. and Petrinic, N. (2005). *Introduction to Computational Plasticity*. Oxford, UK: Oxford University Press.

Elguedj, T., Gravouil, A., and Combescure, A. (2006). Appropriate extended functions for X-FEM simulation of plastic fracture mechanics. *Computer Methods in Applied Mechanics and Engineering* 195 (7–8): 501–515.

Eshelby, J.D. (1956). The continuum theory of lattice defects. *Solid State Physics* 3 (C): 79–144.

Fleck, N.A. and Hutchinson, J.W. (1997). Strain gradient plasticity. *Advances in Applied Mechanics* 33: 295–361.

Fleck, N.A. and Hutchinson, J.W. (2001). A reformulation of strain gradient plasticity. *Journal of the Mechanics and Physics of Solids* 49 (10): 2245–2271.

Fleck, N.A., Muller, G.M., Ashby, M.F., and Hutchinson, J.W. (1994). Strain gradient plasticity: Theory and experiment. *Acta Metallurgica et Materialia* 42 (2): 475–487.

Fries, T.-P. (2008). A corrected XFEM approximation without problems in blending elements. *International Journal for Numerical Methods in Engineering* 75: 503–532.

Fries, T.-P. and Baydoun, M. (2012). Crack propagation with the extended finite element method and a hybrid explicit–implicit crack description. *International Journal for Numerical Methods in Engineering* 89: 1527–1558.

Fuentes-Alonso, S. and Martínez-Pañeda, E. (2020). Fracture in distortion gradient plasticity. *International Journal of Engineering Science* 156: 103369.

Gao, H., Hang, Y., Nix, W.D., and Hutchinson, J.W. (1999). Mechanism-based strain gradient plasticity - I. Theory. *Journal of the Mechanics and Physics of Solids* 47 (6): 1239–1263.

Gracie, R., Wang, H., and Belytschko, T. (2008). Blending in the extended finite element method by discontinuous Galerkin and assumed strain methods. *International Journal for Numerical Methods in Engineering* 74: 1645–1669.

Gurtin, M.E. (2004). A gradient theory of small-deformation isotropic plasticity that accounts for the Burgers vector and for dissipation due to plastic spin. *Journal of the Mechanics and Physics of Solids* 52 (11): 2545–2568.

Gurtin, M.E. and Anand, L. (2005). A theory of strain-gradient plasticity for isotropic, plastically irrotational materials. Part I: Small deformations. *Journal of the Mechanics and Physics of Solids* 53: 1624–1649.

Gudmundson, P. (2004). A unified treatment of strain gradient plasticity. *Journal of the Mechanics and Physics of Solids* 52 (6): 1379–1406.

Guo, S., He, Y., Lei, J., et al. (2017). Individual strain gradient effect on torsional strength of electropolished microscale copper wires. *Scripta Materialia* 130: 124–127.

Huang, Y., Qu, S., Hwang, K.C., et al. (2004). A conventional theory of mechanism-based strain gradient plasticity. *International Journal of Plasticity* 20 (4–5): 753–782.

Hutchinson, J.W. (1968). Singular behaviour at the end of a tensile crack in a hardening material. *Journal of the Mechanics and Physics of Solids* 16 (1): 13–31.

Hutchinson, J.W. (1997). Linking scales in fracture mechanics. *Advances in Fracture Research Proceedings of ICF10* 1–14.

Ji, H., Chopp, D., and Dolbow, J.E. (2002). A hybrid extended finite element/level set method for modeling phase transformations. *International Journal for Numerical Methods in Engineering* 54 (8): 1209–1233.

Jiang, H., Huang, Y., Zhuang, Z., and Hwang, K.C. (2001). Fracture in mechanism based strain gradient plasticity. *Journal of the Mechanics and Physics of Solids* 49 (5): 979–993.

Jiang, Y., Wei, Y., Smith, J.R., et al. (2010). First principles based predictions of the toughness of a metal/oxide interface. *International Journal of Materials Research* 101: 1–8.

Juul, K.J., Martínez-Pañeda, E., Nielsen, K.L., and Niordson, C.F. (2019). Steady-state fracture toughness of elastic-plastic solids: Isotropic versus kinematic hardening. *Engineering Fracture Mechanics* 207: 254–268.

Kok, S., Beaudoin, A.J., and Tortorelli, D.A. (2002). A polycrystal plasticity model based on the mechanical threshold. *International Journal of Plasticity* 18 (5–6): 715–741.

Komaragiri, U., Agnew, S.R., Gangloff, R.P., and Begley, M.R. (2008). The role of macroscopic hardening and individual length-scales on crack tip stress elevation from phenomenological strain gradient plasticity. *Journal of the Mechanics and Physics of Solids* 56 (12): 3527–3540.

Kristensen, P.K., Niordson, C.F., and Martínez-Pañeda, E. (2020). A phase field model for elastic-gradient-plastic solids undergoing hydrogen embrittlement. *Journal of the Mechanics and Physics of Solids* 143: 104093.

Laborde, P., Pommier, J., Renard, Y., and Salaün, M. (2005). High-order extended finite element method for cracked domains. *International Journal for Numerical Methods in Engineering* 64 (3): 354–381.

Legay, A., Wang, H.W., and Belytschko, T. (2005). Strong and weak arbitrary discontinuities in spectral finite elements. *International Journal for Numerical Methods in Engineering* 64 (8): 991–1008.

Li, H., Li, J., and Yuan, H. (2018). A review of the extended finite element method on macrocrack and microcrack growth simulations. *Theoretical and Applied Fracture Mechanics* 97: 236–249.

Martin, A., Esnault, J.B., and Massin, P. (2015). About the use of standard integration schemes for X-FEM in solid mechanics plasticity. *Computer Methods in Applied Mechanics and Engineering* 283: 551–572.

Martínez-Pañeda, E. and Betegón, C. (2015). Modeling damage and fracture within strain-gradient plasticity. *International Journal of Solids and Structures* 59: 208–215.

Martínez-Pañeda, E. and Fleck, N.A. (2018). Crack growth resistance in metallic alloys: The role of isotropic versus kinematic hardening. *Journal of Applied Mechanics* 85: 11002 (6 pages).

Martínez-Pañeda, E. and Fleck, N.A. (2019). Mode I crack tip fields: Strain gradient plasticity theory versus J2 flow theory. *European Journal of Mechanics - A/Solids* 75: 381–388.

Martínez-Pañeda, E., del Busto, S., and Betegón, C. (2017). Non-local plasticity effects on notch fracture mechanics. *Theoretical and Applied Fracture Mechanics* 92: 276–287.

Martínez-Pañeda, E., del Busto, S., Niordson, C.F., and Betegón, C. (2016). Strain gradient plasticity modeling of hydrogen diffusion to the crack tip. *International Journal of Hydrogen Energy* 41 (24): 10265–10274.

Martínez-Pañeda, E.., Deshpande, V.S., Niordson, C.F., and Fleck, N.A. (2019). The 36 role of plastic strain gradients in the crack growth resistance of metals. *Journal of the Mechanics and Physics of Solids* 126: 136–150.

Martínez-Pañeda, E., Díaz, A., Wright, L., and Turnbull, A. (2020). Generalised boundary conditions for hydrogen transport at crack tips. *Corrosion Science* 173: 108698.

Martínez-Pañeda, E., Fuentes-Alonso, S., and Betegón, C. (2019). Gradient-enhanced statistical analysis of cleavage fracture. *European Journal of Mechanics - A/Solids* 77: 103785.

Martínez-Pañeda, E., Natarajan, S., and Bordas, S. (2017). Gradient plasticity crack tip characterization by means of the extended finite element method. *Computational Mechanics* 59: 831–842.

Martínez-Pañeda, E. and Niordson, C.F. (2016). On fracture in finite strain gradient plasticity. *International Journal of Plasticity* 80: 154–167.

Martínez-Pañeda, E., Niordson, C.F., and Bardella, L. (2016). A finite element framework for distortion gradient plasticity with applications to bending of thin foils. *International Journal of Solids and Structures* 96: 288–299.

Martínez-Pañeda, E., Niordson, C.F., and Gangloff, R.P. (2016). Strain gradient plasticity-based modeling of hydrogen environment assisted cracking. *Acta Materialia* 117: 321–332.

Mathew, T.V., Natarajan, S., and Martínez-Pañeda, E. (2018). Size effects in elastic-plastic functionally graded materials. *Composite Structures* 204: 43–51.

Moës, N., Dolbow, J., and Belytschko, T. (1999). A finite element method for crack growth without remeshing. *International Journal for Numerical Methods in Engineering* 46 (1): 131–150.

Moumnassi, M., Belouettar, S., Béchet, É., et al. (2011). Finite element analysis on implicitly defined domains: An accurate representation based on arbitrary parametric surfaces. *Computer Methods in Applied Mechanics and Engineering* 200 (5–8): 774–796.

Mousavi, S. E. and Sukumar, N. (2010). Generalized Gaussian quadrature rules for discontinuities and crack singularities in the extended finite element method. *Computer Methods in Applied Mechanics and Engineering* 199 (49–52): 3237–3249.

Mu, Y., Hutchinson, J.W., and Meng, W.J. (2014). Micro-pillar measurements of plasticity in confined Cu thin films. *Extreme Mechanics Letters* 1: 62–69.

Natarajan, S., Bordas, S., and Mahapatra, D.R. (2009). Numerical integration over arbitrary polygonal domains based on Schwarz–Christoffel conformal mapping. *International Journal for Numerical Methods in Engineering* 80: 103–134.

Natarajan, S. and Song, C. (2013). Representation of singular fields without asymptotic enrichment in the extended finite element method. *International Journal for Numerical Methods in Engineering* 96: 813–841.

Nix, W.D. and Gao, H.J. (1998). Indentation size effects in crystalline materials: A law for strain gradient plasticity. *Journal of the Mechanics and Physics of Solids* 46 (3): 411–425.

Pan, X. and Yuan, H. (2011a). Applications of the element-free Galerkin method for singular stress analysis under strain gradient plasticity theories. *Engineering Fracture Mechanics* 78 (3): 452–461.

Pan, X. and Yuan, H. (2011b). Computational assessment of cracks under strain-gradient plasticity. *International Journal of Fracture* 167 (2): 235– 248.

Poole, W.J., Ashby, M.F., and Fleck, N.A. (1996). Micro-hardness of annealed and work-hardened copper polycrystals. *Scripta Materialia* 34 (4): 559–564.

Pribe, J.D., Siegmund, T., Tomar, V., and Kruzic, J.J. (2019). Plastic strain gradients and transient fatigue crack growth: A computational study. *International Journal of Fatigue* 120: 283–293.

Qian, X., Zhang, S., and Swaddiwudhipong, S. (2011). Calibration of Weibull parameters using the conventional mechanism-based strain gradient plasticity. *Engineering Fracture Mechanics* 78 (9): 1928–1944.

Rao, B.N. and Rahman, S. (2004). An enriched meshless method for non-linear fracture mechanics. *International Journal for Numerical Methods in Engineering* 59 (2): 197–223.

Réthoré, J., Roux, S., and Hild, F. (2010). Hybrid analytical and extended finite element method (HAX-FEM): A new enrichment procedure for cracked solids Julien. *International Journal for Numerical Methods in Engineering* 81: 269–285.

Rice, J.R. and Rosengren, G.F. (1968). Plane strain deformation near a crack tip in a power-law hardening material. *Journal of the Mechanics and Physics of Solids* 16 (1): 1–12.

Shi, M., Huang, Y., Jiang, H., et al. (2001). The boundary-layer effect on the crack tip field in mechanism-based strain gradient plasticity. *International Journal of Fracture* 112 (1): 23–41.

Shlyannikov, V., Martínez-Pañeda, E., Tumanov, A., and Tartygasheva, A. (2021). Crack tip fields and fracture resistance parameters based on strain gradient plasticity. *International Journal of Solids and Structures* 208–209: 63–82.

Simo, J.C. and Hughes, T.J.R. (2006). *Computational Inelasticity, Vol. 7*. New York, NY: Springer Science & Business Media.

Srinivasan, K. Huang, Y., Kolednik, O., and Siegmund, T. (2008). The size dependence of micro-toughness in ductile fracture. *Journal of the Mechanics and Physics of Solids* 56 (8): 2707–2726.

Stölken, J.S. and Evans, A.G. (1998). A micro-bend test method for measuring the plasticity length scale. *Acta Materialia* 46 (14): 5109–5115.

Symington, M., Ortiz, M., and Shih, C.F. (1990). A finite element method for determining the angular variation of asymptotic crack tip fields. *International Journal of Fracture* 45 (1): 51–64.

Taylor, G.I. (1938). Plastic strain in metals. *Journal of the Institute of Metals* 62: 307–324.

Turnbull, A. (1993). Modelling of environment assisted cracking. *Corrosion Science* 34 (6): 921–960.

Tvergaard, V. and Niordson, C.F. (2004). Nonlocal plasticity effects on interaction of different size voids. *International Journal of Plasticity* 20 (1): 107–120.

Ventura, G. (2006). On the elimination of quadrature subcells for discontinuous functions in the eXtended finite-element method. *International Journal for Numerical Methods in Engineering* 66 (5): 761–795.

Ventura, G., Gracie, R., and Belytschko, T. (2009). Fast integration and weight function blending in the extended finite element method. *International Journal for Numerical Methods in Engineering* 77: 1–29.

Wei, Y. and Hutchinson, J.W. (1997). Steady-state crack growth and work of fracture for solids characterized by strain gradient plasticity. *Journal of the Mechanics and Physics of Solids* 45 (8): 1253–1273.

Xiao, Q.Z. and Karihaloo, B.L. (2006). Improving the accuracy of XFEM crack tip fields using higher order quadrature and statically admissible stress recovery. *International Journal for Numerical Methods in Engineering* 66 (9): 1378–1410.

Xiao, Q.Z. and Karihaloo, B.L. (2007). Implementation of hybrid crack element on a general finite element mesh and in combination with XFEM. *Computer Methods in Applied Mechanics and Engineering* 196 (13–16): 1864–1873.

12

An Introduction to Multiscale analysis with XFEM

Robert Gracie

Department of Civil and Environmental Engineering, University of Waterloo, Canada

12.1 Introduction

Many physical processes important to engineers and scientists are the result of phenomena occurring over multiscale length scales. For example, plasticity is governed by numerous phenomena simultaneously occurring across multiple scales. At the nanoscale ($<10^{-9}$ m), plasticity is the motion of a crystal defect, called a dislocation, moving by the making and breaking of atomic bonds. At the sub-microscale (between 10^{-8} m and 10^{-6} m), dislocation-dislocation interactions dominate the plastic response of a single crystal. At the microscale (between 10^{-6} m and 10^{-4} m), grain shape and size play a defining role in the plastic response of a poly-crystalline material. At the scale of a component or a structure ($>10^{-3}$ m), plastic response depends upon both the loads on the structure and the geometry of the structure. In order to fully understand the plastic response of structures or mechanical components, we must study processes occurring over length scales spanning more than 10 orders of magnitude.

Simulations at a single scale are not able to effectively span so many length scales. For example, atomistic scale simulations where each atom is explicitly modeled are restricted (due to computational costs) to domains sizes less than 0.01 μm^3 even using the world's largest computers. Simulations based on continuum mechanics, such as those presented in the previous chapters, do not include the physics of atomic bond breaking and formation and so cannot model most nanoscale phenomena. As a general rule, the computational cost for a given domain size using a fine scale model is at least 1000 times that of using the higher scale model. The limitations of single scale simulations have led to the development of multiscale analyses. Multiscale analysis is still very much a subject under development and a clear consensus of what constitutes "best practices" has not been established. The rest of this chapter provides a brief introduction to the subject of continuum-atomistic multiscale models.

12.1.1 Types of Multiscale Analysis

There are two predominant types of multiscale analyses: hierarchical and concurrent. Hybrid types also exist which blur the boundaries between hierarchical and concurrent multiscale analyses, but these are beyond the scope of this chapter.

In a hierarchical multiscale simulation, the material response at a higher scale (e.g., macroscale) is derived from more detailed simulations at a lower scale (e.g., microscale); a two-scale hierarchical simulation is illustrated in Figure 12.1. In this figure, the strain at a Gauss point in a macroscale finite element (FE) model is used to define

Partition of Unity Methods, First Edition. Stéphane P. A. Bordas, Alexander Menk, and Sundararajan Natarajan.
© 2024 John Wiley & Sons Ltd. Published 2024 by John Wiley & Sons Ltd.

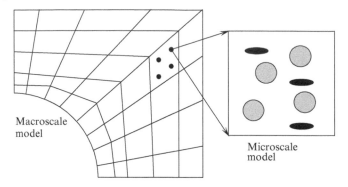

Figure 12.1 Illustration of a hierarchical multiscale model.

displacement boundary conditions for a microscale FE simulation. The microscale model is solved and the average stress in the microscale model is passed back to the Gauss point in the macroscale model. This microscale average stress is then taken as the stress response at the macroscale due to the strain at the Gauss point. Essentially, the microscale simulation plays the role of the constitutive model at the macroscale.

Hierarchical multiscale simulation may link a higher scale continuum model to a lower scale continuum model, a continuum model to an atomistic model, a continuum model to a quantum mechanical model, or an atomistic model to a quantum mechanical model. Furthermore, hierarchical models can span any number of scales by adding additional hierarchical layers. There are many variant of the hierarchical multiscale simulation including classical micromechanics of Mura (1987) and Zohdi and Wriggers (2008); homogenization of Zohdi (2004), Loehnert (2005), and Miehe and Bayreuther (2007); mathematical homogenization of Fish et al. (1997); and the FE^2 method of Feyel (1999), to name just a few.

In a concurrent multiscale analysis, the system being analyzed is decomposed into subdomains. The subdomains where the fine scale physics needs to be simulated and those where it is sufficient to use a higher scale model are identified. In each subdomain, a suitable model is used; constraints on the displacements and tractions on the boundary of the subdomains are introduced to *glue* the different models together. The multiple models are solved simultaneously (i.e., concurrently). A two-scale concurrent multiscale model is illustrated in Figure 12.2, where a microscale model is used in the upper right corner of the domain. In the microscale model, the microstructure of the material is explicitly model while in the rest of the domain a homogenized continuum model is used. Concurrent multiscale simulations may link a higher scale continuum model to a lower scale continuum model, a continuum model to an atomistic model, a continuum model to a quantum mechanical model, or an atomistic model to a quantum mechanical model. It is also possible to couple models at more than two scales, such as the three-scale concurrent quantum-atomistic-continuum simulations of Khare et al. (2008). The rest of this chapter is dedicated to concurrent continuum-atomistic multiscale models.

12.2 Molecular Statics

In this section, the governing equations of a molecular statics (MS) simulation are described. In an MS simulation, a lattice of atoms is modeled as a set of rigid point masses connected by nonlinear springs, as shown in Figure 12.3a. These simulations are quasi-static, i.e., equilibrium of the bond forces acting on each atom is sought. Molecular dynamics (MD) simulations are very similar to MS simulations except inertial forces are also included in the force balance of each atom, making MD simulations transient in time.

Consider a system composed of n_A atoms, as depicted in Figure 12.3b. Let $\mathbf{x}_\alpha$ denote the position of atom α and let $r_{\alpha\beta} = \|\mathbf{x}_\beta - \mathbf{x}_\alpha\|$ denote the distance between atoms β and α. Let the displacement of atom α from a reference position $\mathbf{X}_\alpha$ be denoted as $\mathbf{u}_\alpha^A = \mathbf{x}_\alpha - \mathbf{X}_\alpha$. The total energy of the system is given by the sum of the potential energy stored in the atomic bonds minus the work of external forces:

$$\Pi^A = U^A - W$$

where U is the internal (potential) energy and W is the work done by external forces acting on the atoms. The potential energy, U, is a function of the positions of each atom in the lattice and can be written as the sum of the potential energies of each atom:

Figure 12.2 Illustration of a concurrent multiscale model.

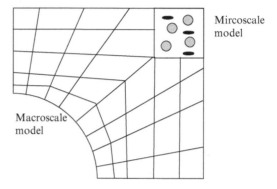

Mircoscale model

Macroscale model

Figure 12.3 (a) Rigid point mass-spring model of an atomic lattice. (b) Illustration of notation for MS model.

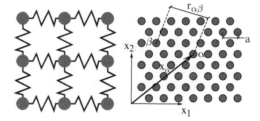

$$U^A \left(\mathbf{x}_1^A, \mathbf{x}_2^A, \dots, \mathbf{x}_{n_A}^A \right) = \sum_{\alpha=1}^{n_A} U_\alpha^A \tag{12.1}$$

where U_α is potential energy of atom α which is related to the positions of atom α and its neighbors. In solids, the potential energy associated with atom α is only a function of the position of atoms within a neighborhood of radius r_{cut} centered on atom α. Atom α only interacts with an atom β if $r_{\alpha\beta} \le r_{cut}$; all such atoms β are called the neighbors of atom α. Typically, the cut-off radius, r_{cut}, of a solid is less than $3a$, where a is the lattice constant.

The equilibrium equations are found by minimizing the total energy (12.1) with respect to the position of each atom, i.e.:

$$\frac{\partial \Pi^A}{\partial \mathbf{x}_\alpha} = \frac{\partial \Pi}{\partial \mathbf{u}_\alpha^A} = \mathbf{f}_\alpha^{int,A} - \mathbf{f}_\alpha^{ext,A} = 0, \ \forall \alpha \tag{12.2}$$

where $\mathbf{f}_\alpha^{int}$ and $\mathbf{f}_\alpha^{ext}$ are the vectors of internal and external force acting on atom α, respectively. The internal forces are given by:

$$\mathbf{f}_\alpha^{int,A} = \frac{\partial U^A}{\partial \mathbf{x}_\alpha} = \sum_{\alpha=1}^{n_A} \frac{\partial U_\alpha^A}{\partial \mathbf{x}_\alpha}$$

In general (12.2) is a set of $n_{nsd} \times n_A$ nonlinear equations which must be solved using a nonlinear solver such as the Newton-Raphson method, where n_{nsd} is the number of space dimensions. For the purpose of clarity, we will restrict our discussions to linear models, in which case we can rewrite $\mathbf{f}^{int,A}$ as:

$$\mathbf{f}^{int,A} = \mathbf{K}^A \mathbf{d}^A$$

where $\mathbf{K}^A$ is the stiffness matrix of the atomistic model and $\mathbf{d}^A = \{\mathbf{u}_1^A, \mathbf{u}_2^A \dots, \mathbf{u}_{n_A}^A\}^\top$. The equilibrium equations become:

$$\mathbf{K}^A \mathbf{d}^A = \mathbf{f}^{ext,A}$$

which is the same form found for the linear finite element method (FEM). The precise expression for $\mathbf{K}^A$ will be derived below.

12.2.1 Atomistic Potentials

In order to solve (12.2), we need to define the atomistic potentials U_α. These functional are not known analytically and must be prescribed by empirical (constitutive) relations. As in continuum mechanics, there exist numerous constitutive models. The selection of an atomistic potential is non-trivial and depends upon the type of atoms involved, upon the lattice types (FCC, BCC, HCP) and upon the physical phenomena being simulated. Each atomistic potential has at least one parameter, though those used in practice often have more than 10; these parameters must be fitted to experimental data or quantum mechanical calculations. In this chapter, we will discuss the harmonic, Lennard Jones (LJ), and embedded atom method (EAM) potentials. Only the harmonic potential leads to a linear system of equilibrium equations. The harmonic and the LJ potentials are generally considered too simple to accurately represent the behavior of a real solid; they are included here for instructive purposes. The EAM potentials are widely used in practize to represent metals. These potentials are generally highly nonlinear making the solution of (12.2) very challenging. The methods presented in this chapter are in principle extendable to the EAM potentials; however, for clarity we will focus our discussion upon the use of the harmonic and LJ potentials.

The harmonic potential is the simplest potential possible. Effectively, it represents the bonds between atoms by linear springs. The potential energy stored in the bond between two atoms is given by:

$$\phi(r) = \frac{1}{2}k(r-a)^2 \tag{12.3}$$

where k is the bond spring constant, r is the deformed bond length, and a is the equilibrium bond length (lattice constant). The potential energy associated with atom α is half the energy of each of the bonds it has with its neighbors, i.e.:

$$U_\alpha^A = \frac{1}{2}\sum_\beta \phi(r_{\alpha\beta}) \tag{12.4}$$

The force acting between atoms α and β is given by:

$$\frac{\partial \phi}{\partial r} = k(r-a) \tag{12.5}$$

The equilibrium equations are obtained using (12.2) and (12.3).

12.2.2 A simple 1D Harmonic Potential Example

Consider a one-dimensional (1D) chain of four atoms, as shown in Figure 12.4. The atoms interact only with their nearest neighbors, so each atom has at most two neighbors, one to the left and one to the right. The atoms are connected by bonds represented by linear springs with spring constants k_1, k_2, and k_3. The equilibrium bond lengths (that when no external forces are applied) are a_1, a_2, and a_3. We will take the reference (undeformed) configuration of the atoms to be such that $X_2^A - X_1^A = a_1$, $X_3^A - X_2^A = a_2$, $X_4^A - X_3^A = a_3$, and $X_1^A < X_2^A < X_3^A < X_4^A$. Let the positions of the four atoms in the deformed configuration be denoted as x_1^A, x_2^A, x_3^A, and x_4^A with $x_1^A < x_2^A < x_3^A < x_4^A$, so that the displacement of each atom is defined by $u_\alpha^A = x_\alpha^A - X_\alpha^A$. The displacement of the left most atom is fixed ($u_1^A = 0$) and a load P is applied to the right most atom. We will now demonstrate the derivation of the governing equation from (12.2) and (12.3) for the lattice shown in Figure 12.4.

The total potential energy is given by:

$$U^A = U_1^A + U_2^A + U_3^A + U_4^A$$

where the potential energies of each atom are given by:

$$U_1^A = \frac{1}{2}\left(\frac{1}{2}k_1\left(r_{12} - a_1\right)^2\right)$$

$$U_2^A = \frac{1}{2}\left(\frac{1}{2}k_1\left(r_{21} - a_1\right)^2\right) + \frac{1}{2}\left(\frac{1}{2}k_2\left(r_{21} - a_2\right)^2\right)$$

$$U_3^A = \frac{1}{2}\left(\frac{1}{2}k_2\left(r_{32} - a_2\right)^2\right) + \frac{1}{2}\left(\frac{1}{2}k_3\left(r_{34} - a_3\right)^2\right)$$

$$U_4^A = \frac{1}{2}\left(\frac{1}{2}k_3\left(r_{43} - a_3\right)^2\right)$$

and

$$r_{\alpha\beta} = abs(x_\beta^A - x_\alpha^A)$$

The equilibrium equations are obtain by (12.2), i.e.:

$$\frac{\partial \Pi}{\partial x_1^A} = \frac{\partial U_1^A}{\partial x_1^A} + \frac{\partial U_2^A}{\partial x_1^A} - f_1^{ext,A} = 0 \tag{12.6}$$

$$\frac{\partial \Pi}{\partial x_2^A} = \frac{\partial U_1^A}{\partial x_2^A} + \frac{\partial U_2^A}{\partial x_2^A} + \frac{\partial U_3^A}{\partial x_2^A} - f_2^{ext,A} = 0 \tag{12.7}$$

$$\frac{\partial \Pi}{\partial x_3^A} = \frac{\partial U_2^A}{\partial x_3^A} + \frac{\partial U_3^A}{\partial x_3^A} + \frac{\partial U_4^A}{\partial x_3^A} - f_3^{ext,A} = 0 \tag{12.8}$$

$$\frac{\partial \Pi}{\partial x_4^A} = \frac{\partial U_3^A}{\partial x_4^A} + \frac{\partial U_4^A}{\partial x_4^A} - f_4^{ext,A} = 0 \tag{12.9}$$

Figure 12.4 Illustration of a chain for four atoms connected by springs.

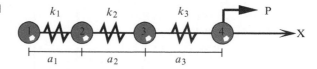

where $f_\alpha^{ext,A}$ is the external force acting on atom α. In this example, $f_1^{ext,A} = f_2^{ext,A} = f_3^{ext,A} = 0$ and $f_4^{ext,A} = P$.

To further clarify the derivation of the equilibrium equations, we will show in detail the steps required to evaluate $\partial\Pi/\partial x_2^A$. Substituting U_1^A to U_3^A into (12.7), we obtain:

$$\frac{\partial\Pi}{\partial x_2^A} = \frac{1}{2}k_1\left(r_{12} - a_1\right)\frac{\partial r_{12}}{\partial x_2^A} + \frac{1}{2}k_1\left(r_{21} - a_1\right)\frac{\partial r_{21}}{\partial x_2^A} + $$
$$\frac{1}{2}k_2\left(r_{23} - a_2\right)\frac{\partial r_{23}}{\partial x_2^A} + \frac{1}{2}k_2\left(r_{32} - a_2\right)\frac{\partial r_{32}}{\partial x_2^A} = 0 \tag{12.10}$$

Now, by definition $r_{\alpha\beta} = r_{\beta\alpha}$ and if we assume that $x_\beta^A > x_\alpha^A$ then $r_{\alpha\beta} = r_{\beta\alpha} = x_\beta^A - x_\alpha^A$; therefore:

$$r_{12} = r_{21} = x_2^A - x_1^A = \left(X_2^A + u_2^A\right) - \left(X_1^A + u_1^A\right) = a_1 + \left(u_2^A - u_1^A\right) \tag{12.11}$$
$$r_{23} = r_{32} = x_3^A - x_2^A = \left(X_3^A + u_3^A\right) - \left(X_2^A + u_2^A\right) = a_2 + \left(u_3^A - u_2^A\right) \tag{12.12}$$
$$r_{34} = r_{43} = x_4^A - x_3^A = \left(X_4^A + u_4^A\right) - \left(X_3^A + u_3^A\right) = a_3 + \left(u_4^A - u_3^A\right) \tag{12.13}$$

Substituting (12.11) and (12.12) into (12.10), we obtain:

$$\frac{\partial\Pi}{\partial x_2^A} = \frac{1}{2}k_1\left(u_2^A - u_1^A\right)\frac{\partial r_{12}}{\partial x_2^A} + \frac{1}{2}k_1\left(u_2^A - u_1^A\right)\frac{\partial r_{21}}{\partial x_2^A} + $$
$$\frac{1}{2}k_2\left(u_3^A - u_2^A\right)\frac{\partial r_{23}}{\partial x_2^A} + \frac{1}{2}k_2\left(u_3^A - u_2^A\right)\frac{\partial r_{32}}{\partial x_2^A} = 0 \tag{12.14}$$

Now,

$$\frac{\partial r_{12}}{\partial x_2^A} = sign\left(x_2^A - x_1^A\right) = 1,$$

$$\frac{\partial r_{21}}{\partial x_2^A} = -sign\left(x_1^A - x_2^A\right) = 1,$$

$$\frac{\partial r_{23}}{\partial x_2^A} = -sign\left(x_3^A - x_2^A\right) = -1, \tag{12.15}$$

$$\frac{\partial r_{32}}{\partial x_2^A} = sign\left(x_2^A - x_3^A\right) = -1.$$

Substituting (12.15) into (12.14) yields:

$$\frac{\partial\Pi}{\partial x_2^A} = 0 = k_1\left(u_2^A - u_1^A\right) + k_2\left(u_3^A - u_2^A\right) \tag{12.16}$$
$$= -k_1 u_1^A + \left(k_1 + k_2\right)u_2^A - k_2 u_3^A$$

Similar logic can be used to derive expressions for (12.6), (12.8), and (12.9) in terms of u_1^A, u_2^A, u_3^A, and u_4^A. Doing so leads to the following linear systems of equations:

$$\begin{bmatrix} (k_1) & -k_1 & 0 & 0 \\ -k_1 & (k_1 + k_2) & -k_2 & 0 \\ 0 & -k_2 & (k_2 + k_3) & -k_3 \\ 0 & 0 & -k_3 & (k_3) \end{bmatrix} \begin{Bmatrix} u_1^A \\ u_2^A \\ u_3^A \\ u_4^A \end{Bmatrix} = \begin{Bmatrix} 0 \\ 0 \\ 0 \\ P \end{Bmatrix} \tag{12.17}$$

which is exactly the same system of equations which would be obtain from a linear FEM analysis of a 1D beam, meshed with elements of length h_1, h_2, and h_3 such that $k_1 = AE/h_1$, $k_2 = AE/h_2$, and $k_3 = AE/h_3$, where A is the cross-sectional area and E is Young's modulus.

12.2.3 The Lennard-Jones Potential

The harmonic potential is too simplistic to represent real atomistic behavior. Processes such as bond creation and bond breakage cannot be simulated using the harmonic potential since the bond force (12.5) increases toward infinity as the distance between atoms increases. Since purely elastic behavior is rarely of interest in atomistic simulations, more complex potentials allowing the simulation of a greater range of physical phenomena must be sought.

One of the simplest atomistic potentials used which captures bonding and de-bonding is the 6-12 LJ potential. The potential energy in the bond between two atoms separated by a distance r is taken to be:

$$\phi^{LJ}(r) = 4\epsilon \left[\left(\frac{\sigma}{r}\right)^{12} - \left(\frac{\sigma}{r}\right)^{6} \right] \tag{12.18}$$

where ϵ and σ are material parameters. ϵ is the energy barrier to overcome bonding and σ is related to the equilibrium bond length, r_0, via $\sigma = 2^{-1/6} r_0$. The bond force is given by:

$$\frac{\partial \phi^{LJ}}{\partial r} = \frac{24\epsilon}{r} \left[\left(\frac{\sigma}{r}\right)^{6} - 2 \left(\frac{\sigma}{r}\right)^{12} \right] \tag{12.19}$$

Setting $\frac{\partial \phi^{LJ}}{\partial r} = 0$ yields $\sigma = 2^{-1/6} r$. As with the harmonic potential, the energy of atom α is given by:

$$U_\alpha^A = \frac{1}{2} \sum_\beta \phi^{LJ}(r_{\alpha\beta}) \tag{12.20}$$

The LJ potential is not very quantitative, as it is only able to fit two properties of a material; however, it is often used because of its high computational efficiency. For example, the one billion atom simulations performed by Abraham et al. (2002) used the LJ potential to qualitatively study ductile and brittle behavior of FCC metals. Two-body potentials, such as the harmonic and LJ potentials, always yield lattices with continuum elastic constants $C_{12} = C_{44}$ which is rarely the case for real materials (Johnson, 1972).

12.2.4 The Embedded Atom Method

The EAM potential is now a standard potential used for the simulation of metallic lattices because it can accurately reproduce a wide range of physical properties such as lattice constants, elastic constants, cohesive energies, stacking fault energies, etc. The EAM potential is significantly more accurate than the LJ potential, at the cost of increase computation complexity. The EAM potential gives the potential energy of atom α as:

$$U_\alpha^A = \phi_\alpha^{core} - \phi^{emb}(\rho_\alpha)$$

where ϕ_α^{core} is the potential energy due to the interaction of the core of atom α with the cores of its neighbors and is given by:

$$\phi_\alpha^{core} = \sum_\beta A \exp(-Br_{\alpha\beta})$$

and $\phi^{emb}(\rho_\alpha)$ is the energy of embedding the core of atom α in an electron gas of density ρ_α. ρ_α is the density of the electron gas which would be present at the location of atom α if atom α were removed from the lattice. For zirconium, Auckland et al. (1995) have found that suitable forms for the embedding function and the electron gas density are given by:

$$\phi^{emb}(\rho) = \sqrt{\rho}$$

and

$$\rho_\alpha = \sum_\beta \sum_{p=1}^{6} \psi_p(r_{\alpha\beta}) c_p$$

where the first summation is over all neighbors of atom α and ψ_p are spline functions. A, B, and c_p are coefficients which are fitted to data from quantum mechanical simulations.

12.3 Hierarchical Multiscale Models of Elastic Behavior – The Cauchy-Born Rule

It this section, we will describe how to define an elastic continuum constitutive model based upon an underlying atomistic model. When a lattice deforms elastically and homogeneously it is possible to map the deformations and forces of the atomistic model to the displacements and stresses of an effective continuum model. This is accomplished via a simple hierarchical multiscale model commonly known as the Cauchy-Born rule, which was made popular by the quasicontinuum method of Tadmor et al. (1996), but was first considered by Johnson (1972) and Milstein (1982).

Consider the deformation of a crystalline lattice by a constant displacement gradient $\nabla \mathbf{u}$, as shown in Figure 12.5. This lattice corresponds to the [111] plane of an FCC lattice. Let the bond vector between atoms α and β in the reference and current configurations be defined by $\mathbf{r}^0 = \mathbf{X}_\beta^A - \mathbf{X}_\alpha^A$ and $\mathbf{r} = \mathbf{x}_\beta^A - \mathbf{x}_\alpha^A$, respectively, and the magnitude of $\mathbf{r}$ is r. When deformations are uniform the components of $\mathbf{r}^0$ are mapped onto $\mathbf{r}$ by:

$$\frac{\partial u_i}{\partial x_j} = \frac{u_{\beta i}^A - u_{\alpha i}^A}{X_{\beta j}^A - X_{\alpha j}^A} = \frac{r_i - r_i^0}{r_j^0}$$

Rearranging we have:

$$r_i = r_i^0 + \frac{\partial u_i}{\partial x_j} r_j^0$$

Let

$$\mathbf{F} = \mathbf{I} + \nabla \mathbf{u}$$

therefore

$$\mathbf{r} = \mathbf{F} \cdot \mathbf{r}^0$$

The distance between any two atoms is thus:

$$r_{\alpha\beta} = \|\mathbf{x}_\alpha^A - \mathbf{x}_\beta^A\| = \|\mathbf{F} \cdot \left(\mathbf{X}_\alpha^A - \mathbf{X}_\beta^A\right)\|$$

Next, consider the potential energy of a unit cell uniformly deformed by $\mathbf{F}$. A unit cell is the smallest repeating building block of a lattice. For a given lattice there is often more than one way to define a unit cell. For example, consider a triangular lattice, illustrated in Figure 12.5, where two possible unit cells are illustrated by the gray subdomains . While the geometry of the different unit cells varies, the number of atoms in each type of unit cell and the volume (area) of each type of unit cell are the same. For the case of a triangular lattice, each unit cell has a single atom.

Figure 12.5 Homogeneous deformation of a lattice by a deformation vector $\mathbf{F} = \mathbf{I} + \nabla\mathbf{u}$. Gray subdomain represents unit cells.

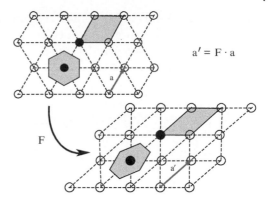

$a' = F \cdot a$

The potential energy of a unit cell, U^{CB}, is defined as the potential energy of the atoms in the unit cell, i.e.:

$$U^{BC} = \sum_{\alpha=1}^{n^A_{cell}} U^A_\alpha$$

where n^A_{cell} is the number of atoms in the unit cell.

The Cauchy-Born rule approximates the potential energy density of an equivalent continuum, deformed at a material point by $\nabla\mathbf{u}$, as the potential energy of a unit cell of a lattice uniform deformed by the same $\nabla\mathbf{u}$ divided by the volume of the unit cell, V_{cell}, i.e.:

$$\delta U^{CB} = \frac{1}{V_{cell}} \sum_{\alpha=1}^{n^A_{cell}} U^A_\alpha$$

For a unit cell with just a single atom and where atomic interactions are modeling by an LJ potential, $U^{CB} = U^A_\alpha$ and by (12.20):

$$\delta U^{CB} = \frac{U^A_\alpha}{V_{cell}} = \frac{1}{V_{cell}} \sum_\beta \phi^{LJ}\left(r_{\alpha\beta}\right) \tag{12.21}$$

The potential energy density (12.21) can be used to define a hyperelastic constitutive model. Assuming small deformations, the Cauchy stress is given by:

$$\sigma_{ij} = \frac{\partial \delta U^{CB}}{\partial \epsilon_{ij}} = \frac{1}{V_{cell}} \sum_\beta \frac{\partial \phi^{LJ}}{\partial r} \frac{\partial r}{\partial F_{ij}}\bigg|_{r=r_{\alpha\beta}} \tag{12.22}$$

where β is summed over all neighbors and

$$\frac{\partial r}{\partial F_{ij}} = \frac{r_i r_j^0}{r}$$

The elasticity tensor ($\sigma_{ij} = C_{ijkl}\epsilon_{kl}$) is defined as:

$$C_{ijkl} = \frac{\partial \sigma_{ij}}{\partial \epsilon_{kl}} = \frac{1}{V_{cell}} \sum_\beta \left[\frac{\partial^2 \phi^{LJ}}{\partial r^2} \frac{\partial r}{\partial F_{ij}} + \frac{\partial \phi^{LJ}}{\partial r} \frac{\partial^2 r}{\partial F_{ij} \partial F_{kl}} \right]\bigg|_{r=r_{\alpha\beta}} \tag{12.23}$$

where:

$$\frac{\partial^2 r}{\partial F_{ij} \partial F_{kl}} = \frac{r_j^0 r_l^0}{r} \left(\delta_{ik} - \frac{r_i r_k}{r^2} \right)$$

Together (12.22) and (12.23) define a continuum elastic constitutive law based on an atomistic model. The continuum constitutive model so defined is a function of the lattice geometry and the chosen atomistic potential used in the underlying MS model. The elasticity and stress tensors inherit all the symmetry and anisotropy of the underlaying atomistic model. A more in-depth presentation of the Cauchy-Born rule is given by Klein (1999).

12.4 Current Multiscale Analysis – The Bridging Domain Method

In this section, the bridging domain method (BDM) of Xiao and Belytschko (2004) and Xiao and Yang (2006) is introduced to couple an MS model with the FEM. This is just one of several concurrent multiscale methods which have been developed over the past two decades. The BDM is presented here because of its ease of implementation and its ability to damp wave reflections in dynamic simulations.

The BDM is an overlapping domain decomposition coupling method. The simulation domain Ω is decomposed into two overlapping subdomains Ω^C and Ω^A, as illustrated in Figure 12.6. Subdomain Ω^A is modeled using an MS model; deformations in this subdomain are expected to be highly nonlinear involving bonding and debonding and are best described by an atomistic model. Subdomain Ω^C is modeled using the FEM (or XFEM – extended finite element method); deformations in this subdomain are expected to be elastic and to be well approximated using a continuum model. Subdomains Ω^C and Ω^A overlap on subdomain Ω^H known as the handshaking (or coupling) domain. It is over Ω^H that *glue* is applied, in the form of compatibility constraints, to couple the atomistic and continuum models. Both the continuum model and the atomistic model co-exist in the handshaking domain Ω^H. Let the contours where Ω^H intersects the interior of Ω^C and Ω^A be denoted by Γ^{HC} and Γ^{HA}, respectively.

The BDM weighs the strain (potential) energy density and work of external forces of the continuum and atomistic models in the handshaking domain, Ω^H, in a complimentary way so that energy is not doubly counted in the coupling domain. The total energy of the system is taken to be:

$$\Pi^{BDM} = \Pi_w^C + \Pi_{1-w}^A - \Pi^\lambda \tag{12.24}$$

where the total energy of the continuum and atomistic subdomain are given by:

$$\Pi_w^C = \int_{\Omega^C} (w) \frac{1}{2} \sigma : \epsilon (\mathbf{u}^C) d\Omega - \int_{\Omega^C} (w) \mathbf{u}^C \cdot \mathbf{b} d\Omega \tag{12.25}$$

and

$$\Pi_{1-w}^A = \sum_{\alpha \in \Omega^A} (1-w) U_\alpha^A - \sum_{\alpha \in \Omega^A} (1-w) \mathbf{u}_\alpha^A \cdot \mathbf{f}_\alpha^{ext,A} \tag{12.26}$$

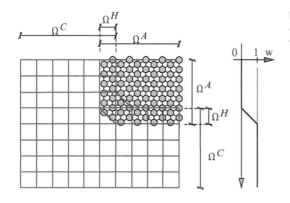

Figure 12.6 Illustration of the decomposition of a domain Ω into overlapping subdomains Ω^C and Ω^A, and the definition of the energy weighting functions *w*.

respectively, where $\mathbf{u}^C$ denotes the continuum displacement field, and w is a scalar weighting function. w has a value of 1 in the purely continuum part of the domain, Ω^C/Ω^H, a value of 0 in the purely atomistic part of the domain Ω^C/Ω^A, and decreases smoothly from 1 on Γ^{HC} to 0 on Γ^{HA}, i.e.:

$$
w(\mathbf{x}) = \begin{cases} 1 & , \text{if } \mathbf{x} \in \Omega^C/\Omega^H \\ 0 & , \text{if } \mathbf{x} \in \Omega^C/\Omega^A \\ 0 \leq d^*/l^* \leq 1 & , \text{if } \mathbf{x} \in \Omega^H \end{cases} \tag{12.27}
$$

Let the Cauchy-Born rule be used to define a constitutive law between $\boldsymbol{\epsilon}$ and $\boldsymbol{\sigma}$, i.e., $\sigma_{ij} = C_{ijkl}\epsilon_{kl}$, where C_{ijkl} is given by (12.23). Let the continuum displacement field, $\mathbf{u}^C$, be approximated by a standard FEM approximation:

$$
\mathbf{u}^C(\mathbf{x}) = \sum_{I=1}^{nn} N_I(\mathbf{x}) \mathbf{u}_I^C = \mathbf{N}\mathbf{d}^C, \ \forall \mathbf{x} \in \Omega^C \tag{12.28}
$$

where $\mathbf{u}_I^C$ are the displacements of the FEM nodes used to discretize the continuum subdomain. Lagrange multipliers are used to enforce compatibility between the displacements of the atoms and the continuum in the handshaking subdomain. More specifically, we wish to enforce the constraint that:

$$
\mathbf{u}^C\left(\mathbf{x}_\gamma^A\right) = \mathbf{u}_\gamma^A, \ \forall \gamma \in \Omega^H \tag{12.29}
$$

i.e., that the atomistic displacements are the same as the continuum displacements in Ω^H. The potential associated with the coupling of the atomistic and continuum subdomains is given by:

$$
\Pi^\lambda = \sum_{\gamma \in \Omega^H} \boldsymbol{\lambda}_\gamma \cdot \left(\mathbf{u}^C\left(\mathbf{x}_\gamma^A\right) - \mathbf{u}_\gamma^A\right) \tag{12.30}
$$

where the summation is over all atoms in the handshaking domain. When the coupling is perfect, i.e., $\mathbf{u}^C\left(\mathbf{x}_\gamma^A\right) - \mathbf{u}_\gamma^A = 0$, Π^λ does not contribute to the total energy of the system. The coupling described above in (12.29) and (12.30), involving only the displacements of the atomistic and continuum models, is known as L2 coupling. Variationally speaking, it is more appropriate to use H1 coupling, involving constraints on both the displacements and the strains; however, experience has shown that there is no significant differences in simulations results obtained using L2 and H1 coupling; see Bauman et al. (2008) for more information.

The equilibrium equations for the BDM are obtained by minimizing (12.24) with respect to $\mathbf{u}_I^C \ \mathbf{u}_\alpha^A$, and $\boldsymbol{\lambda}$:

$$
\frac{\partial \Pi^{BDM}}{\partial \mathbf{u}_I^C} = 0, \ \forall I = 1..nn \tag{12.31}
$$

$$
\frac{\partial \Pi^{BDM}}{\partial \mathbf{u}_\alpha^A} = 0, \ \forall \alpha = 1..n^A \tag{12.32}
$$

$$
\frac{\partial \Pi^{BDM}}{\partial \lambda_\alpha} = 0, \ \forall \alpha \in \Omega^H \tag{12.33}
$$

Substituting (12.28) into (12.31), assuming a linear elastic continuum and switching to matrix notation, we obtain:

$$
\int_{\Omega^C} (w)\,\mathbf{B}_I^\top \mathbf{C}\mathbf{B} d\Omega \mathbf{d}^C - \int_{\Omega^C} (w)\,\mathbf{N}_I^\top \mathbf{b} d\Omega - \sum_{\gamma \in \Omega^H} \boldsymbol{\lambda}_\gamma^\top \left(\mathbf{N}_I\left(\mathbf{x}_\gamma^A\right)\right) = 0, \ \forall I = 1..nn \tag{12.34}
$$

Similarly, assuming atomistic behavior is governed by a harmonic potential (12.3), (12.32) simplifies to:

$$
\sum_{\alpha \in \Omega^A} \sum_\beta \left(1 - w_{\alpha\beta}\right) k_{\alpha\beta}^A \mathbf{u}_\beta^A - \sum_{\alpha \in \Omega^A} \left(1 - w_\alpha\right) \mathbf{f}_\alpha^{ext,A} + \boldsymbol{\lambda}_\alpha^\top = 0, \ \text{if } \alpha \in \Omega^H \tag{12.35}
$$

or

$$\sum_{\alpha \in \Omega^A} \sum_\beta k^A_{\alpha\beta} \mathbf{u}^A_\beta - \sum_{\alpha \in \Omega^A} \mathbf{f}^{ext,A}_\alpha = 0, \text{ if } \alpha \notin \Omega^H \tag{12.36}$$

Equilibrium Equation (12.33) simply yields the original constraint (12.29), i.e.:

$$\mathbf{u}^C \left(\mathbf{x}^A_\alpha \right) - \mathbf{u}^A_\alpha = 0 , \ \forall \alpha \in \Omega^H \tag{12.37}$$

Equations (12.34)–(12.37) can be rewritten in matrix form as:

$$\begin{bmatrix} \mathbf{K}^C & 0 & \mathbf{D}^C \\ 0 & \mathbf{K}^A & \mathbf{D}^A \\ \mathbf{D}^{C^\top} & \mathbf{D}^{A^\top} & 0 \end{bmatrix} \begin{Bmatrix} \mathbf{d}^C \\ \mathbf{d}^A \\ \lambda \end{Bmatrix} = \begin{Bmatrix} \mathbf{f}^{ext,C} \\ \mathbf{f}^{ext,A} \\ 0 \end{Bmatrix} \tag{12.38}$$

where

$$\mathbf{K}^C_{IJ} = \int_{\Omega^C} (w) \, \mathbf{B}^\top_I \mathbf{C} \mathbf{B}_J d\Omega , \forall I, J \tag{12.39}$$

$$\mathbf{K}^A_{\alpha\beta} = - \left(1 - w_{\alpha\beta} \right) \frac{\partial^2 \phi}{\partial \mathbf{x}^A_\alpha \partial \mathbf{x}^A_\beta} , \forall \alpha, \beta \tag{12.40}$$

$$\mathbf{D}^C_{I\gamma} = \begin{cases} -\mathbf{N}_I \left(\mathbf{u}^A_\gamma \right) & , \text{if } \gamma \in \Omega^H \\ 0 & , \text{otherwise} \end{cases} \tag{12.41}$$

$$\mathbf{D}^A_{\alpha\gamma} = \begin{cases} 1 & , \text{if } \alpha = \gamma \text{ and } \gamma \in \Omega^H \\ 0 & , \text{otherwise} \end{cases} \tag{12.42}$$

One should notice that the equilibrium equations for the BDM are not terribly different from those of the FEM. We can easily solve (12.38) by static condensation. Let

$$\mathbf{K} = \begin{bmatrix} \mathbf{K}^C & 0 \\ 0 & \mathbf{K}^A \end{bmatrix}, \mathbf{D} = \begin{bmatrix} \mathbf{D}^C \\ \mathbf{D}^A \end{bmatrix}, \mathbf{d} = \begin{Bmatrix} \mathbf{d}^C \\ \mathbf{d}^A \end{Bmatrix} \text{ and } \mathbf{f} = \begin{Bmatrix} \mathbf{f}^{ext,C} \\ \mathbf{f}^{ext,A} \end{Bmatrix} \tag{12.43}$$

Then (12.38) can be rewritten as:

$$\mathbf{Kd} + \mathbf{D}\lambda = \mathbf{f} \tag{12.44}$$

$$\mathbf{D}^\top \mathbf{d} = 0 \tag{12.45}$$

Solving (12.44) for $\mathbf{d}$ and substituting the result into (12.45), we obtain the following linear system of equations for λ:

$$\mathbf{A}\lambda = \mathbf{D}^\top \mathbf{K}^{-1} \mathbf{f} \tag{12.46}$$

where $\mathbf{A} = \mathbf{D} \mathbf{K}^{-1} \mathbf{D}$. The solution for λ obtained from (12.46) can be substituted back into (12.44) to obtain a solution for $\mathbf{d}$.

12.5 The eXtended Bridging Domain Method

Multiscale analyses which couple atomistic and continuum models are often employed to study crack tip behavior. When a standard multiscale methods such as the BDM is use, it is generally necessary for the whole crack surface to be in the atomistic subdomain. By introducing XFEM at the continuum scale, it is possible to represent a large portion of the crack surface by a continuum model. We will refer to the BDM with XFEM as the eXtended bridging domain method (XFEM-BDM) to differentiate it from the BDM. In the BDM model, the number of atoms is proportional to the length of the crack, while in the XFEM-BDM it is only proportional to the number of crack tips.

The development of the governing equations for the XFEM-BDM follows closely that of the BDM, demonstrated in the previous section. In the XFEM-BDM, the continuum is approximated by:

$$\mathbf{u}^C(\mathbf{x}) = \sum_{I=1}^{nn} N_I(\mathbf{x})\mathbf{u}_I^C + \sum_{J \in S} N_J(\mathbf{x})H(f(\mathbf{x}))\mathbf{a}_J^C = \mathbf{N}\mathbf{d}^C + \bar{\mathbf{N}}\mathbf{a}^C \tag{12.47}$$

where $\mathbf{a}^{C^T} = \{\mathbf{a}_J^C\}, J \in S$ and S is the set of Heaviside enriched nodes, as shown in Figure 12.7. *Note that we do not use the tip enrichment here since the tip region is accurately modeled by an atomistic model.*

We begin with the same energy functional given by (12.24)–(12.26) but with $\mathbf{u}^C$ given by (12.47). The equilibrium equations for the XFEM-BDM are obtained by minimizing (12.24) with respect to $\mathbf{u}_I^C$, $\mathbf{a}_J^C$, $\mathbf{u}_\alpha^A$, and λ :

$$\frac{\partial \Pi^{BDM}}{\partial \mathbf{u}_I^C} = 0, \ \forall I = 1..nn \tag{12.48}$$

$$\frac{\partial \Pi^{BDM}}{\partial \mathbf{a}_I^C J} = 0, \ \forall J \in S \tag{12.49}$$

$$\frac{\partial \Pi^{BDM}}{\partial \mathbf{u}_\alpha^A} = 0, \ \forall \alpha = 1..n^A \tag{12.50}$$

$$\frac{\partial \Pi^{BDM}}{\partial \lambda_\alpha} = 0, \ \forall \alpha \in \Omega^H \tag{12.51}$$

Substituting (12.47) into (12.48) and (12.49), assuming a linear elastic continuum and switching to matrix notation, we obtain:

$$\int_{\Omega^C}(w)\,\mathbf{B}_I^\top \mathbf{C}\mathbf{B}d\Omega\mathbf{d}^C - \int_{\Omega^C}(w)\,\mathbf{B}_I^\top \mathbf{C}\bar{\mathbf{B}}d\Omega\mathbf{a}^C - \int_{\Omega^C}(w)\,\mathbf{N}_I^\top\mathbf{b}d\Omega$$
$$- \sum_{\gamma \in \Omega^H}\lambda_\gamma^\top\left(\mathbf{N}_I\left(\mathbf{x}_\gamma^A\right)\right) = 0, \ \forall I = 1..nn \tag{12.52}$$

$$\int_{\Omega^C}(w)\,\bar{\mathbf{B}}_J^\top \mathbf{C}\mathbf{B}d\Omega\mathbf{d}^C - \int_{\Omega^C}(w)\,\bar{\mathbf{B}}_J^\top \mathbf{C}\bar{\mathbf{B}}d\Omega\mathbf{a}^C - \int_{\Omega^C}(w)\,\bar{\mathbf{N}}_J^\top\mathbf{b}d\Omega$$
$$- \sum_{\gamma \in \Omega^H}\lambda_\gamma^\top\left(\bar{\mathbf{N}}_J\left(\mathbf{x}_\gamma^A\right)\right) = 0, \ \forall J \in S \tag{12.53}$$

Similarly, assuming atomistic behavior is governed by a harmonic potential (12.3), (12.32) simplifies to:

$$\sum_{\alpha \in \Omega^A}\sum_\beta\left(1 - w_{\alpha\beta}\right)k_{\alpha\beta}^A\mathbf{u}_\beta^A - \sum_{\alpha \in \Omega^A}\left(1 - w_\alpha\right)\mathbf{f}_\alpha^{ext,A} + \lambda_\alpha^\top = 0, \ \text{if } \alpha \in \Omega^H \tag{12.54}$$

and

$$\sum_{\alpha \in \Omega^A}\sum_\beta k_{\alpha\beta}^A\mathbf{u}_\beta^A - \sum_{\alpha \in \Omega^A}\mathbf{f}_\alpha^{ext,A} = 0, \ \text{if } \alpha \notin \Omega^H \tag{12.55}$$

which are exactly the same equations obtained with the BDM.

Equation (12.33) simply yields the original constraint (12.29), i.e.:

$$\mathbf{u}^C\left(\mathbf{x}_\alpha^A\right) - \mathbf{u}_\alpha^A = 0 , \; \forall \alpha \in \Omega^H \tag{12.56}$$

The discrete governing Equations (12.52)–(12.56) can be rewritten in matrix form as:

$$
\begin{bmatrix}
\mathbf{K}_{uu}^C & \mathbf{K}_{ua}^C & 0 & \mathbf{D}_u^C \\
\mathbf{K}_{uu}^{C\;\top} & \mathbf{K}_{aa}^C & 0 & \mathbf{D}_a^C \\
0 & 0 & \mathbf{K}^A & \mathbf{D}^A \\
\mathbf{D}_u^{C\;\top} & \mathbf{D}_a^{C\;\top} & \mathbf{D}^{A\;\top} & 0
\end{bmatrix}
\begin{Bmatrix}
\mathbf{d}^C \\
\mathbf{a}^C \\
\mathbf{d}^A \\
\lambda
\end{Bmatrix}
=
\begin{Bmatrix}
\mathbf{f}_u^{ext,C} \\
\mathbf{f}_a^{ext,C} \\
\mathbf{f}^{ext,A} \\
0
\end{Bmatrix}
\tag{12.57}
$$

where

$$\mathbf{K}_{uu,IJ}^C = \int_{\Omega^C} (w)\, \mathbf{B}_I^\top \mathbf{C} \mathbf{B}_J \, d\Omega , \forall I, J \tag{12.58}$$

$$\mathbf{K}_{ua,IJ}^C = \int_{\Omega^C} (w)\, \mathbf{B}_I^\top \mathbf{C} \bar{\mathbf{B}}_J \, d\Omega\Omega, \; J \in S , \forall \alpha, \beta \tag{12.59}$$

$$\mathbf{K}_{aa,IJ}^C = \int_{\Omega^C} (w)\, \bar{\mathbf{B}}_I^\top \mathbf{C} \bar{\mathbf{B}}_J \, d\Omega, \; I, J \in S \tag{12.60}$$

$$\mathbf{K}_{\alpha\beta}^A = -\left(1 - w_{\alpha\beta}\right) \frac{\partial^2 \phi}{\partial \mathbf{x}_\alpha^A \partial \mathbf{x}_\beta^A} \tag{12.61}$$

$$\mathbf{D}_{u,I\gamma}^C = \begin{cases} -\mathbf{N}_I\left(\mathbf{u}_\gamma^A\right) & , \text{if } \gamma \in \Omega^H \\ 0 & , \text{otherwise} \end{cases} \tag{12.62}$$

$$\mathbf{D}_{a,J\gamma}^C = \begin{cases} -\bar{\mathbf{N}}_J\left(\mathbf{u}_\gamma^A\right) & , \text{if } \gamma \in \Omega^H \\ 0 & , \text{otherwise} \end{cases} , \; J \in S \tag{12.63}$$

$$\mathbf{D}_{\alpha\gamma}^A = \begin{cases} 1 & , \text{if } \alpha = \gamma \text{ and } \gamma \in \Omega^H \\ 0 & , \text{otherwise} \end{cases} \tag{12.64}$$

One should notice that the equilibrium equations for the XFEM-BDM are not terribly different from those of the BDM. We can easily solve (12.57) by static condensation as was done for the BDM.

12.5.1 Simulation of a Crack Using XFEM

The example shown in this section was originally published in Gracie and Belytschko (2009). Consider a 2D graphene sheet. The height and width of the sheet are $207A$ and $247A$, respectively. The sheet contains an edge crack with the crack tip located $9A$ above and $10A$ to the left of the center of the sheet. The bottom of the sheet is clamped and displacements of $2.1A$ and $2.5A$ are applied to the top of the sheet in the x and y directions, respectively. The domain is uniformly meshed by 882 triangular elements. A small subdomain around the crack tip is modeled by MS. The rest of the crack is modeled by an XFEM model; the enriched nodes are illustrated in Figure 12.7b

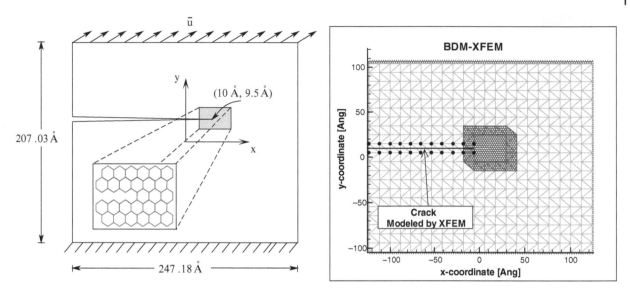

Figure 12.7 Illustration of a multiscale analysis of a crack in a grapheme sheet, Gracie and Belytschko (2009)/John Wiley & Sons. (a) domain geometry, (b) domain decomposition.

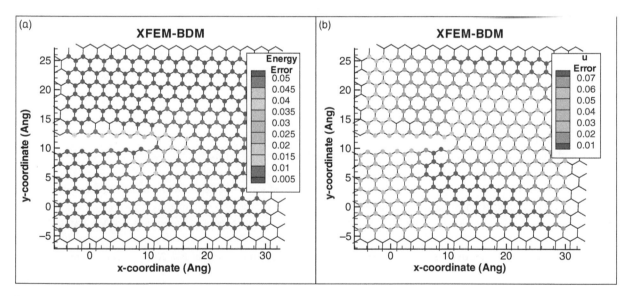

Figure 12.8 Illustration of the error in energy and the displacement error of atoms near the crack tip, after Gracie and Belytschko (2009)/John Wiley & Sons.

Figure 12.8 illustrates the accuracy of the XFEM-BDM relative to a fully atomistic MS simulation. The maximum error in the energy of an atom is about 5%, while the maximum error in the displacements is about 7%. These errors are considered to be small given the saving in computational effort of the XFEM-BDM computation versus the fully MS simulation.

References

Abraham, F., Walkup, R., Gao, H., et al. (2002). Simulating materials failure by using up to one billion atoms and the world's fastest computer: Work-hardening. *Proceeding of the National Academy of Science of the USA* 99 (9): 5777–5782.

Bauman, P., Ben Dhia, H., Elkhodja, N., et al. (2008). On the application of the Arlequin method to the coupling of particle and continuum models. *Computational mechanics* 42 (4): 511–530.

Feyel, F. (1999). Multiscale Fe2 elastoviscoplastic analysis of composite structures. *Computational Materials Science* 16 (1): 344–354.

Fish, J., Shek, K., Pandheeradi, M., and Shephard, M. (1997). Computational plasticity for composite structures based on mathematical homogenization: Theory and practice. *Computer Methods in Applied Mechanics and Engineering* 148 (1): 53–73.

Gracie, R. and Belytschko, T. (2009). Concurrently coupled atomistic and XFEM models for dislocations and cracks. *International Journal for Numerical Methods in Engineering* 78 (3): 354–378.

Johnson, R. (1972). Relationship between two-body interatomic potentials in a lattice model and elastic constants. *Physical Review B* 6 (6): 2094.

Khare, R., Mielke, S., Schatz, G., and Belytschko, T. (2008). Multiscale coupling schemes spanning the quantum mechanical, atomistic forcefield, and continuum regimes. *Computer Methods in Applied Mechanics and Engineering* 197 (41–42): 3190–3202.

Klein, P. (1999). *A Virtual Internal Bond Approach to Modeling Crack Nucleation and Growth*. PhD thesis. Stanford University.

Loehnert, S. (2005). *Computational Homogenization of Microheterogeneous Materials at Finite Strains Including Damage*. PhD thesis. Leibniz University of Hannover-Institut fÃOEr Baumechanik und Numerische Mechanik.

Miehe, C. and Bayreuther, C. (2007). On multiscale Fe analyses of heterogeneous structures: From homogenization to multigrid solvers. *International Journal for Numerical Methods in Engineering* 71 (10): 1135–1180.

Milstein, F. (1982). *Mechanics of Solids*. Pergamon.

Mura, T. (1987). *Micromechanics of Defects in Solids, Vol. 3*. Springer.

Tadmor, E., Ortiz, M., and Philips, R. (1996). Quasi-continuum analysis of defects in solids. *Philosophical Magazine A* 73 (6): 1529–1563.

Xiao, S. and Belytschko, T. (2004). A bridging domain method for coupling continua with molecular dynamics. *Computer Methods in Applied Mechanics and Engineering* 193: 1645–1669.

Xiao, S. and Yang, W. (2006). Temperature-related Cauchy-born rule for multiscale modeling of crystalline solids. *Computational Material Science* 37 (3): 374–379.

Zohdi, T. (2004). *Homogenization Methods and Multiscale Modelling*. Encyclopedia of Computational Mechanics Wiley Online Library.

Zohdi, T. and Wriggers, P. (2008). *An Introduction to Computational Micromechanics, Vol. 20*. Springer Verlag.

Index

Partition of Unity Methods, First Edition. Stéphane P. A. Bordas, Alexander Menk, and Sundararajan Natarajan.
© 2024 John Wiley & Sons Ltd. Published 2024 by John Wiley & Sons Ltd.